Books are to be returned on or before
the last date below.

LIBREX-

UNIFIED THEORY OF CONCRETE STRUCTURES

Thomas T. C. Hsu and Yi-Lung Mo

University of Houston, USA

A John Wiley and Sons, Ltd., Publication

Library of Congress Cataloging-in-Publication Data
Hsu, Thomas T. C. (Thomas Tseng Chuang), 1933-
    Unified theory of concrete structures / Thomas T. C. Hsu and Y. L. Mo.
        p. cm.
    Includes index.
    ISBN 978-0-470-68874-8 (cloth)
    1. Reinforced concrete construction.   I. Mo, Y. L.   II. Title.
    TA683.H73 2010
    624.1′8341–dc22
                                                    2009054418

A catalogue record for this book is available from the British Library.

ISBN 978-0-470-68874-8

Typeset in 10/12pt Times by Aptara Inc., New Delhi, India
Printed in Singapore by Markono Print Media Pte Ltd

Contents

About the Authors

Thomas T. C. Hsu is a John and Rebecca Moores Professor at the University of Houston (UH), Houston, Texas. He received his MS and Ph.D. degrees from Cornell University and joined the Portland Cement Association, Skokie, Illinois, as a structural engineer in 1962. He was a professor and then chairman of the Department of Civil Engineering at the University of Miami, Coral Gables, Florida, 1968–79. After joining UH, he served as the chairman of the Civil and Environmental Engineering Department, 1980–84, built a strong faculty and became the founding director of the Structural Research Laboratory, 1982–2003, which later bears his name. In 2005 he and his wife, Dr. Laura Ling Hsu, established the *"Thomas and Laura Hsu Professorship in Engineering"* at UH.

Dr. Hsu is distinguished by his research in construction materials and in structural engineering. The American Concrete Institute (ACI) awarded him its Wason Medal for Materials Research, 1965; Arthur R. Anderson Research Award, 1990 and Arthur J. Boase Award for Structural Concrete, 2007. Other national awards include the American Society of Engineering Education (ASEE)'s Research Award, 1969, and the American Society of Civil Engineers (ASCE)'s Huber Civil Engineering Research Prize, 1974. In 2009, he was the honoree of the ACI-ASCE co-sponsored *"Thomas T. C. Hsu Symposium on Shear and Torsion in Concrete Structures"* at the ACI fall convention in New Orleans. At UH, Professor Hsu's many honors include the Fluor-Daniel Faculty Excellence Award, 1998; Abraham E. Dukler Distinguished Engineering Faculty Award, 1998; Award for Excellence in Research and Scholarship, 1996; Senior Faculty Research Award, 1992; Halliburton Outstanding Teacher, 1990; Teaching Excellence Award, 1989.

Professor Hsu authored numerous research papers on shear and torsion of reinforced concrete and published two books: *"Unified Theory of Reinforced Concrete"* (1993) and *"Torsion of Reinforced Concrete"* (1984). In this (his third) book *"Unified Theory of Concrete Structures"* (2010), he integrated the action of four major forces (axial load, bending, shear, torsion), in 1,2,3 – dimensions, which culminated into a set of unified theories to analyze and design concrete buildings and infrastructure. Significant parts of Dr. Hsu's work are codified into the ACI Building Code which guides the building industry in the USA and is freely shared worldwide.

Intrinsic to Dr. Hsu's work are two research innovations: (1) the concept that the behavior of whole structures can be derived from studying and integrating their elemental parts, or panels; and (2) the design, construction and use of the *"Universal Panel Tester"* at UH, a unique, million-dollar test rig (NSF grants) that continues to lead the world in producing rigorous, research data on the constitutive models of reinforced concrete, relatable to real-life structures.

In his research on construction materials, Dr. Hsu was the first to visually identify micro-cracks in concrete materials and to correlate this micro-phenomenon to their overt physical properties. His research on fatigue of concrete and fiber-reinforced concrete materials made it possible to interpret the behavior of these structural materials by micro-mechanics.

Among his consulting projects, Dr. Hsu is noted for designing the innovative and cost-saving "*double-T aerial guideways*" for the Dade County Rapid Transit System in Florida; the curved cantilever beams for the Mount Sinai Medical Center Parking Structure in Miami Beach, Florida, and the large transfer girders in the American Hospital Association Buildings, Chicago, Illinois. He is currently a consultant to the US Nuclear Regulatory Commission (NRC).

Dr. Hsu is a fellow of the American Society of Civil Engineers and of the American Concrete Institute. He is a member of ACI Committee 215 (Fatigue), ACI-ASCE joint Committees 343 (Concrete Bridge Design) and 445 (Shear and Torsion). He had also served on ACI Committee 358 (Concrete Guideways), ACI Committee on Publication and ACI Committee on Nomination.

Y. L. Mo is a professor in the Civil and Environmental Engineering Department, University of Houston (UH), and Director of the Thomas T.C. Hsu Structural Research Laboratory. Dr. Mo received his MS degree from National Taiwan University, Taipei, Taiwan and his Ph.D. degree in 1982 from the University of Hannover, Hannover, Germany. He was a structural engineer at Sargent and Lundy Engineers in Chicago, 1984–91, specializing in the design of nuclear power plants. Before joining UH in 2000, Dr. Mo was a professor at the National Cheng Kung University, Tainan, Taiwan.

Professor Mo has more than 27 years of experience in studies of reinforced and prestressed concrete structures subjected to static, reversed cyclic or dynamic loading. In addition to earthquake design of concrete structures, he is an expert in composite and hybrid structures. His outstanding research achievement is in the synergistic merging of structural engineering, earthquake engineering and computer application.

Professor Mo is noted for his innovations in the design of nuclear power plants and is currently a consultant to the US Nuclear Regulatory Commission (NRC). His experience includes developing a monitoring system for structural integrity using the concept of data mining, as well as a small-bore piping design expert system using finite element method and artificial intelligence. Dr. Mo has recently focused on innovative ways to use piezoceramic-based smart aggregates (SAs) to assess the state of health of concrete structures. He also developed carbon nanofiber concrete (CNFC) materials for building infrastructures with improved electrical properties that are required for self health monitoring and damage evaluation.

Professor Mo's wide-ranging consulting work includes seismic performance of shearwalls, optimal analysis of steam curing, effect of casting and slump on ductility of RC beams, effect of welding on ductility of reinforcing bars, early form removal of RC slabs, etc. After the 1999 Taiwan Chi-Chi earthquake, Dr. Mo was selected by Taiwan's National Science Council (NSC) to lead a team of twenty professors to study the damages in concrete structures, to assess causes and to recommend rehabilitation and future research.

Professor Mo is the author of the book "Dynamic Behavior of Concrete Structures" (1994) and is the editor or co-editor of four books. He has written more than 100 technical papers published in national and international journals. For his research and teaching, he received the Alexander von Humboldt Research Fellow Award from Germany in 1995, the Distinguished Research Award from the National Science Council of Taiwan in 1999, the Teaching Excellent

Award from National Cheng Kung University, Taiwan, and the Outstanding Teacher Award from University of Houston.

A fellow of the American Concrete Institute, Professor Mo is also a member of ACI Technical Committees 335 (Composite and Hybrid Structures); 369 (Seismic Repair and Rehabilitation); 374 (Performance-Based Seismic Design of Concrete Building); 444 (Experimental Analysis for Concrete Structures); Joint ACI-ASCE Committee 445 (Shear and Torsion), and 447 (Finite Element Analysis of Reinforced Concrete Structures).

Preface

Concrete structures are subjected to a complex variety of stresses and strains. The four basic actions are: *bending, axial load, shear and torsion*. Each action alone, or in combination with others, may affect structures in different ways under varying conditions. The first two actions – *bending and axial load* – are one-dimensional problems, which were studied in the first six decades of the 20th century, and essentially solved by 1963 when the ultimate strength design was incorporated into the ACI Building Code. The last two actions – *shear and torsion* – are two-dimensional and three-dimensional problems, respectively. These more complicated problems were studied seriously in the second half of the 20th century, and continued into the first decade of the 21st century.

By 1993, a book entitled *Unified Theory of Reinforced Concrete* was published by the first author. At that time, the *unified theory* consisted of five component models: (1) the *struts-and-ties model* for design of local regions; (2) the *equilibrium (plasticity) truss model* for predicting the ultimate strengths of members under all four actions; (3) the *Bernoulli compatibility truss model* for linear and nonlinear theories of bending and axial load; (4) the *Mohr compatibility truss model* for the linear theory of shear and torsion; and (5) the *softened truss model* for the nonlinear theory of shear and torsion.

The first *unified theory* published in 1993 was a milestone in the development of models for reinforced concrete elements. Nevertheless, the ultimate goal must be science-based prediction of the behavior of whole concrete structures. Progress was impeded because the fifth component model, the *softened truss model*, was inadequate for incorporation into the new finite element analysis for whole structures. An innovation in testing facility in 1995 allowed new experimental research to advance the nonlinear theory for shear and torsion. This breakthrough was the installation of a ten-channel servo-control system onto the universal panel tester (UPT) at the University of Houston (UH), which enabled the UPT to perform strain-controlled tests indispensable in establishing more advanced material models.

The expanded testing capabilities opened up a whole new realm of research potentials. One fundamental advance was the understanding of the Poisson effect in cracked reinforced concrete and the recognition of the difference between uniaxial and biaxial strain. The UPT, capable of performing strain-control tests, allowed UH researchers to establish two Hsu/Zhu ratios based on the smeared crack concept, thus laying the foundation for the development of the *softened membrane model*. This new nonlinear model for shear and torsion constitutes the sixth component model of the *unified theory*.

The second advance was the development of the *fixed angle* shear theory, much more powerful than the *rotating angle* shear theory because it can predict the 'contribution of concrete' (V_c). Begun in 1995, the *fixed angle shear theory* gradually evolved into a smooth-operating analytical method by developing a rational shear modulus based on smeared cracks and enrichment of the softened coefficient of concrete. This new fixed angle shear theory serves as a platform to build the *softened membrane model*, even though the term 'fixed angle' is not attached to the name of this model.

The third advance stemming from the expanded capability of the strain-controlled UPT was to obtain the descending branches of the shear stress versus shear strain curves, and to trace the hysteretic loops under reversed cyclic shear. As a result, the constitutive relationships of the cracked reinforced concrete could be established for the whole cyclic loading. These cyclic constitutive relationships, which constitute the *cyclic softened membrane model* (CSMM), opened the door to predicting the behavior of membrane elements under earthquake and other dynamic actions.

A concrete structure can be visualized as an assembly of one-dimensional (1-D) fiber elements subjected to bending/axial load and two-dimensional (2-D) membrane elements subjected to in-plane shear and normal stresses. The behavior of a whole structure can be predicted by integrating the behavior of its component 1-D and 2-D elements. This 'element-based approach' to the prediction of the responses of concrete structures is made possible by the modern electronic computer with its unprecedented speed, and the corresponding rapid development of analytical and numerical tools, such as the nonlinear finite element method.

Finite element method has developed rapidly in the past decade to predict the behavior of structures with nonlinear characteristics, including concrete structures. A nonlinear finite element framework OpenSees, developed during the past decade, is relatively easy to use. By building the constitutive model CSMM of reinforced concrete elements on the platform of OpenSees, a computer program, *Simulation of Concrete Structures* (SCS), was developed at the University of Houston. Program SCS can predict the static, cyclic, and dynamic behavior of concrete structures composed of 1-D frame elements and 2-D wall elements.

The *unified theory* in this 2010 book covers not only the unification of reinforced concrete theories involving bending, axial force, shear and torsion, but also includes the integration of the behavior of 1-D and 2-D elements to reveal the actual behavioral outcome of whole concrete structures with frames and walls. The universal impact of this achievement led to the title for the new book: *Unified Theory of Concrete Structures*, a giant step beyond the scope of *Unified Theory of Reinforced Concrete*. The many challenging goals of this new book are made possible only by the collaboration between the two authors. The first eight chapters were prepared by Thomas T. C. Hsu, and the concluding two chapters by Y. L. Mo.

In closing, this book presents a very comprehensive science-based *unified theory* to design concrete structures and infrastructure for maximum safety and economy. In the USA alone, the value of the concrete construction industry is of the order of two hundred billion dollars a year. Furthermore, the value of this body of work is also reflected by its incalculable human benefit in mitigating the damage caused by earthquakes, hurricanes and other natural or artificial disasters.

With this larger thought, the authors express their deep appreciation to all their colleagues, laboratory staff, and former/current, graduate/undergraduate students, who contributed greatly to the development of the *unified theory*. A special acknowledgment goes to Professor Gregory L. Fenves and his co-workers for the development of the open-domain OpenSees.

Thomas T. C. Hsu and Y. L. Mo
University of Houston
October 7, 2009

Instructors' Guide

This book *Unified Theory of Concrete Structures*, can serve as a comprehensive textbook for teaching and studying a program in concrete structural engineering. Beginning with an undergraduate three-credit course, the program continues on to two graduate-level, three-credit courses. This book can also serve as a reference for researchers and practicing structural engineers who wish to update their current knowledge in order to design unusual or complicated concrete structures.

The undergraduate course could be entitled 'Unified Theory of Concrete Structures I – Beams and Columns' covering Chapters 1, 2 and 3 of the textbook. The course can start with Chapter 3 and the *Bernoulli compatibility truss model* to derive the linear and nonlinear theories of bending for beams and the interaction of bending with axial loads for columns. The derivation process should adhere to Navier's three principles for bending, namely, the 1-D equilibrium condition, the Bernoulli linear compatibility, and the nonsoftened constitutive laws of materials. The course then moves on to Chapter 2, where the *equilibrium (plasticity) truss model* is used to derive the ultimate strengths of the four actions and their interactions. These ultimate strength theories explain the background of bending, shear and torsion in the ACI Building Code, and thus prepare the students to design a concrete beam not only with bending, but also with shear and torsion. Finally, the students are led to Chapter 1, and are introduced to the concept of main regions versus local regions in a structure, and to the *strut-and-ties model* so they can comprehend the equilibrium approach to treating the local regions with disturbed and irregular stresses and strains. Because Chapters 1, 2 and 3 are written in a very concise, 'no-frills' manner, it would be advisable for the instructors of this course to provide a set of additional example problems, and to provide some knowledge of the bond between steel bars and concrete in beams.

The first graduate course could be entitled 'Unified Theory of Concrete Structures II – Shear and Torsion' utilizing Chapter 2, 4, 5, 6 and 7 of the textbook. This first graduate course focuses on shear and torsion, as expressed in the last three component models of the *unified theory* dealing with the *Mohr compatibility truss model,* the *softened truss model* and the *softened membrane model*. These models should be presented in a systematic manner pedagogically and historically, emphasizing the fundamental principles of 2-D equilibrium, Mohr circular compatibility and the softened constitutive laws of materials. A three-credit graduate course taught in this manner was offered in the Spring semester of 2008 at the University of Houston, and in the Fall semester of 2008 at the Hong Kong University of Science and Technology.

The second graduate course could be entitled 'Unified Theory of Concrete Structures III – Finite Element Modeling of Frames and Walls' and covers Chapters 8, 9 and 10 of the textbook. Students who have taken the first graduate course 'Shear and Torsion' and a course in the finite element method could learn to use the finite element framework OpenSees and the UH computer program SCS in Chapter 9. They can first apply these computer programs to the study of beam behavior in Chapter 8, and then expand the application to various forms of concrete structures in Chapter 10. Finally, the students can pursue a research project to study a new form of concrete structure hitherto unexplored.

1

Introduction

1.1 Overview

A reinforced concrete structure may be subjected to four basic types of actions: bending, axial load, shear and torsion. All of these actions can, for the first time, be analyzed and designed by a single unified theory based on the three fundamental principles of mechanics of materials: namely, the stress equilibrium condition, the strain compatibility condition, and the constitutive laws of concrete and steel. Because the compatibility condition is taken into account, this theory can be used to reliably predict the strength of a structure, as well as its load–deformation behavior.

Extensive research of shear action in recent years has resulted in the development of various types of truss model theories. The newest theories for shear can now rigorously satisfy the two-dimensional stress equilibrium, Mohr's two-dimensional circular strain compatibility and the softened biaxial constitutive laws for concrete. In practice, this new information on shear can be used to predict the shear load versus shear deformation histories of reinforced concrete structures, including I-beams, bridge columns and low-rise shear walls. Understanding the interaction of shear and bending is essential to the design of beams, bridge girders, high-rise shear walls, etc.

The simultaneous application of shear and biaxial loads on a two-dimensional (2-D) element produces the important stress state known as 'membrane stresses'. The 2-D element, also known as 'membrane element', represents the basic building block of a large variety of structures made of walls and shells. Such structures, including shear walls, submerged containers, offshore platforms and nuclear containment vessels, can be very large with walls several feet thick. The information in this book provides a rational way to analyze and to design these wall-type and shell-type structures, based on the three fundamental principles of the mechanics of materials for two-dimensional stress and strain states.

The simultaneous application of bending and axial load is also an important stress state prevalent in beams, columns, piers, caissons, etc. The design and analysis of these essential structures are presented in a new light, emphasizing the three principles of mechanics of materials for the parallel stress state, i.e. parallel stress equilibrium, the Bernoulli linear strain compatibility and the uniaxial constitutive laws of materials.

Unified Theory of Concrete Structures   Thomas Hsu and Yi-Lung Mo
© 2010 John Wiley & Sons, Ltd

The three-dimensional (3-D) stress state of a member subjected to torsion must take into account the 2-D shear action in the shear flow zone, as well as the bending action of the concrete struts caused by the warping of the shear flow zone. Since both the 2-D shear action and the bending action can be taken care of by the simultaneous applications of Mohr's compatibility condition and Bernoulli's compatibility condition, the torsional action becomes, for the first time, solvable in a scientific way. This book provides all the necessary information leading up to the rational solution of the problem in torsion.

Because each of the four basic actions experienced by reinforced concrete structures has been found to adhere to the fundamental principles of the mechanics of materials, a unified theory is developed encompassing bending, axial load, shear and torsion in reinforced as well as prestressed concrete structures. This book is devoted to a systematic integration of all the individual theories for the various stress states. As a result of this synthesis, the new rational theories should replace the many empirical formulas currently in use for shear, torsion and membrane stress.

The unified theory is divided into six model components based on the fundamental principles employed and the degree of adherence to the rigorous principles of mechanics of materials. The six models are: (1) the struts-and-ties model; (2) the equilibrium (plasticity) truss model; (3) the Bernoulli compatibility truss model; (4) the Mohr compatibility truss model; (5) the softened truss model; and (6) the softened membrane model. In this book the six models are presented as rational tools for the solution of the four basic actions: bending, axial load, and particularly, 2-D shear and 3-D torsion. Both the four basic actions and the six model components of unified theory are presented in a systematic manner, focusing on the significance of their intrinsic consistencies and their inter-relationships. Because of its inherent rationality, this unified theory of reinforced concrete can serve as the basis for the formulation of a universal and international design code.

In Section 1.2, the position of the unified theory in the field of structural engineering is presented. Then the six components of the unified theory are introduced and defined in Section 1.3, including a historical review of the six model components, and an explanation of how the book's chapters are organized. The conceptual introduction of the first model – the struts-and-ties model – is given in Section 1.4. Detailed study of the struts-and-ties model is not included in this book, but is available in many other textbooks on reinforced concrete.

Chapters 2–7 present a systematic and rigorous study of the last five model components of the unified theory, as rational tools to solve the four basic actions (bending, axial load, shear and torsion) in concrete structures. The last three chapters, 8–10, illustrate the wide applications of the unified theory to prestressed I-beams, ductile frames, various types of framed shear walls, bridge columns, etc., subjected to static, reversed cyclic, dynamic and earthquake loadings.

1.2 Structural Engineering

1.2.1 Structural Analysis

We will now look at the structural engineering of a typical reinforced concrete structure, and will use, for our example, a typical frame-type structure for a manufacturing plant, as shown in Figure 1.1. The main portal frame, with its high ceiling, accommodates the processing work. The columns have protruding corbels to support an overhead crane. The space on the right,

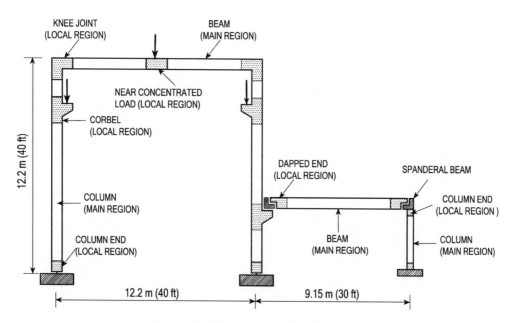

Figure 1.1 A typical frametype reinforced concrete structure

with the low ceiling, serves as offices. The roof beams of the office are supported by spandrel beams which, in turn, are supported by corbels on the left and columns on the right.

The structure in Figure 1.1 is subjected to all four types of basic actions – bending M, axial load N, shear V and torsion T. The columns are subjected to bending and axial load, while the beams are under bending and shear. The spandrel beam carries torsional moment in addition to bending moment and shear force. Torsion frequently occurs in edge beams where the loads are transferred to the beams from one side only. The magnitudes of these four actions are obtained by performing a frame analysis under specified loads. The analysis can be based on either the linear or the nonlinear material laws, and the cross-sections can either be uncracked or cracked. In this way, the four M, N, V and T diagrams are obtained for the whole structure. This process is known as 'structural analysis'.

Table 1.1 illustrates a four-step general scheme in the structural engineering of a reinforced concrete structure. The process of structural analysis is the first step as indicated in row 1 of the table. While this book will not cover structural analysis in details, information on this topics can be found in many standard textbooks on this subject.

1.2.2 Main Regions vs Local Regions

The second step in the structural engineering of a reinforced concrete structure is to recognize the two types of regions in the structure, namely, the 'main regions' and the 'local regions'. The local regions are indicated by the shaded areas in Figure 1.1. They include the ends of a column or a beam, the connections between a beam and a column, the corbels, the region adjacent to a concentrated load, etc. The large unshaded areas, which include the primary portions of each member away from the local regions are called the main regions.

Table 1.1 Unified theory of reinforced concrete structures

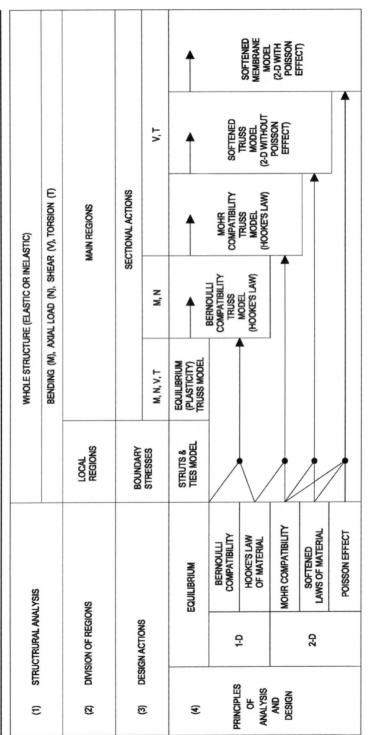

From a scientific point of view, a main region is one where the stresses and strains are distributed so regularly that they can be easily expressed mathematically. That is, the stresses and strains in the main regions are governed by simple equilibrium and compatibility conditions. For columns that are under bending and axial load, the equilibrium equations come from the parallel force equilibrium condition, while the compatibility equations are governed by Bernoulli's hypothesis of the plane section remaining plane. In the case where beams are subjected to shear and torsion, the stresses and strains should satisfy the two-dimensional equilibrium and compatibility conditions, i.e. Mohr's stress and strain circles.

In contrast, a local region is one where the stresses and strains are so disturbed and irregular that they are not amenable to mathematical solution. In particular, the compatibility conditions are difficult to apply. In the design of the local regions the stresses are usually determined by equilibrium condition alone, while the strain conditions are neglected. Numerical analysis by computer (such as the finite element method), can possibly determine the stress and strain distributions in the local regions, but it is seldom employed due to its complexity.

The local region is often referred to as the 'D region'. The prefix D indicates that the stresses and strains in the region are disturbed or that the region is discontinuous. Analogously, the main region is often called the 'B region', noting that the strain condition in this bending region satisfies Bernoulli's compatibility condition. This terminology does not take into account the strain conditions of structures subjected to shear and torsion, which should satisfy Mohr's compatibility condition. Therefore, the term 'B–M region' would be more general and technically more accurate, including both the Bernoulli and the Mohr compatibility conditions. However, since the term 'B region' has been used for a long time, it could be thought of as a simplification of the term 'B–M region'.

The second step of structural engineering is the division of the main regions and local regions in a structure as indicated in row 2 of Table 1.1. On the one hand, the main regions of a structure are designed directly by the four sectional actions, M, N, V and T, according to the four sectional action diagrams obtained from structural analysis. On the other hand, the local regions are designed by stresses acting on the boundaries of the regions. These boundary stresses are calculated from the four action diagrams at the boundary sections. A local region is actually treated as an isolated free body subjected to external boundary stresses.

The third step of structural engineering is the determination of the design actions for the two regions. This third step of finding the sectional actions for main regions and the boundary stresses for local regions is indicated in row 3 of Table 1.1. Once the diagrams of the four actions are determined by structural analysis and the two regions are identified, all the main regions and local regions can be designed.

1.2.3 Member and Joint Design

This fourth step of structural engineering is commonly known as the *member and joint design*. More precisely, it means the *design and analysis of the main and local regions*. By this process the size and the reinforcement of the members as well as the arrangement of reinforcement in the joints are determined.

The unified theory aims to provide this fourth and most important step with a rational method of design and analysis for all of the main and local regions in a typical reinforced concrete structure, such as the one in Figure 1.1. It serves to synthesize all the rational theories and to replace all the empirical design formulas for these regions.

The position of the unified theory in the scheme of structural engineering is shown in row 4 of Table 1.1. The six model components of the unified theory are distinguished by their adherence to the three fundamental principles of the mechanics of materials (the equilibrium condition, the compatibility condition and the constitutive laws of materials). The six models are named to reflect the most significant principle(s) embodied in each as listed in the following section.

1.3 Six Component Models of the Unified Theory

1.3.1 Principles and Applications of the Six Models

As shown in Table 1.1, some of the six models are intended for the main regions and some for the local regions. Others may be particularly suitable for the service load stage or the ultimate load stage. The six models are summarized below, together with their basic principles and the scope of their applications:

1.3.1.1 Struts-and-ties Model

Principles: *Equilibrium* condition only
Applications: Design of local regions

1.3.1.2 Equilibrium (Plasticity) Truss Model

Principles: *Equilibrium* condition and the theory of *plasticity*
Applications: Analysis and design of M, N, V and T in the main regions at the ultimate load stage

1.3.1.3 Bernoulli Compatibility Truss Model

Principles: *1-D Equilibrium* condition, *Bernoulli compatibility* condition and *1-D or uni-axial constitutive law* for concrete and reinforcement. The constitutive laws may be linear or nonlinear
Applications: Analysis and design of M and N in the main regions at both the serviceability and the ultimate load stages

1.3.1.4 Mohr Compatibility Truss Model

Principles: *2-D Equilibrium* condition, *Mohr compatibility* condition and *1-D or uniaxial constitutive law* (*Hooke's Law* is preferred) for both concrete and reinforcement
Applications: Analysis and design of V and T in the main regions at the serviceability load stage

1.3.1.5 Softened Truss Model

Principles: *2-D Equilibrium* condition, *Mohr's compatibility* condition and the *2-D soft-ened constitutive law* for concrete. The constitutive law of reinforcement may be linear or nonlinear

Applications: Analysis and design of *V* and *T* in the main regions at both the serviceability and the ultimate load stages

1.3.1.6 Softened Membrane Model

Principles: *2-D Equilibrium* condition, *Mohr's compatibility* condition and the *2-D soft-ened constitutive law* for concrete. The constitutive law of reinforcement may be linear or nonlinear. The *Poisson effect* is included in the analysis

Applications: Analysis and design of *V* and *T* in the main regions at both the serviceability and the ultimate load stages

1.3.2 Historical Development of Theories for Reinforced Concrete

1.3.2.1 Principles of mechanics of materials

The behavior of a beam subjected to bending was first investigated by Galileo in 1638. In his famous book *Dialogues on Two New Sciences*, he studied the equilibrium of a stone cantilever beam of rectangular section and found that the beam could support twice as much load at the center as at the free end, because a same magnitude of 'bending moment' was produced at the fixed end. Using the rudimentary knowledge of equlibrium, he observed that, for beams 'of equal length but unequal thickness, the resistance to fracture increases in the same ratio as the cube of the thickness (provided the thickness-to-width ratio remains unchanged)'. Galileo's work represented the beginning of a scientific discipline known as the '*mechanics of materials*'.

Since Galileo's beam was considered a rigid body, the deflections of the beam could not be evaluated, thus creating the mystery known as 'Galileo's problem'. The solution to Galileo's problem required two additional sources of information in addition to the principle of equilibrium. The first source came from an understanding of the mechanical properties of materials, summarized as follows: In 1678, Hooke measured the elongations of a long, thin metal wire suspended from a high ceiling at one end and carrying a weight at the bottom end. By systematically varying the weight, he reported that the 'deformation is proportional to force' for wires of various materials under light loads. In 1705, James Bernoulli, a member of a prominent family of Swiss scholars, defined the concept of stress (force divided by area) and strain (displacement divided by original length). This was followed by Euler's postulation in 1727 of 'stress is proportional to strain'. The proportionality constant between stress and strain *E* was measured by Young in 1804 for many materials and was known as Young's modulus. It took 166 years to develop the well-known *Hooke's law*.

The second source of information came from the observation of deformations in beams. In relating the radius of curvature of a beam to the bending moment, Jacob Bernoulli, James' brother, postulated in 1705 the well-known 'Bernoulli's hypothesis', i.e. 'a plane section remains plane'. It should be noted that Jacob Bernoulli misunderstood the neutral axis and took it at the concave surface of the beam. As a result, his derived flexural rigidity *EI* was twice the correct value. Nevertheless, based on Bernoulli's hypothesis and assuming the proportionality between curvature and bending, Euler correctly derived in 1757 the elastic deflection curve of a beam by using the newly developed mathematical tool of calculus. Although Euler was unable to theoretically derive the flexural rigidity, he was able to correctly use *Bernoulli's strain compatibility condition.*

As history bears out, the correct derivation of the flexural rigidity *EI*, the key to the solution of Galileo's problem, requires the integration of all three sources of information on *stress equilibrium, strain compatibility and Hooke's law of materials.* These three principles were put together correctly by a French professor, Navier, in 1826. In his landmark book (Navier, 1826), he systematically and rigorously derived the bending theory using these three principles, thus solving Galileo's problem after almost two centuries. Indeed, Navier's comprehensive book was the first textbook on the mechanics of materials, because these three principles were also applied to shear and torsion (circular sections only). The book showed that a correct load–deformation relationship of a beam must be analyzed according to *Navier's three principles of the mechanics of materials.*

1.3.2.2 Bending Theory in Reinforced Concrete

Reinforced concrete originated four decades after Navier's book. Its birth was credited to Joseph Monier, a French gardener, who obtained a patent in 1867 to reinforce his concrete flower pots with iron wires. The concept of using metal reinforcement to strengthen concrete was quickly used in buildings and bridges, and reinforced concrete became widely accepted in the last quarter of the 19th century. Such growth in applications gave rise to the demand for a rational theory to analyze and design reinforced concrete. By the end of the 19th century a *linear bending theory* for reinforced concrete began to emerge. In this theory, Navier's three principles for one-dimensional stresses and strains were used to analyze the beam. These principles include the equilibrium of parallel, plane stresses, Bernoulli's compatibility of strains, and Hooke's law. Not surprisingly, this theory was developed by French engineers, including Hennibique's firm (Delhumeau, 1999).

The linear bending theory was incorporated into the first ACI Code (1910) and was used for more than half a century. In the 1950s a *nonlinear bending theory* was developed using a nonlinear constitutive law for concrete obtained from tests (Hognestad *et al.*, 1955), rather than Hooke's law. This nonlinear theory was incorporated into the 1963 ACI Code and has been used up to the present time. The analytical tools for both the linear and nonlinear bending theories are given by the *Bernoulli compatibility truss model.* This model can be expanded to analyze and design members subjected to combined bending and axial loads, because it is founded on Navier's three principles of the mechanics of materials.

The linear and nonlinear bending theories, as well as the Bernoulli compatibility truss model, are elaborated in Chapter 3.

1.3.2.3 Struts-and-ties Model (or Truss Model)

Concrete is a material that is very strong in compressive strength but weak in tensile strength. When concrete is used in a structure to carry loads, the tensile regions are expected to crack and, therefore, must be reinforced by materials of high tensile strength, such as steel. The concept of utilizing concrete to resist compression and steel reinforcement to carry tension gave rise to the *struts-and-ties model.* In this model, concrete compression struts and the

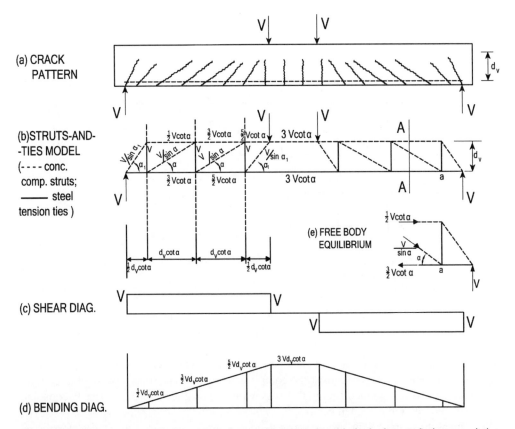

Figure 1.2 Plane truss model of a concrete beam with bottom longidutinal rebars and stirrups resisting shear and bending

steel tension ties form a truss that is capable of resisting applied loadings. The struts-and-ties model has been used, intuitively, by engineers to design concrete structures since the advent of reinforced concrete. At present, it is used primarily for the difficult local regions (D regions). Examples of the struts-and-ties model will be given in Section 1.4.

When the struts-and-ties concept was applied to a main region (B region), it is known as a *truss model*. For a reinforced concrete beam, truss model can be applied, not only to bending and axial loads, but also to shear and torsion. Two examples will be discussed, namely, a beam subjected to shear and bending as shown in Figure 1.2, and a beam subjected to torsion as shown in Figure 1.3.

The first application of the concept of truss model to beam shear was proposed by Ritter (1899) and Morsch (1902) as illustrated in Figure 1.2. In their view, a reinforced concrete beam acts like a parallel-stringer truss to resist bending and shear. Due to the bending moment, the concrete strut near the upper edge serves as the top stringer in a truss and the steel bar near the lower edge assumes the function of the bottom stringer. From shear stresses, the web region would develop diagonal cracks at about 45° inclination to the longitudinal steel. These cracks would separate the concrete into a series of diagonal concrete struts. To resist the applied shear

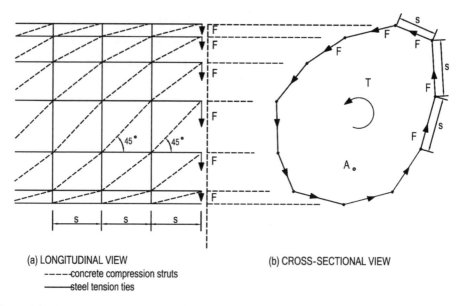

(a) LONGITUDINAL VIEW   (b) CROSS-SECTIONAL VIEW
-----concrete compression struts
———steel tension ties

Figure 1.3   Space truss model of a concrete beam with longitudinal and hoop steel resisting torsional

forces after cracking, the transverse steel bars in the web would carry tensile forces and the diagonal concrete struts would resist the compressive forces. The transverse steel, therefore, serves as the tensile web members in the truss while the diagonal concrete struts become the diagonal compression web members.

The plane truss model for beams was extended to treat members subjected to torsion as shown in Figure 1.3 (Rausch, 1929). In Rausch's concept, a torsional member is idealized as a space truss formed by connecting a series of component plane trusses capable of resisting shear action. The circulatory shear stresses, developed in the cross-section of the space truss, form an internal torsional moment capable of resisting the applied torsional moment.

Although the truss models developed by Ritter (1899), Mörsch (1902) and Rausch (1929) provided a clear concept of how reinforced concrete resists shear and torsion, these models treated the concrete struts and steel ties as lines without cross-sectional dimensions. Consequently, these models did not allow us to treat the beams as a continuous material and to calculate the stresses and strains in the beam. In other words, the precious knowledge developed by the scientific discipline of mechanics of materials could not be applied.

In this book, only a brief, but conceptual, introduction of the struts-and-ties model will be given in Section 1.4.

1.3.2.4 Equilibrium (Plasticity) Truss Model

In the 1960s the truss model of members with dimensionless linear elements to resist shear and torsion was replaced by members made up of more realistic 2-D elements. By treating a 2-D element after cracking as a truss made up of compression concrete struts and tensile steel ties, Nielson (1967) and Lampert and Thurlimann (1968, 1969) derived three equilibrium equations for a 2-D element. The steel and concrete stresses in these three equations should satisfy the Mohr stress circle.

By assuming that all the steel bars in the 2-D element will yield before the crushing of concrete, it is possible to use the three equilibrium equations to calculate the stresses in the steel bars and in concrete struts at the ultimate load stage. This method of analysis and design is called the *equilibrium (plasticity) truss model*.

Since the strain compatibility condition is irrelevant under the plasticity condition, the equilibrium truss model becomes very powerful in two ways: First, it can be easily applied to all four types of actions (bending, axial loads, shear and torsion) and their interactions. The interactive relationship of bending, shear and torsion were elegantly elucidated by Elfgren (1972). Second, this model can easily be incorporated into the strength design codes, such as the ACI Code and the European Code.

Looking at the weakness side of not utilizing the compatibility condition and the constitutive laws of materials, the equilibrium (plasticity) truss model could not be used to derive the load-deformation relationship of RC beams subjected to shear and torsion. More sophisticated theories will have to be developed for shear and torsion that takes care of all three principles of the mechanics of materials.

In this book, the equilibrium (plasticity) truss model will be presented in detail in Chapter 2.

1.3.2.5 Shear Theory

The derivation of three equilibrium equations for 2-D elements was soon followed by the derivation of the three strain compatibility equations by Bauman (1972) and Collins (1973). The steel and concrete strains in these three compatibility equations should satisfy Mohr's strain circle.

Combining the 2-D equilibrium equations, Mohr's compatibility equations, and Hooke's law, a linear shear theory can be developed for a 2-D element. This linear model has been called the *Mohr compatibility truss model*. It could be applied in the elastic range of a 2-D element up to the service load stage. Nonlinear shear theory is required to describe the behavior of 2-D shear elements up to the ultimate load stage.

When an RC membrane element is subjected to shear, it is essentially a 2-D problem because the shear stress can be resolved into a principal tensile stress and a principal compressive stress in the 45° direction. The biaxial constitutive relationship of a 2-D element was a difficult task, because the stresses and strains in two directions affect each other.

The most important phenomenon in a 2-D element subjected to shear was discovered by Robinson and Demorieux (1972). They found that the principal compressive stress was reduced, or 'softened', by the principal tensile stress in the perpendicular direction. However, without the proper equipment to perform biaxial testing of 2-D elements, they were unable to formulate the softened stress-strain relationship of concrete in compression.

Using a biaxial test facility called a 'shear rig', Vecchio and Collins (1981) showed that the softening coefficient of the compressive stress–strain curve of concrete was a function of the principal tensile strain ε_1, rather than the principal tensile stress. Incorporating the equilibrium equations, the compatibility equations, and using the 'softened stress–strain curve' of concrete, Collins and Mitchell (1980) developed a 'compression field theory' (CFT), which could predict the nonlinear shear behavior of an element in the post-cracking region up to the peak point.

Later, Vecchio and Collins (1986) proposed the modified compression field theory (MCFT) which included a constitutive relationship for concrete in tension to better model the post-cracking shear stiffness.

In 1988, a universal panel tester was built at the University of Houston (Hsu, Belarbi and Pang, 1995) to perform biaxial tests on large 2-D elements of $1.4 \times 1.4 \times 0.179$ m ($55 \times 55 \times 7$ in.). By confirming and establishing the softening coefficient as a function of principal tensile strain ε_1, Pang and Hsu (1995) and Belarbi and Hsu (1994, 1995) developed the *rotating-angle softened truss model* (RA-STM). This model made two improvements over the CFT: (1) the tensile stress of concrete was taken into account so that the deformations could be correctly predicted; and (2) the smeared (or average) stress–strain curve of steel bars embedded in concrete was derived on the 'smeared crack level' so that it could be correctly used in the equilibrium and compatibility equations which are based on continuous materials.

Shortly after the development of the rotating-angle model, Pang and Hsu (1996) and Hsu and Zhang (1997) reported the *fixed-angle softened truss model* (FA-STM) that is capable of predicting the 'concrete contribution' V_c by assuming the cracks to be oriented at the fixed angle, rather than the rotating angle. Zhu, Hsu and Lee (2001) derived a rational shear modulus that is a function of the compressive and the tensile stress–strain curves of concrete. Using this simple shear modulus, the solution algorithm of fixed-angle model became greatly simplified.

In 1995, a servo-control system (Hsu, Zhang and Gomez 1995) was installed on the universal panel tester at the University of Houston, so that it could perform strain control tests. Using this new capability, Zhang and Hsu (1998) studied high-strength concrete 2-D elements up to 100 MPa. They found that the softening coefficient was not only a function of the perpendicular tensile strain ε_1, but also a function of the compressive strength of concrete f_c'. More recently, Wang (2006) and Chintrakarn (2001) tested 2-D shear elements with large longitudinal to transverse steel ratios. These tests showed that the softening coefficient was a function of the deviation angle β. Summarizing all three variables, the softening coefficient become a function of ε_1, f_c' and β.

All the above shear theories, based either on rotating-angle or fixed-angle, could predict only the pre-peak branch of the shear stress vs shear strain curve, but not the post-peak branch of the curves, because the Poisson effect of cracked reinforced concrete was neglected. Using the strain-control feature of the universal panel tester, Zhu and Hsu (2002) quantified the Poisson effect and characterized this property by two Hsu/Zhu ratios. Taking into account the Poisson effect, Hsu and Zhu (2002) developed the *softened membrane model* (SMM) which could satisfactorily predict the entire monotonic response of the load–deformation curves, including both the pre-cracking and the post-cracking responses, as well as the ascending and the descending branches.

Mansour and Hsu (2005a,b) extended the SMM for application to reversed cyclic loading. This powerful theory, called the *cyclic softened membrane model* (CSMM), includes new constitutive relationships of concrete and mild steel bars in compressive and tensile directions of cyclic loading, as well as in the unloading and reloading stages. Consequently, CSMM is capable of predicting the hysteretic loops of RC 2-D elements subjected to cyclic loading, particularly their pinching characteristics. Furthermore, CSMM could be used to evaluate the shear stiffness, the shear ductility and the shear energy dissipation of structures subjected to dominant shear (Hsu and Mansour, 2005).

The fundamentals of shear are presented in Chapters 4. The rotating-angle shear theories, including the Mohr compatibility truss model and the rotating-angle softened truss model

(RA-STM), will be treated in Chapter 5. The fixed-angle shear theories are given in Chapter 6, including the fixed-angle softened truss model (FA-STM), the softened membrane model (SMM), and the cyclic softened membrane model (CSMM). CSMM are used in Chapter 9 and 10 to predict the static, dynamic and earthquake behavior of shear-dominant structures, such as framed shear walls, low-rise shear walls, large bridge piers, and wall-type buildings.

1.3.2.6 Torsion Theories

Torsion is a more complicated problem than shear because it is a three-dimensional (3-D) problem involving not only the shear problem of 2-D membrane elements in the tube wall, but also the equilibrium and compatibility of the whole 3-D member and the warping of tube walls that causes bending in the concrete struts. The effective thickness of the tube wall was defined by the shear flow zone (Hsu, 1990) in which the concrete strain varies from zero to a maximum at the edge, thus creating a strong strain gradient.

By incorporating the two compatibility equations of a member (relating angle of twist to the shear strain, and to the curvature of concrete struts), as well as the softened stress–strain curve of concrete, Hsu and Mo (1985a) developed a *rotating-angle softened truss model* (RA-STM) to predict the post-cracking torsional behavior of reinforced concrete members up to the peak point. Hsu and Mo's model predicted all the test results available in the literature very well, and was able to explain why Rausch's model consistently overestimates the experimental ultimate torque. In essence, the softening of the concrete increases the effective thickness of the shear flow zone and decreases the lever arm area. This, in turn, reduces the torsional resistance of the cross-section (Hsu, 1990, 1993).

The *softened membrane model* (SMM) for shear was recently applied to torsion (Jeng and Hsu, 2009). The SMM model for torsion made two improvements. First, it takes into account the bending of a 2-D element in the direction of principal tension, as well as the constitutive relationship of concrete in tension. This allows the pre-cracking torsional response to be predicted. Second, because the SMM takes into account the Poisson effect (Zhu and Hsu, 2002) of the 2-D elements in the direction of principal compression, the post-peak behavior of a torsional member can also be accurately predicted. The Poisson effect, however, must be diluted by 20% to account for the strain gradient caused by the bending of the concrete struts. As a result of these two improvements, this torsion theory become very powerful, capable of predicting the entire torque–twist curve, including the pre-cracking and post-cracking responses, as well as the pre-peak and post-peak behavior.

The rotating-angle softened truss model (RA-STM) for shear will be applied to torsion in Chapter 7. The softened membrane model (SMM), however, will not be included. Readers interested in the application of SMM to torsion are referred to the paper by Jeng and Hsu (2009).

1.4 Struts-and-ties Model

1.4.1 General Description

As discussed in Section 1.2.2, the 'local regions' of a reinforced concrete structure are those areas where stresses and strains are irregularly distributed. These regions include the knee joints, corbels and brackets, deep beams, dapped ends of beams, ledgers of spandrel beams, column ends, anchorage zone of prestressed beams, etc. The struts-and-ties model can be used for the design of local regions by providing a clear concept of the stress flows, following which the reinforcing bars can be arranged. However, the struts-and-ties model does not provide a unique solution. Since this model provides multiple solutions, the best solution may elude an engineer. The best solution is usually the one that best ensures the serviceability of the local region and its ultimate strength. Such service and ultimate behavior of a local region are difficult to predict, because they are strongly affected by the cracking and the bond slipping between the reinforcing bars and the concrete.

Since the 1980s, the struts-and-ties model has received considerable interest. Much research was carried out to study the various types of D regions. As a result, the struts-and-ties model has been considerably refined. The improvements include a better understanding of the stress flow, the behavior of the 'nodes' where the struts and ties intersect, and the dimensioning of the struts and the ties.

In the modern design concept, the local region is isolated as a free body and is subjected to boundary stresses obtained from the four action diagrams (see Section 1.2 and rows 1–3 of Table 1.1) The local region itself is imagined to be a free-form truss composed of compression struts and tension ties. The struts and ties are arranged so that the internal forces are in equilibrium with the boundary forces. In this design method the compatibility condition is not satisfied, and the serviceability criteria may not be assured. Understanding of the stress flows, the bond between the concrete and the reinforcing bars (rebars), and the steel anchorage requirement in a local region can help to improve serviceability and to prevent undesirable premature failures. A good design for a local region depends, to a large degree, on the experience of the engineer.

Proficiency in the application of this design method requires practice. An excellent treatment of the struts-and-ties model is given by Schlaich *et al.* (1987). This 77-page paper provides many examples to illustrate the application of this model. Appendix A of the ACI Code (2008) also provides good guidance for the design of some D regions.

For structures of special importance, design of local regions by the struts-and-ties model may be supplemented by a numerical analysis, such as the finite element method, to satisfy both the compatibility and the equilibrium conditions. The constitutive laws of materials may be linear or nonlinear. Although numerical analysis can clarify the stress flow and improve the serviceability, it is quite tedious, even for a first-order linear analysis.

1.4.2 Struts-and-ties Model for Beams

A struts-and-ties model has been applied to beams to resist bending and shear, as shown in Figure 1.2. Another model simulating beams to resist torsion is given in Figure 1.3. These elegant models convey clearly the message of how the internal forces are mobilized to resist the applied loads.

The cracking pattern of a simply supported beam reinforced with longitudinal bottom bars and vertical stirrups is shown in Figure 1.2(a). The struts-and-ties model for this beam carrying two symmetrical concentrated loads V is given in Figure 1.2(b). The shear and bending moment diagrams are indicated in Figure 1.2(c) and (d). To resist the bending moment, the top and bottom stringers represent the concrete compression struts and the steel tension ties, respectively. The distance between the top and bottom stringers is designated as d_v. To resist the shear forces, the truss also has diagonal concrete compression struts and vertical steel tension ties in the web. The concrete compression struts are inclined at an angle of α, because the diagonal cracks due to shear is assumed to develop at this angle with respect to the longitudinal axis. Each cell of the truss, therefore, has a longitudinal length of $d_v \cot \alpha$, except at the local regions near the concentrated loads where the longitudinal length is $(d_v \cot \alpha)/2$.

The forces in the struts and ties of this idealized truss can be calculated from the equilibrium conditions by various procedures. According to the sectional method, a cut along the section A–A on the right-hand side of the truss will produce a free body, as shown in Figure 1.2(e). Equilibrium assessment of this free body shows that the top and bottom stringers are each sub-jected to a force of $(1/2) V \cot \alpha$ and $(3/2) V \cot \alpha$, respectively. The force in the compression strut is $V/\sin \alpha$. From the vertical equilibrium of the node point a, the force in the vertical tie is V. The results from similar calculations are recorded on the left-hand side of the truss for all the struts and ties.

Figure 1.3 gives a much simplified struts-and-ties model for a beam to resist torsion. The longitudinal and hoop bars are assumed to have the same cross-sectional area and are both spaced at a constant spacing of s. The concrete compression struts are inclined at an angle of 45°. Since each hoop bar is treated as a series of straight ties of length s, a long plane truss is formed in the longitudinal direction between two adjacent longitudinal bars. A series of this kind of identical plane trusses is folded into a space truss with an arbitrary cross-section, as shown in Figure 1.3(b). Because each plane truss is capable of resisting a force F, a series of F thus form a circulatory shear flow, resulting in the torsional resistance T. It has been proven by Rausch (1929), using the equilibrium conditions at the node points, that the force F is related to the torsional moment T by the formula, $F = Ts/2A_o$, where A_o is the cross-sectional area within the truss or the circulatory shear flow (see derivation of Equation (2.46) in Section 2.1.4, and notice the shear flow $q = F/s$).

1.4.3 Struts-and-ties Model for Knee Joints

The knee joint, which connects a beam and a column at the top left-hand corner of the frame, as shown in Figure 1.1, will be used to illustrate how the struts-and-ties model are utilized to help design the reinforcing bars (rebars). Under gravity loads the knee joint is subjected to a closing moment, as shown in Figure 1.4(a). The top rebars in the beam and the outer rebars in the column are stretched by tension, while the bottom portion of the beam and the inner portion of the column are under compression. If the frame is loaded laterally, say by earthquake forces, the knee joint may be subjected to an opening moment, as shown in Figure 1.4(b). In this case the bottom rebars of the beam and the inner rebars of the column are stretched by tension, while the top portion of the beam and the outer portion of the column are under compression. For simplicity, the small shear stresses on the boundaries of the knee joint are neglected.

Three rebar arrangements are shown in Figure 1.4 (c–e). Figure 1.4(c) gives a type of rebar arrangement frequently utilized to resist a closing moment. In this type of arrangement the top and bottom rebars of the beam are connected to form a loop in the joint region. A similar loop is formed by the outer and inner rebars of the column. This type of arrangement is very attractive, because the separation of the beam steel from the column steel makes the construction easy. Unfortunately, tests (Swann, 1969) have shown that the strength of such a joint can be as low as 34% of the strength of the governing member (i. e. the beam or the column, whichever is less).

Two examples of incorrect arrangements of rebars to resist an opening moment are shown in Figure 1.4(d) and (e). In Figure 1.4(d) the bottom tension rebars of the beam are connected

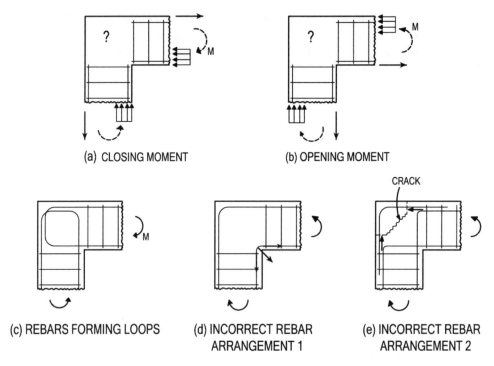

Figure 1.4   Knee joint moments and incorrect reinforcement

to the inner tension rebars of the column, while the top compression steel of the beam is connected to the outer compression steel of the column. This way of connecting the tension rebars of the beam and the column is obviously faulty, because the bottom tensile force of the beam and the inner tensile force of the column would produce a diagonal resultant that tends to straighten the rebar and to tear out a chunk of concrete at the inner corner. In fact, the strength of such a knee joint is only 10% of the strength of the governing member according to Swann's tests.

In Figure 1.4(e), the bottom tension rebars of the beam are connected to the outer compression rebars of the column, while the top compression rebars of the beam are connected to the inner tension rebars of the column. Additional steel bars would be needed along the outer edge to protect the concrete core of the joint region. Such an arrangement also turns out to be flawed. The compression force at the top of the beam and the compression force at the outer portion of the column tend to push out a triangular chunk of concrete at the outer corner of the joint after the appearance of a diagonal crack. The strength of such a knee joint could be as low as 17% of the strength of the governing member (Swann, 1969).

1.4.3.1 Knee Joint under Closing Moment

The struts-and-ties models will now be used to guide the design of rebars at the knee joint. The most important idea in the selection of the struts-and-ties assembly is to recognize the stress flow in the local region. The concrete struts should follow the compression trajectories

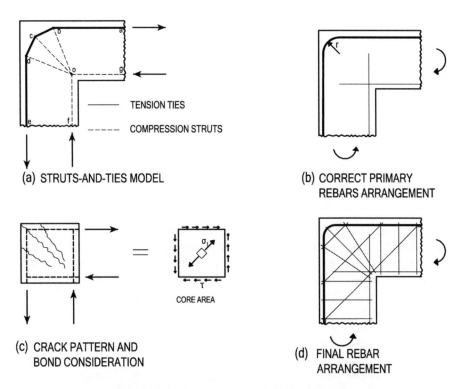

TENSION TIES

COMPRESSION STRUTS

(a) STRUTS-AND-TIES MODEL

(b) CORRECT PRIMARY
    REBARS ARRANGEMENT

CORE AREA

(c) CRACK PATTERN AND
    BOND CONSIDERATION

(d) FINAL REBAR
    ARRANGEMENT

Figure 1.5 Rebar arrangement for closing moment

as closely as possible, and the steel ties should trace the tension trajectories. A model that observes the stress flow pattern is expected to best satisfy the compatibility condition and the serviceability requirements.

Figure 1.5(a) shows the struts-and-ties model for the knee joint subjected to closing moment. First, the top tensile rebars of the beam and the outer tensile rebars of the column are represented by the ties (solid lines), ab and de, respectively. The centroidal lines of the compression zones in the beam and in the column are replaced by the struts (dotted lines), og and fo, respectively. Since the tensile stresses should flow from the top rebar of the beam to the outer rebar of the column, the top tie, ab, is connected to the outer tie, de, by two straight ties, bc and cd, along the outer corner. Because of the changes of angles at the node points, b, c and d, the tensile force in the ties will produce a resultant bearing force on the concrete at each node point directed toward the point o at the inner corner of the joint. Following these compressive stress trajectories, three diagonal struts bo, co and do, are installed. The three compressive forces acting along these diagonal struts and meeting at the node point o must be balanced by the two compression forces in the struts og and fo. As a whole, we have a stable struts-and-ties assembly that is in equilibrium at all the node points. The forces in all the struts and ties can be calculated.

It is clear from this struts-and-ties model that the correct arrangement of the primary tensile rebar should follow the tension ties abcde, as shown in Figure 1.5(b). The radius of bend r of the rebars has a significant effect on the local crushing of concrete under the curved portion, as well as the compression failure of the diagonal struts near the point o. According to an extensive series of full-scale tests by Bai *et al.* (1991), the reinforcement index, $\omega = \rho f_y / f_c'$, of the tension steel at the end section of the beam should be a function of the radius ratio r/d, and should be limited to:

$$\omega = 2.2 \frac{r}{d} \tag{1.1}$$

and

$$\omega = 0.33 + 0.2 \frac{r}{d} \quad \text{for} \quad \frac{r}{d} \leq 0.6 \tag{1.2}$$

where d is the effective depth at the end section of the beam. The compressive strength of the standard cylinder f_c' has been taken as 77% of the compressive strength of the 20 cm cubes used in the tests to define the reinforcement index ω. Equation (1.1) is governed by the local crushing of concrete directly under the curved portion of the tension rebars, and Equation (1.2) by the compression failure of the diagonal struts near the point o.

In addition to the primary tension rebars, the two compression rebars should first follow the compression struts, og and fo, and then each be extended into the joint region for a length sufficient to satisfy the compression anchorage requirement, Figure 1.5(b).

A comparison of Figure 1.4(c) with Figure 1.5(b) reveals why the rebar arrangement in Figure 1.4(c) is deficient. Instead of connecting the top rebar of the beam and the outer rebar of the column by a single steel bar, these two rebars in Figure 1.4(c) are actually spliced together along the edge of the outer corner. This kind of splicing is notoriously weak because the splice is unconfined along the edge of the outer corner and its length is limited. If splices are desired, they should be located away from the joint region and be placed in a well-confined region of the member, either inside the column or inside the beam.

In addition to the primary rebars, we must now consider the secondary rebar arrangement. Secondary rebars are provided for two purposes. First, they are designed to control cracks, and second, they are added to prevent premature failure. The crack pattern of a knee joint under a closing moment is shown in Figure 1.5(c). The direction of the cracks is determined by the stress state in the joint region. To understand this stress state, we notice that the four tension and compression rebars introduce, through bonding, the shear stresses τ around the core area. The shear stress produces the principal tensile stress σ_1, which determines the direction of the diagonal cracks as indicated. To control these diagonal cracks, a set of opposing diagonal rebars perpendicular to the diagonal cracks are added, as shown in Figure 1.5(d).

Figure 1.5(d) also includes a set of inclined closed stirrups which radiate from the inner corner toward the outer corner. This set of closed stirrups is added to prevent two possible types of premature failure. First, since the curved portion of a primary tension rebar in Figure 1.5(b) exerts a severe bearing pressure on the concrete, it could split the concrete directly beneath, along the plane of the frame. The outer horizontal branches of the inclined stirrups, which are perpendicular to the plane of the frame, would serve to prevent such premature failures. Second, concrete in the vicinity of point o is subjected to extremely high compression stresses from all directions. The three remaining branches of the inclined closed stirrups would serve to confine the concrete in this area.

1.4.3.2 Knee Joint under Opening Moment

A struts-and-ties model (model 1) for the knee joint subjected to an opening moment is shown in Figure 1.6(a). This model is essentially a reverse case of the model for a closing moment (Figure 1.5a). This means that the ties and the struts are interchanged. Such a model should also be stable and in equilibrium. According to this model, the correct arrangement of the primary tensile rebars should follow the tension ties, as shown in Figure 1.6(b). It is noted that this set of radially oriented rebars should, in reality, be designed as closed stirrups because of anchorage requirements.

A second struts-and-ties model (model 2) for opening moment is given in Figure 1.6(c). The rebar arrangement according to this model is given in Figure 1.6(d). This model explains why the bottom tension rebar of the beam should be connected to the inner tension rebar of the column by first forming a big loop around the core of the joint area. The connection of the rebars may be achieved by welding.

Design of rebars according to a combination of models 1 and 2 is given in Figure 1.6(f). The big loop of the tension rebar is shown to be formed by splicing rather than welding. The closed stirrups in the radial direction could also serve to control cracking. The direction of the crack is shown in Figure 1.6(e). The final rebar arrangement also includes the diagonal rebars, ab, perpendicular to the diagonal line connecting the inner and outer corners. The effectiveness of such diagonal rebars is explained by the two struts-and-ties models in Figure 1.7. The first one (Figure 1.7a), is an extension of model 1 in Figure 1.6(a), while the second (Figure 1.7b), is a generalization of model 2 in Figure 1.6(c).

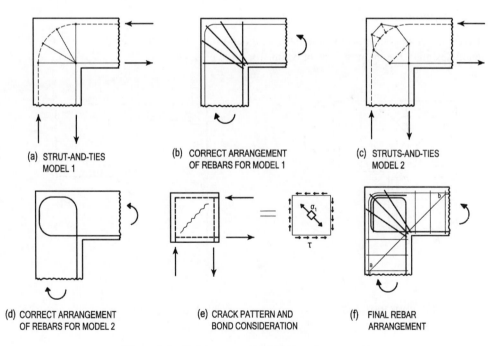

(a) STRUT-AND-TIES MODEL 1

(b) CORRECT ARRANGEMENT OF REBARS FOR MODEL 1

(c) STRUTS-AND-TIES MODEL 2

(d) CORRECT ARRANGEMENT OF REBARS FOR MODEL 2

(e) CRACK PATTERN AND BOND CONSIDERATION

(f) FINAL REBAR ARRANGEMENT

Figure 1.6   Rebar arrangement for opening moment

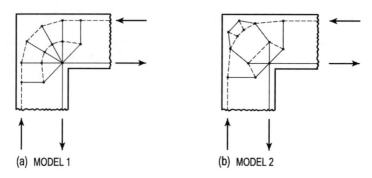

(a) MODEL 1   (b) MODEL 2

Figure 1.7   Struts-and-ties models showing effectiveness of diagonal rebars

1.4.3.3 Knee Joint under both Closing and Opening Moments

The rebar arrangement in a knee joint subjected to both closing and opening moments, as in earthquake loadings, is shown in Figure 1.8. This arrangement is a combination of the two schemes designed separately for a closing moment (Figure 1.5d) and for an opening moment (Figure 1.6f).

Knee joints can also be strengthened by adding a fillet, as shown in Figure 1.8(b). The rebar arrangement could be simplified in such a joint, because the stresses at the joint region are significantly reduced.

1.4.4 Comments

Despite the clear concepts inherent in the struts-and-ties models, they are difficult to use in the actual design of the main regions of beams for three reasons.

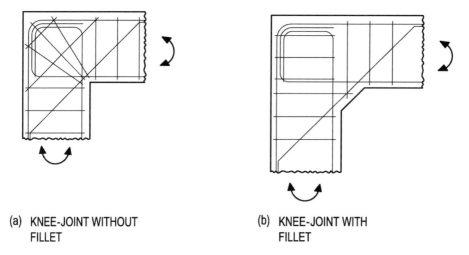

(a) KNEE-JOINT WITHOUT   (b) KNEE-JOINT WITH
     FILLET   FILLET

Figure 1.8   Knee-joint rebar arrangement for both closing and opening moments (earthquake loading)

First, the design can be very tedious, since the forces in every strut and tie must be calculated and scaled proportionally. Instead of treating a beam as a whole to simplify the design, the struts-and-ties model is heading in the opposite direction by treating a beam as an assembly of a large number of struts and ties.

Second, there are no definite objective criteria for the selection and the proportioning of the struts and ties, even though recommendations have been made.

Third, and probably of most importance, there are indeed better and more sophisticated models for the design of beams, which can take into account the compatibility condition and the constitutive laws of materials. As mentioned in Section 1.2.2, Bernoulli's linear compatibility can be used for bending and axial loads in the main regions, and Mohr's circular compatibility can be used for shear and torsion. Detailed study of these better models is the main objective of this book.

Nevertheless, struts-and-ties models are well suited to guide the design of local regions, such as knee joints, for two reasons. First, the strain compatibility conditions in the local regions are usually too complicated to be employed. We have no choice but to rely on the stress equilibrium condition. Second, the local regions are of limited lengths compared with the main regions (see Figure 1.1). Since the number of struts and ties in a local region is limited, the process of designing struts and ties is relatively easy.

2

Equilibrium (Plasticity) Truss Model

2.1 Basic Equilibrium Equations

The basic principles of the equilibrium (plasticity) truss model are stated in Section 1.3.1.2. Because the steel bars are assumed to yield at failure, this model is applicable to all four actions (moment, axial load, shear and torsion) at the ultimate load stage. In this section, we will derive the equilibrium equations for bending, element shear, beam shear and torsion. In Section 2.2 the interactions of these four actions will be presented. Since ACI shear and torsion provisions are based on the equilibrium (plasticity) truss model, we will study these ACI provisions in Section 2.3, including an example problem in Section 2.3.5.

2.1.1 Equilibrium in Bending

The theory of bending will be carefully studied in Chapter 3. In this chapter we will introduce only the truss concept of bending so that it can be conceptually integrated with shear, torsion and axial loads.

In a member subjected to bending, the external moment M acting on the cross-section (Figure 2.1a), is resisted by an internal couple (Figure 2.1b). The internal couple consists of the compression force C and the steel tensile force $T = A_s f_s$. The distance between C and T is the lever arm jd. Taking the equilibrium of the external moment and the internal moment gives:

$$M = A_s f_s(jd) \tag{2.1}$$

In Chapter 3, we will derive $j = (1 - k_2 c/d)$, where c is the depth of the compression zone and is determined by Bernoulli's compatibility condition. The coefficient k_2 is a function of the shape of the stress–strain curve of concrete.

The internal couple concept gives rise to the truss model concept for flexure shown in Figure 2.1(c). In this model, the tensile steel bars are concentrated at the geometric centroid and constitute the tension stringer. This tension stringer is capable of resisting the tensile force

Unified Theory of Concrete Structures   Thomas Hsu and Yi-Lung Mo
© 2010 John Wiley & Sons, Ltd

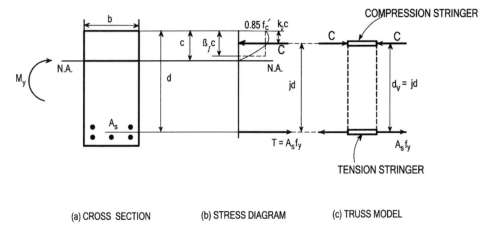

(a) CROSS SECTION   (b) STRESS DIAGRAM   (c) TRUSS MODEL

Figure 2.1   Truss model in bending

$A_s f_s$. The concrete area within the compression stress block is also assumed to concentrate at the location of the resultant C. This concrete element is known as the compression stringer and is capable of resisting the resultant C. The distance between the tension and compression stringers is designated d_v, which, of course, is equal to jd.

For under-reinforced members, the steel yields before the concrete crushes. At the yield condition the steel stress f_s becomes f_y, where f_y is the yield strength of the tensile steel. To simplify the analysis, the ACI Code allows the replacement of the curved concrete stress block by a rectangular one, shown by the dotted line in Figure 2.1(b). The stress of the rectangular stress block is $0.85 f_c'$ and the depth is $\beta_1 c$. The coefficient β_1 is taken as 0.85 for $f_c' = 27.6$ MPa (4000 psi) or less, and decreases by 0.05 for every increment of 6.9 MPa (1000 psi) beyond $f = 27.6$ MPa (4000 psi). Based on this equivalent rectangular stress block, the magnitude of the resultant $C = 0.85 f_c' b \beta_1 c$ and the coefficient $k_2 = \beta_1/2$. Moment equilibrium shown in Figure 2.1(b) gives:

$$M = A_s f_y d \left(1 - \frac{\beta_1 c}{2d} \right) \tag{2.2}$$

where c is obtained from the force equilibrium, $T = C$:

$$c = \frac{A_s f_y}{0.85 f_c' b \beta_1} \tag{2.3}$$

Equations (2.2) and (2.3) show that the lever arm jd is equal to $(1 - \beta_1 c/2d)d$. For normal flexural members with steel reinforcement of 1–1.5%, jd or d_v is approximately $0.9d$.

2.1.2 Equilibrium in Element Shear

2.1.2.1 Element Shear Equations

A 2-D element subjected to a shear flow q is shown in Figure 2.2(a). The element has a thickness of h and is square, with unit length in both directions. The reinforcing steel bars are

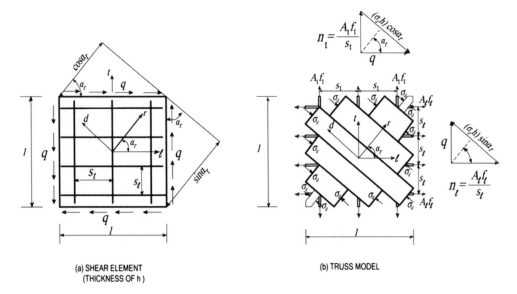

(a) SHEAR ELEMENT
(THICKNESS OF h)

(b) TRUSS MODEL

Figure 2.2   Equilibrium in element shear

defined in the $\ell - t$ coordinate, where the ℓ axis (horizontal axis) is in the direction of the longitudinal steel bars with a uniform spacing of s_ℓ. The transverse steel bars are arranged in the t axis (vertical axis) with a uniform spacing of s_t. After cracking, the concrete is separated by diagonal cracks into a series of concrete struts, as shown in Figure 2.2(b). The cracks are defined in the $r - d$ coordinate, where the r axis is normal to the cracks and the d axis is in the direction of the diagonal concrete struts. The $r - d$ coordinate is oriented at an angle α_r with respect to the $\ell - t$ coordinate. The diagonal concrete struts, the longitudinal steel bars and the transverse steel bars form a truss which is capable of resisting the shear flow q.

Equilibrium in the longitudinal direction is shown by the force triangle on the right face of the shear element, Figure 2.2(b). The shear flow q pointing upward is resisted jointly by a longitudinal steel force n_ℓ and a diagonal concrete force, $(\sigma_d h) \sin \alpha_r$. The steel force n_ℓ is defined as the longitudinal steel force per unit length, $A_\ell f_\ell / s_\ell$, where A_ℓ is the cross-sectional area of one longitudinal steel bar and f_ℓ is the stress in the longitudinal steel bars. The concrete force $(\sigma_d h) \sin \alpha_r$ represents the diagonal concrete stress σ_d acting on a thickness of h and a width of $\sin \alpha_r$. The $\sin \alpha_r$ relationship is shown by the geometry in Figure 2.2(a). From this force triangle the shear flow q can be related to the longitudinal steel force n_ℓ by the geometry:

$$q = n_\ell \cot \alpha_r \tag{2.4}$$

Similarly, equilibrium in the transverse direction is shown by the force triangle on the top face of the shear element, Figure 2.2(b). The shear flow q pointing to the right is resisted jointly by a transverse steel force n_t and a diagonal concrete force $(\sigma_d h) \cos \alpha_r$. The steel force n_t is defined as the transverse steel force per unit length, $A_t f_t / s_t$, where A_t is the cross-sectional area of one transverse steel bar and f_t is the stress in the transverse steel bars. The concrete force $(\sigma_d h) \cos \alpha_r$ represents the diagonal concrete stress σ_d, acting on a thickness of h and a width of $\cos \alpha_r$. The $\cos \alpha_r$ relationship is also shown by the geometry in Figure 2.2(a). From this force triangle the shear flow q can be related to the transverse steel force n_t by the geometry:

$$q = n_t \tan \alpha_r \tag{2.5}$$

The shear flow q can be related to the diagonal concrete stress σ_d using either the force triangle in the longitudinal direction or the force triangle in the transverse direction. From the geometry of the triangles we obtain:

$$q = (\sigma_d h) \sin \alpha_r \cos \alpha_r \tag{2.6}$$

Assuming that yielding occurs in both the longitudinal and transverse steel, then $n_\ell = n_{\ell y} = A_\ell f_{\ell y}/s_\ell$ and $n_t = n_{ty} = A_t f_{ty}/s_t$, where $n_{\ell y}$ and n_{ty} are the longitudinal and transverse yield force per unit length, respectively, and $f_{\ell y}$ and f_{ty} are the longitudinal and transverse yield stress, respectively. Combining Equations (2.4) and (2.5) to eliminate α_r we obtain:

$$q_y = \sqrt{n_{\ell y} n_{ty}} \tag{2.7}$$

where q_y is the shear flow at yielding. Also, combining Equations (2.4) and (2.5) to eliminate q, we have

$$\tan \alpha_r = \sqrt{\frac{n_{\ell y}}{n_{ty}}} \tag{2.8}$$

Equation (2.7) states that the shear flow at yielding q_y is the square-root-of-the-product average of the steel yield forces in the two directions. Equation (2.8) shows that the angle α_r at yield depends on the ratio of the longitudinal to transverse steel yield forces, $n_{\ell y}/n_{ty}$.

In design, the shear flow at yield is usually given. The aim of the design is to find the yield reinforcement in both directions, $n_{\ell y}$ and n_{ty}, and to check the diagonal concrete stress σ_d, so that the concrete will not crush before the yielding of steel. For this purpose Equations (2.4)–(2.6) can be written in the following forms:

$$n_{\ell y} = q_y \tan \alpha_r \tag{2.9}$$
$$n_{ty} = q_y \cot \alpha_r \tag{2.10}$$

$$\sigma_d = \frac{q_y}{h \sin \alpha_r \cos \alpha_r} \tag{2.11}$$

2.1.2.2 Geometric Relationships of Equilibrium Equations

The equations in Section 2.1.2.1. are expressed in terms of q, n_ℓ and n_t, which represent forces per unit length. In order to express these equations in terms of stresses, we divide q, n_ℓ and n_t by h, the thickness of the element, and define these three stress terms as follow:

$$\tau_{\ell t} = \frac{q}{h} = \text{smeared shear stress}$$

$$\rho_\ell f_\ell = \frac{A_\ell f_\ell}{s_\ell h} = \frac{n_\ell}{h} = \text{smeared steel stress in longitudinal direction}$$

$$\rho_t f_t = \frac{A_t f_t}{s_t h} = \frac{n_t}{h} = \text{smeared steel stress in transverse direction}$$

The three equilibrium equations (2.4)–(2-6) can then be written as:

$$\rho_\ell f_\ell = \tau_{\ell t} \tan \alpha_r \tag{2.12}$$

$$\rho_t f_t = \tau_{\ell t} \cot \alpha_r \tag{2.13}$$

$$\sigma_d = \tau_{\ell t} \frac{1}{\sin \alpha_r \cos \alpha_r} = \tau_{\ell t} (\tan \alpha_r + \cot \alpha_r) \tag{2.14}$$

Equations (2.12)–(2,14) can also be expressed in terms of the diagonal concrete stress σ_d by substituting $\tau_{\ell t}$ from Equation (2.14) into Equations (2.12) and (2.13):

$$\rho_\ell f_\ell = \sigma_d \sin^2 \alpha_r \tag{2.15}$$

$$\rho_t f_t = \sigma_d \cos^2 \alpha_r \tag{2.16}$$

$$\tau_{\ell t} = \sigma_d \sin \alpha_r \cos \alpha_r \tag{2.17}$$

These two sets of equations (2.12)–(2.14) and (2.15)–(2.17) each satisfy the Mohr circle, as illustrated in Figure 2.3. Figure 2.3(a) shows the Mohr circle in a $\sigma - \tau$ coordinate system, where σ is the normal stress and τ the shear stress. Point A on the Mohr circle represents the longitudinal face (the direction of a face is defined by its normal axis). Acting on this face are a shear stress $\tau_{\ell t}$, indicated by the vertical coordinate, and a longitudinal steel stress $\rho_\ell f_\ell$, indicated by the horizontal coordinate. At 180° from point A, we have point B which represents the transverse face in the shear element. Acting on this face are a shear stress $-\tau_{\ell t}$ (vertical coordinate) and a transverse steel stress $\rho_t f_t$, (horizontal coordinate). Point C on the σ axis of the Mohr circle represents the principal face in the r direction, i.e. the face of cracks. On this crack face the normal stress is assumed to be zero. The incident angle $2\alpha_r$ between points A and C in the Mohr circle is twice the angle α_r between the ℓ axis and the r axis in the shear element. At 180° from point C, we have point D which represents a face on which only the diagonal concrete stress σ_d is acting. The above explanation of the application of the Mohr circle shows its importance in the analysis of stresses in various directions. Detailed derivation and discussion of Mohr circles will be provided in Chapter 4.

The introduction of the Mohr circle at this point is intended to illustrate the geometric relationships of Equations (2.12)–(2.17). The geometric relationship of the half Mohr circle defined by ACD are illustrated in Figure 2.3(b) and (c). If CD in Figure 2.3(b) is taken as unity, then CE $= \sin^2 \alpha_r$, ED $= \cos^2 \alpha_r$, and AE $= \sin \alpha_r \cos \alpha_r$. These three trigonometric values are actually the ratios of the three stresses $\rho_\ell f_\ell$, $\rho_t f_t$ and $\tau_{\ell t}$, respectively, divided by the diagonal concrete stress σ_d, in the Mohr circle of Figure 2.3(a). Hence, the set of three equilibrium equations, (2.15)–(2.17), simply states the geometric relationships illustrated in Figure 2.3(b). This set of three equilibrium equations will be called the *first type* of expression for Mohr stress circles in Section 5.1.2.

Similarly, if AE in Figure 2.3(c) is taken as unity, then CE $= \tan \alpha_r$, ED $= \cot \alpha_r$, and CD $= (\tan \alpha_r + \cot \alpha_r) = 1/\sin \alpha_r \cos \alpha_r$. These three trigonometric values are actually the ratios of the three stresses $\rho_\ell f_\ell$, $\rho_t f_t$ and σ_d, respectively, divided by the shear stress $\tau_{\ell t}$, in the Mohr circle of Figure 2.3(a). Hence, the set of three equilibrium equations, (2.12)–(2.14), simply states the geometric relationships illustrated in Figure 2.3(c). This set of three

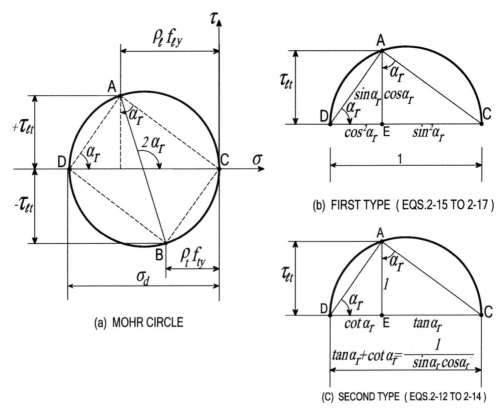

Figure 2.3 Geometric relationships in mohr circle

equilibrium equations will be called the *second type* of expression for Mohr stress circles in Section 5.1.3.

2.1.2.3 Yielding of Reinforcement in 2-D Elements

The plasticity truss model is based on the assumption that both the longitudinal and the transverse steel must yield before failure. In order to ensure this mode of failure, the shear elements are divided into two types: under-reinforced and over-reinforced. In an under-reinforced element the yielding of steel in both directions occurs before the crushing of concrete, thus satisfying the assumption of the plasticity truss model. In an over-reinforced element, however, concrete crushes before the yielding of either the longitudinal or the transverse steel, or both, thus violating the basic assumption. The state of stresses that divide the under-reinforced mode of failure from the over-reinforced will be called the balanced condition. Under the balanced condition, one of the two yielded steels just reach yield point when the concrete crushes at an effective stress of $\zeta f'_c$. The softening coefficient ζ varies from about 0.2 to 0.6 (this will be studied in detail in Chapters 5 and 6). In this chapter we will take ζ as a constant 0.4, a crude average value.

Balanced Condition

The criterion for the balanced condition in the plasticity truss model was defined by Nielsen and Braestrup (1975). Adding Equations (2.15) and (2.16) and noticing that $\sin^2 \alpha_r + \cos^2 \alpha_r = 1$ gives:

$$\rho_\ell f_\ell + \rho_t f_t = \sigma_d \tag{2.18}$$

At the balanced condition both the steel yields ($f_\ell = f_{\ell y}$, $f_t = f_{ty}$) while the concrete crushes at an effective stress of $\zeta f_c'$. Equation (2.18) then becomes

$$\rho_\ell f_{\ell y} + \rho_t f_{ty} = \zeta f_c' \tag{2.19}$$

Define

$$\omega_\ell = \frac{\rho_\ell f_{\ell y}}{\zeta f_c'} = \text{longitudinal reinforcement index}$$

$$\omega_t = \frac{\rho_t f_{ty}}{\zeta f_c'} = \text{transverse reinforcement index}$$

The balanced condition Equation (2.19), can be written in a nondimensional form:

$$\omega_\ell + \omega_t = 1 \tag{2.20}$$

Under-reinforced Elements

When $\omega_\ell + \omega_t < 1$, we have an under-reinforced element where both the steel yields ($f_\ell = f_{\ell y}$, $f_t = f_{ty}$) before the crushing of concrete. The shear stress $\tau_{\ell t}$ becomes $\tau_{\ell t y}$. Substituting $\sin^2 \alpha_r$ from Equation (2.15) and $\cos^2 \alpha_r$ from Equation (2.16) into Equation (2.17) results in

$$\tau_{\ell t y} = \sqrt{(\rho_\ell f_{\ell y})(\rho_t f_{ty})} \tag{2.21}$$

The angle α_r for this under-reinforced element can be obtained by dividing Equation (2.15) by Equation (2.16):

$$\tan \alpha_r = \sqrt{\frac{\rho_{\ell y} f_{\ell y}}{\rho_{ty} f_{ty}}} \tag{2.22}$$

Dividing both sides of Equation (2.21) by $\zeta f_c'$ we obtain

$$\frac{\tau_{\ell t y}}{\zeta f_c'} = \sqrt{\omega_\ell \omega_t} \tag{2.23}$$

The nondimensional ratio, $\tau_{\ell t y}/\zeta f_c'$, will be called the shear stress ratio. Dividing both the numerator and denominator in the square root by $\zeta f_c'$ Equation (2.22) can also be written as

$$\tan \alpha_r = \sqrt{\frac{\omega_\ell}{\omega_t}} \tag{2.24}$$

The three pairs of equations: Equations (2.7) and (2.8); Equations (2.21) and (2.22); and Equations (2.23) and (2.24), are actually the same, except that they are expressed in terms of different units. Equations (2.7) and (2.8) are in terms of force per unit length (q_y, $n_{\ell y}$, n_{ty}).

Equations (2.21) and (2.22) are in terms of stresses ($\tau_{\ell ty}$, $\rho_\ell f_{\ell y}$, $\rho_t f_{ty}$). Equations (2.23) and (2.24) are in terms of the nondimensional indices $\tau_{\ell ty}/\zeta f_c'$, ω_ℓ, ω_t.

Over-reinforced Elements

When $\omega_\ell + \omega_t > 1$, we have an over-reinforced element where the concrete crushes before the yielding of steel in one or both directions. Since this failure mode violates the basic assumption of the plasticity truss model, the design of such an element is unacceptable and no further discussion is necessary.

Three Cases of the Balanced Condition

The balanced condition, $\omega_\ell + \omega_t = 1$, can be divided into three cases:

Case (1): $\omega_\ell = \omega_t$

When $\omega_\ell = \omega_t$, the yielding of both the longitudinal and the transverse steel occur simultaneously with the crushing of concrete. The balanced condition in this case gives $\omega_\ell = \omega_t = 0.5$, resulting in

$$\frac{\tau_{\ell ty}}{\zeta f_c'} = 0.5 \tag{2.25}$$

$$\alpha_r = 45° \tag{2.26}$$

Case (2): $\omega_t < 0.5$

In this case the transverse steel has yielded, but the concrete crushes simultaneously with the yielding of the longitudinal steel. The longitudinal reinforcement index becomes $\omega_\ell = 1 - \omega_t$ according to the balanced condition, and Equation (2.23) becomes

$$\frac{\tau_{\ell ty}}{\zeta f_c'} = \sqrt{\omega_t(1 - \omega_t)} \tag{2.27}$$

The corresponding angle α_r from Equation (2.24) is

$$\tan \alpha_r = \sqrt{\frac{(1 - \omega_t)}{\omega_t}} > 1, \ \alpha_r > 45° \tag{2.28}$$

Case (3): $\omega_\ell < 0.5$

In this case the longitudinal steel has yielded, but the concrete crushes simultaneously with the yielding of the transverse steel. The transverse steel will then be determined by the balanced condition, $\omega_t = 1 - \omega_\ell$, and Equation (2.23) becomes

$$\frac{\tau_{\ell ty}}{\zeta f_c'} = \sqrt{\omega_\ell(1 - \omega_\ell)} \tag{2.29}$$

The corresponding angle α_r is

$$\tan \alpha_r = \sqrt{\frac{\omega_\ell}{(1 - \omega_\ell)}} < 1, \ \alpha_r < 45° \tag{2.30}$$

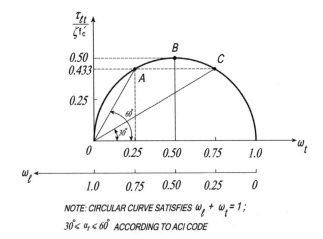

Figure 2.4   Circular relationship between shear stress ratio ($\tau_{\ell t}/\zeta f'_c$) and steel stress indices (ω_ℓ, ω_t)

2.1.2.4 Design Limitations

The balanced condition can also be expressed graphically by a semicircular curve in a $\tau_{\ell ty}/\zeta f'_c$ vs ω_t diagram, as shown in Figure 2.4. This diagram is derived in the following manner. Squaring both sides of Equation (2.27) gives

$$\left(\frac{\tau_{\ell ty}}{\zeta f'_c}\right)^2 + \omega_t^2 - \omega_t = 0 \tag{2.31}$$

Adding 0.5^2 to both sides of Equation (2.31) results in

$$\left(\frac{\tau_{\ell ty}}{\zeta f'_c}\right)^2 + (\omega_t - 0.5)^2 = 0.5^2 \tag{2.32}$$

When $\tau_{\ell ty}/\zeta f'_c$ is plotted against ω_t in Figure 2.4, Equation (2.32) represents a circle with radius 0.5 and center located on the ω_t axis at $\omega_t = 0.5$. This circle gives the nondimensional relationship between the shear stress $\tau_{\ell t}$ and the transverse steel stress, $\rho_t f_{ty}$. When the longitudinal steel is chosen to satisfy the balanced condition, $\omega_\ell + \omega_t = 1$, an axis pointing to the left is also drawn for ω_ℓ.

Substituting ω_ℓ from Equation (2.23) into Equation (2.24) gives

$$\tan \alpha_r = \frac{(\tau_{\ell ty}/\zeta f'_c)}{\omega_t} \tag{2.33}$$

Equation (2.33) shows that the angle α_r represents the slope of a straight line through the origin in Figure 2.4. In the ACI Code, the angle α_r is limited to a range of 30° to 60°, represented by the straight line OC and OA. At these limits, the maximum shear stress ratio $\tau_{\ell ty}/\zeta f'_c$ is 0.433.

Design of reinforcement within the semicircle will give an under-reinforced element, while the region outside the semicircle represents over-reinforcement. Design within the fan-shaped area OABC further satisfies the ACI provisions that limit the range of α_r. On the semicircular curve which expresses the balanced condition, point B represents case (1) of the balanced condition, while the arcs AB and BC give case (2) and case (3), respectively.

In summary, the design of reinforcement in 2-D elements to resist shear is subjected to two limitations. First, the range of $30° \leq \alpha_r \leq 60°$ is specified to prevent excessive cracking. This purpose will be further discussed in Section 5.2.4, after the strain compatibility conditions are introduced. Second, the balanced condition is imposed to ensure the yielding of steel in both directions at failure. To achieve this purpose, a more rigorous method will be elaborated in Section 5.4.7.

2.1.2.5 Minimum Reinforcement

Before cracking, an element subjected to shear stress $\tau_{\ell t}$ will produce a principal tensile stress σ_1, which is equal in magnitude ($\tau_{\ell t} = \sigma_1$) and inclined at an angle of 45°. At cracking, σ_1 is assumed to reach the tensile strength of concrete f_t'. Substituting $\tau_{\ell t} = f_t'$ into Equations (2.12) and (2.13) we have

$$\rho_\ell f_\ell = f_t' \tan \alpha_r \qquad (2.34)$$
$$\rho_t f_t = f_t' \cot \alpha_r \qquad (2.35)$$

Summing Equations (2.34) and (2.35) gives

$$\rho_\ell f_\ell + \rho_t f_t = \frac{f_t'}{\sin \alpha_r \cos \alpha_r} \qquad (2.36)$$

A minimum amount of reinforcement to ensure that the steel will not yield immediately at cracking can be obtained by assuming the yielding of the steel $f_\ell = f_t = f_y$, and by noticing that $\sin \alpha_r \cos \alpha_r$ is very close to 0.5 when α_r is in the vicinity of 45°. Then Equation (2.36) becomes:

$$\rho_\ell + \rho_t = \frac{2f_t'}{f_y} \qquad (2.37)$$

For a typical case when $f_y = 413$ MPa (60 000 psi) and $f_t' = 2.07$ MPa (300 psi), Equation (2.37) gives the minimum total reinforcement $(\rho_\ell + \rho_t)_{min}$ of 1%. If the total reinforcement is equally distributed in the longitudinal and transverse directions, each direction will require a minimum steel percentage of 0.5%.

In the ACI Code the minimum horizontal reinforcement in walls is specified to be in the range of 0.20–0.25%, and the vertical reinforcement in the range of 0.12–0.15%. These ACI requirements provide only about 1/4 to 1/2 the minimum reinforcement required by Equation (2.37). Because most walls in use are designed with steel percentages close to the ACI minimum requirement, it is obvious that these walls will fail in a brittle manner upon cracking. The common notion that shear failure of walls is brittle stems primarily from this practice of supplying insufficient amount of steel in the walls. If a minimum steel requirement was provided according to Equation (2.37), then the wall is expected to behave in a ductile manner.

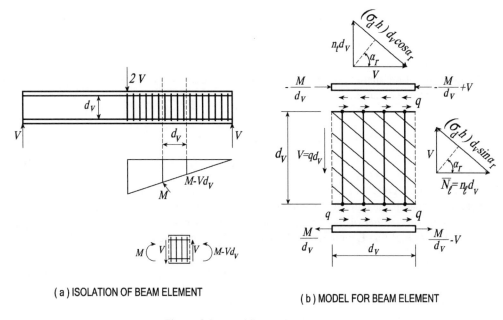

(a) ISOLATION OF BEAM ELEMENT

(b) MODEL FOR BEAM ELEMENT

Figure 2.5   Equilibrium in beam shear

2.1.3 Equilibrium in Beam Shear

A beam subjected to a concentrated load $2V$ at midspan is shown in Figure 2.5(a). Since the reaction is V, the shear force is a constant V throughout one-half of the beam, and the moment diagram is a straight line. When a beam element of length d_v is isolated and the moment on the left face is defined as M, then the moment on the right face is $M - Vd_v$. The shear forces on both the left and right faces are, of course, equal to V.

A model of the isolated beam element is shown in Figure 2.5(b). The top and bottom stringers are separated from the main body of the beam element, so that the mechanism to resist shear can be separated from the mechanism to resist moment. The stringers are resisting the bending moment and the main body is carrying the shear force. Two assumptions are made in the establishment of the model:

1. The shear flow q in the main body is distributed uniformly over the depth d_v (i.e. in the transverse direction). Since q is a constant over the depth, $V = \int q \mathrm{d}(d_v) = q d_v$.
2. The shear flow q in the main body is also distributed uniformly along the length of the main body (i.e. in the longitudinal direction). Hence, the transverse steel stresses (f_t) and the stresses in the diagonal concrete struts (σ_d) vary uniformly along their lengths.

Based on these two assumptions the main body can now be treated as a large shear element discussed in Section 2.1.2. The only difference between the shear element in Figure 2.2 and the main body of the beam in Figure 2.5(b) is that the longitudinal steel is uniformly distributed in the former, but is concentrated at the top and bottom stringers in the latter. This difference, however, does not alter the equilibrium condition, as shown by the two force triangles in the longitudinal and transverse directions. To take care of the nonuniform distribution of the longitudinal steel we simply define $\bar{N}_\ell = n_\ell d_v$, which is the total force in the longitudinal steel to resist shear.

Equilibrium of the main body in the longitudinal, transverse and diagonal directions gives the following three equations:

$$V = \bar{N}_\ell \cot \alpha_r \tag{2.38}$$

$$V = n_t d_v \tan \alpha_r \tag{2.39}$$

$$V = (\sigma_d h) d_v \sin \alpha_r \cos \alpha_r \tag{2.40}$$

This set of three Equations (2.38)–(2.40) is identical to the set of three Equations (2.4)–(2.6) for element shear, if the latter three equations are multiplied by the length d_v.

Assuming the yielding of the longitudinal and transverse steel, then $\bar{N}_\ell = \bar{N}_{\ell y}$, $n_t = n_{ty}$, and $V = V_y$. Multiplying Equations (2.38) and (2.39) to eliminate α_r gives

$$V_y = d_v \sqrt{(\bar{N}_{\ell y}/d_v)n_{ty}} \tag{2.41}$$

Combining Equations (2.38) and (2.39) to eliminate V we have

$$\tan \alpha_r = \sqrt{\frac{\bar{N}_{\ell y}/d_v}{n_{ty}}} \tag{2.42}$$

In design, the total longitudinal steel force $\bar{N}_{\ell y}$ is divided equally between the top and bottom stringers. For the design of the bottom steel bars, this bottom tensile force of $\bar{N}_{\ell y}/2$ due to shear is added to the longitudinal tensile force M/d_v due to bending. The design of the top steel bars, however, is less certain. In theory, the top tensile force of $\bar{N}_{\ell y}/2$ due to shear could be subtracted from the longitudinal compressive force M/d_v due to bending. Such measure would be nonconservative. Perhaps a better approach is to select the larger of the two forces M/d_v or $\bar{N}_{\ell y}/2$, for design purposes.

In the design of beams, it is also cost effective to select an α_r value greater than 45°, because the transverse steel, in the form of stirrups, is more costly than the longitudinal steel bars.

2.1.4 Equilibrium in Torsion

2.1.4.1 Bredt's Formula Relating T and q

A hollow prismatic member of arbitrary bulky cross section and variable thickness is subjected to torsion, as shown in Figure 2.6(a). According to St. Venant's theory the twisting deformation will have two characteristics. First, the cross-sectional shape will remain unchanged after the twisting; and second, the warping deformation perpendicular to the cross-section will be identical throughout the length of the member. Such deformations imply that the in-plane normal stresses in the wall of the tube member should vanish. The only stress component in the wall is the in-plane shear stress, which appears as a circulating shear flow q on the cross-section. The shear flow q is the resultant of the shear stresses in the wall thickness and

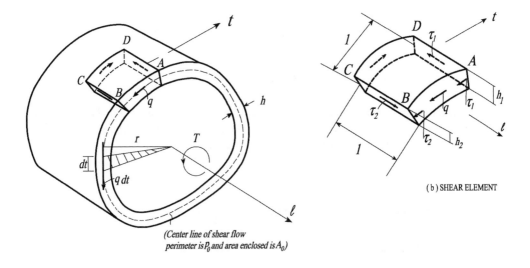

(a) TUBE IN TORSION

Figure 2.6   Equilibrium in torsion

is located on the dotted loop shown in Figure 2.6(a). This dotted loop is defined as the center line of the shear flow, which may or may not lie in the mid-depth of the wall thickness.

A 2-D wall element ABCD is isolated and shown in Figure 2.6(b). It is subjected to pure shear on all four faces. Let us denote the shear stress on face AD as τ_1 and that on face BC as τ_2. The thicknesses at faces AD and BC are designated h_1 and h_2, respectively. Taking equilibrium of forces on the element in the longitudinal ℓ direction we have

$$\tau_1 h_1 = \tau_2 h_2 \tag{2.43}$$

Since shear stresses on mutually perpendicular planes must be equal, the shear stresses on face AB must be τ_1 at point A and τ_2 at point B. Equation (2.43), therefore, means that τh on face AB must be equal at points A and B. Since we define $q = \tau h$ as the shear flow, q must be equal at points A and B. Notice also that the two faces AD and BC of the element can be selected at an arbitrary distance apart without violating the equilibrium condition in the longitudinal direction. It follows that the shear flow q must be constant throughout the cross-section.

The relationship between T and q can be derived directly from the equilibrium of moments about the ℓ axis. As shown in Figure 2.6 (a), the shear force along a length of wall element dt is $q\,dt$. The contribution of this element to the torsional resistance is $q\,dt(r)$, where r is the distance from the center of twist (ℓ axis) to the shear force $q\,dt$. Since q is a constant, integration along the whole loop of the center line of shear flow gives the total torsional resistance:

$$T = q \oint r\,dt \tag{2.44}$$

From Figure 2.6 (a) it can be seen that $r \, \mathrm{d}t$ in the integral is equal to twice the area of the shaded triangle formed by r and $\mathrm{d}t$. Summing these areas around the whole cross-section results in:

$$\oint r \, \mathrm{d}t = 2A_o \qquad (2.45)$$

where A_o is the cross-sectional area bounded by the center line of the shear flow. This parameter A_o is a measure of the lever arm of the circulating shear flow and will be called the lever arm area. Substituting $2A_o$ from Equation (2.45) into Equation (2.44) gives:

$$q = \frac{T}{2A_o} \qquad (2.46)$$

Equation (2.46) was first derived by Bredt (1896).

2.1.4.2 Torsion Equations

A shear element isolated from the wall of a tube of bulky cross-section, Figure 2.6(b), may be subjected to a warping action in addition to the pure shear action discussed above. This warping action will be taken into account in Chapter 7. If the warping action is neglected, then this shear element becomes identical to the shear element in Figure 2.2 which is subjected to pure shear only. As a result, the three equilibrium Equations (2.4)–(2.6) derived for the element shear in Figure 2.2 become valid. Substituting q from Equation (2.46) into Equations (2.4)–(2-6), we obtain the three equilibrium equations for torsion:

$$T = \frac{\bar{N}_\ell}{p_o}(2A_o) \cot \alpha_r \qquad (2.47)$$

$$T = n_t(2A_o) \tan \alpha_r \qquad (2.48)$$

$$T = (\sigma_d h)(2A_o) \sin \alpha_r \cos \alpha_r \qquad (2.49)$$

Notice in Equation (2.47) that $\bar{N}_\ell = n_\ell p_o$. This is because n_ℓ, which is the longitudinal force per unit length, must be multiplied by the whole perimeter of the shear flow p_o to arrive at the total longitudinal force due to torsion $\bar{N}_\ell$.

Assuming the yielding of the longitudinal and transverse steel, then $\bar{N}_\ell = \bar{N}_{\ell y}$, $n_t = n_{ty}$, and $T = T_y$. Multiplying Equations (2.47) and (2.48) to eliminate α_r gives

$$T_y = (2A_o)\sqrt{(\bar{N}_{\ell y}/p_o)n_{ty}} \qquad (2.50)$$

Combining Equations (2.47) and (2.48) to eliminate T we have

$$\tan \alpha_r = \sqrt{\frac{(\bar{N}_{\ell y}/p_o)}{n_{ty}}} \qquad (2.51)$$

2.1.5 Summary of Basic Equilibrium Equations

A summary of the basic equilibrium equations for bending, element shear, beam shear and torsion is given in Table 2.1. The table includes 18 equations, three for bending and five each

Table 2.1   Summary of basic equilibrium equations

	Basic equations	At yield

BENDING

$$M = A_s f_s(jd)$$

$$\alpha_r = 0°$$
$$M_y = A_s f_y(jd)$$

ELEMENT SHEAR (q = shear flow)

$$q = n_\ell \cot \alpha_r$$
$$q = n_t \tan \alpha_r$$
$$q = (\sigma_d h)\sin \alpha_r \cos \alpha_r$$

$$q_y = \sqrt{n_{\ell y} n_{ty}}$$
$$\tan \alpha_r = \sqrt{\frac{n_{\ell y}}{n_{ty}}}$$

BEAM SHEAR ($V = q d_v$ and $\bar{N}_\ell = n_\ell d_v$)

$$V = \bar{N}_\ell \cot \alpha_r$$
$$V = n_t d_v \tan \alpha_r$$
$$V = (\sigma_d h) d_v \sin \alpha_r \cos \alpha_r$$

$$V_y = d_v \sqrt{(\bar{N}_{\ell y}/d_v) n_{ty}}$$
$$\tan \alpha_r = \sqrt{\frac{\bar{N}_{\ell y}/d_v}{n_{ty}}}$$

TORSION ($T = q(2A_o)$ and $\bar{N}_\ell = n_\ell p_o$)

$$T = \frac{\bar{N}_\ell}{p_o}(2A_o)\cot \alpha_r$$
$$T = n_t(2A_o)\tan \alpha_r$$
$$T = (\sigma_d h)(2A_o)\sin \alpha_r \cos \alpha_r$$

$$T_y = (2A_o)\sqrt{(\bar{N}_{\ell y}/p_o) n_{ty}}$$
$$\tan \alpha_r = \sqrt{\frac{(\bar{N}_{\ell y}/p_o)}{n_{ty}}}$$

for the three cases of element shear, beam shear and torsion. Comparison of the three sets of five equations for element shear, beam shear and torsion shows that they are basically the same. The five equations for beam shear are simply the five equations for element shear multiplied by a length dv. The five equations for torsion are simply those for element shear multiplied by the area $2A_o$. Hence it is only necessary to understand the geometric and algebraic relationships of the set of five equations for element shear.

Within each set of five equations for element shear, beam shear, and torsion the three in the left-hand column are derived directly from the three equilibrium conditions. The geometric relationships of these three equations are described by the second type of expression for Mohr circle in Figure 2.3(c).

The two equations of each set in the right-hand column are derived from the basic equations in the left-hand column, assuming that the steels in both the longitudinal and transverse directions have yielded. In the typical case of element shear the angle α_r at yield is obtained by eliminating q from the first two basic equations, while the shear flow at yield, q_y, is obtained by eliminating α_r.

The equations shown in Table 2.1 for pure bending, pure shear and pure torsion are all derived in a consistent and logical manner. Such clarity in concept makes the interaction of these actions relatively simple. These interaction relationships will be enunciated in the next Section 2.2.

2.2 Interaction Relationships

2.2.1 Shear–Bending Interaction

A model of beam subjected to shear and bending is shown in Figure 2.7(a). The moment M creates a tensile force M/d_v in the bottom stringer and an equal compressive force in the top stringer. The shear force V, however, is acting on a shear element as shown in Figure 2.7(d). It induces a total tensile force of $V \tan\alpha_r$ in the longitudinal steel, Figure 2.7(c). Due to symmetry, the top and bottom stringers should each resist one-half of the tensile force, $(V \tan\alpha_r)/2$. In the transverse direction, the shear force V will produce a transverse force $n_t d_v$, in the transverse steel, Figure 2.7(b).

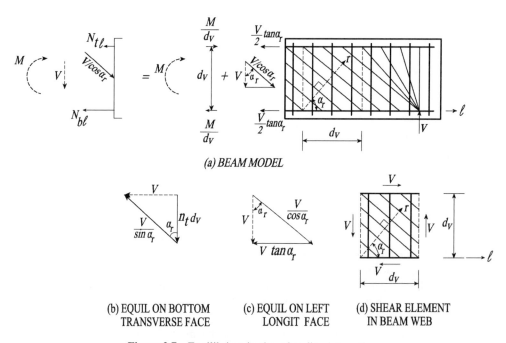

(a) BEAM MODEL

(b) EQUIL ON BOTTOM TRANSVERSE FACE   (c) EQUIL ON LEFT LONGIT FACE   (d) SHEAR ELEMENT IN BEAM WEB

Figure 2.7   Equilibrium in shear-bending interaction

A beam under shear and bending may fail in two modes, because failure may be caused by the yielding of bottom stringer or by the yielding of top stringer. The first and second modes of failure are presented in Sections 2.2.1.1 and 2.2.1.2, respectively.

2.2.1.1 First Failure Mode

In the first failure mode, failure occurs due to yielding of the *bottom stringer* and the *transverse steel*. Looking at Figure 2.7(a), the force in the bottom stringer, $N_{b\ell}$, and in the force per unit length in the transverse steel, n_t, are:

$$N_{b\ell} = \frac{M}{d_v} + \frac{V}{2} \tan \alpha_r \tag{2.52}$$

$$n_t = \frac{V}{d_v} \cot \alpha_r \tag{2.53}$$

Substituting α_r from Equation (2.53) into Equation (2.52) to eliminate α_r, we have

$$\frac{M}{N_{b\ell} d_v} + \frac{V^2}{d_v^2 (2N_{b\ell}/d_v) n_t} = 1 \tag{2.54}$$

Equation (2.54) expresses the interaction relationship between M and V. Assume that yielding occurs in the bottom stringer and in the transverse steel, then $N_{b\ell} = N_{b\ell y}$ and $n_t = n_{ty}$. Also define the pure bending strength M_o and the pure shear strength, V_o, according to Table 2.1 as follows:

$$M_o = N_{b\ell y} d_v \tag{2.55}$$

$$V_o = d_v \sqrt{\left(\frac{2N_{t\ell y}}{d_v}\right) n_{ty}} \tag{2.56}$$

where
  $N_{t\ell y} =$ force in the top stringer at yielding,
  $N_{b\ell y} =$ force in the bottom stringer at yielding,

Notice that the pure shear strength V_o is defined based on the top stringer force at yield $N_{t\ell y}$, rather than the bottom stringer force at yield $N_{b\ell y}$. This definition gives the lowest positive value for V_o, assuming the top stringer force at yield is less than the bottom stringer force at yield. The total longitudinal force due to shear, $\bar{N}_\ell$, is then equal to $2N_{t\ell y}$. To replace $N_{b\ell y}$ by $N_{t\ell y}$ in Equation (2.54) we introduce the ratio R:

$$R = \frac{N_{t\ell y}}{N_{b\ell y}} \tag{2.57}$$

Substituting the definitions of M_o, V_o and R from Equations (2.55)–(2.57) into Equation (2.54), the interaction equation for M and V is now expressed in a nondimensional form:

$$\frac{M}{M_o} + \left(\frac{V}{V_o}\right)^2 R = 1 \tag{2.58}$$

Equation (2.58) is plotted in Figure 2.8 for $R = 0.25, 0.5$ and 1.

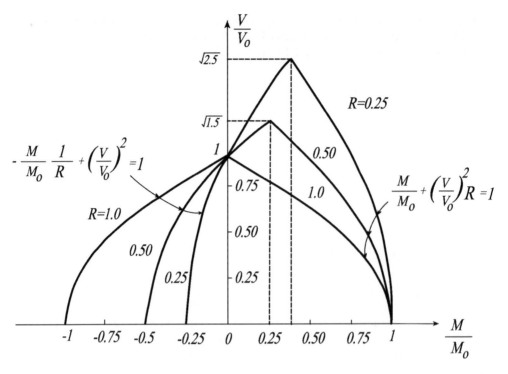

Figure 2.8   Shear-bending interaction curves

2.2.1.2 Second Failure Mode

In the second failure mode, failure occurs due to yielding of the *top stringer* and the *transverse steel*. Looking at Figure 2.7(a), the force in the top stringer $N_{t\ell}$ and the force per unit length in the transverse steel n_t are:

$$N_{t\ell} = -\frac{M}{d_v} + \frac{V}{2}\tan\alpha_r \tag{2.59}$$

$$n_t = \frac{V}{d_v}\cot\alpha_r \tag{2.60}$$

Substituting α_r from Equation (2.60) into Equation (2.59) to eliminate α_r, we have

$$-\frac{M}{N_{t\ell}d_v} + \frac{V^2}{d_v^2\,(2N_{t\ell}/d_v)\,n_t} = 1 \tag{2.61}$$

Assuming that yielding occurs in the top stringer and in the transverse steel, then $N_{t\ell} = N_{t\ell y}$ and $n_t = n_{ty}$. Substituting the definitions of M_o, V_o, and R from Equations (2.55)–(2.57) into Equation (2.61), the nondimensional interaction equation for M and V becomes

$$-\frac{M}{M_o}\frac{1}{R} + \left(\frac{V}{V_o}\right)^2 = 1 \tag{2.62}$$

Equation (2.62) is also plotted in Figure 2.8 for $R = 0.25, 0.5$ and 1. A series of complete interaction curves for shear and bending is now clearly illustrated in Figure 2.8.

Take, for example, the case of $R = 0.5$ where the top stringer has one-half the yield capacity of the bottom stringer. The moment will then vary from $M/M_o = -0.5$ to 1 depending on the magnitude of V/V_o. When $M/M_o = -0.5$, this negative moment introduces a tensile force in the top stringer equal to the yield force. Consequently, the shear strength V which is based on the top stringer strength becomes zero. When the moment is increased (in an algebraic sense toward the right), the tensile force in the top stringer decreases. The remaining tensile capacity in the top stringer is now available to resist shear, resulting in an increase of the shear strength. When the moment is increased to zero (i.e. $M/M_o = 0$), the full capacity of the top stringer is available to resist shear, and the shear strength V becomes the pure shear strength V_o (i.e. $V/V_o = 1$).

When the moment becomes positive, it induces a compressive stress in the top stringer, which can be used to reduce the tensile stress due to shear. As a result, the shear strength continues to increase until the second mode of failure (yielding of top stringer) is changed into the first mode of failure (yielding of bottom stringer). When $M/M_o = 0.25$, the condition is reached where the bottom stringer and the top stringer yield simultaneously. This peak point provides the highest possible shear strength. Beyond this point, the shear strength decreases with increasing moment, because failure is now caused by the tensile yielding of the bottom stringer. In the bottom stringer the tensile stresses due to shear and bending are additive. When the moment reaches the pure bending strength ($M/M_o = 1$), the bottom stringer will yield under the moment itself, leaving no capacity for shear. The shear strength then becomes zero ($V/V_o = 0$).

The point of maximum shear, which corresponds to the simultaneous yielding of top and bottom stringers, is the intersection point of the two curves for the first and second modes of failure. The locations of these peak points can be obtained by solving the two interaction equations, Equations (2.58) and (2.62). Multiplying Equation (2.62) by R and adding it to Equation (2.58), we derive V/V_o by eliminating M/M_o:

$$\frac{V}{V_o} = \sqrt{\frac{1+R}{2R}} \tag{2.63}$$

Multiplying Equation (2.62) by R and subtracting it from Equation (2.58), we derive M/M_o by eliminating V/V_o:

$$\frac{M}{M_o} = \frac{1-R}{2} \tag{2.64}$$

For the case of $R = 0.5$, Equations (2.64) and (2.63) illustrate that the peak point is located at $M/M_o = 0.25$ and $V/V_o = \sqrt{1.5}$. These values are indicated in Figure 2.8.

2.2.2 Torsion–Bending Interaction

A model of beam subjected to torsion and bending is shown in Figure 2.9. The moment M creates a tensile force M/d_v in the bottom stringer and an equal compressive force in the top stringer. The torsional moment, however, induces a total tensile force of $(T p_o/2A_o) \tan \alpha_r$ in the longitudinal steel. Due to symmetry, the top and bottom stringers should each resist

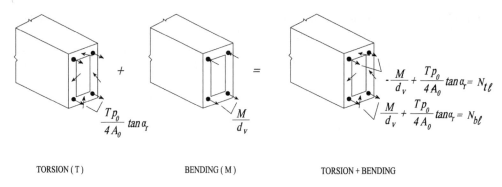

$$-\frac{M}{d_v} + \frac{Tp_0}{4A_0}\tan\alpha_r = N_{t\ell}$$

$$\frac{M}{d_v} + \frac{Tp_0}{4A_0}\tan\alpha_r = N_{b\ell}$$

TORSION (T)   BENDING (M)   TORSION + BENDING

Figure 2.9   Superposition of stringer forces due to torsion and bending

one-half of the total tensile force, $(Tp_o/4A_o)\tan\alpha_r$. In the transverse direction, the torsional moment will also produce a transverse force per unit length, $(T/2A_o)\cot\alpha_r$, in the hoop steel.

2.2.2.1 First Failure Mode

Failure of a beam under torsion and bending also occurs in two modes. In the first failure mode, failure occurs due to yielding of the *bottom stringer* and the *transverse steel*. Reviewing Figure 2.9 reveals that the force in the bottom stringer $N_{b\ell}$ and the force per unit length in the transverse steel n_t, are:

$$N_{b\ell} = \frac{M}{d_v} + \frac{Tp_o}{4A_o}\tan\alpha_r \tag{2.65}$$

$$n_t = \frac{T}{2A_o}\cot\alpha_r \tag{2.66}$$

Substituting α_r from Equation (2.66) into Equation (2.65) to eliminate α_r results in

$$\frac{M}{N_{b\ell}d_v} + \frac{T^2}{4A_o^2\,(2N_{b\ell}/p_o)\,n_t} = 1 \tag{2.67}$$

Equation (2.67) expresses the interaction relationship between M and T. Assume that yielding occurs in the bottom stringer and in the transverse steel, then $N_{b\ell} = N_{b\ell y}$ and $n_t = n_{ty}$. Also define the pure torsional strength T_o, according to Table 2.1 as follows:

$$T_o = 2A_o\sqrt{\left(\frac{2N_{t\ell y}}{p_o}\right)n_{ty}} \tag{2.68}$$

Notice that the pure torsional strength T_o is defined based on the top stringer force at yield $N_{t\ell y}$, rather than the bottom stringer force at yield $N_{b\ell y}$. This definition gives the lowest positive value for T_o, assuming the top stringer force at yield is less than the bottom stringer force at yield. The total longitudinal force due to torsion $\bar{N}_\ell$ is then equal to $2N_{t\ell y}$.

Substituting the definitions of M_o, R, and T_o from Equations (2.55), (2.57) and (2.68) into Equation (2.67), the interaction equation for M and T is expressed in a nondimensional form:

$$\frac{M}{M_o} + \left(\frac{T}{T_o}\right)^2 R = 1 \tag{2.69}$$

2.2.2.2 Second Failure Mode

In the second failure mode, failure occurs due to yielding of the *top stringer* and the *transverse steel*.

Looking at Figure 2.9, the force in the top stringer $N_{t\ell}$ and the force per unit length in transverse steel n_t, are:

$$N_{t\ell} = -\frac{M}{d_v} + \frac{Tp_o}{4A_o}\tan\alpha_r \tag{2.70}$$

$$n_t = \frac{T}{2A_o}\cot\alpha_r \tag{2.71}$$

Substituting α_r from Equation (2.71) into Equation (2.70) to eliminate α_r, we have

$$-\frac{M}{N_{t\ell}d_v} + \frac{T^2}{4A_o^2(2N_{t\ell}/p_o)n_t} = 1 \tag{2.72}$$

Assuming that yielding occurs in the top stringer and in the transverse steel, then $N_{t\ell} = N_{t\ell y}$ and $n_t = n_{ty}$. Substituting the definitions of M_o, R, and T_o from Equations (2.55), (2.57) and (2.68) into Equation (2.72), the nondimensional interaction equation for M and T becomes

$$-\frac{M}{M_o}\frac{1}{R} + \left(\frac{T}{T_o}\right)^2 = 1 \tag{2.73}$$

When the nondimensional interaction equation for the first mode of failure in torsion and moment (Equation 2.69), is compared with that in shear and moment (Equation (2.58), it can be seen that they are identical, if the nondimensional ratio T/T_o is replaced by V/V_o. Similar observation can also be seen when the equations for the second mode of failure, Equations (2.73) and (2.62), are compared. Therefore, the interaction curves for torsion and moment can be illustrated by Figure 2.8, if the axis V/V_o is replaced by an axis T/T_o. Both the torsion–bending interaction curves and the shear–bending interaction curves are plotted in Figure 2.10 using the three axes, V/V_o, T/T_o and M/M_o. The shear–bending curves are shown in the vertical plane and the torsion–bending curves in the horizontal plane.

In Section 2.2.1 we have discussed the peak point of maximum shear. Similar logic can be used to find the peak point of maximum torsional moment, which corresponds also to the simultaneous yielding of top and bottom stringers. This intersection point of two curves for the first and second modes of failure is located at $T/T_o = \sqrt{(1+R)/2R}$ and $M/M_o = (1-R)/2$. For the case of $R = 0.5$, this peak point is located at $T/T_o = \sqrt{1.5}$ and $M/M_o = 0.25$, which are indicated in Figure 2.10.

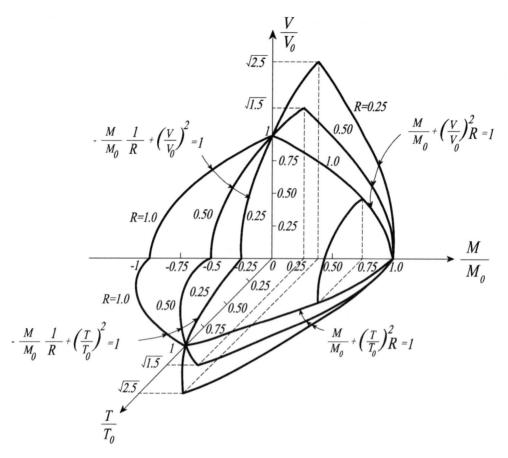

Figure 2.10   Torsion–bending interaction curves (include also V–M curves from Fig. 2.8)

Now that the two sets of shear/bending and torsion/bending interaction curves are obtained, we can derive the interaction relationships of shear, torsion and bending (Elfgren, 1972). These relationships can be represented by a series of interaction surfaces bridging the two sets of interaction curves in Figure 2.10.

2.2.3 Shear–Torsion–Bending Interaction

The interaction relationship for shear, torsion and bending will be derived for a box section as shown in Figure 2.11. When both shear and torsion are present, the shear flows q on the four walls of the box section will be different. The shear flow q_v on each of the two webs due to the applied shear force V is $V/2d_v$, Figure 2.11(a), and the shear flow q_T on each wall due to the applied torsional moment T is $T/2A_o$, Figure 2.11(b). The shear flows due to shear and torsion are additive on the left wall and are subtractive on the right wall. On the top and bottom walls, only shear flow due to torsion exists. Utilizing the subscripts ℓw, rw, tw and bw to indicate left wall, right wall, top wall and bottom wall, respectively, the shear flows in

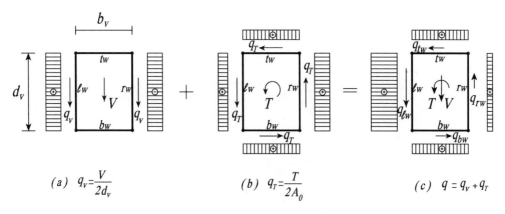

(a) $q_v = \dfrac{V}{2d_v}$ (b) $q_T = \dfrac{T}{2A_0}$ (c) $q = q_v + q_T$

Figure 2.11   Shear flow due to shear and torsion in a box section

the four walls are:

$$q_{\ell w} = \frac{V}{2d_v} + \frac{T}{2A_o} \tag{2.74}$$

$$q_{rw} = -\frac{V}{2d_v} + \frac{T}{2A_o} \tag{2.75}$$

$$q_{tw} = q_{bw} = \frac{T}{2A_o} \tag{2.76}$$

The steel cage of the box section is simplified as shown in Figure 2.12. It is assumed that the centroids of the four longitudinal corner bars are located at the intersection points of the center line of the hoop bars. It is also assumed that the center line of shear flow coincides with the center line of the hoop bars, as well as the center line of the longitudinal bars. The angle α_r of the concrete struts should also be different in the four walls. According to Equation (2.5) the value of α_r is:

$$\tan \alpha_r = \frac{q}{n_t} \tag{2.77}$$

Substituting the four q's from Equations (2.74)–(2.76) into Equation (2.77) we have

$$\tan \alpha_{\ell w} = \frac{1}{n_t}\left(\frac{V}{2d_v} + \frac{T}{2A_o}\right) \tag{2.78}$$

$$\tan \alpha_{rw} = \frac{1}{n_t}\left(-\frac{V}{2d_v} + \frac{T}{2A_o}\right) \tag{2.79}$$

$$\tan \alpha_{tw} = \tan \alpha_{bw} = \frac{1}{n_t}\left(\frac{T}{2A_o}\right) \tag{2.80}$$

These four angles are also shown in Fig. 2.12.

A rectangular box section subjected to shear, torsion and bending may fail in three modes. They are presented in the following three sections.

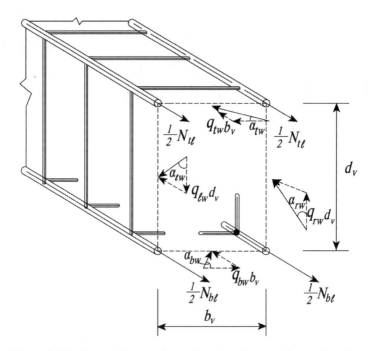

Figure 2.12   Forces in box section subjected to shear, torsion, and bending

2.2.3.1 First Failure Mode

In the first failure mode, failure is caused by yielding in the *bottom stringer* and in the *transverse reinforcement* on the side where shear flows due to shear and torsion are additive (i.e. left wall).

Equating the external moment to the internal moment about the top wall (Figure 2.12) gives:

$$M = N_{b\ell}d_v - q_{\ell w}d_v \tan\alpha_{\ell w}\frac{d_v}{2} - q_{rw}d_v \tan\alpha_{rw}\frac{d_v}{2} - q_{bw}b_v \tan\alpha_{bw}d_v \qquad (2.81)$$

Substituting $q_{\ell w}$, q_{rw}, q_{bw} from Equations (2.74)–(2.76) and $\tan\alpha_{\ell w}$, $\tan\alpha_{rw}$, $\tan\alpha_{bw}$ from Equations (2.78)–(2.80) into Equation (2.81) gives

$$M = N_{b\ell}d_v - \frac{d_v^2}{2n_t}\left(\frac{V}{2d_v} + \frac{T}{2A_o}\right)^2 - \frac{d_v^2}{2n_t}\left(-\frac{V}{2d_v} + \frac{T}{2A_o}\right)^2 - \frac{b_vd_v}{n_t}\left(\frac{T}{2A_o}\right)^2 \qquad (2.82)$$

When the square terms are multiplied in Equation (2.82) we notice that the two mixed terms of VT can be cancelled. The two V^2 terms and the three T^2 terms can be grouped. Making these simplifications and dividing the whole equation by $N_{b\ell}d_v$ results in:

$$\frac{M}{N_{b\ell}d_v} + \left(\frac{V}{2d_v}\right)^2\frac{d_v}{N_{b\ell}}\frac{1}{n_t} + \left(\frac{T}{2A_o}\right)^2\frac{(b_v + d_v)}{N_{b\ell}}\frac{1}{n_t} = 1 \qquad (2.83)$$

Assume yielding in the bottom stringer and in the transverse steel, then $N_{b\ell} = N_{b\ell y}$ and $n_t = n_{ty}$. Also recall the definitions:

$$M_o = N_{b\ell y} d_v$$

$$V_o = 2d_v \sqrt{\left(\frac{2N_{t\ell y}}{d_v}\right) n_{ty}} \qquad \text{for two webs of a box}$$

$$T_o = 2A_o \sqrt{\left(\frac{2N_{t\ell y}}{p_o}\right) n_{ty}} \qquad \text{noting } d_v + b_v = p_o/2$$

$$R = \frac{N_{t\ell y}}{N_{b\ell y}}$$

Substituting these definitions of M_o, V_o, T_o and R into Equation (2.83) we obtain a nondimensional interaction relationship for M, V and T in the first mode:

$$\frac{M}{M_o} + \left(\frac{V}{V_o}\right)^2 R + \left(\frac{T}{T_o}\right)^2 R = 1 \qquad (2.84)$$

Equation (2.84) shows that the interaction of V and T is circular for a constant M. A typical V–T interaction curve for $M/M_o = 0.75$ and $R = 1$ is shown in Figure 2.10. When M is varied, Equation (2.84) represents an interaction surface. This surface intersects the vertical V–M plane to form an interaction curve expressed by Equation (2.58) and intersects the horizontal T–M plane to form an interaction curve expressed by Equation (2.69).

2.2.3.2 Second Failure Mode

In the second failure mode, failure is caused by yielding in the *top stringer* and in the *transverse reinforcement* on the side where shear flows due to shear and torsion are additive (i.e. left-hand side).

Equating the external moment to the internal moment about the bottom wall (Figure 2.12) gives:

$$-M = N_{t\ell} d_v - q_{\ell w} d_v \tan \alpha_{\ell w} \frac{d_v}{2} - q_{rw} d_v \tan \alpha_{rw} \frac{d_v}{2} - q_{tw} b_v \tan \alpha_{tw} d_v \qquad (2.85)$$

Substituting $q_{\ell w}$, q_{rw}, q_{tw} from Equation (2.74)–(2.76) and $\tan \alpha_{\ell w}$, $\tan \alpha_{rw}$, $\tan \alpha_{tw}$ from Equations (2.78)–(2.80) into Equation (2.85) and simplifying gives

$$-\frac{M}{N_{t\ell} d_v} + \left(\frac{V}{2d_v}\right)^2 \frac{d_v}{N_{t\ell}} \frac{1}{n_t} + \left(\frac{T}{2A_o}\right)^2 \frac{(b_v + d_v)}{N_{t\ell}} \frac{1}{n_t} = 1 \qquad (2.86)$$

Assume yielding in the top stringer and in the transverse steel, then $N_{t\ell} = N_{t\ell y}$ and $n_t = n_{ty}$. Substituting the definitions of M_o, V_o, T_o and R into Equation (2.86) we obtain a nondimensional interaction relationship for M, V and T in the second mode:

$$-\left(\frac{M}{M_o}\right) \frac{1}{R} + \left(\frac{V}{V_o}\right)^2 + \left(\frac{T}{T_o}\right)^2 = 1 \qquad (2.87)$$

Equation (2.87) shows that the interaction of V and T is also circular for a constant M in the second mode of failure, Figure 2.10. This equation represents an interaction surface which intersects the vertical V–M plane to form an interaction curve expressed by Equation (2.62) and intersects the horizontal T–M plane to form an interaction curve expressed by Equation (2.73).

The intersection of the two failure surfaces for the first and second modes of failure will form a peak interaction curve between V and T. The expression of this peak curve can be obtained by solving Equations (2.87) and (2.84) to eliminate the M term:

$$\left(\frac{V}{V_o}\right)^2 + \left(\frac{T}{T_o}\right)^2 = \frac{1+R}{2R} \tag{2.88}$$

The location of this curve can be obtained by solving Equations (2.87) and (2.84) to eliminate the V and T terms:

$$\frac{M}{M_o} = \frac{1-R}{2} \tag{2.89}$$

Equation (2.89) is, of course, the same as Equation (2.64). The plane formed by the peak curve at $M/M_o = (1-R)/2$ will be designated as the peak plane. Equation (2.88) for $R = 0.25, 0.5$ and 1 on the peak planes is plotted as a series of dotted curves in Figure 2.13.

2.2.3.3 Third Failure Mode

In the third failure mode, failure is caused by yielding in the *top bar* (not the top stringer), in the *bottom bar* and in the *transverse reinforcement*, all on the side where shear flows due to shear and torsion are additive (i.e. Left-hand side).

Taking moments about the right side wall, Figure 2.12, where shear flows due to shear and torsion are subtractive, will furnish the following equilibrium equation:

$$0 = \frac{1}{2}(N_{b\ell} + N_{t\ell})b_v - q_{\ell w}d_v \tan\alpha_{\ell w}b_v - q_{tw}b_v \tan\alpha_{tw}\frac{b_v}{2} - q_{bw}b_v \tan\alpha_{bw}\frac{b_v}{2} \tag{2.90}$$

Substituting $q_{\ell w}, q_{tw}, q_{bw}$ from Equations (2.74)–(2.76) and $\tan\alpha_{\ell w}, \tan\alpha_{tw}, \tan\alpha_{bw}$ from Equations (2.78)–(2.80) into Equation (2.90) and simplifying gives

$$\frac{N_{b\ell} + N_{t\ell}}{2} = \frac{d_v}{n_t}\left(\frac{V}{2d_v} + \frac{T}{2A_o}\right)^2 + \frac{b_v}{n_t}\left(\frac{T}{2A_o}\right)^2 \tag{2.91}$$

When the square term is multiplied in Equation (2.91) we notice the appearance of a mixed term for VT. Grouping the two T^2 terms and dividing the whole equation by $N_{t\ell}$ results in:

$$\frac{N_{b\ell} + N_{t\ell}}{2N_{t\ell}} = \left(\frac{V}{2d_v}\right)^2\frac{d_v}{N_{t\ell}}\frac{1}{n_t} + \left(\frac{T}{2A_o}\right)^2\frac{(b_v + d_v)}{N_{t\ell}}\frac{1}{n_t} + 2\left(\frac{V}{2d_v}\right)\left(\frac{T}{2A_o}\right)\frac{d_v}{N_{t\ell}}\frac{1}{n_t} \tag{2.92}$$

Assume the yielding of the top bar, the bottom bar and the transverse steel in the left wall, then $N_{t\ell} = N_{t\ell y}$, $N_{b\ell} = N_{b\ell y}$, and $n_t = n_{ty}$. Substituting the definitions of V_o and T_o into

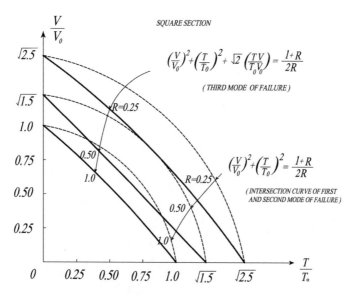

Figure 2.13   Interaction of shear and torsion on peak plane

Equation (2.92) and noticing that

$$\frac{N_{b\ell} + N_{t\ell}}{2N_{t\ell}} = \frac{1 + R}{2R} \quad \text{and} \quad b_v + d_v = \frac{p_o}{2}$$

we obtain a nondimensional interaction relationship for M, V and T in the third mode:

$$\left(\frac{V}{V_o}\right)^2 + \left(\frac{T}{T_o}\right)^2 + \left(\frac{VT}{V_oT_o}\right) 2\sqrt{\frac{2d_v}{p_o}} = \frac{1 + R}{2R} \tag{2.93}$$

It should be pointed out that the third mode of failure is independent of the bending moment M. The mixed VT term is a function of the shape of the cross-section. For a square section, $2\sqrt{2d_v/p_o} = \sqrt{2}$ and Equation (2.93) becomes

$$\left(\frac{V}{V_o}\right)^2 + \left(\frac{T}{T_o}\right)^2 + \sqrt{2}\left(\frac{VT}{V_oT_o}\right) = \frac{1 + R}{2R} \tag{2.94}$$

Equation (2.94) represents a series of cylindrical interaction surfaces perpendicular to the $V - T$ plane. The intersection of each cylindrical surface with its peak plane will produce a curve, which is plotted as a solid curve in Figure 2.13. This solid curve, representing the third interaction surface, is much lower than the corresponding dotted peak interaction curve formed by the intersection of the first and second interaction surfaces. It can be concluded, therefore, that the third mode of failure will always govern in the vicinity of the peak planes.

Figure 2.14 illustrates the three interaction surfaces in a perspective manner for the case of $R = 1/3$. The shaded area in the vicinity of the peak plane is where the third interaction surface

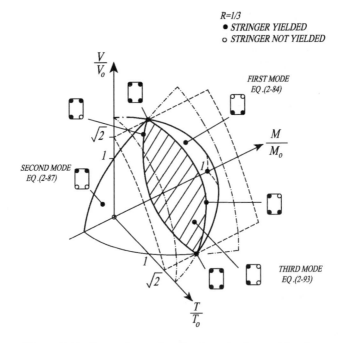

Figure 2.14   Interaction surface for shear, torsion, and bending

replaces the first two interaction surfaces and, therefore, Equation (2.94) governs. Figure 2.14 also shows that the longitudinal steel in one wall will yield when the member fails on one of the three interaction surfaces. On the interaction curves of any two interaction surfaces, however, the longitudinal steel in two adjacent walls will yield. At the two peak points where all three surfaces intersect, all the longitudinal steel will yield.

It should be pointed out that the relationship between shear and torsion is not linear. This is because the plasticity truss model allows the α_r angle to vary in each of the four walls of a member, such that both the longitudinal and transverse steel will yield and the capacity of the member is maximized. As a result, the plasticity truss model always provides an upper bound solution.

If the compatibility condition of a RC member is considered, some of the steel may not yield and the predicted capacity may not be reached. Such situations will be discussed in subsequent chapters. However, tests have shown that the interaction relationships predicted by the plasticity truss model could be conservative, even when the steel does not yield in one direction. Perhaps this theoretical nonconservative behavior is counteracted by the strain hardening of the yielded mild steel bars.

It should also be pointed out that the interaction surfaces based on the plasticity truss model have been shown by tests to be nonconservative near the region of pure torsion. The theoretical torque may overestimate the test results by 20%. This nonconservative behavior is caused by the overestimation of the lever arm area A_o, if it is calculated according to the centerline of the hoop bars. Detailed discussion of this problem will be given in Chapter 7.

2.2.4 Axial Tension–Shear–Bending Interaction

The effect of axial tension on the yield strength of a member can easily be included in the interaction relationship. Only a brief discussion is required in this section. The effect of axial tension will be illustrated by its interaction with shear and bending. In the truss model, it is assumed that the axial tension is resisted only by the longitudinal steel bars. Consequently, the axial tension does not destroy the internal equilibrium of the beam truss action under shear and bending. The only addition to the equilibrium condition is the tensile forces in the bottom and top stringers. In other words, Equations (2.52) and (2.59) for shear and bending interaction should include a simple new term due to the axial tension force P:

$$N_{b\ell} = \frac{P}{2} + \frac{M}{d_v} + \frac{V}{2}\tan\alpha_r \tag{2.95}$$

$$N_{t\ell} = \frac{P}{2} - \frac{M}{d_v} + \frac{V}{2}\tan\alpha_r \tag{2.96}$$

The other equilibrium equations remain valid for the forces in the transverse steel bars and in the concrete struts.

Using the same rational approach as in Section 2.2.1., but including the new term for P, the $P - M - V$ interaction relationships for the two modes of failure can be derived. For the first mode of failure we have

$$\frac{P}{P_o} + \frac{M}{M_o} + \left(\frac{V}{V_o}\right)^2 R = 1 \tag{2.97}$$

The second mode of failure will give

$$\frac{P}{P_o}\left(\frac{1}{R}\right) - \frac{M}{M_o}\left(\frac{1}{R}\right) + \left(\frac{V}{V_o}\right)^2 = 1 \tag{2.98}$$

Equations (2.97) and (2.98) are shown in Figure 2.15 as two interaction surfaces for $R = 0.5$. It can be seen that P and M have a straight line relationship, but the relationship between P and V is nonlinear.

2.3 ACI Shear and Torsion Provisions

The shear and torsion provisions of the ACI Code were significantly revised in 1995 (ACI 318-95). With minor changes, it continued to be used up to the present day (ACI 318-08). The code provisions after 1995 were based on the equilibrium (plasticity) truss model derived in Sections 2.1 and 2.2. The background of the 1995 ACI Code provisions was explained in a paper by Hsu (1997). In this section, we will elaborate on the ACI design procedures and will provide a design example.

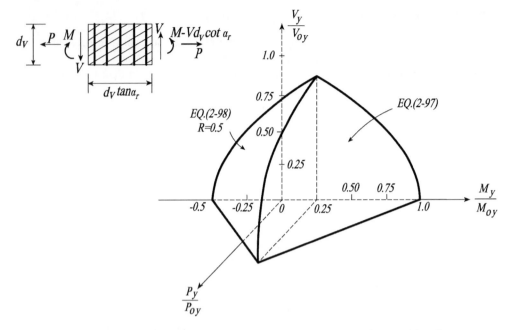

Figure 2.15 Nondimensional interaction surface for axial load, shear, and bending

The ACI code uses some notations that are quite different from the symbols used in this book. First, the angle that defines the direction of cracking is quite different. In the ACI Code, the angle θ is defined as the angle between the direction of concrete struts (i.e. d axis) and the direction of longitudinal steel bars (i.e. ℓ axis). In this book, however, the angle α_r is defined as the angle between the direction of cracking (i.e. r axis) and the direction of longitudinal steel bars (i.e. ℓ axis). Consequently, $\theta = 90° - \alpha_r$ and $\cot \theta = \tan \alpha_r$.

Second, the steel areas are defined differently to resist shear and torsion: A_v is the total area of transverse steel bar for shear; and A_t is one leg of a hoop steel bar for torsion. However, the steel yield stresses f_{yt} are for transverse steel bar for shear, as well as hoop steel bars for torsion. The spacing s is for transverse steel only, and no symbol of spacing is given for longitudinal steel.

2.3.1 Torsional Steel Design

The ACI methodology for the design of torsional steel is based on the three equilibrium equations derived in Section 2.1.4.2, specifically, Equations (2.47)–(2.49). These three equilibrium equations for torsion have also been summarized in Table 2.1. Equation (2.47) is used to size the longitudinal torsional steel. Equation (2.48) is used to design the transverse torsional steel. Equation (2.49) is used to check the stresses in the concrete struts in order to avoid the concrete crushing before the steel yielding.

2.3.1.1 Transverse Torsional Steel

Assuming yielding of the steel, the symbols n_t in Equation (2.48) become $n_t = A_t f_{yt}/s$, and the symbol T becomes $T_n = T_u/\phi$. The torsional transverse steel can be directly designed according to Equation (2.48):

$$\frac{A_t}{s} = \frac{T_u}{\phi 2 A_o f_{yt} \cot\theta} \tag{2.99}$$

In Equation (2.99) the angle θ is limited to a range of $30° < \theta < 60°$ in order to control cracking. It will be shown in Section 5.2.4 and Figure 5.7 (Chapter 5) that crack width increases very rapidly when θ moves away from this range. An angle of $\theta = 45°$ is recommended for reinforced concrete because this angle represents the best crack control. For prestressed concrete though, the ACI code uses an angle for crack control of $37.5°$. This value is based on the past research of MacGregor and Ghoneim (1995). However, current research (Laskar, 2009) indicates that, even for prestressed concrete, $\theta = 45°$ provides optimal crack control. This topic is discussed further in Section 8.3.2.4 and Figure 8.13 (Chapter 8).

The lever arm area A_o in Eq. (2.99) depends on the thickness of the shear flow zone t_d, which, in turn, is a function of the applied torsional moment T_n. The larger the torsional moment T_n, the larger the shear flow zone t_d and the smaller the lever arm area A_o. These relationships can be derived theoretically from the warping compatibility condition of the wall as shown in Section 7.1.2 (Chapter 7). For design practice, simplified expressions are given for t_d and A_o by Equations (7.82) and (7.74), respectively:

$$t_d = \frac{4 T_u}{\phi f'_c A_{cp}} \tag{2.100}$$

$$A_o = A_{cp} - \frac{t_d}{2} p_{cp} \tag{2.101}$$

where A_{cp} is the area enclosed by the outside perimeter of concrete cross-section, and p_{cp} is the outside perimeter of the concrete cross-section. Substituting t_d from Equation (2.100) into Equation (2.101), A_o becomes:

$$A_o = A_{cp} - \frac{2 T_u p_{cp}}{\phi f'_c A_{cp}} \tag{2.102}$$

For structural members commonly employed in buildings (Figure 2.16a), the ACI Code suggests a simpler but less accurate expression:

$$A_o = 0.85 A_{oh} \tag{2.103}$$

where A_{oh} is the area enclosed by the centerline of the outermost closed transverse torsional reinforcement. Equation (2.103) may underestimate the torsional strength of lightly reinforced small members by up to 40% and overestimate the torsional strength of heavily reinforced large members by up to 20%.

The transverse torsional bars required by Equations (2.99) and (2.102) or (2.103) should be in the form of hoops or closed stirrups. They should meet the maximum spacing requirement of $s \leq p_h/8$ or 300 mm (12 in.).

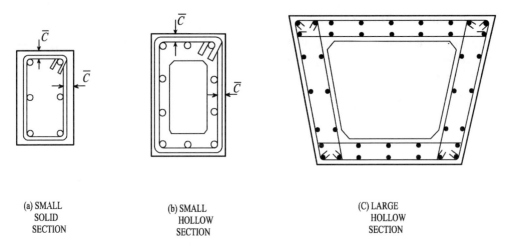

(a) SMALL
SOLID
SECTION

(b) SMALL
HOLLOW
SECTION

(C) LARGE
HOLLOW
SECTION

Figure 2.16   Various cross sections

2.3.1.2 Torsional Longitudinal Steel

The torsional longitudinal steel can be designed according to Equation (2.47) assuming yielding of steel, $f_\ell = f_{y\ell}$, where $f_{y\ell}$ is yield strength of longitudinal torsional reinforcement. However, a more convenient equation relating the torsional longitudinal steel to the torsional transverse steel can be derived by equating Equation (2.47) to Equation (2.48). Noticing in Equation (2.47) that $\bar{N}_\ell = A_\ell f_{y\ell}$, where A_ℓ is now defined as the total area of torsional longitudinal steel in the cross-section. Also assuming that $p_o = p_h$, where p_h is defined as the perimeter of the centerline of the outermost hoop bars, we derive the ACI equation:

$$A_\ell = \left(\frac{A_t}{s} \right) p_h \frac{f_{yt}}{f_{y\ell}} \cot^2 \theta \qquad (2.104)$$

The angle θ is the same one used for transverse torsional steel in Equation (2.99).

2.3.1.3 Minimum Longitudinal Torsional Steel

In order to avoid a brittle torsional failure, a minimum amount of torsional reinforcement (including both transverse and longitudinal steel) is required in a member subjected to torsion. The basic criterion for determining this minimum torsional reinforcement is to equate the post-cracking strength T_n to the cracking strength T_{cr}. The formula derived by Hsu (1997) is:

$$A_{\ell,\min} = \frac{0.42\sqrt{f_c'(MPa)}A_g}{f_{y\ell}} - \left(\frac{A_t}{s} \right) p_h \frac{f_{yt}}{f_{y\ell}} \qquad (2.105)$$

where $A_{\ell,\min}$ is the total area of minimum longitudinal steel; A_g is the cross-sectional area of the hollow section, considering only the concrete and not including the hole(s). A_g becomes A_{cp} for solid sections, where A_{cp} is the area of the same hollow section including the hole(s). Detailed derivation of Equation (2.105) is given in Section 7.2.6 (Chapter 7).

To limit the value of $A_{\ell,\text{min}}$, the transverse steel area per unit length A_t/s in the second term on the right-hand side of Equation (2.105) must not be taken less than $0.17(\text{MPa})b_w/f_{yt}$.

Two equations are required for the design of longitudinal torsional steel. Equation (2.104) is needed to determine the torsional strength, while Equation (2.105) is needed to ensure ductility. Both equations are applicable to solid and hollow sections, as well as nonprestressed and prestressed concrete. The longitudinal torsional bars required by these two equations should be distributed uniformly along the perimeter of the cross-section. They should meet the maximum spacing requirement of $s_\ell \le 305$ mm (12 in.) and the minimum bar diameter of $d_b \ge s/16$ or 9.5 mm (3/8 in.).

2.3.2 Shear Steel Design

The design of shear steel could be based on the three equilibrium equations derived in Section 2.1.3, specifically, Equations (2.38)–(2.40). These three equilibrium equations for beam shear have also been summarized in Table 2.1. Equations (2.38) and (2.39) could be used to size the longitudinal torsional steel and the transverse torsional steel, respectively. Equation (2.40) could be used to check the stresses in the concrete struts in order to avoid the concrete crushing before the steel yielding.

The ACI methodology for the design of shear steel, however, is less conservative. In the ACI Code, the nominal shear resistance V_n is assumed to be made up of two terms:

$$V_n = V_c + V_s \tag{2.106}$$

where V_c is the contributed of concrete, and V_s is the contributed of steel. Only the steel contribution V_s will satisfy the beam shear equilibrium equation (2.39).

2.3.2.1 Concrete Contribution V_c

The empirical expressions of concrete contribution V_c in ACI318-08 are given as follows:
For normal-weight reinforced concrete [ACI Equation (11-3)]

$$V_c = 0.166\sqrt{f_c'(\text{MPa})}b_w d \tag{2.107}$$

For normal-weight prestressed concrete [ACI Equation (11-9)]

$$V_c = \left(0.05\sqrt{f_c'(\text{MPa})} + 4.82(\text{MPa})\frac{V_u d}{M_u}\right) b_w d \tag{2.108}$$

where $0.166\sqrt{f_c'(\text{MPa})}b_w d \le V_c \le 0.42\sqrt{f_c'(\text{MPa})}b_w d$ and $V_u d/M_u \le 1$.

2.3.2.2 Transverse Shear Steel

The transverse shear steel in the beam web is designed according to Equation (2.39) using a shear force of $V_s = (V_u/\phi) - V_c$. Taking $A_t = A_v$ and $d_v = d$, and assuming the yielding of steel $f_t = f_{yt}$ and $\theta = 45°$:

$$\frac{A_v}{s} = \frac{V_s}{d f_{yt}} = \frac{V_u - \phi V_c}{\phi d f_{yt}} \tag{2.109}$$

The transverse shear steel calculated by Equation (2.109) is required in the vertical legs of the cross-section. The spacing s is limited to $d/2$ when $V_s \leq 0.33\sqrt{f_c'(\text{MPa})}b_w d$, and $d/4$ when $V_s > 0.33\sqrt{f_c'(\text{MPa})}b_w d$. The angle θ is taken as $45°$ in the ACI code for shear. It would be more logical to use the same $\theta = 45°$ in Equation (2.99) for torsion, especially in the case of combined shear and torsion.

2.3.2.3 Longitudinal Shear Steel

According to the equilibrium (plasticity) truss model, shear stress also demands longitudinal shear steel according to Equation (2.38). In the 2008 ACI Code, however, longitudinal shear steel continues to be designed indirectly by the so-called shift rule in Section 8.1.1.3 (Chapter 8). In this indirect method, the bending moment diagram is shifted toward the support by a distance d, the effective depth. As a result, the longitudinal bars required by bending are each extended by a length d to take care of the longitudinal shear steel.

2.3.3 Maximum Shear and Torsional Strengths

2.3.3.1 Maximum Shear Strength

The maximum shear strength of a beam cross-section can be derived from Equation (2.40) by taking $h = b_w$, $d_v = 0.9d$, and $\theta = 30°$. The compressive strength of concrete σ_d in Equation (2.40) was found to be softened by the principal tensile strain in the perpendicular direction. Quantifying the softening effect by a softening coefficient ζ, then $\sigma_d = \zeta f_c'$ and the maximum shear strength becomes:

$$V_{n,\max} = 0.39\zeta f_c' b_w d \tag{2.110}$$

The softening coefficient ζ is shown in Section 6.1.7.2 and in Figure 6.9 (Chapter 6) to be inversely proportional to $\sqrt{f_c'}$ for concrete strength up to $f_c' = 100$ MPa (15 000 psi). Assuming a very conservative value of $\zeta = 2.13/\sqrt{f_c'(\text{MPa})}$, we arrive at the long-time ACI provision for maximum shear stress of nonprestressed members:

$$\frac{V_{n,\max}}{b_w d} = 0.83\sqrt{f_c'(\text{MPa})} \quad \text{or} \quad 10\sqrt{f_c'(\text{psi})} \tag{2.111}$$

2.3.3.2 Maximum Torsional Strength

The maximum torsional strength of a cross-section can be derived from Equation (2.49) by taking $h = 0.9A_{oh}/p_h$, $A_o = 0.85A_{oh}$, $\theta = 30°$ and $\sigma_d = \zeta f_c'$:

$$T_{n,\max} = 0.66\zeta f_c' \frac{A_{oh}^2}{p_h} \tag{2.112}$$

Again, assuming a very conservative value of $\zeta = 2.13/\sqrt{f_c'(\text{MPa})}$ gives the maximum torsional strength:

$$\frac{T_{n,\max} p_h}{A_{oh}^2} = 1.41\sqrt{f_c'(\text{MPa})} \quad \text{or} \quad 17\sqrt{f_c'(\text{psi})} \tag{2.113}$$

Equation (2.113) was verified by the PCA torsion tests (Hsu, 1968a, b) with concrete strengths from 14.5 MPa (2 100 psi) to 44.8 MPa (6 500 psi). The shear panel tests shown in Section 6.1.7.2 and in Figure 6.9 (Chapter 6) indicated that Equation (2.113) is also conservative for concrete strength up to 100 MPa (15 000 psi).

2.3.3.3 Interaction of Shear and Torsion

Hollow Sections
Large hollow box sections are typically used as girders in bridges and guideways, as shown in Figure 2.16(c). The vertical stress due to beam shear and the circumferential stress due to torsion will be additive in one of the vertical walls and subtractive in the other vertical wall of a box section. The wall with the additive stresses will generally control the design. The third interaction mode of combined shear, torsion and bending interaction relationship as derived in Equation (2.94) will dominate. The shear–torsion interaction curve near the peak plane, as plotted in Figure 2.13, is very close to a straight line. Consequently, a linear interaction relationship between the shear stress in Equation (2.111) and the torsional stress in Equation (2.113) is adopted by the ACI Code:

$$\left(\frac{V_u}{b_w d}\right) + \left(\frac{T_u p_h}{1.7 A_{oh}^2}\right) \leq \phi \left(\frac{V_c}{b_w d} + 0.667\sqrt{f_c'(\text{MPa})}\right) \tag{2.114}$$

Equation (2.114) is applicable to both nonprestressed and prestressed beams. The term $V_c/b_w d$ on the right-hand side of Equation (2.114) can be conservatively taken as $0.166\sqrt{f_c'(\text{MPa})}$ for nonprestressed members, giving the right-hand side to be $\phi 0.83\sqrt{f_c'(\text{MPa})}$.

The maximum thickness of the shear flow zone (corresponding to the maximum torsional resistance of a cross-section) was found to be $0.8 A_{cp}/p_{cp}$ (see Equation (7.85), Section 7.2.4). This required thickness is taken conservatively as A_{oh}/p_h in the ACI Code. Therefore, if the actual wall thickness t is less than A_{oh}/p_h, the torsional stress in Equation (2.114) should be increased proportionally by substituting $A_{oh}/p_h = t$ in the second term that results in $T_u/1.7 A_{oh} t$. The thickness t is taken at the location where the stresses are being checked.

Solid Sections
In a member with solid cross-section as shown in Figure 2.16(a), the core of the cross-section could be used to resist shear stress due to shear, leaving the outer ring area to resist shear stress due to torsion. Therefore, the stresses due to shear and torsion need not be additive. This favorable condition is reflected in the ACI Code by using a circular interaction relationship between the shear stress in Equation (2.111) and the torsional stress in Equation (2.113):

$$\sqrt{\left(\frac{V_u}{b_w d}\right)^2 + \left(\frac{T_u p_h}{1.7 A_{oh}^2}\right)^2} \leq \phi \left(\frac{V_c}{b_w d} + 0.667\sqrt{f_c'(\text{MPa})}\right) \tag{2.115}$$

Eqs. (2.114) and (2.115) are used to check the cross-section of a member. If these conditions are not satisfied, the cross-section must be enlarged.

2.3.4 Other Design Considerations

2.3.4.1 Compatibility Torsion

In the case of compatibility torsion, where a torsional moment in a statically indeterminate structure can be redistributed to other adjoining members after the formation of a plastic hinge, the ACI code allows the torsional moment to be reduced to the cracking torsional moment under combined loadings. For nonprestressed members, the cracking torque of solid sections subjected to combined torsion, shear and bending T_{cr} has been suggested by Hsu and Burton (1974) and Hsu and Hwang (1977):

$$T_{cr} = 0.33\sqrt{f_c'}\;(\text{MPa})\;\frac{A_{cp}^2}{p_{cp}} \quad \text{or} \quad 4\sqrt{f_c'}\;(\text{psi})\;\frac{A_{cp}^2}{p_{cp}} \tag{2.116}$$

where f_c' and $\sqrt{f_c'}$ have the same stress units (MPa or psi). Detailed derivation of the cracking torsional moment (Equation 2.116), is given in Section 7.2.5 (Chapter 7).

In the case of prestressed concrete, however, the expression on the right-hand side of Equation (2.116) must be multiplied by a factor which reflects the increase of cracking strength by the longitudinal prestress. Using the well-known square-root factor derived from either the Mohr stress circle (Hsu, 1984) or from the skew bending theory (Hsu, 1968b), the cracking torsional moment of solid prestressed members is expressed as follows:

$$T_u = \phi T_{cr} = \phi 0.33\sqrt{f_c'}\;(\text{MPa})\;\frac{A_{cp}^2}{p_{cp}}\sqrt{1 + \frac{f_{pc}}{0.33\sqrt{f_c'}}} \tag{2.117}$$

where f_{pc} in the square-root factor is the compressive stress of concrete at the centroid of cross-section due to effective prestress after allowing for all losses, or at the junction of web and flange when the centroid lies within the flange. In the case of nonprestressed solid members, $f_{pc} = 0$ and the square-root factor becomes unity.

2.3.4.2 Threshold Torque

In order to simplify the design processes, the ACI Code allows a small torsional moment in a structure to be neglected. The limit of this small torsional moment was taken for solid sections as 25% of the cracking torque, Equation (2.117), which results in:

$$T_u = \phi 0.083\sqrt{f_c'}\;(\text{MPa})\;\frac{A_{cp}^2}{p_{cp}}\sqrt{1 + \frac{f_{pc}}{0.33\sqrt{f_c'}}} \tag{2.118}$$

In the case of hollow sections, Mattock (1995) suggested a simple relationship between the cracking torque of a hollow section $(T_{cr})_{hollow}$ and that of a solid section with the same outer dimensions $(T_{cr})_{solid}$:

$$\frac{(T_{cr})_{hollow}}{(T_{cr})_{solid}} = \frac{A_g}{A_{cp}} \tag{2.119}$$

where A_g is the cross-sectional area of the concrete only and not including the hole(s), while A_{cp} is the area of the same hollow section including the hole(s). For solid sections, $A_g = A_{cp}$. This very simple relation expressed by Equation (2.119) has been validated by tests.

In view of Equation (2.119), the ACI Code allows A_{cp} in Equation (2.118) for the threshold torque to be replaced by A_g in the case of hollow sections. In Equation (2.117) for compatibility torsion, however, A_{cp} is not allowed to be replaced by A_g in the case of hollow sections, because of the lack of experimental studies and the lack of need for such an application.

2.3.4.3 Location of Torsional Steel in Shear Flow Zone

The internal torsional moment of a member is contributed by the circulatory shear stresses acting along the centerline of shear flow (Figure 2.6). To be theoretically correct, the centroidal line of the steel cage should be designed to coincide with the centerline of shear flow. Because a steel cage is made up of hoop bars and longitudinal bars, the centroidal line of the steel cage is best represented by the inner edge of the hoop bars, (Figure 2.16a, b). Define $\bar{c}$ as the distance measured from the outer face of the cross-section to the inner edge of the hoop bars. When the centerline of the steel cage defined by $\bar{c}$ lies in the middle of the shear flow zone with a thickness t_d, then the theoretically correct case of design is $\bar{c} = 0.5t_d$. If t_d is taken conservatively as the maximum thickness, $t_{d,\max} = 0.8A_c/p_c$ in Equation (7.85) (Section 7.2.4, Chapter 7), then $\bar{c} = 0.4A_c/p_c$.

In the case of a hollow beam as shown in Figure 2.16(b), the inner concrete cover measured from the inside face of wall to the inner edge of hoop bars should also be $0.5t_d$ in theory. This theoretical requirement of the inner concrete cover is replaced in the ACI Code by a provision specifying that the distance from the inside face of wall to the centerline of the hoop bars shall be not less than $0.5A_{oh}/p_h$. This requirement is conservative because the thickness $t_d = A_{oh}/p_h$ represents the maximum thickness required to resist a maximum torque for the given outer cross-section. It is obvious that this provision is intended for application to the hollow section with one layer of hoop steel as shown in Figure 2.16(b), not to the inner cover of large hollow cross-sections with two layers of hoop steel as shown in Figure 2.16(c).

2.3.4.4 Box Sections with Outstanding Flanges

When the outstanding flanges of a box section are very thin compared with the height of the section, the parameter A_{cp}^2/p_{cp} for the box section with flanges may be less than the same parameter for the box section without flanges. This is conceptually wrong because the addition of flanges is supposed to increase the torsional resistance. Physically, this inconsistency means that the cross-section is not 'bulky' enough and that St. Venant torsional stresses, as expressed by the parameter A_{cp}^2/p_{cp}, can not flow into the flanges. When this happens, the outstanding flanges should be neglected in the calculation of the cross-sectional properties of A_{cp}, A_g and p_{cp} (see 'Check Outstanding Flanges' in the Design Example, Section 2.3.5).

2.3.4.5 Limitation of Angle θ to Prevent Excessive Cracking

When the angle θ deviates too much from 45°, cracking will become excessive. In order to control the crack widths, the ACI Code limits the range of θ to

$$30° \leq \theta \leq 60° \qquad (2.120)$$

The validity of this range of limitations is demonstrated in Section 5.2.4.3, particularly by Figure 5.7. In Chapter 5, α_r is used and $\alpha_r = 90° - \theta$.

2.3.5 Design Example

The ACI design procedures for shear and torsion are given in a flow chart in Figure 2.17. This flow chart procedure can be used to design the shear and torsional reinforcement of a guideway girder. A 3.66-m-wide (12 ft) and 1.27-m-deep (4 ft. 2 in.) box girder with overhanging flanges (Figure 2.18a), was designed as an alternative to the double-tee girder used in the Dade County Rapid Transit System, Florida. The standard prestressed girder in this 35.4 km (22 mile) guideway is simply supported and 24.4 m (80 ft) long. It is prestressed with sixty-two 270 K, 12.7 mm (1/2 in.), seven-wire strands as shown in Figure 2.18(b). The total prestress force is 6076 kN (1366 kips) after prestress loss. The design of flexural steel is omitted for simplicity. The net concrete cover is 3.81 cm (1.5 in.) and the material strengths are $f'_c = 48.2$ MPa (7000 psi) and $f_y = 413$ MPa (60 000 psi).

Sectional Properties

$$L = 24.08 \text{ m (79 ft)}$$
$$h = 1.270 \text{ m (50 in.)}$$
$$d = 1.016 \text{ m (40 in.) at } 0.3L \text{ from support}$$
$$t = 251 \text{ mm (9.875 in.) average of stem width}$$
$$b_w = 502 \text{ mm (19.75 in.) for two stems}$$
$$A = 1523 \times 10^3 \text{ mm}^2 \text{ (2361.4 in.}^2)$$
$$I = 319\,800 \times 10^6 \text{ mm}^4 \text{ (768 336 in.}^4)$$
$$y_t = 517 \text{ mm (20.34 in.)}$$
$$y_b = 753 \text{ mm (29.66 in.)}$$

Loading Criteria

The standard girders are designed to carry a train of cars, each 22.86 m (75 ft) long. Each car has two trucks with a center-to-center distance of 16.46 m (54 ft). Each truck consists of two axles 1.981 m (6 ft 6 in.) apart. The crush live load of each car is 513.7 kN (115.5 kips). The maximum amount of web reinforcement was obtained at the section $0.3L$ from the support under a derailment load, which consists of two truck loads located symmetrically at a distance 3.20 m (10 ft) from the midspan (Figure 2.19b). Each axle load is taken as one-fourth of the crush live load with 100% impact and a maximum sideshift of 0.914 m (3 ft). The load factor is taken as 1.4.

The girder is also subjected to a superimposed dead load caused by the weight of the track rails, rail plinth pads, power rail, guard rail, cableway, acoustic barrier, etc. At derailment, this superimposed dead load is assumed to produce a uniform vertical load of 12.84 kN/m (0.88 kip/ft) and a uniformly distributed torque of 3.158 kN m/m (0.71 kip ft/ft). This torque is neglected in the calculation because the magnitude of the distributed torque is small and because the torque is acting in a direction opposite to the derailment torque.

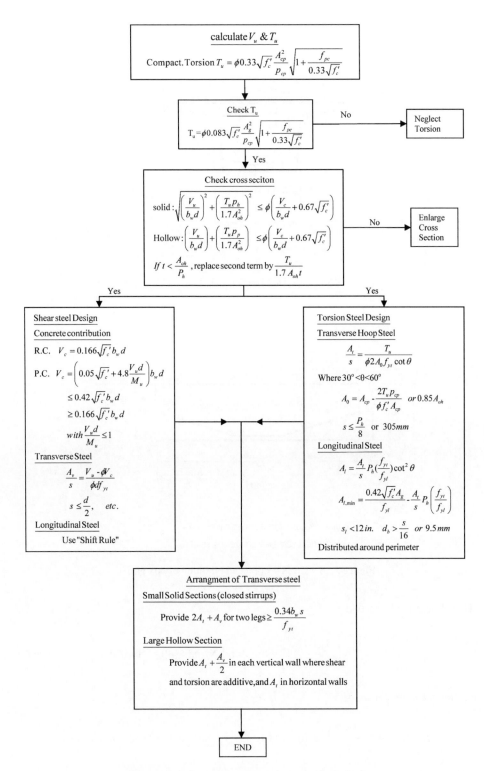

Figure 2.17 ACI shear & torsion design procedures

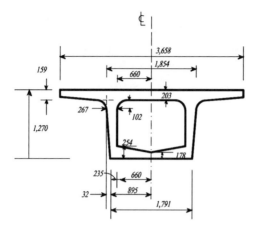

(a) CROSS SECTION (mm)

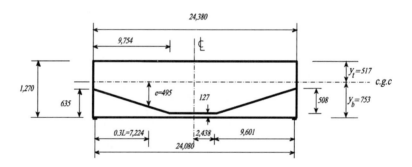

(b) ELEVATION AND PRESTRESS PROFILE (mm)

(VERTICAL SCALE =4.8 HORIZONTAL SCALE)

Figure 2.18   Cross section and elevation of box girder

Derailment load per axle

$$P_{u,L} = 1.4 \left(\frac{513.7}{4} \right) (2) = 359.6 \text{ kN/axle (80.85 kip/axle)}$$

Derailment torque per axle

$$T_{u,L} = 1.4 \left(\frac{513.7}{4} \right) (2)(0.914) = 328.7 \text{ kN m/axle (242.5 kip ft/axle)}$$

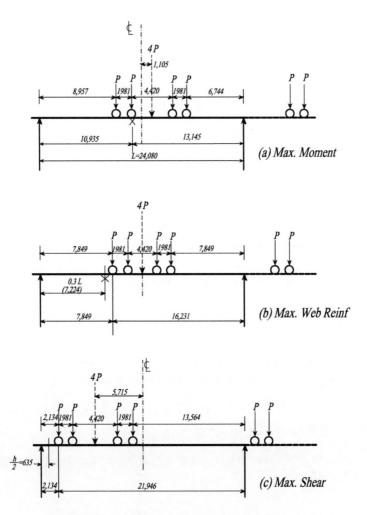

Figure 2.19 Loading conditions and section for design

Girder weight (assumed concrete density of 22.62 kN/m³ (144 lb/ft³) and load factor of 1.4)

$$w_{u,G} = 1.4\,(1.523)(22.62) = 48.23 \text{ kN/m (3.31 kip/ft)}$$

Superimposed dead weight (load factor of 1.4)

$$w_{u,S} = 1.4(12.84) = 17.98 \text{ kN/m (1.23 kip/ft)}$$

Factored Shear, Torque and Bending Moment
Refer to Figure 2.19(b), V_u, T_u and M_u at $0.3L$ from support are:

$$V_u = (w_{u,G} + w_{u,S})(0.2L) + 2P_{u,L} = (48.23 + 17.98)(0.2)(24.08) + 2(359.6)$$
$$= 1038 \text{ kN (233.4 kip)}$$

$$T_u = 2T_{u,L} = 2(328.7) = 657.4 \text{ kN m (485 kip ft)}$$

$$M_u = \frac{1}{2}(w_{u,G} + w_{u,S})L(0.3L) - \frac{1}{2}(w_{u,G} + w_{u,S})(0.3L)^2 + 2P_{u,L}(0.3L)$$

$$= \frac{1}{2}(48.23 + 17.98)(24.08)(7.224) - \frac{1}{2}(48.23 + 17.98)(7.224)^2$$

$$+ 2(359.6)(7.224) = 9227 \text{ kN m (6807 kip ft)}$$

Check Outstanding Flanges

Refer to Figure 2.18(a), the parameter A_{cp}^2/p_{cp} is determined as follows:

Neglect overhanging flanges

$$A_{cp} = 1854(203)\frac{1}{2}(1854 + 1791)(1270 - 203) = 2321 \times 10^3 \text{ mm}^2 \text{ (3597 in.}^2)$$

$$p_{cp} = 1854 + 1791 + 2(1270) = 6185 \text{ mm (243.5 in.)}$$

$$\frac{A_{cp}^2}{p_{cp}} = \frac{2321^2 \times 10^6}{6185} = 871 \times 10^6 \text{ mm}^3 \text{ (53,135 in.}^3)$$

Include overhanging flanges

$$A_{cp} = 2321 \times 10^3 \text{ mm}^2 + (1804)(181) = 2647 \times 10^3 \text{ mm}^2 \text{ (4103 in.}^2)$$

$$p_{cp} = 3658 + 1791 + 2(159) + 2(2000) = 9767 \text{ mm (384.5 in.)}$$

$$\frac{A_{cp}^2}{p_{cp}} = \frac{2647^2 \times 10^6}{9767} = 717 \times 10^6 \text{ mm}^3 \text{ (43 783 in.}^3) < 871 \times 10^6 \text{ mm}^3 \text{ NG.}$$

Neglect outstanding flanges and use $A_{cp} = 2321 \times 10^3$ mm^2; $p_{cp} = 6185$ mm; and $A_{cp}^2/p_{cp} = 871 \times 10^6$ mm^3.

Check Threshold Torque

$$A_g = A_{cp} - A_{hole} = 2321 \times 10^3 - (1320 \times 851) = 1197 \times 10^3 \text{ mm}^2 \text{ (1855 in.}^2)$$

$$f_{pc} = \frac{P}{A_g} = \frac{6076 \times 10^3}{1197 \times 10^3} = 5.076 \text{ MPa (736 psi)}$$

$$T_u = \phi 0.083\sqrt{f_c'(MPa)}\frac{A_g^2}{p_{cp}}\sqrt{1 + \frac{f_{pc}}{0.33\sqrt{f_c'}}}$$

$$= (0.90)0.083\sqrt{48.2}\frac{1197^2 \times 10^6}{6185}\sqrt{1 + \frac{5.076}{0.33\sqrt{48.2}}}$$

$$= 215.1 \text{ kN m (158.7 kip ft)} < 657.4 \text{ kN m (485 kip ft).}$$

Factored torsional moment needs to be considered in design.

Check Cross-section

Assume a clear concrete cover of 38 mm (1.5 in.) and No. 4 bars (d_b = 13 mm) for web reinforcement:

$$A_{oh} = \frac{1}{2}[(1854 - 89) + (1791 - 89)](1270 - 89) = 2047 \times 10^3 \text{ mm}^2 \ (3{,}174 \text{ in.}^2)$$

$$p_h = (1854 - 89) + (1791 - 89) + 2(1270 - 89) = 5829 \text{ mm } (229.5 \text{ in.})$$

$$e = (753 - 127) - \frac{9754 - 7224}{9754} 508 = 494 \text{ mm } (19.47 \text{ in.}) \text{ at } 0.3L \text{ from support}$$

$$d = y_t + e = 517 + 494 = 1011 \text{ mm } (39.81 \text{ in.}) \text{ at } 0.3L \text{ from support}$$

$$d = 0.8h = 0.8(1270) = 1016 \text{ mm } (40 \text{ in.}) \quad \text{governs}$$

$$t = \frac{1}{2}(267 + 235) = 251 \text{ mm } (9.875 \text{ in.})$$

$$b_w = 2t = 2(251) = 502 \text{ mm } (19.75 \text{ in.})$$

$$b_w d = 502(1016) = 510 \times 10^3 \text{ mm}^2 \ (790 \text{ in.}^2)$$

The interaction equation for hollow box sections is

$$\left(\frac{V_u}{b_w d}\right) + \left(\frac{T_u p_h}{1.7 A_{oh}^2}\right) \leq \phi \left(\frac{V_c}{b_w d} + 0.667 \sqrt{f_c'(\text{MPa})}\right)$$

$$\frac{A_{oh}}{p_h} = \frac{2047 \times 10^3}{5829} = 351 \text{ mm } (13.8 \text{ in.}) > t = 251 \text{ mm } (9.875 \text{ in.}).$$

$t = 251$ mm governs

$$\left(\frac{V_u}{b_w d}\right) + \left(\frac{T_u}{1.7 A_{oh} t}\right) = \frac{1038 \times 10^3}{510 \times 10^3} + \frac{657.4 \times 10^6}{1.7(2047 \times 10^3)(251)} = 2.79 \text{ MPa } (404 \text{ psi})$$

$$V_c = \left(0.05\sqrt{f_c'(\text{MPa})} + 4.82(\text{MPa})\frac{V_u d}{M_u}\right) b_w d$$

where

$$\frac{V_u d}{M_u} \leq 1$$

$$\frac{V_u d}{M_u} = \frac{1038 \times 10^3 (1016)}{9227 \times 10^6} = 0.1143 < 1 \text{ OK}$$

$$V_c = [0.05\sqrt{48.2} + 4.82(0.1143)](510 \times 10^3) = 458 \text{ kN } (102.9 \text{ kips})$$

$$V_{c,min} = 0.166\sqrt{f_c' \ (\text{MPa})} b_w d = 0.166\sqrt{48.2}(510 \times 10^3) = 588 \text{ kN } (132.2 \text{ kips})$$

$$> 458 \text{ kN} \quad V_{c,min} = 588 \text{ kN governs}$$

$$\phi \left(\frac{V_c}{b_w d} + 0.667\sqrt{f_c' \ (\text{MPa})}\right) = 0.90 \left(\frac{588 \times 10^3}{510 \times 10^3} + 0.667\sqrt{48.2}\right) = 0.90(5.784)$$

$$= 5.20 \text{ MPa } (753 \text{ psi}) > 2.79 \text{ MPa } (404 \text{ psi}) \text{ OK}$$

Design of Torsional Hoop Steel

$$A_o = A_{cp} - \frac{2T_u p_{cp}}{\phi f_c' A_{cp}} = 2321 \times 10^3 - \frac{2(657.4 \times 10^6)(6185)}{0.90(48.2)(2321 \times 10^3)}$$

$$= 2240 \times 10^3 \text{ mm}^2 \ (3472 \text{ in.}^2)$$

Assume $\theta = 37.5°$ as recommended by the Code provision for prestressed members:

$$\frac{A_t}{s} = \frac{T_u}{\phi 2A_o f_{yt} \cot \theta} = \frac{657.4 \times 10^6}{(0.90)(2)(2240 \times 10^3)(413)(1.303)}$$

$$= 0.303 \text{ mm}^2/\text{mm} \ (0.0119 \text{ in.}^2/\text{in.})$$

$$s_{max} = \frac{p_h}{8} = \frac{5829}{8} = 729 \text{ mm} \ (28.7 \text{ in.}) > 305 \text{ mm} \ (12 \text{ in.})$$

305 mm (12 in.) spacing governs.

Design of Torsional Longitudinal Steel

$$A_\ell = \left(\frac{A_t}{s}\right) p_h \frac{f_{yt}}{f_{y\ell}} \cot^2 \theta = (0.303)(5829)(1)(1.303^2) = 2998 \text{ mm}^2 \ (4.64 \text{ in.}^2)$$

Check minimum limitation for $(A_t/s)_{min}$:

$$\left(\frac{A_t}{s}\right)_{min} = \frac{0.172(\text{MPa})b_w}{f_{yt}} = \frac{0.172(502)}{413} = 0.209 \text{ mm}^2/\text{mm} \ (0.00823 \text{ in.}^2/\text{in.})$$

$$< 0.303 \text{ mm}^2/\text{mm} \ (0.0119 \text{ in.}^2/\text{in.}). \quad \text{OK}$$

$$A_{\ell,min} = \frac{0.42\sqrt{f_c'(\text{MPa})}A_g}{f_{y\ell}} - \left(\frac{A_t}{s}\right) p_h \frac{f_{yt}}{f_{y\ell}} = \frac{0.42\sqrt{48.2}(1197 \times 10^3)}{413}$$

$$- (0.303)(5829)(1) = 8451 - 1766 = 6685 \text{ mm}^2 \ (10.36 \text{ in.}^2) \quad \text{governs}$$

Select 36 No. 5 longitudinal bars: $A_\ell = 36(200) = 7200 \text{ mm}^2 > 6685 \text{ mm}^2$ OK.

Design of Shear Steel

$$V_c = V_{c,min} = 588 \text{ kN} \ (132.2 \text{ kips})$$

$$\frac{A_v}{s} = \frac{V_u - \phi V_c}{\phi d f_{yt}} = \frac{1038 \times 10^3 - 0.90(588 \times 10^3)}{0.90(1016)(413)}$$

$$= 1.347 \text{ mm}^2/\text{mm} \ (0.0530 \text{ in.}^2/\text{in.})$$

$$s_{max} = 305 \text{ mm} \ (12 \text{ in.}) \text{ for torsion governs}$$

Transverse Steel for Vertical Walls
Transverse steel in the vertical walls is contributed by both shear and torsion:

$$\frac{A_t}{s} + \frac{1}{2}\frac{A_v}{s} = 0.303 + \frac{1}{2}(1.347) = 0.977 \text{ mm}^2/\text{mm } (0.0384 \text{ in.}^2/\text{in.})$$

$$\left(\frac{A_t}{s} + \frac{1}{2}\frac{A_v}{s}\right)_{min} = \frac{0.172(502)}{413} = 0.209 \text{ mm}^2/\text{mm } (0.00823 \text{ in.}^2/\text{in.})$$

$$< 0.977 \text{ mm}^2/\text{mm } (0.0384 \text{ in.}^2/\text{in.}) \text{ OK}$$

Select two layers of No. 5 bars in each vertical wall at 305 mm (12 in.) spacing:

$$\frac{2(200)}{305} = 1.311 \text{ mm } (0.0516 \text{ in.}^2/\text{in.}) > 0.977 \text{ mm}^2/\text{mm } (0.0384 \text{ in.}^2/\text{in.}) \quad \text{OK}$$

Transverse Steel for Horizontal Walls
Transverse steel in the horizontal walls is contributed by torsion only:

$$\frac{A_t}{s} = 0.303 \text{ mm}^2/\text{mm } (0.0119 \text{ in.}^2/\text{in.})$$

Select two layers of No. 3 bars in each horizontal wall at 305 mm (12 in.) spacing:

$$\frac{2(71)}{305} = 0.466 \text{ mm}^2/\text{mm } (0.0183 \text{ in.}^2/\text{in.}) > 0.303 \text{ mm}^2/\text{mm. OK}$$

The transverse steel in the top wall should be added to the flexural steel required in the top flange acting as a transverse continuous slab.

Arrangement of Reinforcing Bars
The arrangement of the reinforcing bars for shear and torsion is shown in Figure 2.20. This steel arrangement could be conservatively used throughout the length of the girder.

2.4 Comments on the Equilibrium (Plasticity) Truss Model

The equilibrium (plasticity) truss model presented in this chapter summarizes the major advances in reinforced concrete theory achieved during the 1970s. A milestone in this development was the adoption of this model by the 1978 European Code (CEB-FIP, 1978). This model was incorporated into the 1995 ACI Code (ACI 318-95).

However, serious weaknesses exist in these provisions due to its historical limitations. The difficulties stemmed primarily from a lack of understanding of the strain compatibility condition for shear and torsion, as well as the biaxial behavior of reinforced concrete 2-D elements. These problems had been extensively studied in the 1980s, 1990s up to the present time. The new information is presented in Chapters 4–8.

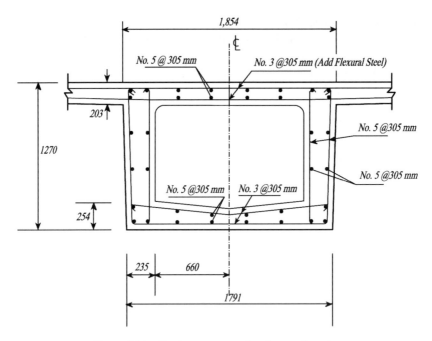

Figure 2.20   Steel arrangement for shear and torsion

The advantages and deficiencies of the equilibrium (plasticity) truss model are summarized below:

Advantages
1. The equilibrium truss model theory satisfies completely the equilibrium condition. It provides three equilibrium equations which are conceptionally identical in element shear, beam shear and torsion.
2. From a design point of view, the three equilibrium equations can be used directly to design the three components of the truss model, namely, the transverse steel, the longitudinal steel and the diagonal concrete struts.
3. The three equilibrium equations satisfy the Mohr circle.
4. The model provides a very clear concept of the interaction of bending, shear, torsion and axial load.
5. Equilibrium (plasticity) truss model is applied to the analysis of beams in Section 8.1 (Chapter 8). The model provides a rational explanation of the 'shift rule' as described in Section 8.1.1.3.

Deficiencies
1. The equilibrium truss model does not take into account the strain compatibility condition. As a result, it can not predict the shear or torsional deformations of a RC member.
2. The model can not predict the strains in the steel or concrete. Consequently, the yielding of steel or the crushing of concrete can not be rationally determined, and the modes of failure can not be discerned.

3. The model is unable to produce a method to correctly determine the thickness of the shear flow zone t_d in a torsional member. As such, the lever arm area, A_o, and the perimeter of the shear flow p_o can not be logically defined.
4. The staggering concept of shear design in beams is discussed in Section 8.1.2.1 and shown in Figure 8.2(d). This concept may not be conservative.

3

Bending and Axial Loads

3.1 Linear Bending Theory

3.1.1 Bernoulli Compatibility Truss Model

3.1.1.1 Basic Principles

The analysis of a prismatic member subjected to bending, or flexure, is illustrated in Figure 3.1. On a rectangular cross-section with width b and height h (Figure 3.1a), a bending moment M creates compression stresses in the top part of the cross-section and tensile stresses in the bottom part. Since concrete is weak in tension, the bottom part of the cross-section will crack and the tensile stresses will be picked up by the reinforcing bars (rebar, in short) indicated by the area A_s. The basic concept of reinforced concrete is to utilize the high compressive strength of concrete to resist the compression at the top, and the high tensile strength of steel reinforcement to resist tension at the bottom. This action is similar to a truss, where the top and bottom chords resist the compressive and tensile forces, respectively.

The linear bending theory described in this section is based on *Bernoulli compatibility truss model* as stated in Section 1.3.1.3. This theory is rational and rigorous because it satisfies Navier's three principles of the mechanics of deformable bodies. First, the stresses in the concrete and the reinforcement, Figure 3.1(c), satisfy the equilibrium condition. Second, the linear strain distribution as shown in Figure 3.1(b) satisfies Bernoulli's compatibility hypothesis, which assumes a plane-cross section to remain plane after the bending deformation. Third, Hooke's linear law is applicable to both concrete and reinforcing bar as shown in Figure 3.1(e) and (f). As a result, rigorous solutions can be obtained not only up to the yield strength of a flexural member, but also the flexural deformations.

Hooke's constitutive law is generally assumed to be applicable to both concrete and rebars up to the yielding of steel and, therefore, up to the *service load stage*. Bending theory based on Hooke's law is called the *linear bending theory* and is studied in this section. The constitutive laws of concrete and reinforcement are definitely nonlinear at the ultimate load stage. Bending theory based on a nonlinear stress–strain relationship of concrete will be called the *nonlinear bending theory* and will be studied in Section 3.2.

A reinforced concrete member subjected to bending is expected to crack before the service load stage. The stresses and strains in the rebars at the cracked sections should be greater than

Unified Theory of Concrete Structures   Thomas Hsu and Yi-Lung Mo
© 2010 John Wiley & Sons, Ltd

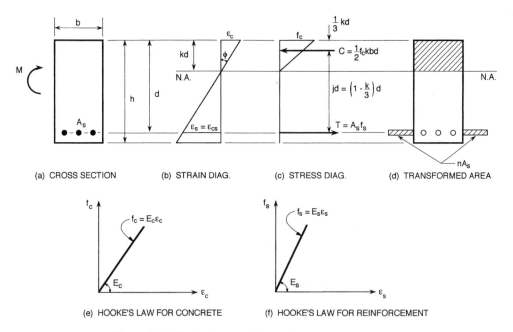

Figure 3.1 Cracked rectangular sections (singly reinforced)

those between the cracked sections. Therefore, the application of Bernoulli's compatibility hypothesis to cracked reinforced concrete members should theoretically be based on the *smeared stresses* and *smeared strains* of mild steel bars embedded in concrete. An accurate relationship between the *smeared stress* and the *smeared strain* is presented in Section 6.1.9 (Chapter 6).

In the linear bending theory of reinforced concrete, however, two assumptions are made to simplify the bending analysis and design. First, the stress–strain relationship of steel bars is based on the bare bars, not on the steel bars embedded in concrete. In other words, the *local* stress–strain relationship of steel at the cracks is assumed, rather than the *smeared* stress–strain relationship. Second, the tensile stress of concrete is neglected in the bending analysis. Since these two assumptions are physically correct at the cracks, the linear bending theory is accurate in the prediction of yield moment. However, the omission of tensile stress of concrete will cause the bending rigidity and stiffness to be significantly under-predicted, and the yield curvature to be significantly over-predicted. For the calculation of bending deflection in Section 3.1.5, this weakness is remedied by using an 'effective bending rigidity' which has been adjusted to fit the deflection test data.

3.1.1.2 Moment and Curvature

The characteristic of a bending member is represented by its moment–curvature relationship. Consequently, the crux of the flexural analysis is how to find the moment and the curvature using Navier's three principles of stress equilibrium, strain compatibility and stress–strain relationships of materials. For a bending member with materials that satisfy Hooke's law, the stress equilibrium condition can be established using the stress diagram in Figure 3.1(c). Bernoiulli's strain compatibility condition is represented by the strain diagram in Figure 3.1(b).

According to Bernoulli linear strain distribution, as shown in Figure 3.1(b), a bending moment M will induce a maximum compressive strain ε_c at the top surface and the tensile strain ε_s in the rebar. The neutral axis (NA), which indicates the level of zero strain, is located at a distance kd from the top surface. Assuming a perfect bond exists between the rebars and the concrete, a compatibility equation can be established in terms of the three variables, ε_c, ε_s and k. Using the geometric relationship of similar triangles we have:

$$\frac{\varepsilon_s}{\varepsilon_c} = \frac{1-k}{k} \tag{3.1}$$

In the stress diagram of Figure 3.1(c), the linear stress distribution in the compression zone is obtained from the linear strain distribution (Figure 3.1b), according to Hooke's law. The stresses vary linearly from the maximum of f_c at the top surface to zero at the neutral axis. Furthermore, the resultant force $C = (1/2) f_c kbd$ is located at a distance $(1/3)kd$ from the top surface. By neglecting the tensile stresses of concrete below the neutral axis, the tensile resistance is concentrated in the rebar tensile force $T = A_s f_s$. Equilibrium of $T = C$ gives:

$$A_s f_s = \frac{1}{2} f_c kbd \tag{3.2}$$

In Figure 3.1(c), the rebar tension force T and the concrete compression force C constitute an internal couple. The lever arm of the couple jd is $(1 - k/3)d$. Equilibrium of internal and external moments gives:

$$M = A_s f_s d \left(1 - \frac{k}{3} \right) \tag{3.3}$$

Hooke's laws for concrete and steel, as shown in Figure 3.1(e) and (f), furnish two additional equations:

$$f_c = E_c \varepsilon_c \tag{3.4}$$

$$f_s = E_s \varepsilon_s \tag{3.5}$$

Using these five equations, (3.1)–(3.5), we can solve five unknowns, f_s, f_c, ε_s, ε_c and k. The solution will allow us to calculate the curvature ϕ using the compatibility condition:

$$\phi = \frac{\varepsilon_c}{kd} \text{ or } \frac{\varepsilon_s}{(1-k)d} \tag{3.6}$$

Equations (3.3) and (3.6) establish the moment curvature relationship (M–ϕ curve).

The detailed solution procedures of the five equations depend on the type of bending problems described in the next section.

3.1.1.3 Types of Bending Problem

In a singly reinforced concrete member, the bending action involves nine variables, b, d, A_s, M, f_s, f_c, ε_s, ε_c and k, as indicated in Figure 3.1(a)–(c). The five equations (3.1)–(3.5), derived from Navier's three principles, can only be used to solve five unknown variables. Therefore, four of the nine variables must be given before the remaining five unknown variables can be solved. Depending on the given variables, the problems are generally categorized into five types as shown in the following table:

Type of problem	Given variables	Unknown variables
(1) First type of analysis:	b d A_s M	f_s f_c ε_s ε_c k
(2) Second type of analysis:	b d A_s f_s (or f_c)	M f_c (or f_s) ε_s ε_c k
(3) Balanced design:	b d f_s f_c	M A_s ε_s ε_c k
(4) First type of design:	b d M f_s (or f_c)	A_s f_c (or f_s) ε_s ε_c k
(5) Second type of design:	b (or d) M f_s f_c	d (or b) A_s ε_s ε_c k

The first two types of problem are called analysis, because the cross-sectional properties of concrete and rebars, b, d and A_s, are given. The last three types are called design, because at least one of these three cross-sectional properties, primarily A_s, is an unknown.

The last three types of problem were known as the allowable stress design method, prevalent prior to 1971. The allowable stress design method had been made obsolete by the ultimate strength design method and, therefore, will not be treated in this book. However, the two types of analysis problems are still very relevant at present for checking the serviceability criteria, i.e. the deflections and the crack widths.

The problems in the analysis and design of flexural members boils down to finding the most efficient way to solve the five available equations for each type of problem. The methodology of the solution process is demonstrated in the next section, where the five equations are applied to the two types of analysis problems.

3.1.1.4 Solution for First Type of Analysis Problem

We will first look at the first type of analysis problem, which is posed as follows:

Given four variables: b, d, A_s, M
Find five unknown variables: f_s, f_c, ε_s, ε_c, k

The nine variables, b, d, A_s, M, f_s, f_c, ε_s, ε_c and k, are defined in Figure. 3.1 for a singly reinforced flexural member. Because four variables are given, the remaining five unknown variables can be solved by the five available equations. These equations and their unknown variables are summarized as follows:

Type of equations	Equations	Unknowns		
Equilibrium of forces	$A_s f_s = \frac{1}{2} f_c k b d$	f_s f_c	k	(3.7)
Equilibrium of moments	$M = A_s f_s \left(1 - \frac{k}{3}\right) d$	f_s	k	(3.8)
Bernoulli compatibility	$\frac{\varepsilon_s}{\varepsilon_c} = \frac{1-k}{k}$	ε_s ε_c k		(3.9)
Hooke's law for rebar	$f_s = E_s \varepsilon_s$	f_s ε_s		(3.10)
Hooke's law for concrete	$f_c = E_c \varepsilon_c$	f_c ε_c		(3.11)

It is interesting to point out that the equilibrium of a parallel coplanar force system will furnish two independent equations. The three simplest forms of equilibrium equations are: (1) equilibrium of forces in the longitudinal direction; (2) equilibrium of moments about the resultant compression force of concrete; and (3) equilibrium of moments about the tensile force of rebars. The selection of two of these three equilibrium equations is strictly a matter of convenience. It must be emphasized that the equilibrium condition in bending can only be used to solve two unknowns, even if all three equilibrium equations are used.

The solution of the five unknowns (f_s, f_c, ε_s, ε_c and k) by the five equations, (3.7)–(3.11), is facilitated by identifying the unknown variables in each equation. These variables are shown after each equation under the column heading *Unknowns*.

In the set of five equations, (3.7)–(3.11), the first two equilibrium equations deal with unknown stresses, while the third equation expresses Bernoulli's compatibility condition in terms of unknown strains. Substituting the stress–strain relationships of Equations (3.10) and (3.11) into Equation (3.9) we obtain Bernoulli's compatibility equation in terms of stresses:

$$\frac{f_s}{n f_c} = \frac{1-k}{k} \tag{3.12}$$

where $n = E_s/E_c$ = modulus ratio. With this simple maneuvering, we have now reduced a set of five equations to a set of three equations, (3.7), (3.8) and (3.12), which involve only three unknowns, f_s, f_c and k.

Of the three equations (3.7), (3.8) and (3.12), the first and the last have the unknown k and the unknown stress ratio f_s/f_c. Substituting the ratio f_s/f_c from Equation (3.7) into Equation (3.12) gives an equation with only one unknown k:

$$\frac{kbd}{2n A_s} = \frac{1-k}{k} \tag{3.13}$$

Defining $\rho = A_s/bd$ = percentage of rebars, Equation (3.13) becomes

$$\frac{k}{2n\rho} = \frac{1-k}{k} \tag{3.14}$$

The unknown k can be solved by Equation (3.14), using whatever method that is convenient. Two methods are generally used. The first is the trial-and-error method. A value of k is assumed and inserted into Equation (3.14). If the equation is satisfied, the assumed k is the solution. If the equation is not satisfied, another k value is assumed and the process is repeated. This trial-and-error method could be quite efficient if the k value could be closely estimated in the first trial. It is, therefore, convenient for engineers with experience.

The second method is to rewrite Equation (3.14) in the form of a second-order equation:

$$k^2 + (2n\rho)k - 2n\rho = 0 \tag{3.15}$$

From Equation (3.15), k can be determined by the formula for quadratic equation:

$$k = \sqrt{(n\rho)^2 + 2n\rho} - n\rho \tag{3.16}$$

Now that the position of the neutral axis, represented by k, is solved, the rebar stresses f_s can be obtained from the equilibrium of moments about the resultant C using Equation (3.8):

$$f_s = \frac{M}{A_s(1 - \frac{k}{3})d} \tag{3.17}$$

and the concrete stress f_c can be obtained either from the equilibrium of forces (Equation 3.10), or more directly in this case (since M is given) from the equilibrium of moments about the tensile force T:

$$f_c = \frac{M}{\frac{1}{2}k(1 - \frac{k}{3})bd^2} \tag{3.18}$$

Once the stresses, f_s and f_c, are found, the strains, ε_s and ε_c, can be calculated from Hooke's law by Equations (3.10) and (3.11). Knowing k and ε_c, the curvature ϕ can be calculated from the compatibility equation (3.6).

3.1.1.5 Solution for Second Type of Analysis Problem

The solution of the second type of analysis is similar to the first type described above. The similarity can be observed by examining the five unknown variables in the five available equations. The problem posed is:

> Given four variables: b, d, A_s, f_s (or f_c)
> Find five unknown variables: M, f_c (or f_s), ε_s, ε_c, k

The variables f_c and f_s in the parentheses should be understood as follows. If the given variable f_s is replaced by f_c in the parenthesis, then the unknown variables f_c must be replaced by f_s in the parenthesis.

The five equations, (3.7)–(3.11), are still valid in this case, but the list of unknowns under the column heading *Unknowns* should be revised. The stress f_s (or f_c) becomes a known value and should be replaced by the new unknown M. The most efficient algorithm of solution still consists of the following three steps:

Step 1: Utilize the stress–strain relationships of rebar and concrete to express Bernoulli's compatibility equation in terms of stresses, thus arriving at Equation (3.12). In this way a set of five equations is reduced to a set of three equations, (3.7), (3.8) and (3.12), which involve only three unknowns, M, f_c (or f_s) and k.

Step 2: Of the three equations, both Equation (3.7) and (3.12) contain the same two unknowns, f_c (or f_s) and k. Simultaneous solution of these two equations results in Equation (3.16), expressing the unknown variable k.

Step 3: Once the position of the neutral axis, k, is determined, the unknowns f_c (or f_s) can be calculated from the force equilibrium equation, and the unknown M from any one of the two moment equilibrium equations. The strains, ε_s and ε_c, can easily be calculated from the stresses f_s and f_c by Hooke's laws.

It can be seen that the solution of the second type of analysis problems is the same as the first type in steps 1 and 2, i.e. in solving the position of the neutral axis. The only small difference is in step 3 in the calculation of different unknowns by equilibrium equations.

3.1.2 Transformed Area for Reinforcing Bars

The key problem in the linear bending analysis (Section 3.1.2) is to find the location of the neutral axis, represented by k, using Bernoulli's compatibility condition, the force equilibrium condition and Hooke's law. A short-cut method to find k will now be introduced. The method utilizes the concept of the transformed area for the rebars.

Based on the assumption that perfect bond exists between the rebar and the concrete, the rebar strain ε_s should be equal to the concrete strain at the rebar level ε_{cs} (Figure 3.1b):

$$\varepsilon_s = \varepsilon_{cs} \tag{3.19}$$

Applying Hooke's law, $f_s = E_s \varepsilon_s$ and $f_{cs} = E_c \varepsilon_{cs}$, to Eq. (3.19) gives:

$$f_s = \frac{E_s}{E_c} f_{cs} = n f_{cs} \tag{3.20}$$

where f_{cs} is the concrete stress at the rebar level.

Using Equation (3.20) the tensile force T can be written as

$$T = A_s f_s = (n A_s) f_{cs} \tag{3.21}$$

Equation (3.21) states that the tensile force T can be thought of as being supplied by a concrete area of $n A_s$ in connection with a concrete stress of f_{cs}. Physically, this means that the rebar area A_s can be transformed into a concrete area $n A_s$, as long as the rebar stress f_s is simultaneously converted to the concrete stress f_{cs}. A cross-section with the transformed rebar area is shown in Figure 3.1(d).

The transformation of the rebar area A_s to a concrete area $n A_s$ has a profound significance. A flexural beam, which is made up of two materials, steel and concrete, can now be thought of as a homogeneous elastic material made of concrete only. For such a homogeneous elastic beam, the neutral axis coincides with the centroidal axis. So, instead of solving the neutral axis from the five equations of equilibrium, compatibility and constitutive laws, as illustrated in Section 3.1.1.4, we can now locate the neutral axis using the simple and well-known method of finding the centroidal axis of a homogeneous beam. In other words, the static moments of the stressed areas, shaded in Figure 3.1(d), about the centroidal axis must be equal to zero. Hence,

$$\frac{1}{2} b(kd)^2 - n A_s(d - kd) = 0 \tag{3.22}$$

The variable k in Equation (3.22) can be solved by two methods. The first method is the trial-and-error procedure. A value of k is assumed and is inserted into Equation (3.22). If the equation is satisfied, the assumed k is the solution. If the equation is not satisfied, another k value is assumed and the process repeated. This trial-and-error method is very efficient for an experienced engineer, who can closely estimate the k value in the first trial.

The second method is to solve Equation (3.22) by the formula of quadratic equation, giving

$$k = \sqrt{(n\rho)^2 + 2n\rho} \; - \; n\rho \tag{3.23}$$

Equation (3.23) is, of course, the same as (3.16). This method requires an engineer to remember Equation (3.23) or to have this formula at hand.

Once the neutral axis, represented by k, is known, the moment can be calculated directly from the equilibrium equation (3.3) and the curvature from the compatibility equation (3.6).

3.1.3 Bending Rigidities of Cracked Sections

3.1.3.1 Singly Reinforced Rectangular Sections

For a homogeneous elastic member, the bending rigidity EI is the product of the modulus of elasticity of the material E and the moment of inertia of the cross section I. For a cracked concrete member, we will define the bending rigidity $E_c I_{cr}$ as the product of the modulus of elasticity of concrete E_c and a cracked moment of inertia I_{cr}. In this section we will derive I_{cr} for singly reinforced rectangular sections.

By definition, the bending rigidity of a member is the bending moment per unit curvature M/ϕ. Hence, we can write

$$E_c I_{cr} = \frac{M}{\phi} \tag{3.24}$$

From the strain diagram in Figure 3.1(b) the curvature ϕ can be expressed by

$$\phi = \frac{\varepsilon_c}{kd} \tag{3.25}$$

From the stress diagram in Figure 3.1(c) the moment M can be obtained by taking moment about the tensile rebar:

$$M = \frac{1}{2} f_c k j b d^2 \tag{3.26}$$

Substituting ϕ and M from Equations (3.25) and (3.26) into Equation (3.24) and noticing that $f_c = E_c \varepsilon_c$ and $j = (1 - k/3)$ results in:

$$I_{cr} = bd^3 \left[\frac{1}{2} k^2 \left(1 - \frac{k}{3} \right) \right] \tag{3.27}$$

Equation (3.27) shows that the cracked moment of inertia I_{cr} can be calculated if the k value is determined from Equation (3.23), or directly from Equation (3.22).

3.1.3.2 Doubly Reinforced Rectangular Sections

The transformed area concept can be used to great advantage in finding the cracked moments of inertia for doubly reinforced sections and flanged sections. These types of sections occur regularly in continuous beams, as shown in Figure 3.2. In a continuous beam the cross-sections are usually doubly reinforced at the column faces, while those at the midspan are often designed as flanged sections.

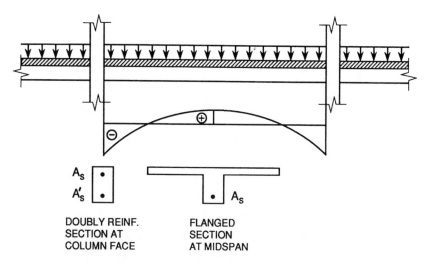

Figure 3.2 Doubly reinforced and flanged sections in continuous beams

Figure 3.3(a) shows a doubly reinforced rectangular section of height h and width b. The tensile rebar area A_s is located at a distance d'' from the bottom surface, and the compression rebar area A'_s at a distance d' from the top surface. The effective depth d is equal to $h - d''$.

The strains in the tensile and compressive rebars are designated ε_s and ε'_s, respectively, as shown in Figure 3.3 (b). The neutral axis is assumed to be located at a distance kd from the top surface. The stresses in the tensile and compressive rebars are denoted as f_s and f'_s, respectively, in Figure 3.3(c).

The transformed area for tensile rebar is indicated by nA_s in Figure 3.3(d). The transformed area for compression rebar is $mA'_s = (n-1)A'_s$. The subtraction of an area of compression rebar A'_s in the expression mA'_s is to compensate for the same area included in the concrete area $b(kd)$. This is strictly for convenience in the calculation.

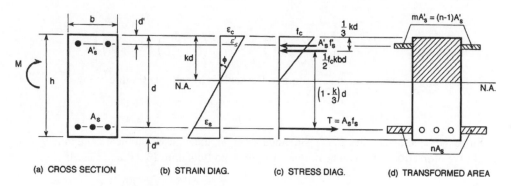

Figure 3.3 Cracked rectangular sections (Doubly reinforced)

The position of the neutral axis can be determined by taking the static moments of the shaded areas (Figure 3.3d), about the centroidal axis (same as neutral axis):

$$\frac{1}{2}b(kd)^2 + mA'_s(kd - d') = nA_s(d - kd)$$ (3.28)

Dividing Equation (3.28) by bd^2 and denoting $\rho = A_s/bd$ and $\rho' = A'_s/bd$ results in:

$$k^2 + 2(n\rho + m\rho')k - 2\left(n\rho + m\rho'\frac{d'}{d}\right) = 0$$ (3.29)

Let

$$\beta_c = \frac{m\rho'}{n\rho}$$ (3.30)

and solving Equation (3.29) by the quadratic equation formula gives:

$$k = \sqrt{(n\rho)^2(1 + \beta_c)^2 + 2n\rho\left(1 + \beta_c\frac{d'}{d}\right)} - n\rho(1 + \beta_c)$$ (3.31)

Equation (3.31) determines the location of the neutral axis for doubly reinforced cracked beams. For the special case of singly reinforced cracked beams, $\beta_c = 0$, and Equation (3.31) degenerates into Equation (3.23).

The moment resistance of the cross-section M can be calculated from the stress diagram in Figure 3.3(c) by taking moments about the tensile force:

$$M = \frac{1}{2}f_c k\left(1 - \frac{k}{3}\right)bd^2 + A'_s f'_s(d - d')$$ (3.32)

From Hooke's law and Bernoulli's strain compatibility, (Figure 3.3b), we can express the compressive rebar stress f'_s as

$$f'_s = E_s\varepsilon'_s = (mE_c)\left(\varepsilon_c\frac{kd - d'}{kd}\right) = m(E_c\varepsilon_c)\frac{kd - d'}{kd}$$ (3.33)

Substituting f'_s from Equation (3.33), $A'_s = \rho'bd$ and $m\rho' = n\rho\beta_c$ into Equation (3.32) gives:

$$M = bd^2(E_c\varepsilon_c)\left[\frac{1}{2}k\left(1 - \frac{k}{3}\right) + n\rho\beta_c\frac{1}{k}\left(k - \frac{d'}{d}\right)\left(1 - \frac{d'}{d}\right)\right]$$ (3.34)

The curvature can be written according to Bernoulli's hypothesis (Figure 3.3b), as

$$\phi = \frac{\varepsilon_c}{kd}$$ (3.35)

Since the bending rigidity $E_c I_{cr}$ is defined as M/ϕ, the cracked moment of inertia I_{cr} for doubly reinforced sections can be derived according to Equations (3.34) and (3.35):

$$I_{cr} = bd^3\left[\frac{1}{2}k^2\left(1 - \frac{k}{3}\right) + n\rho\beta_c\left(k - \frac{d'}{d}\right)\left(1 - \frac{d'}{d}\right)\right]$$ (3.36)

The coefficient k in Equation (3.36) is determined from Equation (3.31). In the absence of compression steel ($\beta_c = 0$), Equation (3.36) degenerates into Equation (3.27).

The cracked moment of inertia can be written simply as $I_{cr} = K_{i2}bd^3$, where K_{i2} denotes the expression in the bracket in Equation (3.36). K_{i2} is a function of $n\rho$, β_c and d'/d and is tabulated in ACI Special Publication SP-17(73).

3.1.3.3 Flanged Sections

A typical flanged section (T-section) is shown in Figure 3.4(a), together with the transformed area in Figure 3.4(b). Comparing this transformed area for the T-section with the transformed area for the doubly rectangular reinforced section in Figure 3.3(d), we observe that they are identical except for two minor differences. First, the area of the top flange is $(b - b_w)h_f$ rather than mA_s'. Second, the centroidal axis of the top flange is located at a distance $h_f/2$ from the top surface rather than at a distance d'.

Therefore, the formulas for cracked moment of inertia of doubly reinforced sections (Equations 3.36 and 3.31), are also applicable to T-sections, if these two differences are taken care of as follows:

(1) Redefine the original definition of the symbol β_c in Equation (3.30) by substituting $(b - b_w)h_f/b_w d$ for $m\rho'$:

$$\beta_c = \frac{(b - b_w)h_f}{(n\rho)b_w d} \tag{3.37}$$

(2) Replace

$$\frac{d'}{d} \text{ by } \frac{h_f}{2d}$$

In conclusion, the cracked moments of inertia I_{cr} for flanged sections are calculated by Equations (3.36) and (3.31), with the symbol β_c defined by Equation (3.37) and the ratio d'/d replaced by $h_f/2d$.

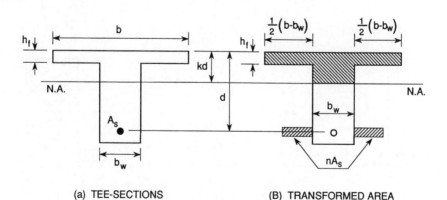

(a) TEE-SECTIONS   (B) TRANSFORMED AREA

Figure 3.4   Cracked flanged sections

3.1.4 Bending Rigidities of Uncracked Sections

In a flexural member at service load, some parts of the member will be cracked and some parts will remain uncracked, depending on the bending moment diagram. The deflection of the member will be a function of both the cracked and the uncracked portions. The bending rigidities of the *cracked* sections have been derived in Section 3.1.3 for singly reinforced sections, doubly reinforced sections and flanged sections. The bending rigidity of *uncracked* sections will be derived in this section.

3.1.4.1 Rectangular Sections

The bending rigidities of uncracked rectangular sections (Figure 3.5a), will first be derived. Before the cracking of a flexural member, the strain and the stress distributions in the cross-section are shown in Figure 3.5 (b) and (c), respectively. First, the concrete stresses below the neutral axis can no longer be neglected. Second, the full height of the cross-section h becomes effective, rather than the effective depth d. The transformed area of the cross-section is shown in Figure 3.5(d).

The internal resistance to the applied moment is contributed primarily by the concrete before cracking. Assuming that the rebars have no effect on the position of the neutral axis, then the neutral axis will lie at the mid-depth of the cross-section. As a result, $kd = h/2$, $jd = (2/3)h$ and $C = T = (1/4)f_c bh$. Taking moments about the neutral axis,

$$M = \frac{1}{6}bh^2 f_c + mA_s f_{cs}(d - h/2) \tag{3.38}$$

where $m = n - 1$, because the original rebar area A_s has been included in the first term for concrete and must be subtracted from the transformed rebar area in the second term.

Observing that $A_s = \rho bd$ and $f_{cs} = f_c(d - h/2)/(h/2)$, Equation (3.38) becomes

$$M = \frac{1}{6}bh^2 f_c \left[1 + 3m\rho \left(\frac{d}{h}\right) \left(\frac{d - (h/2)}{(h/2)}\right)^2\right] \tag{3.39}$$

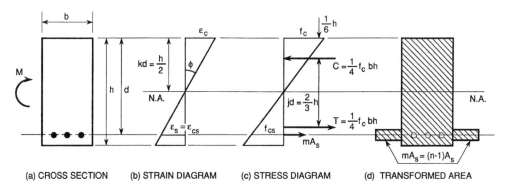

(a) CROSS SECTION   (b) STRAIN DIAGRAM   (c) STRESS DIAGRAM   (d) TRANSFORMED AREA

Figure 3.5   Uncracked rectangular sections (singly reinf.)

Assuming $d/h = 0.9$ and $(d - h/2)/(h/2) = 0.8$, then Equation (3.39) gives

$$M = \frac{1}{6}bh^2 f_c [1 + 1.7m\rho] \tag{3.40}$$

In Equation (3.40) $m\rho$ is the transformed area percentage of tensile rebars. The steel ratio ρ is usually in the range of 1–1.5% and the modulus ratio m varies from 5 to 7 for conventional concrete. Therefore, the second term contributed by rebars, $1.7m\rho$ in the bracket, is roughly 10–15% of the first term (unity) contributed by concrete.

The curvature before cracking is

$$\phi = \frac{\varepsilon_c}{0.5h} \tag{3.41}$$

The uncracked flexural rigidity $E_c I_{uncr}$ is defined by the ratio M/ϕ according to Equations (3.40) and (3.41). Dividing M by ϕ and noticing that $f_c = E_c \varepsilon_c$, the uncracked moment of inertia I_{uncr} becomes:

$$I_{uncr} = \frac{1}{12}bh^3 [1 + 1.7m\rho] \tag{3.42}$$

If the effect of the reinforcement is neglected before cracking, we have the uncracked moment of inertia for the gross section I_g. Setting $\rho = 0$ in Equation (3.42) gives:

$$I_g = \frac{1}{12}bh^3 \tag{3.43}$$

Equation (3.43) is the well-known formula for the moment of inertia of a rectangular cross-section in the case of a homogeneous beam. In design practice, the effect of rebars is frequently neglected, and the uncracked moment of inertia is calculated from the gross section (Equation 3.43).

The moment of inertia of a gross section can also be obtained from the moment of inertia for a cracked section (Equation 3.27). Substituting $k = (1/2)h/d$ and $j = (2/3)h/d$ into Equation (3.27), the resulting equation is identical to Equation (3.43). Hence, the transition of flexural behavior from an uncracked stage to a cracked stage can be mathematically modeled by a unified set of equations, rather than by different sets of formulas.

3.1.4.2 Flanged Sections

The calculation of the moment of inertia of uncracked T-sections is quite straightforward, but rather tedious. In practice, I_g for gross T-sections are obtained from graphs in handbooks, where the rebar areas are neglected. The equation for plotting the graphs is briefly derived here.

The T-section shown in Figure 3.6 has a height of h and a width of b. The thickness of the flange is h_f and the width of the web is b_w. The first step is to find the position of the centroidal axis defined by the distance x measured from the top surface of the cross-section. Taking the static moments of the web $b_w h$ and the outstanding flange $(b - b_w)h_f$ about an axis lying on

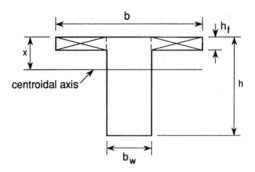

Figure 3.6   Uncracked T-sections

the top surface gives:

$$X = \frac{h}{2} \frac{1 + \left(\frac{b}{b_w} - 1\right)\left(\frac{h_f}{h}\right)^2}{1 + \left(\frac{b}{b_w} - 1\right)\left(\frac{h_f}{h}\right)} \tag{3.44}$$

The second step is to find the gross moment of inertia about the centroidal axis by the transfer axis theorem:

$$I_g = \frac{b_w h^3}{12} + b_w h \left(\frac{h}{2} - x\right)^2 + (b - b_w) h_f \left(x - \frac{h_f}{2}\right)^2 + \frac{(b - b_w) h_f^3}{12} \tag{3.45}$$

Inserting x from Equation (3.44) into (3.45), simplifying and grouping the terms results in:

$$I_g = \frac{b_w h^3}{12} \left[1 + \left(\frac{b}{b_w} - 1\right)\left(\frac{h_f}{h}\right)^3 + \frac{3\left(\frac{b}{b_w} - 1\right)\left(\frac{h_f}{h}\right)\left(1 - \frac{h_f}{h}\right)^2}{1 + \left(\frac{b}{b_w} - 1\right)\left(\frac{h_f}{h}\right)} \right] \tag{3.46}$$

Let the expression within the bracket of Equation (3.46) be denoted as K_{i4}, then Equation (3.46) can be written simply as $I_g = K_{i4}(b_w h^3/12)$, where K_{i4} is a function of b/b_w and h_f/h. K_{i4} is plotted as a function of b/b_w and h_f/h in a graph in the ACI Special Publication SP-17(73) and in other textbooks.

3.1.5 Bending Deflections of Reinforced Concrete Members

3.1.5.1 Deflections of Homogeneous Elastic Members

The bending deformations, i.e. deflections and rotations, of a member are, in general, calculated from the basic moment–curvature relationships of the sections along the member, as shown in Figure 3.7. In the simple case of a homogeneous elastic member, however, the curvature ϕ is proportional to the moment M and the proportionality constant is the bending rigidity EI. Since the curvature ϕ is defined as the bending rotation per unit length $(d\Phi/dx)$, the

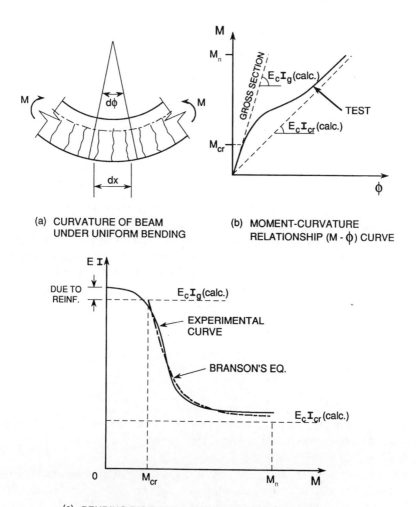

(a) CURVATURE OF BEAM
UNDER UNIFORM BENDING

(b) MOMENT-CURVATURE
RELATIONSHIP (M - ϕ) CURVE

(c) BENDING RIGIDITY E$\mathbf{I}$ AS A FUNCTION OF MOMENT M

Figure 3.7   Variation of bending rigidity with moment

differential bending rotation dΦ is expressed for a differential length dx as

$$d\Phi = \frac{M}{EI}dx \qquad (3.47)$$

The deflections and rotations of a member can be calculated from Equation (3.47) by integration if the moment diagram is given analytically. Other forms of integration, such as moment area method, elastic load method and conjugate beam method are available in textbooks on structural analysis.

In design practice, the deflections and rotations of a prismatic member are calculated from formulas available in handbooks and textbooks. A method of organizing the various formulas is given in the 1983 ACI Commentary 9.5.2.4 (ACI Committee 318, 1983) and in the ACI Special Publication SP-43 (ACI Committee 435, 1974). In this method the deflection Δ of a member at the midspan of simple and continuous beams or at the end of a cantilevers is

$$\Delta = K \frac{5M\ell^2}{48EI} \tag{3.48}$$

where

$M =$ moment at midspan for simple and continuous beams or moment at support for cantilever beams;

$\ell =$ clear span of a beam;

$K = 1$ for simple beam, 12/5 for cantilever beam, $1.20 - 0.20 M_o/M_c$ for continuous beam, where $M_o =$ simple span moment $= w\ell^2/8$, and M_c is the net midspan moment.

For a beam with both ends fixed $M_c = w\ell^2/24$ and $K = 0.60$; and for a beam with one end fixed and the other end hinged $M_c = w\ell^2/16$ and $K = 0.8$.

The difficulty of applying Equation (3.48) to reinforced concrete beams with both cracked and uncracked sections is the evaluation of the bending rigidity EI. To overcome this difficulty we introduce an effective bending rigidity $E_c I_e$, which can be used in connection with Equation (3.48) to calculate the deflections.

3.1.5.2 Effective Bending Rigidities

The effective bending rigidity $E_c I_e$ is the product of two quantities E_c and I_e. E_c is the modulus of elasticity of concrete defined by the ACI Code (ACI 318-08) to be

$$E_c = 44w^{1.5}\sqrt{f_c'(MPa)} \text{ or } E_c = 33w^{1.5}\sqrt{f_c'(psi)} \tag{3.49}$$

where E_c and f_c' are in MPa (or psi) and the unit weight w is in kN/m^3 (or lb/ft^3). In the case of normal weight concrete w is taken as 22.6 kN/m^3 (or 144 lb/ft^3) and $E_c = 4730\sqrt{f_c'(MPa)}$ (or $57000\sqrt{f_c'(psi)}$).

The effective moment of inertia I_e will now be derived. A uniform moment M acting on a length of beam is expected to create a curvature $\phi = d\Phi/dx$, as shown in Figure 3.7(a). The moment–curvature relationship (M–ϕ curve) is plotted in Figure 3.7(b). Up to the cracking moment M_{cr}, the curve follows approximately a slope calculated from the bending rigidity $E_c I_g$ of the gross uncracked section. After cracking, however, the curve changes direction drastically. When the nominal moment M_n is reached, the slope approaches the bending rigidity $E_c I_{cr}$, calculated by the cracked section.

The trend described above can be clearly illustrated by plotting the bending rigidity EI against the moment M in Figure 3.7(c). Below the cracking moment M_{cr}, the experimental curve is roughly horizontal at the level of the calculated $E_c I_g$. After cracking, however, the curve drops drastically. When the nominal moment M_n is approached, the curve becomes asymptotic to the horizontal line of the calculated $E_c I_{cr}$. A theoretical equation to express the moment of inertia, therefore, must vary from I_g at cracking to I_{cr} at ultimate.

The trend of the experimental curve, Figure 3.7(c), can be closely approximated by a theoretical curve suggested by Branson (1965):

$$I = \left(\frac{M_{cr}}{M}\right)^4 I_g + \left[1 - \left(\frac{M_{cr}}{M}\right)^4\right] I_{cr} \leq I_g \tag{3.50}$$

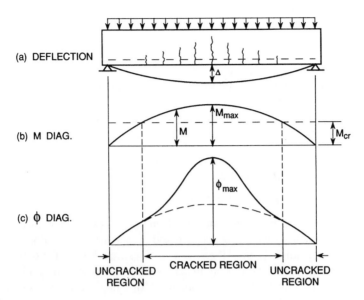

Figure 3.8 Moment and curvature diagrams for simply supported beams under uniform load

Equation (3.50) satisfies the two limiting cases. First, $I = I_g$ when $M_{cr}/M = 1$. Second, $I = I_{cr}$ when M_{cr}/M approaches zero. The moment of inertia I obviously has a value between I_g and I_{cr}. The power of 4 for (M_{cr}/M) in Equation (3.50) was selected to best fit the experimental curve.

When a beam is subjected to lateral loads, the moment is no longer uniform along the length of the beam. Figure 3.8(a) shows a simply supported beam under uniform load. According to the parabolic bending moment diagram (Figure 3.8b), the moment in the central region of the beam exceeds the cracking moment M_{cr}, causing cracks to develop and the curvature ϕ to increase rapidly (Figure 3.8c). Near the ends, however, the regions are still uncracked since the moment is still less than the cracking moment. For such a beam with both cracked and uncracked regions, a rigorous calculation of the deflection using numerical integration would be very tedious. For practical purposes, an effective moment of inertia I_e was proposed by Branson (1977) and adopted by the ACI Code, Section 9.5.2.3 (ACI 318-08):

$$I_e = \left(\frac{M_{cr}}{M_a}\right)^3 I_g + \left[1 - \left(\frac{M_{cr}}{M_a}\right)^3\right] I_{cr} \leq I_g \tag{3.51}$$

where M_a is the maximum moment at midspan, and the cracking moment is

$$M_{cr} = \frac{f_r I_g}{y_t} \tag{3.52}$$

In Equation (3.52) y_t is the distance from the centroidal axis of gross section to the extreme fiber in tension. f_r is the modulus of rupture defined for normal-weight concrete by

$$f_r = 0.63\sqrt{f_c'(MPa)} \ \text{ or } \ f_r = 7.5\sqrt{f_c'(psi)} \tag{3.53}$$

Equation (3.51) is a modification of Equation (3.50). The power of 3 for (M_{cr}/M_a) in Equation (3.51) is chosen to best fit the test results obtained for simply supported beams. Equation (3.51) is also found to be applicable to cantilever beams if M_a is the maximum moment at the support. For continuous beams the ACI Code allows the effective moment of inertia to be taken as the average of the values obtained from Equation (3.51) for the critical positive and negative moment sections.

The effective bending rigidity $E_c I_e$, determined by Equation (3.49) and Equations (3.51)–(3.53), is used to calculate the deflections in conjunction with Equation (3.48).

3.2 Nonlinear Bending Theory

3.2.1 Bernoulli Compatibility Truss Model

Linear bending theory has been presented in Section 3.1. This linear theory utilizes the Bernoulli compatibility truss model, which is based on Navier's three principles of equilibrium, Bernoulli's compatibility, and Hooke' *linear* constitutive laws for both concrete and steel reinforcement. In this section, a nonlinear bending theory will be studied which also utilizes the Bernoulli compatibility truss model. In this case, however, *nonlinear* constitutive laws of concrete and steel will be used, in addition to equilibrium and Bernoulli's compatibility.

The stress–strain relationship of mild steel is assumed to be linear up to yielding, followed by a yield plateau. This elastic–perfectly plastic type of stress–strain relationship imparts a distinct method of analysis and design for concrete members reinforced with mild steel. The nonlinear bending theory presented in this section is, therefore, applicable only to *mild steel* reinforced concrete members at *ultimate* load stage.

Navier's three principles will first be applied to singly reinforced rectangular beams in Section 3.2.2, and then to doubly reinforced rectangular beams and flanged beams in Sections 3.2.3 and 3.2.4, respectively. Finally, the moment–curvature relationship of mild steel reinforced beams will be discussed in Section 3.2.5, using both the nonlinear bending theory and the linear bending theory.

3.2.1.1 Equilibrium and Compatibility Conditions

A singly reinforced rectangular beam is subjected to a nominal bending moment M_n at ultimate load stage (Figure 3.9a). The moment M_n will induce an ultimate curvature ϕ_u, which is defined as the bending rotation per unit length of the member. The fundamental assumption relating M_n and ϕ_u is the well-known Bernoulli compatibility hypothesis, which states that a plane section before bending will remain plane after bending. In other words, the strains along the depth of the cross-section will be distributed linearly, as shown in Figure 3.9(b). The maximum compressive strain of concrete at the top surface is ε_u. The tensile strain ε_s at the centroid of the rebars is located at a distance d from the top surface. The neutral axis, where the strain is zero, is located at a distance c from the top surface. The compatibility equation can then be established in terms of c, ε_u and ε_s. Using the geometric relationship of similar triangles we have:

$$\frac{c}{d} = \frac{\varepsilon_u}{\varepsilon_u + \varepsilon_s} \tag{3.54}$$

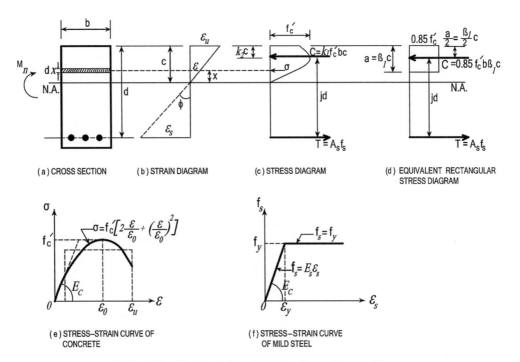

(a) CROSS SECTION   (b) STRAIN DIAGRAM   (c) STRESS DIAGRAM   (d) EQUIVALENT RECTANGULAR STRESS DIAGRAM

(e) STRESS–STRAIN CURVE OF CONCRETE   (f) STRESS–STRAIN CURVE OF MILD STEEL

Figure 3.9 Singly reinforced rectangular sections at ultimate

The equilibrium condition can be derived from the distribution of stresses along the depth of the cross-section, as given in Figure 3.9(c). Above the neutral axis is a compression stress block sketched by a curve. The stresses in the compression stress block are related to the strain in Figure 3.9(b) through the stress–strain relationship of the concrete. The stress–strain relationship of concrete is assumed to be identical to the stress–strain curve obtained from the compression test of a standard concrete cylinder, shown in Figure 3.9(e).

The compression stress block can be replaced statically by a resultant C in the equilibrium equations. This resultant has a magnitude of $C = k_1 f_c' bc$, where k_1 is a coefficient representing the ratio of the average stress of the compression stress block to the maximum stress of concrete f_c'. The resultant C also has a location defined by a distance $k_2 c$, where k_2 is coefficient representing the ratio of the depth of the resultant C from the extreme fiber to the depth of the compression zone. These two coefficients k_1 and k_2 will be determined in Section 3.2.1.2.

Below the neutral axis in Figure 3.9(c), the small tensile stress of concrete adjacent to the neutral axis is neglected. All the tensile stresses in the rebars are assumed to be concentrated at the centroid of the reinforcing bars with a total area of A_s. The stress in the rebar f_s is assumed to be related to the strain ε_s in Figure 3.9(b) through the stress–strain curve obtained from the tension tests of bare reinforcing bars, Figure 3.9(f). For mild steel reinforcing bars, the stress–strain curve is first linear, with a slope of E_s up to the yielding strength f_y, then followed by a yield plateau, represented by a long horizontal straight line up to failure. Before steel yielding, the total tensile force is denoted as $T = A_s f_s$. After steel yielding, the total tensile force is denoted as $T = A_s f_y$.

From the stress diagram in Figure 3.9(c), we recognize that the external moment M_n is resisted by the internal moment, consisting of a pair of equal, opposite and parallel forces, T and C, located at a distance jd from each other. This idea of bending resistance is frequently referred to as the *internal couple concept*. From the force equilibrium $T = C$ we have

$$A_s f_s = k_1 f'_c bc \tag{3.55}$$

From the moment equilibrium, the two most convenient forms are: $M_n = T(jd)$ or $M_n = C(jd)$, resulting in:

$$M_n = A_s f_s (jd) = A_s f_s (d - k_2 c) \tag{3.56}$$

$$M_n = k_1 f'_c bc(jd) = k_1 f'_c bc (d - k_2 c) \tag{3.57}$$

Although we have written three equilibrium equations (3.55)–(3.57), only two of these equations are independent, meaning that they could only be used to solve two unknown variables.

Using two of the three equilibrium equations, the compatibility equation (3.54), and the two stress–strain curves of concrete and steel, we can solve the variable c and determine the location of the neutral axis.

Finally, from the linear strain distribution given in Figure 3.9(b) the ultimate curvature ϕ_u is expressed as

$$\phi_u = \frac{\varepsilon_u}{c} \tag{3.58}$$

Equations (3.56) and (3.58) establish the moment–curvature relationship (M_n and ϕ_u) at ultimate strength.

The nonlinear constitutive relationships of concrete will be elaborated in Section 3.2.1.2. The detailed solution procedures using Equations (3.54)–(3.57) and the nonlinear constitutive laws of concrete and steel will be explained in Sections 3.2.2, 3.2.3 and 3.2.4 for singly reinforced, doubly reinforced and flanged sections, respectively.

3.2.1.2 Nonlinear Constitutive Relationship of Concrete

Theoretical Derivation
The distribution of stresses in the concrete compression stress block (Figure 3.9c), is assumed to have the same shape as the stress–strain curve of concrete (Figure 3.9e), obtained from the tests of standard 6×12 in. concrete cylinders. The stress-strain curve of concrete is frequently expressed by a parabolic equation:

$$\sigma = f'_c \left[2 \left(\frac{\varepsilon}{\varepsilon_o} \right) - \left(\frac{\varepsilon}{\varepsilon_o} \right)^2 \right] \tag{3.59}$$

where
    $\sigma =$ compressive stress of concrete;
    $\varepsilon =$ compressive strain of concrete;
    $f'_c =$ maximum compressive stress of concrete obtained from standard cylinders;
    $\varepsilon_o =$ strain at the maximum stress f'_c, usually taken as 0.002.

The magnitude of the compression resultant C can be expressed in the following manner. Let x be the distance from the neutral axis to a level where the strain is ε and the stress is σ (Figure 3.9b and c). From similar triangle of the strain diagram, $x = (c/\varepsilon_u)\varepsilon$. Differentiating x gives $dx = (c/\varepsilon_u)d\varepsilon$ and the resultant C can be expressed by integrating the compression stress block:

$$C = b \int_0^c \sigma \, dx = b \left(\frac{c}{\varepsilon_u}\right) \int_0^{\varepsilon_u} \sigma \, d\varepsilon \tag{3.60}$$

Substituting $C = k_1 f'_c bc$ into Equation (3.60) gives the expression for k_1:

$$k_1 = \frac{1}{f'_c \varepsilon_u} \int_0^{\varepsilon_u} \sigma \, d\varepsilon \tag{3.61}$$

Substituting the concrete stress σ from Equation (3.59) into Equation (3.61) and integrating results in

$$k_1 = \frac{\varepsilon_u}{\varepsilon_o} \left(1 - \frac{1}{3}\frac{\varepsilon_u}{\varepsilon_o}\right) \tag{3.62}$$

Inserting the ACI value of $\varepsilon_u = 0.003$ and $\varepsilon_o = 0.002$ into Equation (3.62) gives $k_1 = 0.75$. The location of the resultant C is defined by $k_2 c$ measured from the top surface of the cross-section. The coefficient k_2 is determined by taking moment about the neutral axis:

$$C(c - k_2 c) = b \int_0^c \sigma x \, dx = b \left(\frac{c}{\varepsilon_u}\right)^2 \int_0^{\varepsilon_u} \sigma \varepsilon \, d\varepsilon \tag{3.63}$$

Substituting C from Equation (3.60) into Equation (3.63) gives

$$k_2 = 1 - \frac{1}{\varepsilon_u} \frac{\int_0^{\varepsilon_u} \sigma \varepsilon \, d\varepsilon}{\int_0^{\varepsilon_u} \sigma \, d\varepsilon} \tag{3.64}$$

Again, substituting the concrete stress σ from Equation (3.59) into Equation (3.64) and integrating results in

$$k_2 = 1 - \frac{2}{3}\frac{1 - \frac{3}{8}(\varepsilon_u/\varepsilon_o)}{1 - \frac{1}{3}(\varepsilon_u/\varepsilon_o)} \tag{3.65}$$

Inserting the ACI value of $\varepsilon_u = 0.003$ and $\varepsilon_o = 0.002$ into Equation (3.65) gives $k_2 = 0.4167$.

The values of $k_1 = 0.75$ and $k_2 = 0.4167$ can be inserted into Equations (3.55)–(3.57), so that these equilibrium equations can be used in connection with the compatibility equation and the constitutive laws.

Hognestad et al's tests

The shape of the concrete stress–strain curve was found by the 'dog bone' tests of Hognestad *et al.* (1955) to have two additional characteristics. First, the coefficients k_1 and k_2 depend on the compressive strength of the concrete f_c'. Apparently, the stress–strain curve becomes more linear in the ascending portion and more steep in the descending portion when the concrete strength is increased. In other words, the coefficients k_1 and k_2 should decrease with an increase of f_c'. Second, the maximum stress of concrete in the stress block of a beam is somewhat less than the maximum stress of a 152×305 mm (6 × 12 in.) standard concrete cylinder, f_c', because of size effect, shape effect, loading rate effect, etc. The maximum stress of concrete in the stress block is defined as $k_3 f_c'$, where k_3 is also function of f_c'.

The three coefficients, k_1, k_2 and k_3, are given for concrete with $f_c' < 55$ MPa (8000 psi) as follows:

$$k_1 = 0.94 - \frac{f_c'(\text{MPa})}{179} \quad \text{or} \quad k_1 = 0.94 - \frac{f_c'(\text{psi})}{26\,000} \tag{3.66}$$

$$k_2 = 0.50 - \frac{f_c'(\text{MPa})}{552} \quad \text{or} \quad k_2 = 0.50 - \frac{f_c'(\text{psi})}{80\,000} \tag{3.67}$$

$$k_3 = \frac{3.9 + 0.051\,f_c'(\text{MPa})}{3.0 + 0.115\,f_c'(\text{MPa}) - 0.00081\,f_c'^2(\text{MPa})}$$

$$\text{or} \quad \frac{3,900 + 0.35\,f_c'(\text{psi})}{3\,000 + 0.82\,f_c'(\text{psi}) - 0.000038\,f_c'^2(\text{psi})} \tag{3.68}$$

Whitney's equivalent rectangular stress block

To simplify the analysis and design, ACI Code (ACI 318-08) allows the curved concrete stress block to be replaced by an equivalent rectangular stress block as shown in Figure 3.9(d). The replacement is such that the magnitude and the location of the resultant C remain unchanged. The ACI rectangular stress block has a uniform stress of $0.85\,f_c'$ and a depth, $a = \beta_1 c$, where β_1 is determined to be:

f_c'	β_1
≤ 27.6 MPa (4000 psi)	0.85
34.5 MPa (5000 psi)	0.80
41.4 MPa (6000 psi)	0.75
48.3 MPa (7000 psi)	0.70
≥ 55.2 MPa (8000 psi)	0.65

In terms of the US conventional units, this table shows that β_1 is 0.85 when f_c' is less than 4000 psi, and reduces by 0.05 for every 1000 psi, up to 8000 psi.

Based on this rectangular stress block ($a = \beta_1 c$), the coefficients $k_1 = 0.85\beta_1$ and $k_2 = \beta_1/2$. The magnitude of the resultant $C = 0.85\,f_c'ba = 0.85\,f_c'b\beta_1 c$. The force equilibrium equation (3.55) and the moment equilibrium equation (3.56) become

$$A_s f_s = 0.85\,f_c'ba \tag{3.69}$$

$$M_u = \varphi A_s f_s \left(d - \frac{a}{2} \right) \tag{3.70}$$

where M_u is the factored moment $= \varphi M_n$, and φ is the reduction factor for material. $\varphi = 0.9$ for bending in the ACI Code.

3.2.2 Singly Reinforced Rectangular Beams

A singly reinforced rectangular section Figure 3.9(a), is subjected to a nominal bending moment M_n at the ultimate load stage. The cross-section has a width of b, and an effective depth of d. The mild steel reinforcing bars have a total area of A_s. The moment M_n induces a linear distribution of strains, as indicated in the strain diagram of Figure 3.9(b). The maximum concrete strain at the top surface ε_u is specified by the ACI Code (ACI 318-08) to be 0.003 at ultimate. The moment M_n is resisted by a $C - T$ couple as shown in Figure 3.9(c) and (d). The resultant C is statically equivalent to the compression stress block with real stress–strain curve shown in Figure 3.9(c) and to an equivalent stress block of rectangular shape shown in Figure 3.9(d).

3.2.2.1 Modes of Failure

A singly reinforced beam may fail in three different modes:
Insufficient amount of steel
When the beam is reinforced with a very small amount of steel, the steel will yield at cracking and the beam will collapse suddenly. To exclude this undesirable mode of failure, the ACI Code provides a minimum amount of flexural steel as follows:

$$\rho_{\min} = \frac{0.25\sqrt{f_c'(\text{MPa})}}{f_y} \quad \text{or} \quad \rho_{\min} = \frac{3\sqrt{f_c'(\text{psi})}}{f_y} \tag{3.71}$$

Under-reinforced beams
When the beam is reinforced with a moderate amount of steel, the steel will yield first, followed by a secondary crushing of concrete. Consequently, failure is preceded by a large deflection, and the failure mode is ductile and desirable.
Over-reinforced beams
When the beam is reinforced with an excessive amount of steel, the concrete will crush first, before the yielding of steel. Consequently, failure is preceded by a very small deflection, and the failure mode is brittle and undesirable.

To differentiate the under-reinforced beams from the over-reinforced beams, we will derive a balanced percentage of steel ρ_b, which is defined as the percentage of steel that causes the yielding of the steel and the crushing of concrete to occur simultaneously. Therefore:

$\rho < \rho_b$ gives under-reinforced beams
$\rho > \rho_b$ gives over-reinforced beams

3.2.2.2 Balanced Condition

The 'balanced condition' is the condition when the steel reaches the yield point, i.e. $\varepsilon_s = \varepsilon_y$, simultaneously with the crushing of concrete, i.e. $\varepsilon_u = 0.003$.

The bending of singly reinforced beams involves nine variables b, d, A_s, M_u, f_s, f_c', ε_s, ε_u, and c (or a), as shown in Figure 3.9(a)–(d). The coefficients β_1, are not considered variables because they are determined from the given compression stress–strain curve (Figure 3.9e), and listed in Section 3.2.1.2. A total of four equations are available from the Bernoulli's compatibility truss model, i.e. two from equilibrium, one from Bernoulli compatibility and one from the constitutive law of mild steel (Figure 3.9f). Therefore, five variables must be given and the remaining four unknown variables can be solved by the four equations.

The problem posed for solving the balanced condition is:

$$\text{Given: } b, d, f_c', \varepsilon_s = \varepsilon_y, \varepsilon_u = 0.003$$
$$\text{Find: } A_s, M_u, f_s, \text{ and } a \text{ (or } c\text{)}$$

The four available equations and their unknowns are summarized as follows:

Type of equations	Equations	Unknowns				
Equilibrium of forces	$A_s f_s = 0.85 f_c' ba$	A_s			a	(3.72)
Equilibrium of moments	$M_u = \varphi A_s f_s \left(d - \dfrac{a}{2}\right)$	A_s	M_u	f_s	a	(3.73)
Bernoulli compatibility	$\dfrac{a}{\beta_1 d} = \dfrac{\varepsilon_u}{\varepsilon_u + \varepsilon_y}$				a	(3.74)
Constitutive law for steel	$f_s = E_s \varepsilon_s \quad \text{for } \varepsilon_s \le \varepsilon_y$			f_s		(3.75a)
	$f_s = f_y \quad \text{for } \varepsilon_s > \varepsilon_y$			f_s		(3.75b)

The four unknown variables, A_s, M_u, f_s, and a, are indicated for each equation under the column heading *Unknowns*. Because the steel strain ε_s is given to be the yield strain ε_y, the steel stress f_s should obviously be equal to f_y based on the stress–strain relationship of either Equations (3.75a) or (3.75b). Also, the unknown depth a of the equivalent rectangular stress block can be solved directly from the compatibility condition (Equation 3.74). Once the depth a and the steel stress $f_s = f_y$ are determined, the steel area A_s can be obtained from the force equilibrium (Equation 3.72), and the factored moment M_u from the moment equilibrium (Equation 3.73). In this particular case of the balanced condition there is no need to solve simultaneous equations.

In this case of balanced condition we will now add a subscript b to the three unknown quantities: A_{sb} for balanced steel area, M_{ub} for balanced factored moment and a_b for balanced depth of equivalent rectangular stress block. Substituting $\varepsilon_u = 0.003$ and $\varepsilon_y = f_y(\text{MPa})/200\,000$ into the compatibility equation (3.74), the balanced depth a_b is expressed in terms of yield stress f_y (MPa) as

$$a_b = \beta_1 d \frac{600}{600 + f_y(\text{MPa})} \tag{3.76}$$

Substituting a_b into the force equilibrium (Equation 3.72), the balanced steel area A_{sb} is

$$A_{sb} = 0.85 \beta_1 bd \frac{f_c'}{f_y} \frac{600}{600 + f_y(\text{MPa})} \tag{3.77}$$

Dividing Equation (3.77) by bd and defining $\rho_b = A_{sb}/bd$ as the balanced percentage of steel gives

$$\rho_b = 0.85 \beta_1 \frac{f_c'}{f_y} \frac{600}{600 + f_y(\text{MPa})} \tag{3.78}$$

Inserting a_b from Equation (3.76), A_{sb} from Equation (3.77), $f_s = f_y$ and the reduction factor φ into the moment equilibrium equation (3.73), results in

$$M_{ub} = \left[\varphi 0.85\beta_1 \frac{600}{600 + f_y(\text{MPa})} \left(1 - 0.5\beta_1 \frac{600}{600 + f_y(\text{MPa})}\right) f_c'\right] bd^2 \qquad (3.79)$$

Equation (3.79) can be written simply as $M_{ub} = R_b\, bd^2$, where R_b is the expression within the bracket. The coefficient R_b is a function of the material properties f_c' and f_y (tabulated in some textbooks).

The balanced condition expressed by a_b (Equation 3.76), ρ_b (Equation 3.78) or M_{ub} (Equation 3.79), divides under-reinforced beams from over-reinforced beams. The balanced percentage of steel ρ_b is useful for the analysis type of problems when the cross-sections of concrete and steel are given. The balanced factored moment M_{ub} is convenient for the design type of problems when the moment is given. The balanced depth of rectangular stress block a_b could be used either in analysis or design, but mostly in design in lieu of M_{ub}.

The application of the balanced condition discussed above can be summarized in the following table:

Types of problem	Under-reinforced beams	Over-reinforced beams
Analysis	$\rho < \rho_b$	$\rho > \rho_b$
Design	$M_u < M_{ub}$ or $a < a_b$	$M_u > M_{ub}$ or $a > a_b$

3.2.2.3 Over-reinforced Beams

If the percentage of steel of a beam is greater than the balanced percentage ($\rho > \rho_b$) or the factored moment is greater than the balanced factored moment ($M_u > M_{ub}$), then the beam is over-reinforced. The problem posed for the analysis of over-reinforced beams is:

Given: $b, d, A_s, f_c', \varepsilon_u = 0.003$
Find: $M_u, f_s, \varepsilon_s < \varepsilon_y$, and a (or c)

The four available equations and their unknowns are:

Type of equation	Equations	Unknowns		
Equilibrium of forces	$A_s f_s = 0.85 f_c' ba$	f_s	a	(3.80)
Equilibrium of moments	$M_u = \varphi A_s f_s \left(d - \frac{a}{2}\right)$	M_u f_s	a	(3.81)
Bernoulli compatibility	$\dfrac{a}{\beta_1 d} = \dfrac{\varepsilon_u}{\varepsilon_u + \varepsilon_y}$	ε_s a		(3.82)
Constitutive law for steel	$f_s = E_s \varepsilon_s$ for $\varepsilon_s \le \varepsilon_y$	f_s ε_s		(3.83)

Based on the unknowns in each equations, the best strategy to solve the above four equations is as follows: (1) Substitute ε_s from the stress–strain relationship of Equation (3.83) into

Equation (3.82) and express the compatibility equation in terms of the steel stress f_s and the depth a. (2) Solve the new compatibility equation simultaneously with the force equilibrium equation (3.80), to determine the unknowns f_s and a. (3) Insert the newly found unknowns, f_s and a, into the moment equilibrium equation (3.81), and calculate the moment M_u. (4) Calculate the steel strain $\varepsilon_s = f_s/E_s$ from the stress–strain relationship of steel. Knowing ε_s allows us to calculate the ultimate curvature $\phi_u = \varepsilon_s/(d - c)$ from the strain compatibility condition, Figure 3.9(b).

Elaboration of the above procedures is not warranted, because over-reinforced beams are seldom encountered. Over-reinforced beams are expected to fail in a brittle manner and to exhibit small deflections. Consequently, the design of over-reinforced beams is either prohibited or penalized by all the building and bridge codes.

3.2.2.4 Ductility of Under-reinforced Beams

In an under-reinforced beam, the steel bars are expected to yield before the crushing of concrete. However, if the steel bars develop only a small plastic strain after yielding, the resulting small deflection may not be adequate. To ensure the beam will develop a sufficiently large deflection before collapse due to secondary crushing of concrete, the tensile steel bars must be able to develop sufficiently large strains. This concept results in the ACI Code (ACI 318-08) specifying a *tensile steel strain limit of 0.005*.

The ACI *tensile strain limit method* to ensure sufficient ductility of flexural beams is illustrated in Figure 3.10. To implement this method, the Code defines a new effective depth d_t, defined as the distance from the extreme compression fiber to the centroid of the extreme

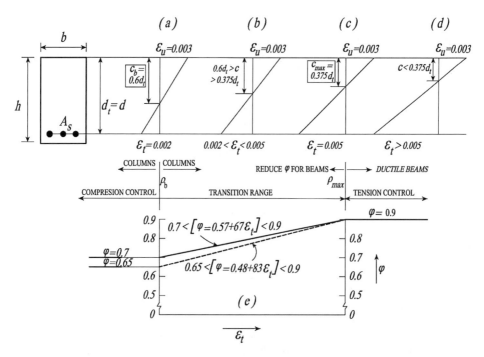

Figure 3.10   Strain limit methods for ductile beams

layer of longitudinal tension steel. The depth d_t is somewhat different from the well-known effective depth d, which is defined as the distance from the extreme compression fiber to the centroid of longitudinal tensile reinforcement. If only one layer of steel bars is used, then $d_t = d$. If more than one layer of steel bars are used, then d_t is somewhat larger than d. For simplicity, one layer of steel bars is used in Figure 3.10 and $d_t = d$.

Figure 3.10 (a), (b), (c) and (d) from left to right show four strain distributions at ultimate load stage of four beams reinforced with decreasing percentages of steel. The first diagram (a) shows the strain distribution of a beam with a balanced steel percentage ρ_b. That means when the concrete reaches the ultimate strain of $\varepsilon_u = 0.003$ specified by the ACI Code, the steel bar reaches the yield strain of 0.002 for a mild steel bar with a yield stress of about $f_y = 413$ MPa (60 000 psi). In this case, the position of the neutral axis gives a balanced value of $c_b = 0.6d_t$. When the steel percentage ρ is greater than ρ_b, the beam is considered to be under 'compression control' because concrete will crush before the yielding of steel. Such behavior is only allowed in columns, which must be designed with a low reduction factor $\varphi = 0.7$ for spiral columns and $\varphi = 0.65$ for tied columns.

The third diagram (c) shows the strain distribution of a beam with the maximum steel percentage ρ_{max}. In this case, when the concrete reaches the ultimate strain of $\varepsilon_u = 0.003$, the steel bar reaches the ACI specified strain limit of $\varepsilon_t = 0.005$. The position of the neutral axis gives a maximum value of $c_{max} = 0.375d_t$, or $a_{max} = 0.375\beta_1 d_t$.

The fourth diagram (d) shows the strain distribution of a beam with the steel percentage ρ less than ρ_{max}. That means that when the concrete reaches the ultimate strain of $\varepsilon_u = 0.003$, the steel bar will have a strain ε_t greater than 0.005, while c is less than $0.375d_t$. A beam located in this region will be called a 'ductile beam', and the behavior of the beam is called 'tension control.' A ductile beam has the privilege of using a high reduction factor of $\varphi = 0.9$.

A typical ductile beam could be designed for $\rho = 0.5\rho_{max}$. In this case, the neutral axis is located at $c = 0.5(0.375d_t) = 0.1875d_t$, and the tensile strain $\varepsilon_t = \varepsilon_u(d_t - c)/c = 0.003[(d_t/c) - 1] = 0.003[(1/0.1875) - 1] = 0.013 > 0.005$, ductility, OK. For a beam with $b = 200$ mm, $d = 500$ mm, and $f_c' = 41.4$ MPa (6000 psi), $a = \beta_1 c = 0.75(0.1875 \times 500) = 70.3$ mm. The factored moment $M_u = \varphi 0.85 f_c' b a (d - a/2) = (0.9)(0.85)(41.4)(200)(70.3)(500 - 70.3/2) = 207$ kN m.

The second diagram (b) shows the strain distribution of a beam with the steel percentage ρ greater than the maximum ρ_{max}, but less than the balanced ρ_b ($\rho_{max} \leq \rho \leq \rho_b$). When the concrete reaches the ultimate strain of $\varepsilon_u = 0.003$, the steel bar reaches a strain of ε_t less than 0.005, but greater than 0.002 ($0.005 \geq \varepsilon_t \geq 0.002$), while c is larger than $0.375d_t$, but less than $0.6d_t$ ($0.375d_t \leq c \leq 0.6d_t$). In this range, the reduction factor φ will decrease from 0.9 to 0.7 (or 0.65) with decreasing ε_t from 0.005 to 0.002. The analytical expressions for this transition are shown in Figure 3.10(e).

In practice, it is recommended to use 'ductile beams', rather than nonductile beams, for two reasons. First, ductile beams are more economical because they can be designed with a φ factor of 0.9 without penalty. Second, ductile beams are much simpler to design because the procedure to penalize the φ factor is complicated.

3.2.2.5 Analysis of Ductile Sections

The analysis and design of ductile beams will be presented separately. The analysis of ductile beams is included in this section, while the design of ductile beams will be discussed in Section 3.2.2.6.

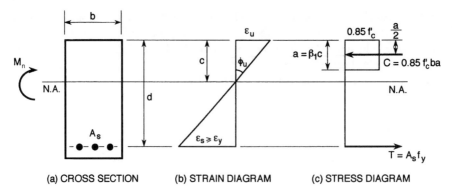

(a) CROSS SECTION   (b) STRAIN DIAGRAM   (c) STRESS DIAGRAM

Figure 3.11   Under-Reinforced sections at ultimate

The strain and stress diagrams for a ductile beam are shown in Figure 3.11(b) and (c), respectively. The analysis and design of ductile beams are considerably simplified for two reasons. First, the tensile steel stress is known to yield, $f_s = f_y$, and the stress–strain curve of steel is not required in the solution. Second, the tensile steel strain is expected to lie within the plastic range, $\varepsilon_s \geq \varepsilon_y$. Bernoulli's compatibility condition, which relates the steel strain ε_s to the maximum concrete strain ε_u at the top surface, becomes irrelevant to the solution of other stress-type variables. Because of these two simplifications, the analysis and design of under-reinforced beams involve only eight variables, b, d, A_s, M_u, f_y, f_c', ε_u and a(or c). The available equations are now down to two, the only two from the equilibrium condition. Therefore, six variables must be given before the two remaining unknown variables can be solved by the two equilibrium equations.

The analysis problem to find the moment is posed as follows:

Given: b, d, A_s, f_y, f_c' and ε_u
Find: M_u and a

As mentioned previously, three forms of equilibrium equations are frequently used in the parallel, coplanar force system of bending action, but only two are independent. However, the three equations are given below for the purpose of finding the most convenient solution. Remember that only two of these three equations will be selected.

Type of equations	Equations	Unknowns	
Equilibrium of forces	$A_s f_y = 0.85 f_c' ba$	a	(3.84)
Equilibrium of moment about C	$M_u = \varphi A_s f_y \left(d - \dfrac{a}{2}\right)$	$M_u \quad a$	(3.85)
Equilibrium of moment about C	$M_u = \varphi 0.85 f_c' ba \left(d - \dfrac{a}{2}\right)$	$M_u \quad a$	(3.86)

The solution for the analysis type of problems turns out to be quite simple when the unknowns in each equation are examined. It can be seen from Equation (3.84) that the depth a can be determined directly from the equilibrium of forces, C and T. Then the moment M_u could be calculated either from the equilibrium of moments about the compressive force C (Equation 3.85), or from the equilibrium of moments about the tensile force T (Equation 3.86).

From the equilibrium of forces:

$$a = \frac{A_s f_y}{0.85 f_c' b} \qquad (3.87)$$

and from the equilibrium of moments about the compression force C:

$$M_u = \varphi A_s f_y \left(d - \frac{a}{2} \right) \qquad (3.88)$$

It should again be emphasized that the depth of the neutral axis c, which is equal to a/β_1, can be determined directly from the equilibrium condition (Equation 3.87), without using the compatibility condition and the stress–strain relationship of steel. This is a special characteristic of mild steel, under-reinforced, concrete beams.

3.2.2.6 Design of Ductile Sections

First Type of Design (find area of steel)
                      Given: b, d, M_u, f_y, f_c' and ε_u
              Find: A_s and a

The three equilibrium equations and their unknowns are:

Type of equation	Equations	Unknowns		
Equilibrium of forces	$A_s f_y = 0.85 f_c' ba$	A_s	a	(3.89)
Equilibrium of moment about C	$M_u = \varphi A_s f_y \left(d - \frac{a}{2} \right)$	A_s	a	(3.90)
Equilibrium of moment about T	$M_u = \varphi 0.85 f_c' ba \left(d - \frac{a}{2} \right)$		a	(3.91)

Examination of the two unknowns A_s and a in the three equations reveals two ways to arrive at the solution:

First way. The depth a is solved from the equilibrium of moment about T (Equation 3.91). Then the depth a is inserted into either Equation (3.89) or Equation (3.90) to solve for the steel area A_s. This way of solution is quite straightforward, but not often used because Equation (3.91) is a quadratic equation for the depth a. The solution of a quadratic equation by hand is somewhat tedious.

Second way. The depth a is obtained by solving Equations (3.89) and (3.90) simultaneously using trial-and-error method. This solution is actually quite simple for an experienced engineer because the depth a can be closely estimated in the first trial. Assuming a depth a_1 to be less than $a_{\max} = 0.375\beta_1 d_t$, and substituting it into Equation (3.89) gives the value of A_{s1} for the first cycle as

$$A_{s1} = \frac{M_u}{\varphi f_y \left(d - \frac{a_1}{2} \right)} \tag{3.92}$$

Then inserting A_{s1} into Equation (3.89) to calculate the depth a_2:

$$a_2 = \frac{A_{s1} f_y}{0.85 f_c' b} \tag{3.93}$$

If $a_2 = a_1$, a solution is found. If $a_2 \neq a_1$, then assume another a_2 value and repeat the cycle. The convergence is usually very rapid, and two or three cycles of trial and error is usually sufficient to give an accurate solution.

Second Type of Design (find cross sections of concrete and steel)

Given: M_u, f_y, f_c' and ε_u
Find: b, d, A_s and a

Notice that four unknowns are shown. Obviously, only two of these four unknowns can be determined by the two available equations. Therefore, two unknowns will have to be assumed in the process of design. It is a characteristic of design problems that the number of unknowns frequently exceeds the available number of equations.

The three equilibrium equations and their unknowns are:

Type of equation	Equations	Unknowns			
Equilibrium of forces	$A_s f_y = 0.85 f_c' b a$	b	A_s	a	(3.94)
Equilibrium of moment about C	$M_u = \varphi A_s f_y d \left(1 - \dfrac{a}{2d} \right)$	d	A_s	a	(3.95)
Equilibrium of moment about T	$M_u = \varphi 0.85 f_c' b d a \left(1 - \dfrac{a}{2d} \right)$	b d		a	(3.96)

Two methods of design have been used:

First Method. Assume a suitable percentage of steel $\rho = A_s/bd$, say $\rho = 0.5\rho_{\max}$. This means that we are specifying the ratio of the two quantities that are being designed.

From equilibrium of forces (Equation 3.94):

$$\frac{a}{d} = \frac{A_s f_y}{0.85 f_c' b d} = \frac{\rho f_y}{0.85 f_c'} \tag{3.97}$$

Inserting a/d from Equation (3.97) into the moment equilibrium equation about C (Equation 3.95), and noticing $A_s = \rho b d$, we can solve for a parameter bd^2 for the concrete cross-section:

$$bd^2 = \frac{M_u}{\varphi \rho f_y \left(1 - 0.59 \frac{\rho f_y}{f_c'} \right)} \tag{3.98}$$

Once bd^2 is obtained, either b or d has to be assumed to determine the remaining concrete dimension. The steel area A_s can then be calculated from ρbd. However, because the dimensions b and d are always adjusted to become round numbers, an accurate calculation of the steel area A_s can be obtained using the 'first type of design' method in which the cross-section of concrete (b and d) is given.

Second Method. Assume a suitable reinforcement index $\omega = \rho f_y / f_c'$. From Equation (3.97) it can be seen that this index ω is directly proportional to the depth ratio, a/d. Substituting $a/d = \omega/0.85$ into the moment equilibrium equation, (3.96), gives

$$bd^2 = \frac{M_u}{\varphi f_c' \omega (1 - 0.59\omega)} \tag{3.99}$$

Equation (3.99) is tabulated and used in the ACI Design Handbook (ACI Committee 340, 1973).

3.2.2.7 Curvature and Tensile Steel Strain

In conclusion, once a mild steel reinforced concrete beam is found to be ductile, it is analyzed and designed by the equilibrium condition without using Bernoulli's compatibility condition and the stress–strain curve of steel. It would appear that the analysis and design of ductile beams are based on the equilibrium (plasticity) truss model presented in Chapter 2, Section 2.1.1. This similarity is true if we are interested only in the bending strength. If we are also interested in deformations, however, we must then rely on Bernoulli's compatibility condition, Figure 3.11(b), to determine the ultimate curvature ϕ_u:

$$\phi_u = \frac{\varepsilon_u}{c} \tag{3.100}$$

where $c = a/\beta_1$ and the depth a is one of the two unknowns solved by the two equilibrium equations in either the analysis or the design problems. If the steel strain ε_s is desired at the ultimate load stage, Bernoulli's compatibility condition is also required to give:

$$\varepsilon_s = \varepsilon_u \frac{d - c}{c} \tag{3.101}$$

3.2.3 Doubly Reinforced Rectangular Beams

A doubly reinforced beam has compression steel A_s' in addition to the tension steel A_s (Figure 3.12a). The compression steel could be employed for various purposes. First, to increase the moment capacity of a beam when the cross-section is limited. Second, in a continuous beam (Figure 3.2) the ACI code requires that a portion of the bottom positive steel in the center region of a beam must be extended into the supports. These extended bars provide the compression steel for the rectangular support sections which are subjected to negative moment. Third, compression steel could be used to reduce deflections. Fourth, the ductility of a beam could be enhanced by adding compression steel (see Section 3.3.5).

The additional compression steel introduces three additional variables, namely, A_s', f_s' and ε_s' for the area, stress and strain of the compression steel, respectively, Figure 3.12(a)–(c). Therefore, the analysis and design of doubly reinforced rectangular concrete beams involve 12 variables, namely, b, d, A_s, A_s', M_u, f_s, f_s', f_c', ε_s, ε_s', ε_u and c (or a). At the same time,

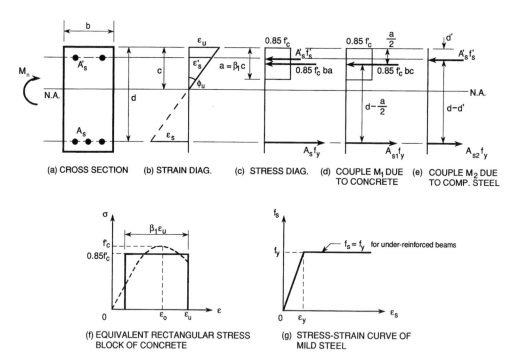

(a) CROSS SECTION   (b) STRAIN DIAG.   (c) STRESS DIAG.   (d) COUPLE M_1 DUE   (e) COUPLE M_2 DUE
                                                           TO CONCRETE   TO COMP. STEEL

(f) EQUIVALENT RECTANGULAR STRESS   (g) STRESS-STRAIN CURVE OF
    BLOCK OF CONCRETE   MILD STEEL

Figure 3.12   Doubly reinforced rectangular sections at ultimate

two additional equations are available, one for the compatibility of compression steel and the other for the stress–strain relationship of the compression steel.

The analysis and design of doubly reinforced rectangular sections will be limited to ductile beams in this section, using the stress–strain relationships of concrete and steel shown in Figure 3.12(f) and (g). In this type of problems the tensile steel will be in the yield range, $f_s = f_y$, and the tensile steel strain, ε_s, is irrelevant to the solution of stress-type variables. Correspondingly, the compatibility equation for the tensile steel and the stress–strain relationship of the tensile steel are not required. As a result, we now have eleven variables, b, d, A_s, A'_s, M_u, f_y, f'_s, f'_c, ε'_s, ε_u and c (or a), and four available equations. Two of these four equations come from the equilibrium condition and the other two from the compatibility and stress–strain relationship of compression steel. If seven variables are given, the remaining four unknown variables can be solved by the four available equations.

The analysis and design of doubly reinforced beams will now be treated separately.

3.2.3.1 Analysis of Ductile Sections

Analysis problems to find moment are posed as follows:

Given: b, d, A_s, A'_s, f_y, f'_c, ε_u
Find: M_u, f'_s, ε'_s and a (or $c = a/\beta_1$)

The four available equations and their unknowns are:

Type of equation	Equations	Unknowns		
Equilibrium of forces	$A_s f_y = 0.85 f'_c ba + A'_s f'_s$	f'_s	a	(3.102)
Equilibrium of moment about T	$M_u = \varphi \begin{bmatrix} 0.85 f'_c ba \left(d - \frac{a}{2}\right) \\ + A'_s f'_s (d - d') \end{bmatrix}$	$M_u\ f'_s$	a	(3.103)
Compatibility of compression steel	$\dfrac{\varepsilon'_s}{\varepsilon_u} = \dfrac{c - d'}{c}$ $(c = a/\beta_1)$	ε'_s a		(3.104)
Constitutive law of compression steel	$f'_s = E_s \varepsilon'_s$ for $\varepsilon'_s \leq \varepsilon_y$	$f'_s,\ \varepsilon'_s$		(3.105a)
	$f'_s = f_y$ for $\varepsilon'_s > \varepsilon_y$	$f'_s,\ \varepsilon'_s$		(3.105b)

It should be pointed out that an approximation has been introduced in writing Equations (3.102) and (3.103). The compression steel force $A'_s f'_s$ should have been written as $A'_s (f'_s - 0.85 f'_c)$ to include the negative force attributed to the area of concrete displaced by the compression steel. Since $0.85 f'_c$ is smaller than f'_s by an order of magnitude, this small force $A'_s (0.85 f'_c)$ has been neglected in the two equations.

Examination of the unknowns in the four equations points indicates two methods to solve these equations:

Method 1. (Trial-and-error method)

Step 1: Assume a value of depth a $(c = a/\beta_1)$ and calculate the compression steel strain ε'_s from the compatibility condition of Equation (3.104).

Step 2: Calculate the compression steel stress f'_s from the stress–strain relationship of compression steel (Equation 3.105a or b).

Step 3: Insert f'_s into the force equilibrium equation (3.102) to calculate a new value of the depth a. If the new a is the same as the assumed a, a solution is obtained. If not, assume another value of a and repeat cycle. The convergence is usually quite rapid.

Step 4: Once the depth a and the compression steel stress f'_s is solved, insert them into the moment equilibrium equation (3.103) to calculate the moment M_u.

Method 2. (solve quadratic equation)

Step 1: Insert the compression steel strain ε'_s from Equation (3.105a) into Equation (3.104) and express the compatibility of compression steel by a new equation in terms of the compression steel stress f'_s and the depth a.

Step 2: Solve the new compatibility equation simultaneously with the force equilibrium equation (3.102) to obtain the stress f'_s and the depth a. This is the process of solving a quadratic equation. If $f'_s \geq f_y$, then use $f'_s = f_y$ according to Equation (3.105b) and recalculate the depth a from Equation (3.102).

Step 3: Substitute the stress f'_s and the depth a into Equation (3.103) to determine the moment M_u.

3.2.3.2 Design of Ductile Sections

Design problems to find areas of tension and compression steel are posed as follows:

$$\text{Given: } b, d, M_u, f_y, f_c', \varepsilon_u$$
$$\text{Find: } A_s, A_s', f_s', \varepsilon_s' \text{ and } a (\text{or } c = a/\beta_1)$$

The problem shows five unknowns, but we have only four equations. Therefore, one of the five unknowns must be assumed during the design process. It will be shown below that the most convenient unknown to choose is the depth a.

In the design process it is convenient to take advantage of the two-internal-couples concept as shown in Figure 3.12(d) and (e). The bending resistance of a doubly reinforced beam can be considered to consist of two internal couples. One couple M_1 is contributed by the concrete compression stress block $0.85 f_c' ba$ and the other M_2 by the compression steel area A_s'. The tensile steel area A_s is separated into A_{s1} and A_{s2} for M_1 and M_2, respectively. Because the first couple is identical to a singly reinforced beam which can be easily solved by the equilibrium condition alone, the separation of the two internal couples is very convenient for the design of doubly reinforced beams.

In the separation process we have expanded the two variables, A_s and M_u, into four variables, A_{s1}, A_{s2}, M_1 and M_2, resulting in the addition of two unknowns. At the same time, two additional equations stating these separations are created. These two additional equations are:

$$\text{Separation of } M_u \quad M_u = M_1 + M_2 \tag{3.106}$$

$$\text{Separation of } A_s \quad A_s = A_{s1} + A_{s2} \tag{3.107}$$

The two groups of equations and their unknown variables are summarized as follows:

First Internal Couple M_1

Type of equation	Equations	Unknowns		
Equilibrium of forces	$A_{s1} f_y = 0.85 f_c' ba$	A_{s1}	a	(3.108)
Equilibrium of moments	$M_1 = \varphi 0.85 f_c' ba \left(d - \dfrac{a}{2}\right)$	M_1	a	(3.109)

Second Internal Couple M_2

Type of equation	Equations	Unknowns		
Equilibrium of forces	$A_{s2} f_y = A_s' f_s'$	A_{s2} A_s' f_s'		(3.110)
Equilibrium of moments	$M_2 = \varphi A_{s2} f_y (d - d')$	M_2 A_{s2}		(3.111)
Compatibility of compression steel	$\dfrac{\varepsilon_s'}{\varepsilon_u} = \dfrac{c - d'}{c} (c = a/\beta_1)$		ε_s' a	(3.112)
Constitutive law of compression steel	$f_s' = E_s \varepsilon_s' \quad \text{for } \varepsilon_s' \le \varepsilon_y$		f_s', ε_s'	(3.113a)
	$f_s' = f_y \quad \text{for } \varepsilon_s' > \varepsilon_y$		f_s', ε_s'	(3.113b)

Examination of the two groups of equations and their unknown variables shows that the only unknown variable common to the two groups is the depth a (or c). If the depth a is assumed, the unknowns M_1 and A_{s1} for the first group can be calculated directly from the two equilibrium equations (3.109) and (3.108), respectively. Then the moment M_2 for the second group is determined from Equation (3.106), $M_2 = M_u - M_1$; and the rest of the unknowns in the second groups can be easily solved.

The depth a could be used to control the ductility (see Section 3.2.2.4) of the beam according to the compatibility of strains in Figure 3.12(b). The most economical way to design the beam, however, is to assume the depth $a = a_{max}$. This assumption maximizes the compression stress block in the first internal couple and, therefore, minimizes the compression steel area A'_s for the second couple. If $a = a_{max}$, the procedures for design are as follows:

Step 1: Apply $a = a_{max}$ to the first couple to find A_{s1} and M_1 directly from equilibrium equations, (3.108) and (3.109), respectively. Then compare M_1 with M_u. If $M_u \leq M_1$, a singly reinforced beam would suffice. If $M_u > M_1$, a doubly reinforced beam is required.

Step 2: If a doubly reinforced beam is confirmed, calculate $M_2 = M_u - M_1$ from Equation (3.106) and insert M_2 into the moment equilibrium equation (3.111) of the second couple to find the steel area A_{s2}. The total steel area A_s is the sum of A_{s1} and A_{s2} (Equation 3.107).

Step 3: Apply $a = a_{max}$ to the second couple by inserting $c = a_{max}/\beta_1$ into the compatibility equation (3.112) to determine the compression steel strain ε'_s. Then find the compression steel stress, f'_s, from the stress–strain relationship of Equations (3.113a) or (3.113b).

Step 4: Substitute f'_s and A_{s2} into the force equilibrium equation (3.110) and calculate the compression steel area A'_s.

3.2.4 Flanged Beams

Flanged beams may have cross-sections in the shape of a T, an L, an I or a box. In these sections the flanges are used to enhance the compression forces of the internal couples. The flanges may be purposely added or may be available as parts of a structure, such as a floor system. In a slab-and-beam floor system the slab serves as the flange, while the beam serves as the web. Together they form a T-beam. According to the ACI Code, the effective width of the flange of a T-beam shall not exceed one-quarter of the span length of the beam and the effective overhanging flange width on each side of the web shall not exceed eight times the slab thickness and one-half the clear distance to the next web.

A T-cross-section subjected to a nominal moment M_n is shown in Figure 3.13(a). The strain and stress diagrams are given in Figure 3.13(b) and (c). It can be seen that these strain and stress distributions are identical to those for rectangular sections (Figure 3.11b and c), except that the magnitude and the location of the compression resultant C are difficult to determine. Assuming that the average stress of $0.85 f'_c$ and the coefficient β_1 are still valid for T-sections, the magnitude of the resultant should include all the stresses on the shaded area in Figure 3.13(a). The position of the resultant should be measured from the top surface to the centroidal axis of the shaded area. Such calculations are obviously very tedious.

A simple way to avoid the above difficulties is to employ the two-internal-couples concept previously used in the doubly reinforced beams. The one internal $C - T$ couple shown in Figure 3.13(c) can be separated into two couples, one for the web area shown in Figure 3.13(d) and one for the flange area shown in Figure 3.13(e). Accordingly, we denote the internal couple

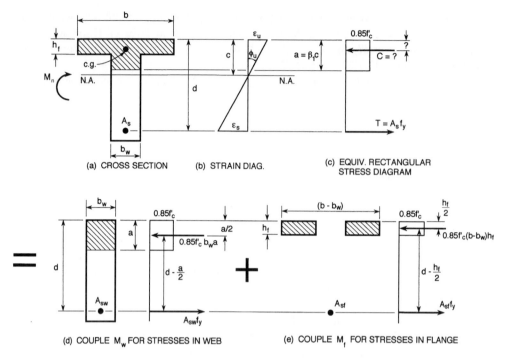

Figure 3.13   Flanged sections at ultimate

and the steel area in the web as M_w and A_{sw}, respectively, and the internal couple and the steel area in the flange as M_f and A_{sf}, respectively.

The analysis and design of T-sections will be limited to ductile beams in this section. As discussed previously in connection with singly and doubly reinforced rectangular beams, the tensile steel will be in the yield range, $f_s = f_y$; and the compatibility equation and the stress–strain relationship for the tensile steel become irrelevant. The only equations required for the solution come from the equilibrium condition. We will first examine the two equilibrium equations corresponding to the internal couple for the flange M_f:

Type of equation	Equations	Unknowns	
Equilibrium of forces (flange)	$A_{sf} f_y = 0.85 f_c'(b - b_w)h_f$	A_{sf}	(3.114)
Equilibrium of moments (flange)	$M_f = \varphi 0.85 f_c'(b - b_w)h_f \left(d - \dfrac{h_f}{2} \right)$	M_f	(3.115)

Because the area of the flange is always considered a given value, the magnitude and the position of the resultant for the flange do not change. The steel area and the moment for the flange A_{sf} and M_f could be determined directly from the equilibrium equations, (3.114) and (3.115).

3.2.4.1 Analysis of Ductile Sections

In the case of an analysis problem when A_s is given, we can first calculate A_{sf} from Equation (3.114) and compare it with A_s. If $A_s \leq A_{sf}$, meaning the flange is not fully utilized, the cross-section should be analyzed as a rectangular cross section with a width b. If $A_s > A_{sf}$, meaning a T-section is confirmed, then we can proceed to make an analysis of the web with a given steel area $A_{sw} = A_s - A_{sf}$.

Analysis of the web (finding moment M_w) is carried out as follows:

$$\text{Given: } b, d, h_f, A_{sw}, f_y, f'_c, \varepsilon_u$$
$$\text{Find: } M_w \text{ and } a \text{ (or } c = a/\beta_1)$$

The equilibrium equations and their unknowns for the web are:

Type of equation	Equations	Unknowns	
Equilibrium of forces (web)	$A_{sw} f_y = 0.85 f'_c b_w a$	a	(3.116)
Equilibrium of moments (web)	$M_w = \varphi A_{sw} f_y \left(d - \dfrac{a}{2} \right)$	$M_w \quad a$	(3.117)

Equations (3.116) and (3.117) are identical to Equations (3.84) and (3.85) for the analysis of a singly reinforced beam. The procedure for solution, which is very easy, is exactly the same as that given in Equations (3.87)–(3.88).

Once the moment for the web M_w is found, we can calculate the total moment $M_u = M_w + M_f$, where the moment for the flange M_f is obtained from Equation (3.115).

3.2.4.2 Design of Ductile Sections

In the case of a design problem when M_u is given, we can first calculate M_f from Equation (3.115) and compare it with M_u. If $M_u \leq M_f$, meaning the flange is not fully utilized, the cross-section should be analyzed as a rectangular cross-section with a width b. If $M_u > M_f$, meaning a T-section is confirmed, then we can proceed to design the web with a given moment $M_w = M_u - M_f$.

Design of the web to find steel area A_{sw} is carried out as follows:

$$\text{Given: } b, d, h_f, M_w, f_y, f'_c, \varepsilon_u$$
$$\text{Find: } A_{sw} \text{ and } a \text{ (or } c = a/\beta_1)$$

The equilibrium equations and their unknowns for the web are:

Type of equation	Equations	Unknowns	
Equilibrium of forces (web)	$A_{sw} f_y = 0.85 f'_c b_w a$	$A_{sw} \quad a$	(3.118)
Equilibrium of moments (web)	$M_w = \varphi A_{sw} f_y \left(d - \dfrac{a}{2} \right)$	$A_{sw} \quad a$	(3.119)

Equationss (3.118) and (3.119) are exactly the same as Equations (3.89) and (3.90) for the design of a singly reinforced beam. The trial-and-error procedure for the solution has been described in conjunction with Equations (3.92)–(3.93).

Once the steel area for the web A_{sw} is found, we can calculate the total steel area $A_s = A_{sw} + A_{sf}$, where the steel area for the flange A_{sf} is obtained from Equation (3.114).

3.2.5 Moment–Curvature (M–φ) Relationships

3.2.5.1 Characteristics of M–φ Curve

We have now studied the linear bending theory in Section 3.1 and the nonlinear bending theory at ultimate load in Section 3.2. These theories will be used to understand and to calculate the fundamental moment–curvature relationships in the bending of reinforced concrete beams.

Figure 3.14(a) gives the moment–curvature (M–$φ$) curve for a singly reinforced beam using mild steel bars. Since mild steel has an elastic–perfectly plastic stress–strain relationship as

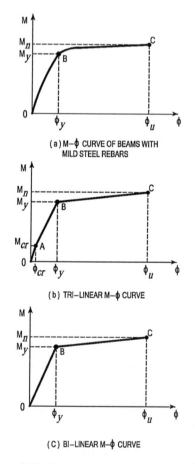

Figure 3.14   Moment–Curvature relationships

shown in Figure 3.9(f), a distinct kink point B reflects the yield point ($\varepsilon_s = \varepsilon_y$). The moment and the curvature at this point are the yield moment, M_y, and the yield curvature, ϕ_y, respectively. Both the curves before the yield point, OB, and after the yield point, BC, are slightly curved, reflecting the nonlinear stress–strain curve of concrete in Figure 3.9(e).

Point C (for ultimate) and point B (for yielding) in Figure 3.14(a) are calculated from cracked sections. Before cracking, however, the member is much stiffer and should be calculated according to the uncracked sections. This uncracked M–ϕ relationship is shown in Figure 3.14(b) at low loads and extended up to the point A, representing the cracking of concrete. The cracking moment and the cracking curvature at this distinctive point are designated as M_{cr} and ϕ_{cr}, respectively. If the four points, O, A, B and C are connected by straight lines, we have the so-called trilinear M–ϕ curve.

If the uncracked region of the M–ϕ curve is neglected and if the stress–strain curve of the concrete is assumed to be linear up to the load stage when the steel reaches the yield point, then the curve OB becomes a straight line as shown in Figure 3.14(c). When points B and C are also jointed by a straight line, the M–ϕ curve becomes bilinear. This bilinear M–ϕ curve captures the basic force–deformation characteristic of bending action, and is very useful in defining the ductility of reinforced concrete members, as explained in the next section.

The two distinctive points B (at yield) and C (at ultimate) in the bilinear M–ϕ curve can be calculated as follows:

Singly Reinforced Beams
The yield moment M_y and the yield curvature ϕ_y, represented by point B in Figure 3.14(c), can be calculated by the linear bending theory given in Sections 3.1.1 and 3.1.2. Changing ε_s to ε_y in the strain diagram of Figure 3.1(b) and f_s to f_y in the stress diagram of Figure 3.1(c), Equations (3.3) and (3.6) give:

$$M_y = A_s f_y d \left(1 - \frac{k}{3} \right) \tag{3.120}$$

$$\phi_y = \frac{\varepsilon_y}{d(1 - k)} \tag{3.121}$$

The coefficient k has been derived in Equation (3,16) by solving five equations, or in Equation (3.23) using the transformed area concept:

$$k = \sqrt{(n\rho)^2 + 2n\rho} \; - \; n\rho \tag{3.122}$$

The nominal moment M_n and the ultimate curvature ϕ_u, represented by the point C in the M–ϕ curve (Figure 3.14c), can be calculated according to the nonlinear bending theory of Sections 3.2.2.5 and 3.2.2.7 for ductile beams. Equations (3.88), (3.87) and (3.100) give M_n and ϕ_u as follows:

$$M_n = A_s f_y \left(d - \frac{a}{2} \right) \tag{3.123}$$

$$\phi_u = \frac{\varepsilon_u}{c} \quad (c = a/\beta_1) \tag{3.124}$$

where

$$a = \frac{A_s f_y}{0.85 f_c' b} \tag{3.125}$$

Since the lever arm $(d - kd/3)$ at yielding (Equation 3.120) must be less than the lever arm $(d - a/2)$ at ultimate (Equation 3.123), then M_y must be less than M_n. Therefore, if M_y is found to be greater than M_n, as in the rare case of large percentage of tension steel, then M_y should be taken as equal to M_n.

Doubly Reinforced Beams
The yield moment M_y and the yield curvature ϕ_y can be calculated by the linear bending theory explained in Section 3.1.3.2. Changing f_s to f_y in the stress diagram of Figure 3.3(c), we can write from the equilibrium of moments about the concrete compression resultant the following equations:

$$M_y = A_s f_y d \left(1 - \frac{k}{3}\right) + A_s' f_s' \left(\frac{kd}{3} - d'\right) \tag{3.126}$$

Changing ε_s to ε_y in the strain diagram of Figure 3.3(b), we can also obtain the following compatibility equation for the compression steel strain ε_s':

$$\varepsilon_s' = \varepsilon_y \frac{kd - d'}{d - kd} \tag{3.127}$$

The compression steel stress f_s' in Equation (3.126) can now be written using Hooke's law:

$$f_s' = E_s \frac{\varepsilon_y}{1 - k} \left(k - \frac{d'}{d}\right) \leq f_y \tag{3.128}$$

The coefficient k in Equations (3.126) and (3.128) is obtained from Equation (3.31). The ratio β_c in Equation (3.31) is defined in Equation (3.30). Once the coefficient k is obtained, the yield curvature ϕ_y is computed from Equation (3.121).

The nominal moment M_n and the ultimate curvature ϕ_u can be computed by the principles enunciated in Section 3.2.3. The moment M_n and the depth a can be obtained by solving Equations (3.102)–(3.105) according to method 1 or method 2. Once the depth a is obtained, the curvature ϕ_u is computed from Equation (3.124).

3.2.5.2 Bending Ductility

Because the bilinear M–ϕ curve in Figure 3.14(c) captures the basic force–deformation characteristic of bending action, it will be used to define the ductility of a reinforced concrete member. The bending ductility ratio, μ, is defined as the ratio of the ultimate curvature ϕ_u to the yield curvature ϕ_y, i.e.

$$\mu = \frac{\phi_u}{\phi_y} \tag{3.129}$$

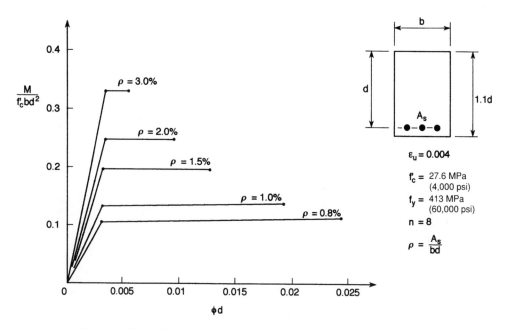

Figure 3.15   Bi-linear M-ϕ curves for various precentages of tension steel

The two curvatures ϕ_y and ϕ_u can be calculated according to Equations (3.121) and (3.124), respectively, for singly reinforced beams. In the case of doubly reinforced beams, these two curvatures can be calculated according to the procedures described in Section 3.2.5.1.

The ultimate curvature ϕ_u, and therefore the ductility ratio μ of a reinforced concrete member is strongly affected by the percentages of steel. Five singly reinforced beams with steel percentages varying from 0.8 to 3.0% are analyzed according to Equations (3.121)–(3.124) and their bilinear $M/f_c'bd^2$–ϕd curves are plotted in Figure 3.15. The curvature ϕ, which has units of radian per unit length, has been made nondimensional by multiplying by the effective depth d. The moment M has also been made nondimensional by dividing by the parameter $f_c'bd^2$. In calculating these curves a more realistic ultimate strain ε_u of 0.004 is assumed rather than the conservative ACI value of 0.003. The normal material properties and geometric location of steel bars are selected, namely, $f_c' = 27.6$ MPa (4000 psi), $f_y = 413$ MPa (60 000 psi), $n = E_s/E_c = 8$ and a concrete cover (center of steel bar to surface) of 0.1d. Figure 3.15 shows clearly that the ultimate rotation $\phi_u d$ decreases rapidly with the increase of the tension steel percentage, ρ. The ductility ratio μ decreases from 8.0 for $\rho = 0.8\%$ to 1.7 for $\rho = 3.0\%$.

The ductility of a bending member can be enhanced by the addition of compression steel as shown by the nondimensional $M/f_c'bd^2 - \phi d$ curves in Figure 3.16. The curves for the doubly reinforced beams in this figure are calculated by the procedures discussed in Section 3.2.5.1. For the three beams with tensile steel percentage ρ of 3%, the compression steel percentages ρ' of 0, 1 and 2% will give ultimate rotations ϕd of 0.0064, 0.0096 and 0.0193, respectively. The corresponding ductility ratio μ increases in the sequence of 1.7, 2.5 and 5.4. For the two beams with tension steel percentages ρ of 1%, an increase of compression steel percentage ρ'

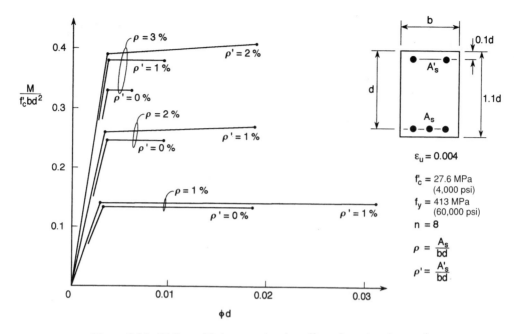

Figure 3.16 Bi-linear M-ϕ curves showing effect of compression steel

from 0 to 1% increases the ductility ratio μ from 6.2 to 10.8. It is clear that the compression steel is indeed very effective in increasing the ductility of flexural beams.

In short, the effects of the tension and compression reinforcement on the ductility of concrete structures are very profound. Such information is crucial in the design of reinforced concrete structures to resist earthquake.

3.3 Combined Bending and Axial Load

In Sections 3.1 and 3.2 we have studied the bending action of reinforced concrete members. The bending analysis and design of such members have been based on the three fundamental principles of parallel stress equilibrium, Bernoulli compatibility and uniaxial constitutive laws of materials. In this section we will apply these same principles to reinforced concrete members subjected to combined bending and axial load. To limit the scope of this presentation, however, only the nonlinear theory dealing with *mild steel* reinforced concrete members at *ultimate* load stage will be included. This section is, in fact, an extension of Section 3.2 to cover combined bending and axial load.

3.3.1 Plastic Centroid and Eccentric Loading

A member subjected to axial compression, with or without bending, is called a column. A concentric load on a column is defined as a load that produces a uniform stress in the column section and induces no bending moment. In the case of a symmetrically reinforced rectangular cross-section a concentric load is located at the geometric centroid of the concrete section.

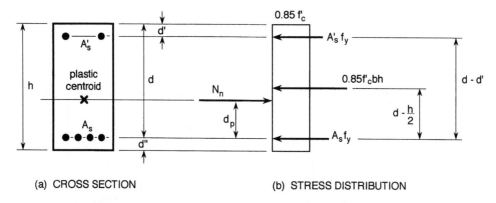

(a) CROSS SECTION   (b) STRESS DISTRIBUTION

Figure 3.17   Plastic centroid for unsymmetrically reinforced sections

However, in the general case of an unsymmetrically reinforced rectangular section, as shown in Figure 3.17(a), a concentric load is said to be located at the *plastic centroid* of the cross-section.

A concentric load N_n at the plastic centroid should, by definition, produce a uniform stress in the concrete as shown in Figure 3.17(b). At maximum load this uniform stress is assumed to be $0.85 f_c'$. At the same time both the top and bottom steel are assumed to reach yielding, $f_s = f_s' = f_y$. The location of the plastic centroid is defined by the distance, d_p, measured from the concentric load to the bottom steel. The load N_n and the distance d_p can be determined from the two equilibrium equations. Equilibrium of forces gives:

$$N_n = 0.85 f_c' bh + A_s f_y + A_s' f_y \qquad (3.130)$$

Note that in writing Equation (3.130) the forces of the concrete areas displaced by the compression steel $0.85 A_s' f_c'$, and by tension steel $0.85 A_s f_c'$, have been neglected. These simplifications will be made for all the equilibrium equations in combined bending and axial load. Taking moments of all the forces about the bottom steel we have

$$N_n d_p = 0.85 f_c' bh \left(d - \frac{h}{2} \right) + A_s' f_y (d - d') \qquad (3.131)$$

Inserting N_n from Equation (3.130) into Equation (3.131) results in

$$d_p = \frac{0.85 f_c' bh \left(d - \frac{h}{2} \right) + A_s' f_y (d - d')}{0.85 f_c' bh + A_s f_y + A_s' f_y} \qquad (3.132)$$

Figure 3.18(a) illustrates an unsymmetrically reinforced column section subjected to a load, N_n, and a bending moment, M_n. The load N_n is located at the plastic centroid, and the moment M_n is taken about an axis through the plastic centroid. This combined bending and axial loading is statically equivalent to a load N_n acting at an eccentricity, $e = M_n / N_n$, measured from the plastic centroid (Figure 3.18b).

An eccentric load N_n is acting on an unsymmetrically reinforced column section, as shown in Figure 3.19(a). The strain and stress distributions due to the eccentric load are illustrated in Figure 3.19(b) and (c) at the ultimate load stage. The compression stress block has been converted to the ACI equivalent rectangular stress block (Figure 3.19d). The load

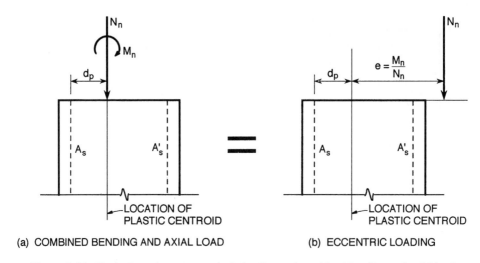

(a) COMBINED BENDING AND AXIAL LOAD   (b) ECCENTRIC LOADING

Figure 3.18   Equivalence between eccentric loading and combined bending and axial load

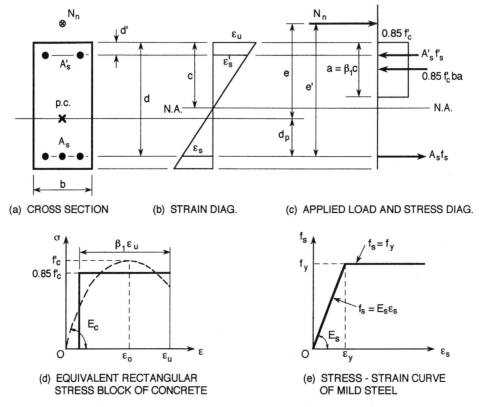

(a) CROSS SECTION   (b) STRAIN DIAG.   (c) APPLIED LOAD AND STRESS DIAG.

(d) EQUIVALENT RECTANGULAR   (e) STRESS - STRAIN CURVE
    STRESS BLOCK OF CONCRETE   OF MILD STEEL

Figure 3.19   Unsymmetrically reinforced rectangular column sections at ultimate

N_n is located at an eccentricity e measured from the plastic centroid or at an eccentricity of e' measured from the centroid of the tensile steel. The latter is frequently more convenient for the analysis. The difference between e and e' is the distance d_p given by Eqution (3.132).

The mild steel has an elastic–perfectly plastic stress–strain curve, as shown in Figure 3.19(e). In a mild steel reinforced column, therefore, the tensile steel may or may not yield at ultimate. If the tension steel is in the plastic range ($\varepsilon_s \geq \varepsilon_y$) when the concrete crushes at ultimate load ($\varepsilon_u = 0.003$), the column is said to fail in tension. If the tension steel is in the elastic range ($\varepsilon_s < \varepsilon_y$) when the concrete crushes, the column is said to fail in compression. In order to divide these two types of columns, we define the 'balanced condition' when the steel reaches the yield point ($\varepsilon_s = \varepsilon_y$) simultaneously with the crushing of concrete ($\varepsilon_u = 0.003$).

The eccentric loading of an unsymmetrically reinforced rectangular column involves thirteen variables $b, d, A_s, A'_s, N_n, e', f_s, f'_s, f'_c, \varepsilon_s, \varepsilon'_s, \varepsilon_u$ and a (or c), as shown in Figure 3.19(a)–(c). The coefficient β_1 is not considered a variable because it has been determined independently from tests. A total of six equations are available, two from equilibrium (Figure 3.19c), two from Bernoulli compatibility (Figure 3.19b), and two from the constitutive law of mild steel (Figure 3.19e). Therefore, seven variables must be given before the remaining six unknown variables can be solved by the six equations.

3.3.2 Balanced Condition

The problem posed for the balanced condition is:

$$\text{Given: } b, d, A_s, A'_s, f'_c, \varepsilon_s = \varepsilon_y, \varepsilon_u = 0.003$$
$$\text{Find: } N_n, e', f_s, f'_s, \varepsilon'_s, a \text{ (or } c = a/\beta_1)$$

The six available equations and their unknowns are:

Type of equation	Equations	Unknowns			
Equilibrium of forces	$N_n = 0.85 f'_c ba - A_s f_s$ $+ A'_s f'_s$	N_n	$f_s \; f'_s$	a	(3.133)
Equilibrium of moment about T	$N_n e' = \begin{bmatrix} 0.85 f'_c ba \left(d - \frac{a}{2}\right) \\ + A'_s f'_s (d - d') \end{bmatrix}$	$N_n \; e'$	f'_s	a	(3.134)
Compatibility of tension steel	$\dfrac{a}{\beta_1 d} = \dfrac{\varepsilon_u}{\varepsilon_u + \varepsilon_y}$			a	(3.135)
Compatibility of compression steel	$\dfrac{\varepsilon'_s}{\varepsilon_u} = \dfrac{c - d'}{c} \; (c - a/\beta_1)$			$\varepsilon'_s \; a$	(3.136)
Consttutive law of tension steel	$f_s = E_s \varepsilon_s \quad \text{for} \quad \varepsilon_s \leq \varepsilon_y$		f_s		(3.137a)
	$f_s = f_y \quad \text{for} \quad \varepsilon_s > \varepsilon_y$		f_s		(3.137b)
Constitutive law of compression steel	$f'_s = E_s \varepsilon'_s \quad \text{for} \quad \varepsilon'_s \leq \varepsilon_y,$		f'_s, ε'_s		(3.138a)
	$f'_s = f_y \quad \text{for} \quad \varepsilon'_s > \varepsilon_y$		f'_s, ε'_s		(3.138b)

Notice that no material reduction factor φ is shown in the equilibrium equations, because φ is included in the symbol N_n. For convenience, the nominal load N_n is defined as N_u/φ.

The six unknown variables, N_n, e', f_s, f'_s, ε'_s, and a (or c) for the six equations are indicated after the equations under the heading *Unknowns*. Since the tension steel strain ε_s is given as the yield strain ε_y, the tension steel stress f_s should obviously be equal to the yield stress f_y based on the stress–strain relationship of either Equations (3.137a) or (3.137b). The unknown depth a of the equivalent rectangular stress block can be solved directly from the compatibility condition for tension steel (Equation 3.135). Once the depth a (and c) is determined, the compression steel strain ε'_s can be obtained from the compatibility equation (3.136) and the compression steel stress f'_s from the stress–strain equation for compression steel (3.138). Substituting a, f_s and f'_s into the force equilibrium equation (3.133) gives the force N_n; and inserting a, f'_s and N_n into the moment equilibrium equation (3.134) results in the eccentricity e'. For this particular case of balanced condition there is no need to solve simultaneous equations.

In this case of the balanced condition we will now add a subscript b to the three unknown quantities: N_{nb} for balanced force, e'_b for balanced eccentricity and a_b for balanced depth of equivalent rectangular stress block. The balanced depth a_b is expressed from Equation (3.135) as

$$a_b = \beta_1 d \frac{\varepsilon_u}{\varepsilon_u + \varepsilon_y} \tag{3.139}$$

Equation (3.139) is, of course, identical to Equation (3.76), derived for pure bending without axial load, except that in Equation (3.76) the ultimate strain ε_u has been taken as 0.003 and $E_s = 200\,000$ MPa (29 000 000 psi).

From Equations (3.133) and (3.134) the balanced force N_{nb} and the balanced eccentricity e'_b are:

$$N_{nb} = 0.85 f'_c b a_b - A_s f_y + A'_s f'_s \tag{3.140}$$

$$e'_b = \frac{1}{N_{nb}} \left[0.85 f'_c b a_b \left(d - \frac{a_b}{2} \right) + A'_s f'_s (d - d') \right] \tag{3.141}$$

The balanced depth a_b in Equations (3.140) and (3.141) is determined from Equation (3.139) and the compression steel stress, f'_s, is obtained from Equations (3.136) and (3.138)

The balanced N_{nb} and e'_b can be used to divide the tension failure and the compression failure of the column as follows:

Types of problem	Tension failure	Compression failure
N_n is given	$N_n < N_{nb}$	$N_n > N_{nb}$
e' is given	$e' > e'_b$	$e' < e'_b$

3.3.3 Tension Failure

The analysis of column sections failing in tension is considerably simplified for the same reasons as in under-reinforced beams (Section 3.2.2.5). First, the tensile steel stress is known to yield $f_s = f_y$ and the stress–strain curve of tension steel is not required in the solution.

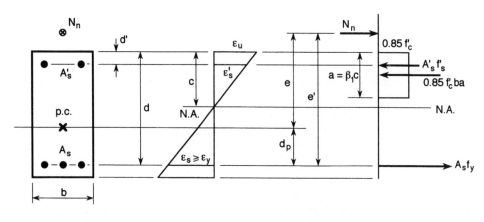

(a) CROSS SECTION   (b) STRAIN DIAG.   (c) APPLIED LOAD AND STRESS DIAG.

Figure 3.20   Tension failure in column sections

Second, the tensile steel strain is expected to lie within the plastic range, $\varepsilon_s \geq \varepsilon_y$. Bernoulli's compatibility condition, which relates the tensile steel strain ε_s to the maximum strain of concrete ε_u becomes irrelevant to the solution of the stress-type variables. Because of these two simplifications, analysis of column sections failing in tension involves only twelve variables, $b, d, A_s, A'_s, N_n, e', f'_s, \varepsilon'_s, f_y, f'_c, \varepsilon_u$ and a (or c), Figure 3.20. The available equations are now down to four, in which two are from the equilibrium condition, one is from the compatibility condition of compression steel and one is from the stress–strain relationship of compression steel. Therefore, eight variables must be given before the four remaining unknown variables can be solved by the four available equations.

The problem posed for the analysis of columns failing in tension is:

$$\text{Given: } b, d, A_s, A'_s, e', f_y, f'_c \text{ and } \quad \varepsilon_u = 0.003$$
$$\text{Find: } N_n, f'_s, \varepsilon'_s \text{ and } a \ (c = a/\beta_1)$$

The four available equations and their unknowns are:

Type of equation	Equations	Unknowns	
Equilibrium of forces	$N_n = 0.85 f'_c b a - A_s f_y + A'_s f'_s$	$N_n \ f'_s \quad a$	(3.142)
Equilibrium of moment about T	$N_n e' = \begin{bmatrix} 0.85 f'_c b a \left(d - \frac{a}{2}\right) \\ + A'_s f'_s (d - d') \end{bmatrix}$	$N_n \ f'_s \quad a$	(3.143)
Compatibility of compression steel	$\dfrac{\varepsilon'_s}{\varepsilon_u} = \dfrac{c - d'}{c} \ (c = a/\beta_1)$	$\varepsilon'_s \ a$	(3.144)
Constitutive law of compression steel	$f'_s = E_s \varepsilon'_s \quad \text{for } \varepsilon'_s \leq \varepsilon_y$	$f'_s \ \varepsilon'_s$	(3.145a)
	$f'_s = f_y \quad \text{for } \varepsilon'_s > \varepsilon_y$	$f'_s \ \varepsilon'_s$	(3.145b)

Examination of the four unknowns in the four equations, (3.142)–(3.145), suggests that the best method of solution is the generic trial-and-error procedure:

Step 1: Check $e' > e'_b$ to confirm the tension failure of the column section.
Step 2: Assume a value of the depth a, which should be less than the balanced a_b, and calculate $c = a/\beta_1$.
Step 3: Determine the compression steel strain ε'_s from the compatibility equation (3.144), and the compression steel stress f'_s from the stress–strain equation (3.145).
Step 4: Calculate the force N_n from the moment equilibrium equation (3.143).
Step 5: Insert f'_s and N_n into the force equilibrium equation (3.142), and solve for a new value of the depth a.
Step 6: If the new depth a is equal to the assumed depth a, a solution is obtained. If not, assume another depth a and repeat the cycle. The convergence is usually quite rapid.

It would be instructive to mention here that the problem posed above is to give the eccentricity e' and to find the force N_n. If the problem is reversed by giving N_n and finding e', the solution procedure will be simplified, because Equation (3.142) will contain only two unknowns rather than three. This simplified solution procedure is left as an exercise for the reader.

3.3.4 Compression Failure

The analysis of column sections failing in compression is more complicated than that of failing in tension, because two types of analysis are required to deal with two very different situations. The first situation occurs when the neutral axis lies within the concrete section $(c \leq h)$ as shown in Figure 3.21. The second situation occurs when the neutral axis lies outside the concrete section $(c > h)$ (Figure 3.22). These two situations will be treated separately below.

When c ≤ h (Figure 3.21)
When the neutral axis lies within the concrete cross-section, the ACI method of converting an actual compression stress block into an equivalent rectangular one remains valid. That is to say, the coefficient β_1 for the depth c and the coefficient 0.85 for the stress f'_c are still applicable. Consequently, the six equations, (3.133)–(3.138), remain available for the analysis of columns failing in compression with $c \leq h$.

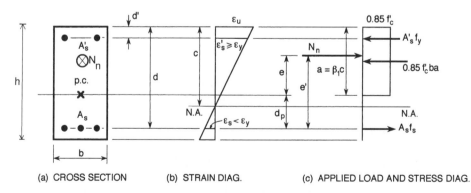

(a) CROSS SECTION   (b) STRAIN DIAG.   (c) APPLIED LOAD AND STRESS DIAG.

Figure 3.21   Compression failure (c≤h) in column sections

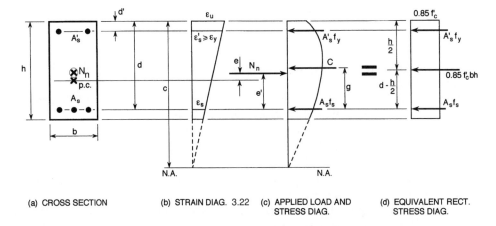

(a) CROSS SECTION   (b) STRAIN DIAG. 3.22   (c) APPLIED LOAD AND   (d) EQUIVALENT RECT.
                                                    STRESS DIAG.   STRESS DIAG.

Figure 3.22   Compression failure (c > h) in column sections

An obvious observation is available, however, to simplify the analysis of a column section failing in compression with $c \leq h$. The compression steel strain is expected to lie within the plastic range, $\varepsilon'_s \geq \varepsilon_y$, because the neutral axis is low enough such that ε'_s is close to the ultimate concrete strain ε_u of 0.003. This is true for all columns of practical size, except those very small ones with height h close to the lowest allowable limit of 305 mm (12 in.), and depth a close to the balanced a_b. Bernoulli's compatibility condition, which relates ε'_s to ε_u, becomes irrelevant to the solution of the stress-type variables. As a result, the compression steel is expected to yield $f'_s = f_y$ and the stress–strain curve of compression steel is not required in the solution. Because of this simplification, analysis of column sections failing in compression involves only twelve variables, b, d, A_s, A'_s, N_n, e', f_s, ε_s, f_y, f'_c, ε_u and c (or a) (Figure 3.21). The number of available equations is reduced to four, in which two are from the equilibrium condition, one is from the compatibility condition of tension steel and one is from the stress–strain relationship of tension steel. Therefore, eight variables must be given before the four remaining unknown variables can be solved by the four available equations.

The problem posed for the analysis of column sections failing in compression with $c \leq h$ is:

$$\text{Given: } b, d, A_s, A'_s, e', f_y, f'_c, \varepsilon_u = 0.003$$
$$\text{Find: } N_n, f_s, \varepsilon_s, \text{ and } a \text{ (or } c = a/\beta_1)$$

The four available equations and their unknowns are:

Type of equation	Equations	Unknowns		
Equilibrium of forces	$N_n = 0.85 f'_c b a - A_s f_s + A'_s f_y$	N_n f_s	a	(3.146)
Equilibrium of moment about T	$N_n e' = \begin{bmatrix} 0.85 f'_c b a \left(d - \frac{a}{2} \right) \\ + A'_s f_y (d - d') \end{bmatrix}$	N_n	a	(3.147)
Compatibility of tension steel	$\dfrac{\varepsilon_s}{\varepsilon_u} = \dfrac{d - c}{c}$ $(c = a/\beta_1)$	ε_s a		(3.148)
Constitutive law of tension steel	$f_s = E_s \varepsilon_s$ for $\varepsilon_s \leq \varepsilon_y$	f_s ε_s		(3.149)

The generic trial-and-error procedure can be used to solve these four equations:

Step 1: Check $e' < e_b'$ to confirm the compression failure of the column section.

Step 2: Assume a value of the depth a, which should be greater than the balanced a_b, and calculate $c = a/\beta_1$.

Step 3: Determine the tensile steel strain, ε_s, from the compatibility equation (3.148) and the tensile steel stress f_s from the stress–strain equation (3.149).

Step 4: Calculate the force N_n from the moment equilibrium equation (3.147).

Step 5: Insert f_s and N_n into the force equilibrium equation (3.146) and solve for a new value of the depth a.

Step 6: If the new depth a is equal to the assumed depth a, a solution is obtained. If not, assume another depth a and repeat the cycle.

Step 7: Check the assumption for the yielding of the compression steel, $\varepsilon_s' \geq \varepsilon_y$, by Equation (3.136). In all practical cases, this step is not necessary.

Incidentally, the problem posed here is to give the eccentricity e' and to find the force N_n. If the problem is reversed by giving N_n and finding e', the solution procedure will be simplified, because Equation (3.146) will contain only two unknowns rather than three. The reader should now be able to work out this simplified solution procedure if required.

When c > h (Figure 3.22)

When the neutral axis lies outside the concrete cross-section, the ACI rule to convert the actual compression stress block to an equivalent rectangular one is no longer valid. In other words, the coefficient β_1 derived from pure bending, is not applicable. Therefore, a column section failing in compression with $c > h$ requires special analysis.

When the neutral axis leaves the bottom surface, failure begins to change from a compression failure at the extreme fiber (say $\varepsilon_u = 0.003$) to a concentric compression failure of the whole cross-section ($\varepsilon_o = 0.002$). In this range of transition no simple method is available to determine the magnitude and the location of the concrete compression stress resultant C (Figure 3.22c). Fortunately, tests have shown that the moment of C about the bottom steel is approximately equal to the moment of a rectangular stress block shown in Figure 3.22(d) about the same bottom steel. The rectangular stress block has a stress of $0.85 f_c'$ throughout the whole cross-section. Hence,

$$Cg = 0.85 f_c' bh \left(d - \frac{h}{2} \right) \tag{3.150}$$

where g is the distance from the compression resultant C to the centroid of the bottom steel. Taking all the moments about the bottom steel gives:

$$N_n e' = 0.85 f_c' bh \left(d - \frac{h}{2} \right) + A_s' f_y (d - d') \tag{3.151}$$

Equation (3.151) shows that $N_n e'$ is no longer a function of the depth c (or a). That is to say, the position of the neutral axis can not be determined.

At the limiting case of concentric loading, $e = 0$ and $e' = d_p$. Inserting $e' = d_p$ from Equation (3.132) into Equation (3.151) gives

$$N_n = 0.85 f'_c bh + A_s f_y + A'_s f_y \qquad (3.152)$$

Equation (3.152) is exactly the same as Equation (3.130), meaning that the approximation made for the equivalent rectangular stress block is exactly correct for the limiting case of concentric loading.

We can now summarize the method of analysis for a column section failing in compression. First, assume $c \le h$ and go through the trial-and-error procedures of Equations (3.146)–(3.149). If c is found to be less than or equal to h, the assumption is correct and we have a solution. Second, if c is found to be greater than h, then the load N_n can be calculated directly from Equation (3.151) with the given eccentricity e'.

Although the approximation of Equation (3.150) allows us to determine the load N_n when $c > h$, the inability to calculate the position of the neutral axis does not permit us to determine the curvature $\phi_u = \varepsilon_u/c$ in this range. Fortunately, when $c > h$, the concrete is uncracked and the curvature is very small. The effect of such curvature would be small on the overall deflection of the member. If the curvature is desired, however, the position of the neutral axis can best be obtained from numerical integration of the stress–strain curve of concrete.

3.3.5 Bending–Axial Load Interaction

The methods of analysis of eccentrically loaded column sections have been presented in Sections 3.3.1–3.3.4. These analyses reveal an interesting trend. With an increase of the eccentricity e, the behavior of a column section changes in the following manner:

$e = 0$:	axial load
$e < e_b$:	compression failure
$e = e_b$:	balanced condition
$e > e_b$:	tension failure
$e = \infty$:	pure bending

This whole range of interaction from axial load to pure bending can be easily visualized by a bending–axial load interaction curve, shown in Figure 3.23.

Figure 3.23 gives a diagram with the bending moment M as the abscissa and the load N as the ordinate. The point on the abscissa denoted as M_o represents pure bending, and the point on the ordinate denoted as N_o represents the concentric load. The point B, representing the balanced condition, has a coordinate of M_{nb} and N_{nb} calculated from Equations (3.139)–(3.141) in Section 3.3.2. In the region between the point M_o and the balanced point B we have the tension failure zone, which can be analyzed by the method in Section 3.3.3. The compression failure zone, which is located between the balanced point B and the point N_o can be analyzed by the method in Section 3.3.4.

Three characteristics can be observed from the bending–axial load interaction curve in Figure 3.23:

1. Any point on the interaction curve, such as point A, represents a pair of values, M_n and N_n, which will cause the column section to fail. Points inside the curve are safe and points outside the curve are unsafe.

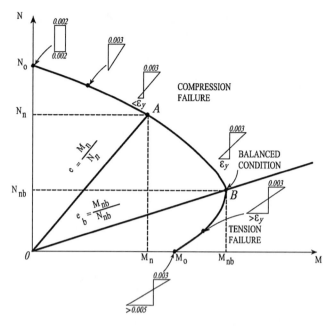

Figure 3.23 Bending–axial load interaction curve

2. Any radial line, such as OA, has a slope whose reciprocal represents the eccentricity $e = M_n/N_n$. A vertical slope has an eccentricity of 0 and the horizontal slope infinity.

3. In the compression zone the moment M_n *decreases* with increasing load N_n, while in the tension zone the moment M_n *increases* with increasing load N_n. This characteristic can be explained as follows. First, in the compression failure zone, failure is caused by the overstraining of concrete in compression ($\varepsilon_u = 0.003$). Since axial compression increases the compression strain, it decreases the capacity of the concrete to resist flexural compression. Second, in the tension failure zone, failure is caused by yielding of the steel in tension. Since axial compression decreases the tensile strain in the steel, it increases the capacity of the steel to resist flexural tension.

4. The balanced point B in Figure 3.23 corresponds to the case of balanced percentage ρ_b in Figure 3.10(e). In the region of compression failure (Figure 3.23), a column should be designed using a reduction factor $\varphi = 0.7$ (tied columns) and $\varphi = 0.65$ (spiral columns) as indicated in the region to the left of ρ_b (Figure 3.10e). In the region of tension failure (Figure 3.23), however, we should use a reduction factor φ that is a linear function of tensile strain ε_t, as given to the right of ρ_b.

3.3.6 *Moment–Axial Load–Curvature (M − N − ϕ) Relationship*

The interaction of bending and axial load is shown in Figure 3.24(a) for a column with 2% of tension steel and 1% of compression steel. All the steel bars are located at a distance $0.1d$ from the surfaces. The material properties are: $\varepsilon_u = 0.004$, $f'_c = 27.6$ MPa (4000 psi),

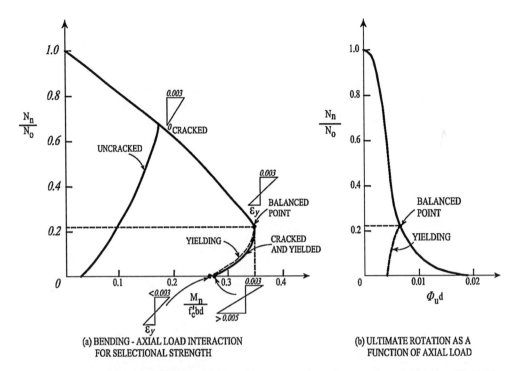

Figure 3.24 M-N-ϕ relationship for a typical column ($\rho = 2\%$, $\rho' = 1\%$, $\varepsilon_u = 0.004$, f$_c'$ = 27.6 MPa (4 000 psi), f_y = 413 MPa (60,000 psi), n = 8, h = 1.1d)

$f_y = 413$ MPa (60 000 psi) and $n = E_s/E_c = 8$. The moment M_n has been made nondimensional by the parameter $f_c'bd^2$, and the axial load N_n has been normalized by the capacity of the concentric load N_o.

The solid curve in this nondimensional interaction diagram represents the sectional strength of the given column under combined bending and axial load. The balanced point is calculated from Equations (3.139)–(3.141). The tension failure zone is computed by solving Equations (3.142)–(3.145); and the compression failure zone is determined either by solving Equations (3.146)–(3.149) or directly by Equation (3.151), depending on the location of the neutral axis. The dotted curve from the balanced point to the point of pure bending represents the yielding of the tension steel. The method of calculation for this dotted curve, which is based on Hooke's laws for both concrete and steel, has not been presented. The reader should be able to derive the equations required to plot this curve using the equilibrium, Bernoulli compatibility and Hooke's constitutive laws.

The nondimensional ultimate rotation, $\phi_u d$, is related to the normalized axial load, N/N_o, by the solid curve in Figure 3.24(b). The ultimate rotation $\phi_u d$ is calculated from the Bernoulli compatibility:

$$\phi_u d = \frac{\beta_1 \varepsilon_u}{\left(\frac{a}{d}\right)} \tag{3.153}$$

Equation (3.153) can be calculated as soon as the depth ratio, a/d, is determined in the solution process. The dotted curve in Figure 3.24(b) also represents the yielding of the tension steel.

Figure 3.24(b) shows that the ultimate rotation $\phi_u d$ decreases rapidly with increasing axial compression, N_n/N_o. The ductility ratio μ drops to unity at the balanced point. In earthquake design, therefore, columns are not designed to develop flexural hinges or to dissipate energy. Flexural hinges should always be designed to occur in the beams of a frame, rather than in the columns. This so-called strong columns–weak beams concept of frame design is fundamental to the earthquake design of reinforced concrete structures.

4

Fundamentals of Shear

4.1 Stresses in 2-D Elements

4.1.1 Stress Transformation

In Chapter 2 an equilibrium (plasticity) truss model was presented for 2-D elements which are subjected to pure shear and in which the tensile strength of concrete is assumed to be zero. In this chapter, however, the 2-D elements will be subjected to in-plane normal stresses in addition to in-plane shear stresses, and the concrete will also exhibit a small average tensile strength, even after cracking. Such elements are called membrane elements and the in-plane stresses are known as the membrane stresses.

The stresses in a 2-D membrane element are best analyzed by the *principle of stress transformation*. Figure 4.1(a) shows an element with a stationary $\ell - t$ coordinate system, defined by the directions of the longitudinal and transverse steel. The three stress components on the element, $\sigma_\ell, \sigma_t, \tau_{\ell t}$ (or $\tau_{t\ell}$) are shown in their positive directions.

For the shear stress $\tau_{\ell t}$ the first subscript ℓ represents the face on which the stress acts, and each face is defined by its outward normal. The second subscript t represents the direction of the stress itself with respect to the $\ell - t$ coordinate. The upward shear stress on the right face is positive because this stress is acting on the positive ℓ-face and pointing in the positive t-direction. The downward shear stress on the left face is also positive because this stress is acting on the negative ℓ-face and pointing in the negative t-direction. In the case of the shear stress, $\tau_{t\ell}$, the rightward stress on the top t-face is positive, because it is acting on the positive t-face and pointing in the positive ℓ-direction. This approach to define the sign of a stress will be called the basic sign convention, Figure 4.1(a).

For the two normal stresses σ_ℓ and σ_t the single subscript represents both the direction of face and the direction of stress. As such, the tensile stress will always be positive, because the direction of the stress always coincides with the outward normal. Compressive stresses will always be negative, since their direction is always opposite to the outward normal.

To find the three stress components in various directions, we introduce a rotating $1 - 2$ coordinate system as shown in Figure 4.1(b). The $1 - 2$ coordinates have been rotated counter-clockwise by an angle of α_1 with respect to the stationary $\ell - t$ axes. The three stress components in this rotating coordinate system are σ_1, σ_2 and τ_{12} (or τ_{21}). They are all shown in their positive directions. The relationship between the rotating stress components σ_1, σ_2

Unified Theory of Concrete Structures   Thomas Hsu and Yi-Lung Mo
© 2010 John Wiley & Sons, Ltd

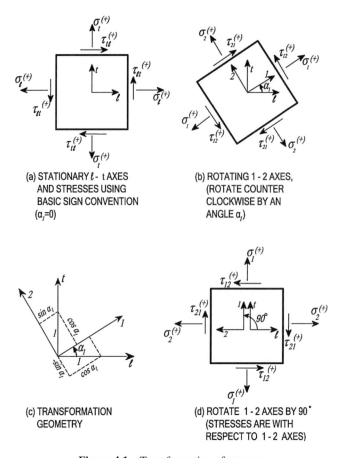

(a) STATIONARY ℓ- t AXES
AND STRESSES USING
BASIC SIGN CONVENTION
(α_1=0)

(b) ROTATING 1 - 2 AXES,
(ROTATE COUNTER
CLOCKWISE BY AN
ANGLE α_1)

(c) TRANSFORMATION
GEOMETRY

(d) ROTATE 1 - 2 AXES BY 90°
(STRESSES ARE WITH
RESPECT TO 1 - 2 AXES)

Figure 4.1   Transformation of stresses

and τ_{12} and the stationary stress components σ_ℓ, σ_t and $\tau_{\ell t}$ is the stress transformation. This relationship is, of course, a function of α_1, the angle between the ℓ-axis and the 1-axis.

The relationship between the rotating 1–2 coordinate and the stationary $\ell - t$ coordinate is shown by the transformation geometry in Figure 4.1(c). A positive unit length on the ℓ-axis will have a projection of $\cos\alpha_1$ on the 1-axis and a projection of $-\sin\alpha_1$ on the 2-axis. A positive unit length on the t-axis, however, should give projections of $\sin\alpha_1$ and $\cos\alpha_1$ on the 1-axis and 2-axis, respectively. Hence, the rotation matrix [R] is

$$[R] = \begin{bmatrix} \cos\alpha_1 & \sin\alpha_1 \\ -\sin\alpha_1 & \cos\alpha_1 \end{bmatrix} \tag{4.1}$$

The relationship between the stresses in the 1–2 coordinate $[\sigma_{12}]$ and the stresses in the $\ell - t$ coordinate $[\sigma_{\ell t}]$ is

$$[\sigma_{12}] = [R][\sigma_{\ell t}][R]^T \tag{4.2}$$

or

$$\begin{bmatrix} \sigma_1 & \tau_{12} \\ \tau_{21} & \sigma_2 \end{bmatrix} = \begin{bmatrix} \cos\alpha_1 & \sin\alpha_1 \\ -\sin\alpha_1 & \cos\alpha_1 \end{bmatrix} \begin{bmatrix} \sigma_\ell & \tau_{\ell t} \\ \tau_{t\ell} & \sigma_t \end{bmatrix} \begin{bmatrix} \cos\alpha_1 & -\sin\alpha_1 \\ \sin\alpha_1 & \cos\alpha_1 \end{bmatrix} \tag{4.3}$$

Performing the matrix multiplications and noticing that $\tau_{t\ell} = \tau_{\ell t}$ and $\tau_{12} = \tau_{21}$ are applicable to equilibrium condition result in the following three equations:

$$\sigma_1 = \sigma_\ell \cos^2\alpha_1 + \sigma_t \sin^2\alpha_1 + \tau_{\ell t}\, 2\,\sin\alpha_1\,\cos\alpha_1 \tag{4.4}$$

$$\sigma_2 = \sigma_\ell \sin^2\alpha_1 + \sigma_t \cos^2\alpha_1 - \tau_{\ell t}\, 2\,\sin\alpha_1\,\cos\alpha_1 \tag{4.5}$$

$$\tau_{12} = (-\sigma_\ell + \sigma_t)\,\sin\alpha_1\,\cos\alpha_1 + \tau_{\ell t}\,(\cos^2\alpha_1 - \sin^2\alpha_1) \tag{4.6}$$

Equations (4.4)–(4.6) can be expressed in the matrix form by one equation:

$$\begin{bmatrix} \sigma_1 \\ \sigma_2 \\ \tau_{12} \end{bmatrix} = \begin{bmatrix} \cos^2\alpha_1 & \sin^2\alpha_1 & 2\sin\alpha_1\cos\alpha_1 \\ \sin^2\alpha_1 & \cos^2\alpha_1 & -2\sin\alpha_1\cos\alpha_1 \\ -\sin\alpha_1\cos\alpha_1 & \sin\alpha_1\cos\alpha_1 & (\cos^2\alpha_1 - \sin^2\alpha_1) \end{bmatrix} \begin{bmatrix} \sigma_\ell \\ \sigma_t \\ \tau_{\ell t} \end{bmatrix} \tag{4.7}$$

This 3×3 matrix in Equation (4.7) is the *transformation matrix* $[T]$ for transforming the stresses in the stationary $\ell - t$ coordinate to the stresses in the rotating 1–2 coordinate. Using the tensor notation, Equation (4.7) becomes:

$$[\sigma_{12}] = [T][\sigma_{\ell t}] \tag{4.8}$$

4.1.2 Mohr Stress Circle

The trigonometric functions of α_1 in Equations (4.4)–(4.6) can be written in terms of double angle $2\alpha_1$ by

$$\cos^2\alpha_1 = \frac{1}{2}(1 + \cos 2\alpha_1) \tag{4.9}$$

$$\sin^2\alpha_1 = \frac{1}{2}(1 - \cos 2\alpha_1) \tag{4.10}$$

$$\sin\alpha_1\cos\alpha_1 = \frac{1}{2}\sin 2\alpha_1 \tag{4.11}$$

$$\cos^2\alpha_1 - \sin^2\alpha_1 = \cos 2\alpha_1 \tag{4.12}$$

Substituting Equations (4.9)–(4.12) into Equations (4.4)–(4.6) gives the three transformation equations in terms of the double angle $2\alpha_1$ as follows:

$$\sigma_1 = \frac{\sigma_\ell + \sigma_t}{2} + \frac{\sigma_\ell - \sigma_t}{2}\cos 2\alpha_1 + \tau_{\ell t}\,\sin 2\alpha_1 \tag{4.13}$$

$$\sigma_2 = \frac{\sigma_\ell + \sigma_t}{2} - \frac{\sigma_\ell - \sigma_t}{2}\cos 2\alpha_1 - \tau_{\ell t}\,\sin 2\alpha_1 \tag{4.14}$$

$$\tau_{12} = -\frac{\sigma_\ell - \sigma_t}{2}\sin 2\alpha_1 + \tau_{\ell t}\,\cos 2\alpha_1 \tag{4.15}$$

Equations (4.13)–(4.15) had been recognized by Otto Mohr to be algebraically analogous to a set of equations describing a circle in the $\sigma - \tau$ coordinate, as shown in Figure 4.2(a).

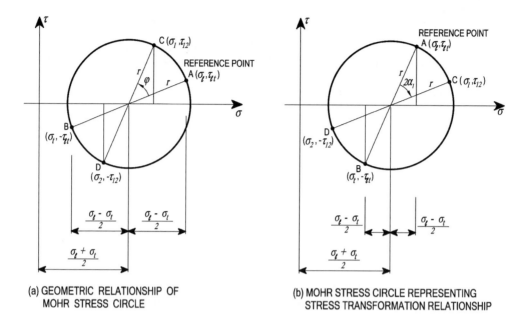

(a) GEOMETRIC RELATIONSHIP OF
   MOHR STRESS CIRCLE

(b) MOHR STRESS CIRCLE REPRESENTING
   STRESS TRANSFORMATION RELATIONSHIP

Figure 4.2   Mohr stress circle

This set of three equations is:

$$\sigma_1 = \frac{\sigma_\ell + \sigma_t}{2} + \frac{\sigma_\ell - \sigma_t}{2} \cos\phi - \tau_{\ell t} \sin\phi \tag{4.16}$$

$$\sigma_2 = \frac{\sigma_\ell + \sigma_t}{2} - \frac{\sigma_\ell - \sigma_t}{2} \cos\phi + \tau_{\ell t} \sin\phi \tag{4.17}$$

$$\tau_{12} = \frac{\sigma_\ell - \sigma_t}{2} \sin\phi + \tau_{\ell t} \cos\phi \tag{4.18}$$

The circular property of Equations (4.16)–(4.18) can be demonstrated as follows: The first two equations, (4.16) and (4.17), can be written as one equation, where $\sigma_{1,2}$ means either σ_1 or σ_2:

$$\left(\sigma_{1,2} - \frac{\sigma_\ell + \sigma_t}{2}\right) = \pm\left(\frac{\sigma_\ell - \sigma_t}{2} \cos\phi - \tau_{\ell t} \sin\phi\right) \tag{4.19}$$

Squaring Equation (4.19) gives

$$\left(\sigma_{1,2} - \frac{\sigma_\ell + \sigma_t}{2}\right)^2 = \left(\frac{\sigma_\ell - \sigma_t}{2}\right)^2 \cos^2\phi + \tau_{\ell t}^2 \sin^2\phi - (\sigma_\ell - \sigma_t)\tau_{\ell t} \sin\phi \cos\phi \tag{4.20}$$

Also squaring Equation (4.18) gives

$$\tau_{12}^2 = \left(\frac{\sigma_\ell - \sigma_t}{2}\right)^2 \sin^2\phi + \tau_{\ell t}^2 \cos^2\phi + (\sigma_\ell - \sigma_t)\tau_{\ell t} \sin\phi \cos\phi \tag{4.21}$$

Adding Equations (4.20) and (4.21) results in

$$\left(\sigma_{1,2} - \frac{\sigma_\ell + \sigma_t}{2}\right)^2 + \tau_{12}^2 = \left(\frac{\sigma_\ell - \sigma_t}{2}\right)^2 + \tau_{\ell t}^2 = r^2 \tag{4.22}$$

Note that a term r^2 is added to Equation (4.22). By equating both the left- and the right-hand sides of Equation (4.22) to r^2, we can see that Equation (4.22) is the equation of a circle with a radius r, centered on the σ-axis at a distance of $(\sigma_\ell + \sigma_t)/2$ from the origin, as shown in Figure 4.2(a). The left-hand side of Equation (4.22) represents points C (σ_1, τ_{12}) and D $(\sigma_2, -\tau_{12})$ on the circle and the right-hand side represents points A $(\sigma_\ell, \tau_{\ell t})$ and B $(\sigma_t, -\tau_{\ell t})$. The angle between radius for C and radius for A is the angle ϕ. Angle ϕ does not appear in Equation (4.22), and, therefore, can have any magnitude.

In summary, Figure 4.2(a) shows that the abscissas and ordinates for points C (σ_1, τ_{12}) and D $(\sigma_2, -\tau_{12})$ on the circle can be expressed in terms of the abscissas and ordinates of points A $(\sigma_\ell, \tau_{\ell t})$ and B $(\sigma_t, -\tau_{\ell t})$ at an angle ϕ apart based on geometric (i.e. circular) relationship. Notice that points C and D represent a set of stresses $(\sigma_1, \sigma_2, \tau_{12})$ in the 1–2 coordinate, and points A and B represent a set of stresses $(\sigma_\ell, \sigma_t, \tau_{\ell t})$ in the $\ell - t$ coordinate. Therefore, the first set of stresses $(\sigma_1, \sigma_2, \tau_{12})$ can be expressed in terms of the second set of stresses $(\sigma_\ell, \sigma_t, \tau_{\ell t})$ by the three equations (4.16)–(4.18).

Now we can compare the stress transformation relationship, (Equations 4.13–4.15), to the geometric (i.e. circular) relationship (Equations 4.16–4.18). Obviously, the two sets equations are similar, but not identical. Two differences can be noted: (1) all the angles in the transformation equations are twice those in the geometric relationship; (2) all the signs of the angles are opposite in the two set of equations.

In order to bring these two sets of equations into direct correspondence, we can eliminate these two differences by taking

$$\phi = -2\alpha_1 \tag{4.23}$$

The resulting Mohr stress circle is shown in Figure 4.2(b).

In Mohr's graphical representation of stress transformation (Figure 4.2b), the angles in Mohr's stress circle are always twice as great and measured in the opposite direction, compared with the actual stress field in an element. Since the angle α_1 in the $\ell - t$ coordinate rotates counter-clockwise, the angle $2\alpha_1$ in the Mohr circle must rotate clockwise. In addition, one must always remember that, although the algebraic analogy is precisely true, the geometric analogy is not, due to the double angle relationship.

In the Mohr stress circle (Figure 4.2b), the point B presents some confusion with regard to the negative sign of the shear stress $\tau_{\ell t}$. This can be explained by rotating the 1–2 axes by 90° as shown in Figure 4.1(d). It can be seen that the shear stress τ_{12} on the top 1-face is positive and pointing leftward. In contrast, the shear stress $\tau_{t\ell}$ on the top t-face as shown in Figure 4.1(a) is positive and pointing rightward. Obviously, in the case of finding a shear stress at 90°, a positive shear stress τ_{12} in the rotating 1–2 coordinate must be treated as a negative shear stress $\tau_{t\ell}$ in the stationary $\ell - t$ coordinate. To be consistent with the sign convention of τ_{12} (calculated from stress transformation), we can define $\tau_{t\ell}$ as

$$\tau_{t\ell} = -\tau_{\ell t} \tag{4.24}$$

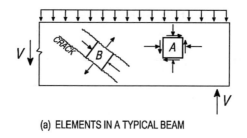

(a) ELEMENTS IN A TYPICAL BEAM

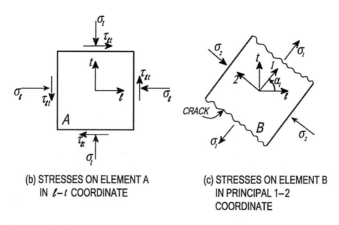

(b) STRESSES ON ELEMENT A
IN $\ell-t$ COORDINATE

(c) STRESSES ON ELEMENT B
IN PRINCIPAL 1–2
COORDINATE

Figure 4.3   Stresses on elements in typical reinforced concrete beams

According to Equation (4.24) in the stationary $\ell - t$ coordinate, a shear stress is considered positive if it causes a counter-clockwise rotation of the element, while shear stress is negative if it causes a clockwise rotation of the element.

In the graphical representation of the Mohr circle in Figure 4.2(b), a rotation of an angle α_1 in an element is represented by a rotation of $2\alpha_1$ from the reference point A to the rotating point C. For a 90° rotation in the element, point C would have rotated 180° from the reference point A to point B. Point B, therefore, should represent a shear stress of negative $\tau_{\ell t}(-\tau_{\ell t})$ as indicated in Figure 4.2(b).

The right-hand support of a typical reinforced concrete beam is shown in Figure 4.3(a). Two concrete elements, A and B, are indicated and are shown separately in Figure 4.3(b) and (c). Element A is taken in the $\ell - t$ coordinate of the steel bars and element B in the principal 1–2 coordinate of the applied stresses. The stresses on these two elements have been indicated in a qualitative sense with regard to sign. We will now find how to draw a Mohr circle to represent the stresses on these two elements.

The stresses on element A are shown in Fig. 4.4(b). The shear stress, $-\tau_{\ell t}$, on the top face is based on the sign convention given in Fig. 4.4(a). Figure 4.4(a) is obtained by applying $\tau_{t\ell} = -\tau_{\ell t}$ to the basic sign convention shown in Fig. 4.1(a). A Mohr stress circle is constructed graphically in Fig. 4.4 (c).

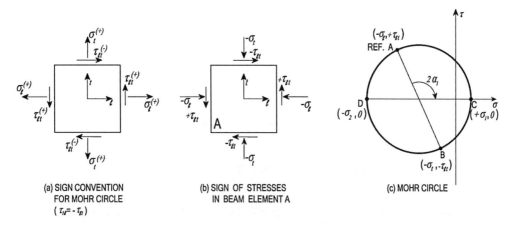

Figure 4.4   Mohr stress circle for a typical beam element

The reference point A in Figure 4.4 (c) represents the stresses on the ℓ-face $(-\sigma_\ell, \tau_{\ell t})$. The point B at 180° away gives the stresses $(-\sigma_t, -\tau_{\ell t})$ on the t-face. The point C, which is located at an angle $2\alpha_1$ from the reference point A, represents the principal tensile stress condition $(\sigma_1, 0)$ on element B. Point D, which is 180° from point C, gives the principal compressive stress condition $(-\sigma_2, 0)$.

4.1.3 Principal Stresses

Principal stresses are defined as the normal stresses on the face oriented in such a way that the shear stress vanishes. To find the principal stresses we will first look at the double angle transformation equations (4.13)–(4.15). The stresses σ_1 and σ_2 in Equations (4.13) and (4.14) become the principal stresses when the shear stress τ_{12} in Equation (4.15) vanishes.

Setting τ_{12} in Equation (4.15) equal to zero gives the angle α_1 which defines the two principal directions:

$$\cot 2\alpha_1 = \frac{\sigma_\ell - \sigma_t}{2\tau_{\ell t}} \tag{4.25}$$

From the geometric relationship shown in Figure 4.5 we find $\sin 2\alpha_1$ and $\cos 2\alpha_1$ as follows:

$$\sin 2\alpha_1 = \frac{\tau_{\ell t}}{\sqrt{\left(\frac{\sigma_\ell - \sigma_t}{2}\right)^2 + \tau_{\ell t}^2}} \tag{4.26}$$

$$\cos 2\alpha_1 = \frac{\left(\frac{\sigma_\ell - \sigma_t}{2}\right)}{\sqrt{\left(\frac{\sigma_\ell - \sigma_t}{2}\right)^2 + \tau_{\ell t}^2}} \tag{4.27}$$

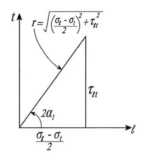

Figure 4.5   Geometric relationship for trignometric function

Substituting $\sin 2\alpha_1$ and $\cos 2\alpha_1$ from Equations (4.26) and (4.27) into Equations (4.13) and (4.14) results in:

$$\sigma_1 = \frac{\sigma_\ell + \sigma_t}{2} + \sqrt{\left(\frac{\sigma_\ell - \sigma_t}{2}\right)^2 + \tau_{\ell t}^2} \qquad (4.28)$$

$$\sigma_2 = \frac{\sigma_\ell + \sigma_t}{2} - \sqrt{\left(\frac{\sigma_\ell - \sigma_t}{2}\right)^2 + \tau_{\ell t}^2} \qquad (4.29)$$

The expressions for these two principal stresses in Equations (4.28) and (4.29) can be combined into one equation:

$$\sigma_{1,2} - \frac{\sigma_\ell + \sigma_t}{2} = \pm \sqrt{\left(\frac{\sigma_\ell - \sigma_t}{2}\right)^2 + \tau_{\ell t}^2} = \pm r \qquad (4.30)$$

Equation (4.30) can actually be obtained directly from Equation (4.22) by setting τ_{12} equal to zero and taking the square root. The square root term in Equation (4.30) is the radius r of the Mohr circle.

Equations (4.28) and (4.29) can be expressed graphically by the Mohr circle in Figure 4.6(c). For this example we choose an element that is subjected to biaxial tension and shear (Figure 4.6a). The principal tensile stress σ_1 is much larger, in magnitude, than the principal compressive stress σ_2 (Figure 4.6b). σ_1 and σ_2 are represented by the two points C and D lying on the σ-axis of the circle. It can be seen that σ_1 is the sum of the average stress $(\sigma_\ell + \sigma_t)/2$ and the radius r, while σ_2 is the difference of these two terms.

4.2 Strains in 2-D Elements

4.2.1 Strain Transformation

In Section 4.1.1 we have studied the principle of transformation for stresses. This same principle will now be applied to strains. A strain is defined as a displacement per unit length. The definitions of strains ε_ℓ, ε_t and $\gamma_{\ell t}$ (or $\gamma_{t\ell}$) in the $\ell - t$ coordinate are illustrated in Figure 4.7(a) and (b). They are indicated as positive, using the basic sign convention described in Section 4.1.1. The strain ε_ℓ indicated in Figure 4.7(a) is positive, because both the displacement

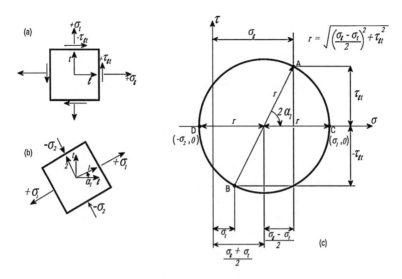

Figure 4.6 Graphical expression of principal stresses

and the unit length are in the positive ℓ-direction. For the same reason, the strain ε_t is positive in the t-direction.

For the shear strain $\gamma_{\ell t}$ in Figure 4.7(b), the first subscript ℓ indicates the original unit length in the ℓ-direction, and the second subscript t indicates the displacement in the t-direction. Since both the unit length and the displacement are in their positive directions, the shear strains $\gamma_{\ell t}$ indicated is positive. Similarly, $\gamma_{t\ell}$ indicated is also positive.

To find the three strain components in various directions, we introduce a rotating 1–2 coordinate system as shown in Figure 4.7(c). The 1–2 coordinate have been rotated counter-clockwise by an angle of α_1 with respect to the stationary $\ell - t$ axes. The three strain components in this rotating coordinate system are ε_1, ε_2 and γ_{12} (or γ_{21}). The relationship between the rotating

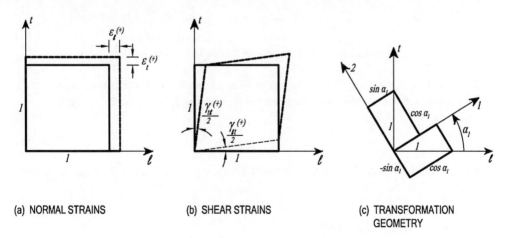

(a) NORMAL STRAINS   (b) SHEAR STRAINS   (c) TRANSFORMATION
                                                          GEOMETRY

Figure 4.7 Definitions of strains and transformation geometry

strain components, ε_1, ε_2 and γ_{12}, and the stationary strain components, ε_ℓ, ε_t and $\gamma_{\ell t}$, is the strain transformation.

In Figure 4.7(c) a positive unit length on the ℓ-axis will have projections of $\cos\alpha_1$ and $-\sin\alpha_1$ on the 1- and 2-axis, respectively. A positive unit length on the t-axis, however, should give projections of $\sin\alpha_1$ and $\cos\alpha_1$ on the 1- and 2-axis, respectively. Hence, the rotation matrix $[R]$ is

$$[R] = \begin{bmatrix} \cos\alpha_1 & \sin\alpha_1 \\ -\sin\alpha_1 & \cos\alpha_1 \end{bmatrix} \tag{4.31}$$

Notice that the rotation matrix $[R]$ for strains in Equation (4.31) is identical to matrix $[R]$ for stresses in Equation (4.1). The relationship between the strains in the 1-2 coordinate $[\varepsilon_{12}]$ and the strain in the $\ell - t$ coordinate $[\varepsilon_{\ell t}]$ is

$$[\varepsilon_{12}] = [R][\varepsilon_{\ell t}][R]^T \tag{4.32}$$

or

$$\begin{bmatrix} \varepsilon_1 & \dfrac{\gamma_{12}}{2} \\ \dfrac{\gamma_{21}}{2} & \varepsilon_2 \end{bmatrix} = \begin{bmatrix} \cos\alpha_1 & \sin\alpha_1 \\ -\sin\alpha_1 & \cos\alpha_1 \end{bmatrix} \begin{bmatrix} \varepsilon_\ell & \dfrac{\gamma_{\ell t}}{2} \\ \dfrac{\gamma_{t\ell}}{2} & \varepsilon_t \end{bmatrix} \begin{bmatrix} \cos\alpha_1 & -\sin\alpha_1 \\ \sin\alpha_1 & \cos\alpha_1 \end{bmatrix} \tag{4.33}$$

Performing the matrix multiplications and noticing that $\gamma_{t\ell} = \gamma_{\ell t}$ and $\gamma_{12} = \gamma_{21}$ results in the following three equations:

$$\varepsilon_1 = \varepsilon_\ell \cos^2\alpha_1 + \varepsilon_t \sin^2\alpha_1 + \frac{\gamma_{\ell t}}{2}(2\sin\alpha_1 \cos\alpha_1) \tag{4.34}$$

$$\varepsilon_2 = \varepsilon_\ell \sin^2\alpha_1 + \varepsilon_t \cos^2\alpha_1 - \frac{\gamma_{\ell t}}{2}(2\sin\alpha_1 \cos\alpha_1) \tag{4.35}$$

$$\frac{\gamma_{12}}{2} = (-\varepsilon_\ell + \varepsilon_t)\sin\alpha_1 \cos\alpha_1 + \frac{\gamma_{\ell t}}{2}\left(\cos^2\alpha_1 - \sin^2\alpha_1\right) \tag{4.36}$$

Equations (4.34)–(4.36) can be expressed in the matrix form by one equation:

$$\begin{bmatrix} \varepsilon_1 \\ \varepsilon_2 \\ \dfrac{\gamma_{12}}{2} \end{bmatrix} = \begin{bmatrix} \cos^2\alpha_1 & \sin^2\alpha_1 & 2\sin\alpha_1 \cos\alpha_1 \\ \sin^2\alpha_1 & \cos^2\alpha_1 & -2\sin\alpha_1 \cos\alpha_1 \\ -\sin\alpha_1 \cos\alpha_1 & \sin\alpha_1 \cos\alpha_1 & (\cos^2\alpha_1 - \sin^2\alpha_1) \end{bmatrix} \begin{bmatrix} \varepsilon_\ell \\ \varepsilon_t \\ \dfrac{\gamma_{\ell t}}{2} \end{bmatrix} \tag{4.37}$$

This 3×3 matrix in Eq. (4.37) is the *transformation matrix* $[T]$ for transforming the strain in the stationary $\ell - t$ coordinate to the strains in the rotating 1–2 coordinate. Using the tensor notation, Equation (4.37) becomes:

$$[\varepsilon_{12}] = [T][\varepsilon_{\ell t}] \tag{4.38}$$

It is interesting to note that the transformation matrix $[T]$ for strain is the same as that for stress.

4.2.2 Geometric Relationships

Equations (4.34)–(4.36) can be illustrated strictly by geometry in Figure 4.8. Figure 4.8(a) gives the geometric relationships between the three strain components ε_1, ε_2 and $\gamma_{12}/2$, in

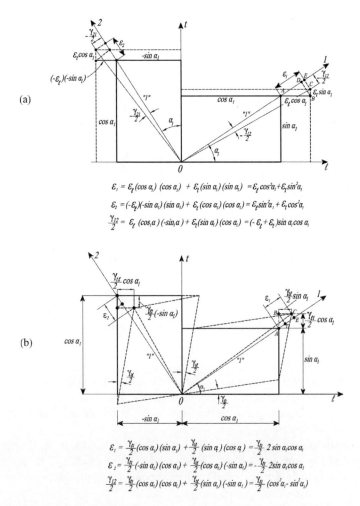

Figure 4.8   Geometric relationships for strain transformation. (a) for normal strains ε_ℓ and ε_t; (b) for shear strain $\gamma_{\ell t}$

the 1–2 coordinate and the two normal strains, ε_ℓ and ε_t, in the $\ell - t$ coordinate. Take ε_1 for example, it is contributed by both ε_ℓ and ε_t. Taking a unit diagonal length OA on the 1-axis, the projections of this unit length on the ℓ- and t-axes are $\cos\alpha_1$ and $\sin\alpha_1$. The displacements along the ℓ- and t-axes are then $\varepsilon_\ell\cos\alpha_1$ (AB) and $\varepsilon_t\sin\alpha_1$ (BC). The projections of these two displacements on the 1-axis are $\varepsilon_\ell\cos^2\alpha_1$ (AD) and $\varepsilon_t\sin^2\alpha_1$ (DE). The sum of these two displacement projections, $\varepsilon_\ell\cos^2\alpha_1 + \varepsilon_t\sin^2\alpha_1$(AE), is the strain ε_1 due to ε_ℓ and ε_t, because it is measured from an original length OA of unity.

   Figure 4.8(b) shows the geometric relationships between the three strain components ε_1, ε_2 and $\gamma_{12}/2$ in the 1–2 coordinate and the shear strain $\gamma_{\ell t}/2$ in the $\ell - t$ coordinate. Again, take ε_1 for example. The projections of the diagonal unit length OA on the ℓ and t axes are $\cos\alpha_1$ and $\sin\alpha_1$, respectively. For a shear strain $\gamma_{\ell t}/2$, indicated by the dotted lines, the angular displacements, AB and BC, along the ℓ- and t-axes are then $(\gamma_{\ell t}/2)\cos\alpha_1$ and $(\gamma_{t\ell}/2)\sin\alpha_1$,

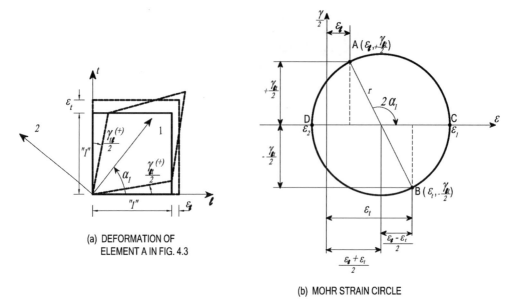

(a) DEFORMATION OF
   ELEMENT A IN FIG. 4.3

(b) MOHR STRAIN CIRCLE

Figure 4.9   Mohr circle for strains

respectively. The projections of these two displacements, AD and DE, on the 1-axis are $(\gamma_{\ell t}/2)\cos\alpha_1(\sin\alpha_1)$ and $(\gamma_{t\ell}/2)\sin\alpha_1(\cos\alpha_1)$, respectively. The sum of these two displacement projections, $AE = (\gamma_{\ell t}/2)2\cos\alpha_1\sin\alpha_1$, is the strain ε_1 due to the shear strain $\gamma_{\ell t}$, because it has an original length OA of unity.

Summing the strain ε_1 due to ε_ℓ and ε_t (Figure 4.8a) and that due to $\gamma_{\ell t}$ (Figure 4.8b), the total ε_1 is expressed by Equation (4.34). The expressions for ε_2 and $\gamma_{12}/2$ in Equations (4.35) and (4.36) can similarly be demonstrated by direct geometric relationships.

4.2.3 Mohr Strain Circle

Referring to element A at the right-hand support of a typical reinforced concrete beam, as shown in Figure 4.3, the deformation of this 2-D element is indicated in Figure 4.9(a), and the transformation of strains in this element is illustrated by a Mohr strain circle in Figure 4.9(b).

Recalling the double angle trigonometric relationships of Equations (4.9)–(4.12), and substituting these equations into Equations (4.34)–(4.36) results in:

$$\varepsilon_1 = \frac{\varepsilon_\ell + \varepsilon_t}{2} + \frac{\varepsilon_\ell - \varepsilon_t}{2}\cos 2\alpha_1 + \frac{\gamma_{\ell t}}{2}\sin 2\alpha_1 \tag{4.39}$$

$$\varepsilon_2 = \frac{\varepsilon_\ell + \varepsilon_t}{2} - \frac{\varepsilon_\ell - \varepsilon_t}{2}\cos 2\alpha_1 - \frac{\gamma_{\ell t}}{2}\sin 2\alpha_1 \tag{4.40}$$

$$\frac{\gamma_{12}}{2} = -\frac{\varepsilon_\ell - \varepsilon_t}{2}\sin 2\alpha_1 + \frac{\gamma_{\ell t}}{2}\cos 2\alpha_1 \tag{4.41}$$

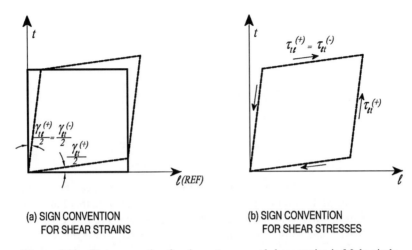

(a) SIGN CONVENTION
FOR SHEAR STRAINS

(b) SIGN CONVENTION
FOR SHEAR STRESSES

Figure 4.10   Sign convention for shear stresses and shear strains in Mohr circle

Comparing Equations (4.39)–(4.41) with Equations (4.13)–(4.15), it can be seen that they are the same, if the strains (ε_1, ε_2, $\gamma_{12}/2$, ε_ℓ, ε_t and $\gamma_{\ell t}/2$) are replaced by the corresponding stresses (σ_1, σ_2, τ_{12}, σ_ℓ, σ_t and $\tau_{\ell t}$). Therefore, Equations (4.39)–(4.41) represent a circle in the $\varepsilon - \gamma/2$ coordinate, as shown in Figure 4.9(b). The Mohr strain circle has its center on the ε-axis at a distance, $(\varepsilon_\ell + \varepsilon_t)/2$, from the origin. The radius r of the circle is

$$r = \sqrt{\left(\frac{\varepsilon_\ell - \varepsilon_t}{2}\right)^2 + \left(\frac{\gamma_{\ell t}}{2}\right)^2} \tag{4.42}$$

Point A on the Mohr circle represents the strain state of the element in the positive ℓ-direction. In this direction the longitudinal normal strain is ε_ℓ and the shear strain is $\gamma_{\ell t}$. The positive t-direction, which is 90° counter-clockwise from the positive ℓ-direction, is represented by point B on Mohr circle, located at 180° clockwise from the reference point A. In this t-direction the transverse normal strain is ε_t and the shear strain is $-\gamma_{\ell t}$. The sign convention for shear strains is based on $\gamma_{t\ell} = -\gamma_{\ell t}$, similar to the sign conversion for shear stresses, $\tau_{t\ell} = -\tau_{\ell t}$. The sign conventions for strains and for stresses are shown in Figure 4.10(a) and (b), respectively.

4.2.4 Principle Strains

The principal tensile strain ε_1 is oriented at an angle α_1 from the reference ℓ-direction, as shown in Figure 4.9(a). The angle α_1 can be found by setting $\gamma_{12} = 0$ in Equation (4.41):

$$\cot 2\alpha_1 = \frac{\varepsilon_\ell - \varepsilon_t}{\gamma_{\ell t}} \tag{4.43}$$

Using the same procedure as in Section 4.1.3 to find principal stresses, we can substitute $2\alpha_1$ from Equation (4.43) into Equations (4.39) and (4.40) to find the principal strains as follows:

$$\varepsilon_1 = \frac{\varepsilon_\ell + \varepsilon_t}{2} + \sqrt{\left(\frac{\varepsilon_\ell - \varepsilon_t}{2}\right)^2 + \left(\frac{\gamma_{\ell t}}{2}\right)^2} \tag{4.44}$$

$$\varepsilon_2 = \frac{\varepsilon_\ell + \varepsilon_t}{2} - \sqrt{\left(\frac{\varepsilon_\ell - \varepsilon_t}{2}\right)^2 + \left(\frac{\gamma_{\ell t}}{2}\right)^2} \tag{4.45}$$

The principal strains ε_1 and ε_2, given by Equations (4.44)–(4.45), can be observed graphically in terms of the Mohr strain circle in Figure 4.9(b). The principal tensile strain ε_1 is represented by point C, and is located at an angle of $2\alpha_1$ from the reference point A. The principal compressive strain ε_2 is represented by point D at 180° from point C.

4.3 Reinforced Concrete 2-D Elements

4.3.1 Stress Condition and Crack Pattern in RC 2-D Elements

A RC 2-D elements subjected to in-plane shear and normal stresses is shown in Figure 4.11(a). The directions of the longitudinal and transverse steel bars are designated as ℓ- and t-axes, respectively, constituting the $\ell - t$ coordinate system. The normal stresses are designated as σ_ℓ and σ_t in the ℓ- and t-directions, respectively, and the shear stresses are represented by $\tau_{\ell t}$ in the $\ell - t$ coordinate system. Based on the basic sign convention for Mohr circles, a positive shear stress $\tau_{\ell t}$ is the one that causes counter-clockwise rotation of a reinforced concrete element.

The RC element shown in Figure 4.11(a) can be separated into a concrete element (Figure 4.11b) and a steel grid element (Figure 4.11c). The normal and shear stresses on the concrete element are designated as σ_ℓ^c, σ_t^c and $\tau_{\ell t}^c$ in the $\ell - t$ coordinate system. The smeared steel stresses on the steel grid element are designated as $\rho_\ell f_\ell$ and $\rho_t f_t$ in the $\ell - t$ coordinate system. The symbols f_ℓ and f_t are the stresses in the steel bars. The steel bars are assumed to take only axial stresses, neglecting any possible dowel action.

Summing the concrete stresses (σ_ℓ^c and σ_t^c) and the steel stresses ($\rho_\ell f_\ell$ and $\rho_t f_t$), respectively, in the ℓ- and t-directions and equating them to the externally applied stresses (σ_ℓ, σ_t) gives Equations (4.46) and (4.47). Also, taking the moment equilibrium of internal shear stress $\tau_{\ell t}^c$ and external shear stresses $\tau_{\ell t}$ gives Equation (4.48):

$$\sigma_\ell = \sigma_\ell^c + \rho_\ell f_\ell \tag{4.46}$$

$$\sigma_t = \sigma_t^c + \rho_t f_t \tag{4.47}$$

$$\tau_{\ell t} = \tau_{\ell t}^c \tag{4.48}$$

The applied principal stresses for the RC element (Figure 4.11a), are defined as σ_1 and σ_2, based on the $1 - 2$ coordinate system as shown in Figure 4.11(d). The angle from the $\ell - t$ coordinate to the $1 - 2$ coordinate is defined as the fixed angle α_1, because this angle does not change when the three in-plane stresses σ_ℓ, σ_t and $\tau_{\ell t}$ increase proportionally. This angle α_1

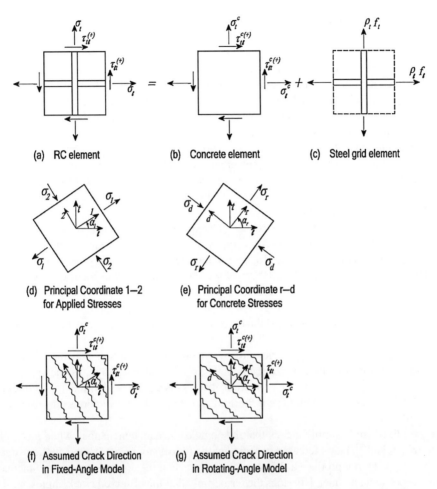

(a) RC element

(b) Concrete element

(c) Steel grid element

(d) Principal Coordinate 1–2 for Applied Stresses

(e) Principal Coordinate r–d for Concrete Stresses

(f) Assumed Crack Direction in Fixed-Angle Model

(g) Assumed Crack Direction in Rotating-Angle Model

Figure 4.11   Reinforced concrete membrane elements subjected to in-plane stress

is also called the steel bar angle because it defines the direction of the steel bars with respect to the applied principal stresses.

The principal stresses for the concrete element (Figure 4.11b), are defined as σ_r and σ_d, based on the $r - d$ coordinate system as shown in Figure 4.11(e). The angle from the $\ell - t$ coordinate to the $r - d$ coordinate is defined as the rotating angle α_r, because this angle will rotate when the three in-plane stresses σ_ℓ, σ_t and $\tau_{\ell t}$ on the RC element increase proportionally.

Before cracking, the steel bars have negligible effect on the behavior of a reinforced concrete element. The principal stresses in the concrete element coincide with the applied principal stresses σ_1 and σ_2. When the principal tensile stress σ_1 reaches the tensile strength of concrete, cracks will form and the concrete will be separated by the cracks into a series of concrete struts in the 2-direction as shown in Figure 4.11(f). If the RC element is reinforced with different amounts of steel in the ℓ- and t-directions, i.e., $\rho_\ell f_\ell \neq \rho_t f_t$ in Figure 4.11(c), the $r - d$ coordinate after cracking will deviate from the $1 - 2$ coordinate of the applied principal stresses on the RC element. The cracking pattern based on the $r - d$ coordinate after cracking is shown in Figure 4.11(g).

In the general case of $\rho_\ell f_\ell \neq \rho_t f_t$, when the set of applied stresses σ_ℓ, σ_t and $\tau_{\ell t}$ is increased proportionally, the $r - d$ coordinate will deviate from the $1 - 2$ coordinate. The deviation angle is defined as $\beta = \alpha_r - \alpha_1$, which will increase until it reaches the ultimate strength. If the RC element is reinforced with the same amounts of steel in the ℓ- and t-directions, i.e., $\rho_\ell f_\ell = \rho_t f_t$ in Figure 4.11(c), the $r - d$ coordinate will coincide with the $1 - 2$ coordinate, and the deviation angle β will become zero.

After cracking and in the subsequent cracking process under increasing proportional loading, however, the direction of the subsequent cracks was observed to deviate from the $1 - 2$ coordinate and to move toward the $r - d$ coordinate. Tests showed that the observed deviation will be less than the theoretical deviation. At ultimate load, the observed β-angle is about one-half of the calculated β-angle.

In view of the fact that the direction of subsequent cracks occurs in between the 1–2 coordinate and the $r - d$ coordinate, two types of theories have been developed, namely, the rotating angle theories (Hsu, 1991a, 1993) and the fixed angle theories (Pang and Hsu, 1996; Hsu and Zhang, 1997; Hsu, 1996, 1998). The rotating angle theories are based on the assumption that the direction of cracks is perpendicular to the principal tensile stress in the concrete element, as shown in Figure 4.11(g). In other words, the direction of cracks is governed by the $r - d$ coordinate, and the derivations of all the equilibrium and compatibility equations are based on the $r - d$ coordinate.

In contrast, the fixed angle theories are based on the assumption that the direction of subsequent cracks is perpendicular to the applied principal tensile stress, as shown in Figure 4.11(f). In the fixed angle theories, the direction of cracks is represented by the $1 - 2$ coordinate system, and the derivations of all the equilibrium and compatibility equations are based on the $1 - 2$ coordinate.

The term fixed angle simply means that the angle α_1 remains unchanged when the applied stresses (σ_ℓ, σ_t and $\tau_{\ell t}$) are increased proportionally. It does not mean that the applied stresses (σ_ℓ, σ_t and $\tau_{\ell t}$) can not produce a different α_1 angle, when the applied stresses are increased in a nonproportional manner. Nor does the term imply that the observed crack angles are fixed in the subsequent cracking process.

4.3.2 Fixed Angle Theory

As in Section 4.1.1, Equation (4.7), we can derive the matrix $[T]$ that transforms a set of concrete stresses in the $\ell - t$ coordinate (σ_ℓ^c, σ_t^c and $\tau_{\ell t}^c$) into a set of concrete stresses in the 1–2 coordinate (σ_1^c, σ_2^c and τ_{12}^c). The superscript c in all six stress symbols emphasizes that these stresses are intended for the concrete element (Figure 4.11b). The transformation process in Section 4.1.1 is shown in Figure 12(a), where the angle α_1 between the two coordinate systems is positive and rotates counter-clockwise. Now, let us reverse the transformation process to find the transformation matrix which transforms the stresses in the 1–2 coordinate (σ_1^c, σ_2^c and τ_{12}^c) into the stresses in the $\ell - t$ coordinate (σ_ℓ^c, σ_t^c and $\tau_{\ell t}^c$). This process is shown in Figure 4.12(b), where the angle α_1 is negative and rotates clockwise.

To reverse the direction of a rotation, it is only necessary to change the angle α_1 into $-\alpha_1$ in all the terms in the matrix of Equation (4.7). It can be seen that all the trigonometric functions remain the same, except that $\sin(-\alpha_1) = -\sin\alpha_1$. Changing the sign of the four terms with

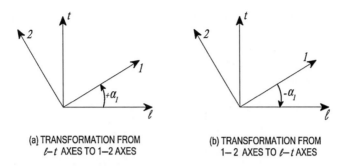

(a) TRANSFORMATION FROM
ℓ−t AXES TO 1−2 AXES

(b) TRANSFORMATION FROM
1−2 AXES TO ℓ−t AXES

Figure 4.12   Reserved transformation

$\sin \alpha_1$ in Equation (4.7) results in:

$$\begin{bmatrix} \sigma_\ell^c \\ \sigma_t^c \\ \tau_{\ell t}^c \end{bmatrix} = \begin{bmatrix} \cos^2 \alpha_1 & \sin^2 \alpha_1 & -2\sin\alpha_1\cos\alpha_1 \\ \sin^2 \alpha_1 & \cos^2 \alpha_1 & 2\sin\alpha_1\cos\alpha_1 \\ \sin\alpha_1\cos\alpha_1 & -\sin\alpha_1\cos\alpha_1 & (\cos^2\alpha_1 - \sin^2\alpha_1) \end{bmatrix} \begin{bmatrix} \sigma_1^c \\ \sigma_2^c \\ \tau_{12}^c \end{bmatrix} \tag{4.49}$$

This 3×3 matrix in Equation (4.49) is the *inverse transformation matrix* $[T]^{-1}$ for transforming a set of stresses in the 1–2 coordinate to a set of stresses in the $\ell - t$ coordinate. Using the tensor notation, Equation (4.49) becomes:

$$[\sigma_{\ell t}^c] = [T]^{-1}[\sigma_{12}^c] \tag{4.50}$$

Substituting Equation (4.49) into Equations (4.46)–(4.48) gives the three equilibrium equations for fixed angle theories as:

$$\sigma_\ell = \sigma_1^c \cos^2 \alpha_1 + \sigma_2^c \sin^2 \alpha_1 - \tau_{12}^c 2\sin\alpha_1\cos\alpha_1 + \rho_\ell f_\ell \tag{4.51}$$

$$\sigma_t = \sigma_1^c \sin^2 \alpha_1 + \sigma_2^c \cos^2 \alpha_1 + \tau_{12}^c 2\sin\alpha_1\cos\alpha_1 + \rho_t f_t \tag{4.52}$$

$$\tau_{\ell t} = (\sigma_1^c - \sigma_2^c)\sin\alpha_1\cos\alpha_1 + \tau_{12}^c(\cos^2\alpha_1 - \sin^2\alpha_1) \tag{4.53}$$

Similarly, in Section 4.2.1, Equation (4.37), we derived the same matrix $[T]$ that transforms a set of strains in the $\ell - t$ coordinate (ε_ℓ, ε_t and $\gamma_{\ell t}$) into a set of strains in the 1–2 coordinate (ε_1, ε_2 and γ_{12}). Using the same process as in stress transformation, we can reverse the process, as shown in Figure 4.12, and use the reverse transformation matrix $[T]^{-1}$ to transform a set of strains in the 1–2 coordinate (ε_1, ε_2 and γ_{12}) into a set of strains in the $\ell - t$ coordinate (ε_ℓ, ε_t and $\gamma_{\ell t}$) as follows:

$$\begin{bmatrix} \varepsilon_\ell \\ \varepsilon_t \\ \dfrac{\gamma_{\ell t}}{2} \end{bmatrix} = \begin{bmatrix} \cos^2 \alpha_1 & \sin^2 \alpha_1 & -2\sin\alpha_1\cos\alpha_1 \\ \sin^2 \alpha_1 & \cos^2 \alpha_1 & 2\sin\alpha_1\cos\alpha_1 \\ \sin\alpha_1\cos\alpha_1 & -\sin\alpha_1\cos\alpha_1 & (\cos^2\alpha_1 - \sin^2\alpha_1) \end{bmatrix} \begin{bmatrix} \varepsilon_1 \\ \varepsilon_2 \\ \dfrac{\gamma_{12}}{2} \end{bmatrix} \tag{4.54}$$

Consequently, the three compatibility equations for fixed-angle theories are:

$$\varepsilon_\ell = \varepsilon_1 \cos^2 \alpha_1 + \varepsilon_2 \sin^2 \alpha_1 - \frac{\gamma_{12}}{2}2\sin\alpha_1\cos\alpha_1 \tag{4.55}$$

$$\varepsilon_t = \varepsilon_1 \sin^2 \alpha_1 + \varepsilon_2 \cos^2 \alpha_1 + \frac{\gamma_{12}}{2}2\sin\alpha_1\cos\alpha_1 \tag{4.56}$$

$$\frac{\gamma_{\ell t}}{2} = (\varepsilon_1 - \varepsilon_2)\sin\alpha_1\cos\alpha_1 + \frac{\gamma_{12}}{2}(\cos^2\alpha_1 - \sin^2\alpha_1) \tag{4.57}$$

In the fixed angle theory, the solution of these three stress equilibrium equations and the three strain compatibility equations requires stress–strain relationships for concrete and for mild steel bars. The three sets of stress–strain curves for concrete are the $\sigma_1^c - \varepsilon_1$ curve, the $\sigma_2^c - \varepsilon_2$ curve, and the $\tau_{12}^c - \gamma_{12}$ curve. In the matrix form, we can summarize these concrete constitutive relationships as:

$$
\begin{bmatrix} \sigma_1^c \\ \sigma_2^c \\ \tau_{12}^c \end{bmatrix} = \begin{bmatrix} E_1 & 0 & 0 \\ 0 & E_2 & 0 \\ 0 & 0 & G_{12} \end{bmatrix} \begin{bmatrix} \varepsilon_1 \\ \varepsilon_2 \\ \dfrac{\gamma_{12}}{2} \end{bmatrix}
\tag{4.58}
$$

where the 3×3 matrix in Equation. (4.58) is called the constitutive matrix for fixed angle theories, and the three nonlinear moduli E_1, E_2 and G_{12} represent the $\sigma_1^c - \varepsilon_1$ curve, the $\sigma_2^c - \varepsilon_2$ curve, and the $\tau_{12}^c - \gamma_{12}$ curve, respectively.

4.3.3 Rotating Angle Theory

In the rotating angle theory the direction of cracks is defined by the rotating angle α_r in the principal $r - d$ coordinate of concrete, as shown previously in Figure 4.11(g). Similar to the derivation of equilibrium equations for the fixed angle theories, we can derive the reverse matrix $[T]^{-1}$ that transforms a set of concrete stresses in the $r - d$ coordinate (σ_r and σ_d) into a set of concrete stresses in the $\ell - t$ coordinate (σ_ℓ^c, σ_t^c and $\tau_{\ell t}^c$). Notice that the shear stress in concrete τ_{rd} must vanish (i.e. $\tau_{rd} = 0$), because the $r - d$ coordinate is a principal coordinate. As a result, the three transformation equations for concrete element are:

$$
\begin{bmatrix} \sigma_\ell^c \\ \sigma_t^c \\ \tau_{\ell t}^c \end{bmatrix} = \begin{bmatrix} \cos^2 \alpha_r & \sin^2 \alpha_r & -2 \sin \alpha_r \cos \alpha_r \\ \sin^2 \alpha_r & \cos^2 \alpha_r & 2 \sin \alpha_r \cos \alpha_r \\ \sin \alpha_r \cos \alpha_r & -\sin \alpha_r \cos \alpha_r & (\cos^2 \alpha_r - \sin^2 \alpha_r) \end{bmatrix} \begin{bmatrix} \sigma_r \\ \sigma_d \\ 0 \end{bmatrix}
\tag{4.59}
$$

Substituting Equation (4.59) for concrete elements into Equations (4.46)–(4.48) for RC elements results in three equilibrium equations for RC elements as follows:

$$
\sigma_\ell = \sigma_r \cos^2 \alpha_r + \sigma_d \sin^2 \alpha_r + \rho_\ell f_\ell
\tag{4.60}
$$

$$
\sigma_t = \sigma_r \sin^2 \alpha_r + \sigma_d \cos^2 \alpha_r + \rho_t f_t
\tag{4.61}
$$

$$
\tau_{\ell t} = (\sigma_r - \sigma_d) \sin \alpha_r \cos \alpha_r
\tag{4.62}
$$

Similarly, the three compatibility equations in the rotating angle theories are expressed as:

$$
\varepsilon_\ell = \varepsilon_r \cos^2 \alpha_r + \varepsilon_d \sin^2 \alpha_r
\tag{4.63}
$$

$$
\varepsilon_t = \varepsilon_r \sin^2 \alpha_r + \varepsilon_d \cos^2 \alpha_r
\tag{4.64}
$$

$$
\frac{\gamma_{\ell t}}{2} = (\varepsilon_r - \varepsilon_d) \sin \alpha_r \cos \alpha_r
\tag{4.65}
$$

Equations (4.63)–(4.65) represent the transformation relationship between the strains (ε_ℓ, ε_t and $\gamma_{\ell t}$) in the $\ell - t$ coordinate of the steel grid element and the principal strains (ε_r and ε_d) in the $r - d$ coordinate of the concrete element.

In the rotating angle theory, the solution of the three equilibrium equations and the three compatibility equations also requires the stress–strain relationships of concrete and mild steel bars. The two sets of stress–strain curves for concrete are the $\sigma_r - \varepsilon_r$ curve and the $\sigma_d - \varepsilon_d$ curve. In matrix form, we can write the constitutive relationships as:

$$\begin{bmatrix} \sigma_r \\ \sigma_d \\ 0 \end{bmatrix} = \begin{bmatrix} E_r & 0 & 0 \\ 0 & E_d & 0 \\ 0 & 0 & G_{rd} \end{bmatrix} \begin{bmatrix} \varepsilon_r \\ \varepsilon_d \\ 0 \end{bmatrix} \tag{4.66}$$

where the 3×3 matrix in Equation (4.66) is the constitutive matrix for the rotating angle theory. The two nonlinear moduli E_r and E_d represent the $\sigma_r - \varepsilon_r$ curve and the $\sigma_d - \varepsilon_d$ curve, respectively. The shear modulus G_{rd} is usually taken as a small value to provide stability in calculation.

4.3.4 'Contribution of Concrete' (V_c)

4.3.4.1 Pure Shear Elements with Equilibrium at Cracks

In order to study the 'contribution of concrete' (V_c), it is most convenient to apply the fixed angle and the rotating angle theories to a pure shear elements with equilibrium at cracks. For such a RC element the two theories can be simplified by assuming: (1) the normal stresses on the RC element $\sigma_\ell = \sigma_t = 0$; and (2) the stresses in the concrete element and in the steel bars are taken at the cracks, which implies that the tensile strength of concrete vanishes (i.e. $\sigma_1 = 0$ in fixed angle theory, and $\sigma_r = 0$ in rotating angle theory), and the stresses in the steel bars will reach its yield strength at yielding of the RC element (i.e. $f_\ell = f_{\ell y}$ and $f_t = f_{ty}$).

4.3.4.2 V_c in Rotating Angle Theory

In the case of a pure shear element with equilibrium at cracks, Equations (4.60)–(4.62) become:

$$\rho_\ell f_{\ell y} = (-\sigma_d) \sin^2 \alpha_r \tag{4.67}$$
$$\rho_t f_{ty} = (-\sigma_d) \cos^2 \alpha_r \tag{4.68}$$
$$\tau_{\ell ty} = (-\sigma_d) \sin \alpha_r \cos \alpha_r \tag{4.69}$$

where $\tau_{\ell ty}$ is the applied shear stress at yielding, and $(-\sigma_d)$ is a positive value. Multiplying Equations (4.67) and (4.68) and taking the square root gives:

$$\sqrt{\rho_\ell f_{\ell y} \rho_t f_{ty}} = (-\sigma_d) \sin \alpha_r \cos \alpha_r \tag{4.70}$$

Combining Equations (4.69) and (4.70) results in a very simple equation:

$$\tau_{\ell ty} = \sqrt{\rho_\ell f_{\ell y} \rho_t f_{ty}} \tag{4.71}$$

Equation (4.71) states simply that the shear stress at yielding is the square-root-of-the-product average of the steel yield stresses in the longitudinal and transverse directions. It also implies that the 'contribution of concrete' (V_c) is equal to zero.

4.3.4.3 V_c in Fixed Angle Theory

In the case of a pure shear element with equilibrium at cracks, Equations (4.51)–(4.53) become:

$$\rho_\ell f_{\ell y} - \tau_{12}^c 2 \sin\alpha_1 \cos\alpha_1 = -\sigma_2^c \sin^2\alpha_1 \tag{4.72}$$

$$\rho_t f_{ty} + \tau_{12}^c 2 \sin\alpha_1 \cos\alpha_1 = -\sigma_2^c \cos^2\alpha_1 \tag{4.73}$$

$$\tau_{\ell t y} - \tau_{12}^c(\cos^2\alpha_1 - \sin^2\alpha_1) = -\sigma_2^c \sin\alpha_1 \cos\alpha_1 \tag{4.74}$$

The following four steps are used to derive an explicit expression for $\tau_{\ell t y}$. First, the compressive stress σ_2^c is eliminated from Equations (4.72) and (4.74) by multiplying Equation (4.72) by $\cos\alpha_1$ and Equation (4.74) by $\sin\alpha_1$, and then subtracting the latter from the former. After simplifying, the resulting equation is:

$$(\tau_{\ell t y} + \tau_{12}^c) = \rho_\ell f_{\ell y} \cot\alpha_1 \tag{4.75}$$

Second, multiply Equation (4.73) by $\sin\alpha_1$ and Equation (4.74) by $\cos\alpha_1$, and then subtract the latter from the former. After simplifying, the resulting equation is:

$$(\tau_{\ell t y} - \tau_{12}^c) = \rho_t f_{ty} \tan\alpha_1 \tag{4.76}$$

Third, multiplying Equations (4.75) and (4.76) and then simplifying gives:

$$\tau_{\ell t y} = \sqrt{(\tau_{12}^c)^2 + \rho_\ell f_{\ell y} \rho_t f_{ty}} \tag{4.77}$$

Fourth, expanding Equation (4.77) into a Taylor series and taking the first two terms of the series gives, to a good approximation:

$$\tau_{\ell t y} = \frac{(\tau_{12}^c)^2}{2\sqrt{\rho_\ell f_{\ell y} \rho_t f_{ty}}} + \sqrt{\rho_\ell f_{\ell y} \rho_t f_{ty}} \tag{4.78}$$

Equation (4.78) was first derived by Pang and Hsu (1996). The second term on the right-hand side of Equation (4.78) is the 'contribution of steel' (V_s), the same as that derived in Equation (4.71) for rotating angle theory. The first term is the 'contribution of concrete' (V_c). It can be seen that V_c is contributed by τ_{12}^c, the concrete shear stress along the cracks.

4.3.4.4 Summary for V_c

Comparison of Equation (4.78) for fixed angle theory and Equation (4.71) for rotating angle theory shows clearly that the 'contribution of concrete' (V_c) is caused by defining the cracks in the principal 1–2 coordinate of the externally applied stresses (or fixed angle). In the fixed angle theory, a concrete strut is subjected not only to the axial compressive stress σ_2^c, but also to a concrete shear stress τ_{12}^c, in the direction of the cracks. This concrete shear stress τ_{12}^c is the source of the V_c term observed in tests.

When the cracks are assumed to be governed by the principal $r - d$ coordinate of the concrete element, a concrete strut is subjected only to an axial compressive stress σ_d. In the rotating angle theory, the $r - d$ coordinate rotates with increasing proportional loading in such a way that the concrete shear stress τ_{12}^c vanishes. As a result, the V_c term is always zero.

In summary, both the rotating angle theory and the fixed angle theory are useful. The fixed angle theory is more accurate because it can predict the V_c term observed in tests. However, it

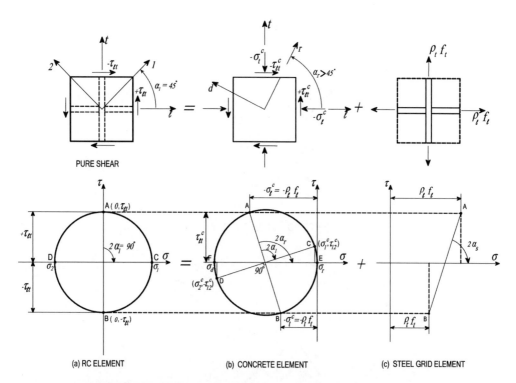

Figure 4.13   RC element subjected to pure shear (assuming $\sigma_t = 0$)

is more complicated because the equilibrium equations must involve the concrete shear stress τ_{12}^c. In contrast, the rotating angle theory has the advantage of simplicity and can be used to study the effect of steel reinforcement through the V_s term, without the complication induced by the concrete shear stress τ_{12}^c.

4.3.5 Mohr Stress Circles for RC Shear Elements

The state of stresses of a RC 2-D elements subjected to pure shear is illustrated in Figure 4.13 in terms of Mohr stress circles. These Mohr circles will further help our understanding of the rotating angle theory and the fixed angle theory.

Figure 4.13(a) shows a RC element subjected only to an applied shear stress $\tau_{\ell t}$. Consequently, the $1 - 2$ principal coordinate is oriented at $45°$ to the $\ell - t$ coordinate of the steel bars, giving a fixed angle α_1 of $45°$. On the Mohr circle of applied stresses in Figure 4.13(a), points A and B represent the vertical face ($\alpha_1 = 0°$) with a shear stress of $+\tau_{\ell t}$ and the horizontal faces ($\alpha_1 = 90°$) with a shear stress of $-\tau_{\ell t}$, respectively. Points C and D represent the principal 1-face ($\alpha_1 = 45°$) with a normal stress of σ_1 and the principal 2-face ($\alpha_1 = 135°$) with a normal stress of σ_2, respectively. The first crack is expected to develop perpendicular to the principal tension direction (1-direction) with an α_1 angle of $45°$ to the $\ell - t$ coordinate.

After cracking, the smeared steel stresses, $\rho_\ell f_\ell$ and $\rho_t f_t$, are activated. The post-cracking principal $r - d$ coordinate of the concrete element begins to deviate from the $1 - 2$ coordinate

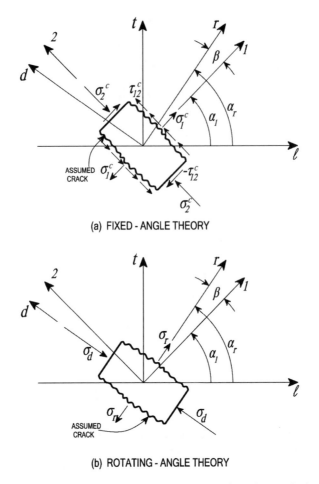

(a) FIXED - ANGLE THEORY

(b) ROTATING - ANGLE THEORY

Figure 4.14   Stress states in fixed-angle theory and rotating-angle theory

of the applied stresses. Assuming that the longitudinal steel ratio ρ_ℓ is greater than the transverse steel ratio ρ_t, the rotating angle α_r will be greater than the fixed angle α_1, as shown in Figure 4.13(b). The deviation angle $\beta = \alpha_r - \alpha_1$ will increase under increasing applied shear stresses $\tau_{\ell t}$. The angle α_r will rotate until it reaches a maximum angle that corresponds to the yielding of steel in both ℓ- and t-directions.

The state of stresses in the concrete element is shown by the Mohr circle in Figure 4.13(b). This Mohr circle is constructed from points A and B, representing the stress conditions on the vertical and horizontal face of the concrete element, respectively. The vertical face (point A), normal to ℓ-axis, is subjected not only to the shear stress $\tau_{\ell t}^c$ but also to a normal compressive stress, $-\sigma_\ell^c$. From the horizontal force equilibrium between the concrete stresses in Figure 4.13(b) and the steel stresses in Figure 4.13(c), $-\sigma_\ell^c = -\rho_\ell f_\ell$. Similarly, the horizontal faces (point B), normal to the t-axis, is subjected not only to the shear stress, $-\tau_{\ell t}^c$, but also to a normal compressive stress, $-\sigma_t^c$. From the vertical force equilibrium, $-\sigma_t^c$ is equal to $-\rho_t f_t$.

Figure 4.13(b) also shows that point C on the Mohr circle, which is $90°$ $(2\alpha_1)$ from point A, represents a face normal to the 1-axis and inclined at the fixed angle $\alpha_1 = 45°$. Such a $45°$ diagonal face is subjected to a tensile stress σ_1^c and a shear stress τ_{12}^c. On the opposite side of the Mohr circle, $180°$ away, point D represents a face normal to the 2-axis and inclined at a fixed angle $\alpha_1 = 135°$. Such a diagonal face is subjected to a compressive stress σ_2^c and a shear stress $-\tau_{12}^c$. The stress state represented by points C and D is illustrated in Figure 4.14(a) and serves as the basis of the *fixed angle theory*.

Figure 4.13(b) further shows that point E on the horizontal axis of the Mohr circle represents a face normal to the r-axis and inclined at a rotating angle α_r. This inclined face is subjected to a principal tensile stress σ_r (assumed to be zero at cracked sections) and a zero shear stress. On the opposite side of the Mohr circle, $180°$ away, point F represents a face normal to the d-axis and inclined at a rotation angle $90° + \alpha_r$. This inclined face is subjected to a principal compressive stress σ_d and a zero shear stress. The stress state represented by points E and F is illustrated in Figure 4.14(b) and serves as the basis of the *rotating angle theory*.

The rotating angle theory will be studied in Chapter 5, and the fixed angle theory in Chapter 6.

5

Rotating Angle Shear Theories

5.1 Stress Equilibrium of RC 2-D Elements

5.1.1 Transformation Type of Equilibrium Equations

In Chapter 4, Section 4.3.3, we derived three equilibrium equations, (4.60)–(4.62), for the rotating angle theory of a reinforced concrete 2-D element as follows:

$$\sigma_\ell = \sigma_r \cos^2 \alpha_r + \sigma_d \sin^2 \alpha_r + \rho_\ell f_\ell \tag{5.1}$$

$$\sigma_t = \sigma_r \sin^2 \alpha_r + \sigma_d \cos^2 \alpha_r + \rho_t f_t \tag{5.2}$$

$$\tau_{\ell t} = (\sigma_r - \sigma_d) \sin \alpha_r \cos \alpha_r \tag{5.3}$$

The state of stress in such a RC 2-D element subjected to pure shear $\tau_{\ell t}$ was illustrated by Mohr stress circles in Figure 4.13. Now we will deal with the general case of an element subjected not only to shear, $\tau_{\ell t}$, but also to normal stresses σ_ℓ and σ_t. The understanding of the state of stress in a RC element should focus on three aspects:

1. We can look at the reinforced concrete element as a whole. The state of stress due to the three applied external stresses σ_ℓ, σ_t and $\tau_{\ell t}$, is shown by the Mohr circle in Figure 5.1(a). It can be seen that point A represents the reference ℓ-face with stresses of σ_ℓ and $\tau_{\ell t}$, while point B represents the t-face with stresses of σ_t and $-\tau_{\ell t}$. The principal tensile stress, σ_1, is denoted as point C which is located at an angle of $2\alpha_1$ away from point A. The principal compressive stress, σ_2, is denoted as point D located at an angle of $180°$ away from point C.
2. The axial smeared stresses of the steel grid element are shown in Figure 5.1(c). The longitudinal and transverse stresses $\rho_\ell f_\ell$ and $\rho_t f_t$ are plotted at the level of points A and B, respectively. There is no Mohr circle for steel stresses, because the steel bars are assumed to be incapable of resisting shear (or dowel) stresses.
3. The state of stress in the concrete element is shown in Figure 5.1(b). The stresses in the ℓ- and t-faces are represented by points A and B, respectively. Point A has a normal stresses of $\sigma_\ell^c = \sigma_\ell - \rho_\ell f_\ell$ and a shear stress of $\tau_{\ell t}$. Point B has a normal stresses of $\sigma_t^c = \sigma_t - \rho_t f_t$, and a shear stress of $-\tau_{\ell t}$. The principal tensile stress σ_r is represented by point C located at an angle of $2\alpha_r$ away from point A. The principal compressive stress σ_d, is represented by point D located at an angle of $180°$ away from point C.

Unified Theory of Concrete Structures   Thomas Hsu and Yi-Lung Mo
© 2010 John Wiley & Sons, Ltd

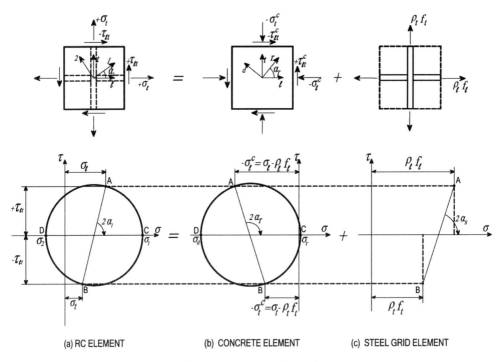

(a) RC ELEMENT   (b) CONCRETE ELEMENT   (c) STEEL GRID ELEMENT

Figure 5.1   Stress state in reinforced concrete

Adding Equations (5.1) and (5.2) gives

$$\sigma_r + \sigma_d = \sigma_\ell + \sigma_t - (\rho_\ell f_\ell + \rho_t f_t) \qquad (5.4)$$

Since the tensile stress of concrete, σ_r, is small and can be considered a given value, the compressive stress in the concrete element σ_d can be calculated from the externally applied stresses σ_ℓ and σ_t, as well as the steel stresses $\rho_\ell f_\ell$ and $\rho_t f_t$, according to Equation (5.4).

The three equilibrium equations (5.1)–(5.3), derived from the transformation, are convenient for computer analysis. These equations can be expressed in two ways that are convenient for calculation by hand and/or calculator. The first type given in Section 5.1.2 is convenient for analysis, and the second type given in Section 5.1.3 is convenient for design.

5.1.2 First Type of Equilibrium Equations

In the three equilibrium equations, (5.1)–(5.3), the tensile stress of concrete σ_r is smaller by an order of magnitude when compared with the other internal stresses σ_d, $\rho_\ell f_\ell$ and $\rho_t f_t$. Therefore, our focus will be on the relationships among the stresses $(\sigma_\ell - \rho_\ell f_\ell)$, $(\sigma_t - \rho_t f_t)$, $\tau_{\ell t}$ and σ_d, while considering σ_r as a small value of secondary importance. With this in mind, we can relate the three stresses $(\sigma_\ell - \rho_\ell f_\ell)$, $(\sigma_t - \rho_t f_t)$ and $\tau_{\ell t}$, to the compressive stress in the concrete struts σ_d.

Inserting $\sigma_r \cos^2 \alpha_r = \sigma_r - \sigma_r \sin^2 \alpha_r$ into Equation (5.1) gives

$$-\sigma_\ell + \rho_\ell f_\ell + \sigma_r = (\sigma_r - \sigma_d) \sin^2 \alpha_r \qquad (5.5)$$

Similarly, inserting $\sigma_r \sin^2 \alpha_r = \sigma_r - \sigma_r \cos^2 \alpha_r$ into Equation (5.2) gives

$$-\sigma_t + \rho_t f_t + \sigma_r = (\sigma_r - \sigma_d) \cos^2 \alpha_r \qquad (5.6)$$

Equation (5.3) remains the same

$$\tau_{\ell t} = (\sigma_r - \sigma_d) \sin \alpha_r \cos \alpha_r \qquad (5.7)$$

Equations (5.5)–(5.7) are the first type of expression for the equilibrium condition. These three equations are convenient for the analysis of RC 2-D elements.

Equations (5.5)–(5.7) represent the first type of geometric relationship in the Mohr circle as shown in Figure 5.2(a). The geometric relationship of the half Mohr circle defined by ACD is illustrated in Figure 5.2(b). If CD in Figure 5.2(b) is taken as unity, then $CE = \sin^2 \alpha_r$, $ED = \cos^2 \alpha_r$, and $AE = \sin \alpha_r \cos \alpha_r$. These three trigonometric values are actually the ratios of the three stresses $(-\sigma_\ell + \rho_\ell f_\ell + \sigma_r)$, $(-\sigma_t + \rho_t f_t + \sigma_r)$ and $\tau_{\ell t}$, respectively, divided by the sum of the principal stresses $(\sigma_r - \sigma_d)$.

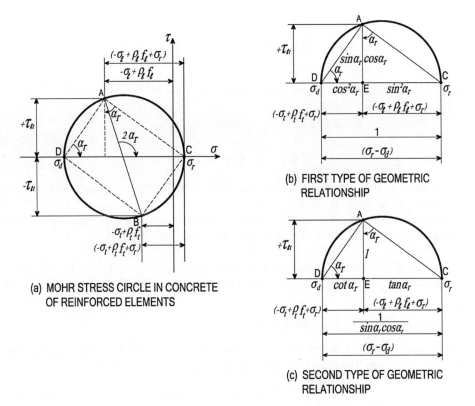

(a) MOHR STRESS CIRCLE IN CONCRETE OF REINFORCED ELEMENTS

(b) FIRST TYPE OF GEOMETRIC RELATIONSHIP

(c) SECOND TYPE OF GEOMETRIC RELATIONSHIP

Figure 5.2 Geometric relationship in Mohr stress circle for concrete in reinforced elements

Combining Equations (5.5) and (5.6) gives:

$$(-\sigma_\ell + \rho_\ell f_\ell + \sigma_r)(-\sigma_t + \rho_t f_t + \sigma_r) = (\sigma_r - \sigma_d)^2 \sin^2 \alpha_r \cos^2 \alpha_r \tag{5.8}$$

Squaring Equation (5.7) gives

$$\tau_{\ell t}^2 = (\sigma_r - \sigma_d)^2 \sin^2 \alpha_r \cos^2 \alpha_r \tag{5.9}$$

Equating Equations (5.8) and (5.9) and taking square root results in:

$$\tau_{\ell t} = \pm\sqrt{(-\sigma_\ell + \rho_\ell f_\ell + \sigma_r)(-\sigma_t + \rho_t f_t + \sigma_r)} \tag{5.10}$$

It can be seen that the applied shear stress is simply the square-root-of-the-product average of the two stress quantities, $(-\sigma_\ell + \rho_\ell f_\ell + \sigma_r)$ and $(-\sigma_t + \rho_t f_t + \sigma_r)$.

Dividing Equation (5.5) by Equation (5.6) we have:

$$\tan^2 \alpha_r = \frac{-\sigma_\ell + \rho_\ell f_\ell + \sigma_r}{-\sigma_t + \rho_t f_t + \sigma_r} \tag{5.11}$$

Neglecting the tensile strength of concrete by taking $\sigma_r = 0$, Equations (5.5)–(5.7) are simplified as follows:

$$-\sigma_\ell + \rho_\ell f_\ell = (-\sigma_d) \sin^2 \alpha_r \tag{5.12}$$
$$-\sigma_t + \rho_t f_t = (-\sigma_d) \cos^2 \alpha_r \tag{5.13}$$
$$\tau_{\ell t} = (-\sigma_d) \sin \alpha_r \cos \alpha_r \tag{5.14}$$

In the case of a reinforced concrete element subjected to pure shear $\sigma_\ell = \sigma_t = 0$. Then

$$\rho_\ell f_\ell = (-\sigma_d) \sin^2 \alpha_r \tag{5.15}$$
$$\rho_t f_t = (-\sigma_d) \cos^2 \alpha_r \tag{5.16}$$
$$\tau_{\ell t} = (-\sigma_d) \sin \alpha_r \cos \alpha_r \tag{5.17}$$

Equations (5.15)–(5.17) are identical to Equations (2.15)–(2.17) in Chapter 2. The difference in the sign of σ_d is due to the difference in sign conventions. In Equations (5.15)–(5.17) the sign convention in Section 4.1.1 requires the compressive stress in the concrete element to be negative. In Equations (2.15)–(2.17), however, all stresses have been taken as absolute values without sign.

When a RC 2-D element is subjected to pure shear ($\sigma_\ell = \sigma_t = 0$), and assuming zero tensile stress ($\sigma_r = 0$), Figure 5.2(a) and (b) become the same as Figure 2.3(a) and (b), respectively.

5.1.3 Second Type of Equilibrium Equations

The three equilibrium equations, (5.1)–(5.3), can also be expressed in another form. Substituting $(\sigma_r - \sigma_d)$ from Equation (5.7) into Equations (5.5) and (5.6) gives

$$(-\sigma_\ell + \rho_\ell f_\ell + \sigma_r) = \tau_{\ell t} \tan \alpha_r \tag{5.18}$$
$$(-\sigma_t + \rho_t f_t + \sigma_r) = \tau_{\ell t} \cot \alpha_r \tag{5.19}$$

Equation (5.7) itself can be written as

$$(\sigma_r - \sigma_d) = \tau_{\ell t} \frac{1}{\sin \alpha_r \cos \alpha_r} \tag{5.20}$$

Equations (5.18)–(5.20) are the second type of expression for the equilibrium condition. They are convenient for the design of RC 2-D elements.

Equations (5.18)–(5.20) represent the second type of geometric relationship in the Mohr circle, as shown in Figure 5.2(c). When AE in Figure 5.2(c) is taken as unity, then EC $=$ $\tan \alpha_r$, ED $= \cot \alpha_r$, and CD $= 1/\sin \alpha_r \cos \alpha_r$. These three trigonometric values are actually the ratios of the three stresses $(-\sigma_\ell + \rho_\ell f_\ell + \sigma_r)$, $(-\sigma_t + \rho_t f_t + \sigma_r)$ and $(\sigma_r - \sigma_d)$, divided by the shear stress $\tau_{\ell t}$, respectively.

Multiplying Equation (5.18) by Equation (5.19) gives Equation (5.10), and dividing Equation (5.18) by Equation (519) produces Equation (5.11).

In design, the small tensile stress of concrete is often neglected, i.e. $\sigma_r = 0$. Then Equations (5.18)–(5.20) become

$$\rho_\ell f_\ell = \sigma_\ell + \tau_{\ell t} \tan \alpha_r \tag{5.21}$$

$$\rho_t f_t = \sigma_t + \tau_{\ell t} \cot \alpha_r \tag{5.22}$$

$$(-\sigma_d) = \tau_{\ell t} \frac{1}{\sin \alpha_r \cos \alpha_r} \tag{5.23}$$

Furthermore, for the case of pure shear, $\sigma_\ell = \sigma_t = 0$. Then

$$\rho_\ell f_\ell = \tau_{\ell t} \tan \alpha_r \tag{5.24}$$

$$\rho_t f_t = \tau_{\ell t} \cot \alpha_r \tag{5.25}$$

$$(-\sigma_d) = \tau_{\ell t} \frac{1}{\sin \alpha_r \cos \alpha_r} \tag{5.26}$$

Equations (5.24)–(5.26) are identical to Equations (2.12)–(2.14). When the tensile stress of concrete is neglected ($\sigma_r = 0$) and for the case of pure shear elements ($\sigma_\ell = \sigma_t = 0$), Figure 5.2(a) and (c) become the same as Figure 2.3(a) and (c), respectively.

5.1.4 Equilibrium Equations in Terms of Double Angle

The equilibrium condition shown in Figure 5.1(a), (b) and (c) can also be expressed in terms of the relationships between the fixed angle α_1, the rotating angle α_r, and an imaginary angle α_s to be defined. Subtracting Equation (5.22) from Equation (5.21) gives

$$(\sigma_\ell - \sigma_t) = -\tau_{\ell t}(\tan \alpha_r - \cot \alpha_r) + (\rho_\ell f_\ell - \rho_t f_t) \tag{5.27}$$

Dividing Equation (5.27) by $2\tau_{\ell t}$ gives

$$\frac{(\sigma_\ell - \sigma_t)}{2\tau_{\ell t}} = -\frac{1}{2}(\tan \alpha_r - \cot \alpha_r) + \frac{(\rho_\ell f_\ell - \rho_t f_t)}{2\tau_{\ell t}} \tag{5.28}$$

Notice in Equation (5.28) that the term on the left-hand side can be written according to Equation (4.25) in Chapter 4 as:

$$\frac{(\sigma_\ell - \sigma_t)}{2\tau_{\ell t}} = \cot 2\alpha_1 \qquad (5.29)$$

The angle in Equation (5.29) is α_1, because the externally applied stresses σ_ℓ, σ_t and $\tau_{\ell t}$ are acting on the RC element as a whole, as shown in Figure 5.1(a).

The first term on the right-hand side of Equation (5.28) is

$$-\frac{1}{2}(\tan \alpha_r - \cot \alpha_r) = \cot 2\alpha_r \qquad (5.30)$$

The angle in Equation (5.30) is α_r, because the internal stresses of concrete are acting on the concrete element, as shown in Figure 5.1(b).

Let us now define an imaginary angle α_s for the second term on the right-hand side of Equation (5.28), such that

$$\frac{(\rho_\ell f_l - \rho_t f_t)}{2\tau_{\ell t}} = \cot 2\alpha_s \qquad (5.31)$$

The angle in Equation (5.31) is α_s, because the internal stresses of steel bars are acting on the steel grid element, as shown in Figure 5.1(c).

Substituting Equations (5.29)–(5.31) into Equation (5.28) we arrived at a very simple equation, relating the three angles α_1, α_r and α_s:

$$\cot 2\alpha_1 = \cot 2\alpha_r + \cot 2\alpha_s \qquad (5.32)$$

Equation (5.32) is very useful in understanding the relationship among the three diagrams in Figure 5.1(a), (b) and (c). It states that the cotangent of the angle $2\alpha_1$ for the RC element is simply the sum of the cotangent of $2\alpha_r$ for the concrete element and the cotangent of $2\alpha_s$ for the steel grid element.

It should again be emphasized that the angle α_s is imaginary, because the steel grid element can not resist an applied shear stress $\tau_{\ell t}$ in the $\ell - t$ coordinate system. The vertical τ-axis in Figure 5.1(c) does not have a real physical meaning and no Mohr circle can be prepared for the steel grid. However, the invention of the α_s angle provides a very convenient way to understand and to check the three diagrams in Figure 5.1(a), (b) and (c).

5.1.5 Example Problem 5.1 Using Equilibrium (Plasticity) Truss Model

5.1.5.1 Statement of Problem

A RC 2-D element is subjected to a set of three membrane stresses, as shown in Figure 5.3(a), i.e. $\sigma_\ell = 2.13$ MPa (tension), $\sigma_t = -2.13$ MPa (compression) and $\tau_{\ell t} = 3.69$ MPa. Design the reinforcement in the longitudinal and transverse directions according to the equilibrium (plasticity) truss model, i.e. assuming the yielding of steel in the two directions, $f_{\ell y} = f_{ty} = 413$ MPa. Calculate the stresses in the principal directions of the RC element (σ_1 and σ_2), the smeared stresses in the steel grid ($\rho_\ell f_{\ell y}$ and $\rho_t f_{ty}$), and the principal compressive stress in

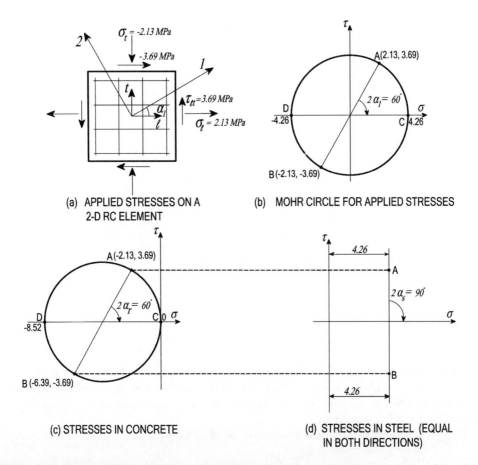

(a) APPLIED STRESSES ON A
    2-D RC ELEMENT

(b) MOHR CIRCLE FOR APPLIED STRESSES

(c) STRESSES IN CONCRETE

(d) STRESSES IN STEEL (EQUAL
    IN BOTH DIRECTIONS)

Figure 5.3   Stresses (MPa) in RC 2-D element with equal steel in two directions (example problem 5.1)

the concrete element (σ_d). Express all these stresses in terms of Mohr circles. Assume equal percentages of steel in both directions, i.e. $\rho_\ell = \rho_t$, and neglect the tensile strength of concrete ($\sigma_r = 0$).

5.1.5.2 Basic Principles for Solution

In the equilibrium (plasticity) truss model, the design of reinforced concrete membrane elements is based on the yielding of both the longitudinal and transverse steel. Assuming $f_\ell = f_{\ell y}$ and $f_t = f_{ty}$, and neglecting the tensile stress of concrete ($\sigma_r = 0$), the three equilibrium equations, Equations (5.21)–(5.23), become

$$\rho_\ell f_{\ell y} = \sigma_\ell + \tau_{\ell t}\ \tan\alpha_r \tag{5.33}$$

$$\rho_t f_{ty} = \sigma_t + \tau_{\ell t}\ \cot\alpha_r \tag{5.34}$$

$$(-\sigma_d) = \tau_{\ell t}\ \frac{1}{\sin\alpha_r \cos\alpha_r} \tag{5.35}$$

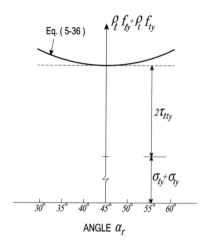

Figure 5.4   Total reinforcement in element as a function of angle α_r

Adding Equations (5.33) and (5.34) and recalling $\tan \alpha_r + \cot \alpha_r = 1/\sin \alpha_r \cos \alpha_r$ gives

$$\rho_\ell f_{\ell y} + \rho_t f_{ty} = \sigma_\ell + \sigma_t + \tau_{\ell t} \frac{1}{\sin \alpha_r \cos \alpha_r} \tag{5.36}$$

The total steel requirement in two directions, $\rho_\ell f_{\ell y} + \rho_t f_{ty}$, is plotted against the angle α_r in Figure 5.4 according to Equation (5.36). It can be seen that a minimum total steel can be achieved by designing α_r equal to 45°. This angle of 45° also provides the best crack control as will be studied in Section 5.2.4.

Designing a membrane element based on $\alpha_r = 45°$, Equations (5.33) and (5.34) become

$$\rho_\ell f_{\ell y} = \sigma_\ell + \tau_{\ell t} \tag{5.37}$$
$$\rho_t f_{ty} = \sigma_t + \tau_{\ell t} \tag{5.38}$$

Equations (5.37) and (5.38) show that the steel bars in either the ℓ- or t-direction can simply be designed to resist the normal stress plus the shear stress.

In Figure 5.3(a) the RC 2-D element is subjected to an applied transverse compressive stress, $\sigma_t = -2.13$ MPa, which is less than the applied shear stress $\tau_{\ell t} = 3.69$ MPa in magnitude. Equation (5.38) will give $\rho_t f_{ty} = 1.56$ MPa > 0. That means that tensile steel is required in the transverse direction.

If the applied transverse compressive stress, say $\sigma_t = -5.00$ MPa, which is greater in magnitude than the applied shear stress $\tau_{\ell t}$, Equation (5.38) gives a negative $\rho_t f_{ty} = -5.00 + 3.69 = -1.31$ MPa, i.e. steel is required for compression. Such design is, of course, not valid, because concrete has not cracked. The best we can do is to set $\rho_t f_{ty} = 0$. In theory, the extra compressive stress, -1.31 MPa, in the transverse direction can be utilized to reduce the required steel in the longitudinal direction as follows. Inserting $\sigma_t = -500$ MPa and $\rho_t f_{ty} = 0$ into Equation (5.34) gives

$$\cot \alpha_r = \frac{(-\sigma_t)}{\tau_{\ell t}} = \frac{5.00}{3.69} = 1.355 > 1, \text{ and } \alpha_r = 36.43° < 45° \tag{5.39}$$

Substituting this $\alpha_r = 36.43°$ into Equation (5.33) we obtain the required longitudinal steel:

$$\rho_\ell f_{\ell y} = \sigma_\ell + \tau_{\ell t}(\tan\ 36.43°) \tag{5.40}$$

Since $\tau_{\ell t}$ in the second term is multiplied by $\tan 36.43° < 1$, $\rho_\ell f_{\ell y}$ will be less than that if $\alpha_r = 45°$, thus resulting in a more economical design. However, when α_r is designed to be less than $45°$, the crack widths are expected to be larger.

5.1.5.3 Solutions

The fixed angle α_1 and the principal applied stresses, σ_1 and σ_2, for the RC 2-D element can be calculated directly from the applied membrane stresses σ_ℓ, σ_t and $\tau_{\ell t}$. From Equation (4.25) in Chapter 4:

$$\cot 2\alpha_1 = \frac{(\sigma_\ell - \sigma_t)}{2\tau_{\ell t}} = \frac{2.13 - (-2.13)}{2(3.69)} = 0.577$$
$$2\alpha_1 = 60°, \text{ and } \alpha_1 = 30°.$$

From Equations (4.28) and (4.29) in Chapter 4:

$$\sigma_1 = \frac{\sigma_\ell + \sigma_t}{2} + \sqrt{\left(\frac{\sigma_\ell - \sigma_t}{2}\right)^2 + \tau_{\ell t}^2} = \frac{2.13 - 2.13}{2} + \sqrt{\left(\frac{2.13 + 2.13}{2}\right)^2 + 3.69^2}$$
$$= 0 + 4.26 = 4.26 \text{ MPa}$$
$$\sigma_2 = \frac{\sigma_\ell + \sigma_t}{2} - \sqrt{\left(\frac{\sigma_\ell - \sigma_t}{2}\right)^2 + \tau_{\ell t}^2} = 0 - 4.26 = -4.26 \text{ MPa}.$$

The Mohr circle for the applied stresses is shown in Figure 5.3(b). Indicated in this circle are the angle $2\alpha_1 = 60°$, $\sigma_1 = 4.26$ MPa, and $\sigma_2 = -4.26$ MPa.

The reinforcement can be designed according to the basic equations of the equilibrium (plasticity) truss model (Equations 5.33 and 5.34). Since $\rho_\ell f_{\ell y} = \rho_t f_{ty}$, the rotating angle α_r can be found by equating these two equations:

$$\sigma_\ell + \tau_{\ell t}\ \tan\alpha_r = \sigma_t + \tau_{\ell t}\ \cot\alpha_r$$
$$\tau_{\ell t}(\tan\alpha_r - \cot\alpha_r) = -(\sigma_\ell - \sigma_t)$$

Noticing that $(\tan\alpha_r - \cot\alpha_r)$ is equal to $-2\cot 2\alpha_r$, the angle α_r is expressed as:

$$\cot 2\alpha_r = \frac{\sigma_\ell - \sigma_t}{2\tau_{lt}} = \frac{2.13 - (-2.13)}{2(3.69)} = 0.577$$

$2\alpha_r = 60°$, and $\alpha_r = 30°$. Also, $\tan\alpha_r = 0.577$ and $\cot\alpha_r = 1.732$
Substituting α_r back into Equations (5.33) and (5.34), we have

$$\rho_\ell f_{\ell y} = \sigma_\ell + \tau_{\ell t}\ \tan\alpha_r = 2.13 + 3.69(0.577) = 4.26 \text{ MPa}$$
$$\rho_t f_{ty} = \sigma_t + \tau_{\ell t}\ \cot\alpha_r = -2.13 + 3.69(1.732) = 4.26 \text{ MPa}$$
$$\rho_\ell = \rho_t = \rho = \frac{4.26}{413} = 0.0103$$

If the RC element has a thickness h of 305 mm., the steel area per unit length A_ℓ/s and A_t/s, is:

$$\frac{A_\ell}{s} = \frac{A_t}{s} = \rho h = 0.0103(305) = 3.14 \text{ mm}^2/\text{mm}$$

We could use two layers of No. 7 bars at 245 mm spacing in both directions. $A_\ell/s = A_t/s = 2(387)/245 = 3.16 \text{ mm}^2/\text{mm} > 3.14 \text{ mm}^2/\text{mm}$, OK.

The smeared steel stresses $\rho_\ell f_{\ell y}$ and $\rho_t f_{ty}$, are shown in Figure 5.3(d), in which $\rho_\ell f_{\ell y} = \rho_t f_{ty} = 4.26$ MPa. The stresses in the concrete element are shown by the Mohr circle in Figure 5.3(c) using the following additional values:

$$\sigma_\ell - \rho_\ell f_{\ell y} = 2.13 - 4.26 = -2.13 \text{ MPa}$$
$$\sigma_t - \rho_t f_{ty} = -2.13 - 4.26 = -6.39 \text{ MPa}$$

Equation (5.35) gives:

$$\sigma_d = \frac{-\tau_{\ell t}}{\sin\alpha_r \cos\alpha_r} = \frac{-3.69}{(0.500)(0.866)} = -8.52 \text{ MPa}$$

In conclusion, Figure 5.3(b), (c) and (d) clearly illustrate that the rotating angle α_r is equal to the fixed angle α_1 when the smeared steel stresses in both directions $\rho_\ell f_{\ell y}$ and $\rho_t f_{ty}$ are equal. This interesting case is a direct consequence of Equation (5.32), which relates the double angles of the steel grid element, the concrete element and the RC element. When $\cot 2\alpha_s$ for the steel grid is zero (i.e. $2\alpha_s = 90°$), Equation (5.32) requires that $\cot 2\alpha_r$ for the concrete element is equal to $\cot 2\alpha_1$ for the whole RC element.

5.2 Strain Compatibility of RC 2-D Elements

5.2.1 Transformation Type of Compatibility Equations

In Chapter 4, Section 4.3.3, we derived the three compatibility equations, (4.63)–(4.65), for the rotating angle theory of a reinforced concrete 2-D element as follows:

$$\varepsilon_\ell = \varepsilon_r \cos^2\alpha_r + \varepsilon_d \sin^2\alpha_r \tag{5.41}$$
$$\varepsilon_t = \varepsilon_r \sin^2\alpha_r + \varepsilon_d \cos^2\alpha_r \tag{5.42}$$
$$\frac{\gamma_{\ell t}}{2} = (\varepsilon_r - \varepsilon_d)\sin\alpha_r \cos\alpha_r \tag{5.43}$$

The compatibility condition, defined by Equations (5.41)–(5.43), contains six variables, namely, $\varepsilon_\ell, \varepsilon_t, \gamma_{\ell t}, \varepsilon_r, \varepsilon_d$ and α_r. When any three of the six variables are given, the other three can be solved by these three compatibility equations.

It is interesting to note that the four normal strains $\varepsilon_\ell, \varepsilon_t, \varepsilon_r$ and ε_d have a simple relationship. Adding Equations (5.41) and (5.42) gives

$$\varepsilon_\ell + \varepsilon_t = \varepsilon_r + \varepsilon_d \tag{5.44}$$

Equation (5.44) is the principle of first invariance for strains.
From Eq. (5.44), the cracking strain ε_r can be expressed as:

$$\varepsilon_r = \varepsilon_\ell + \varepsilon_t - \varepsilon_d \tag{5.45}$$

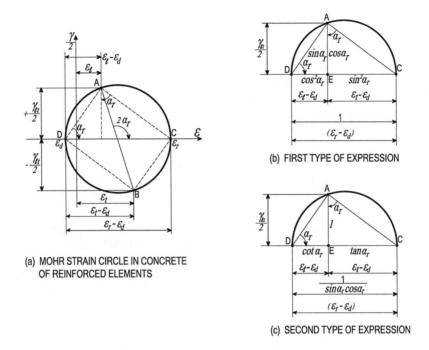

(a) MOHR STRAIN CIRCLE IN CONCRETE
OF REINFORCED ELEMENTS

(b) FIRST TYPE OF EXPRESSION

(c) SECOND TYPE OF EXPRESSION

Figure 5.5   Geometric relationship in Mohr stress circle

It should be noted that the magnitude of ε_d is an order of magnitude smaller than the other three strains, ε_r, ε_ℓ, and ε_t, because of cracking. If we neglect ε_d, Equation (5.45) states that the principal tensile strain (or cracking strain) ε_r is the sum of the steel strains in the longitudinal and transverse directions ($\varepsilon_\ell + \varepsilon_t$).

The state of strain in a RC 2-D element can be illustrated by Mohr strain circles in Figure 5.5(a). In the $\ell - t$ coordinate, point A represents the reference ℓ-direction with strains of ε_ℓ and $\gamma_{\ell t}/2$, while point B represents the t-direction with strains of ε_t and $-\gamma_{\ell t}/2$. The principal tensile strain, ε_r, is denoted as point C which is located at an angle of $2\alpha_r$ away from point A. The principal compressive strain, ε_d, is denoted as point D located at an angle of $180°$ away from point C.

5.2.2 First Type of Compatibility Equations

In view of the fact that ε_d is much smaller than the other four strains, our interest will naturally be focused on the relationship among the four strains, ε_ℓ, ε_t, ε_r and $\gamma_{\ell t}$, while considering ε_d as a small secondary strain. With this aim in mind, the three transformation equations, (5.41)–(5.43), can be changed into three explicit relationships among the four strains ε_ℓ, ε_t, $\gamma_{\ell t}$ and ε_r.

Inserting $\varepsilon_d \sin^2 \alpha_r = \varepsilon_d - \varepsilon_d \cos^2 \alpha_r$ into Equation (5.41) gives

$$(\varepsilon_\ell - \varepsilon_d) = (\varepsilon_r - \varepsilon_d)\cos^2 \alpha_r \qquad (5.46)$$

Similarly, inserting $\varepsilon_d \cos^2 \alpha_r = \varepsilon_d - \varepsilon_d \sin^2 \alpha_r$ into Equation (5.42) gives

$$(\varepsilon_t - \varepsilon_d) = (\varepsilon_r - \varepsilon_d)\, \sin^2 \alpha_r \qquad (5.47)$$

Equation (5.43) remains the same

$$\frac{\gamma_{\ell t}}{2} = (\varepsilon_r - \varepsilon_d)\, \sin \alpha_r \cos \alpha_r \qquad (5.48)$$

Equations (5.46)–(5.48) are the first type of expression for the compatibility condition. In this type of expression, the three strains in the $\ell - t$ coordinate, ε_ℓ, ε_t and $\gamma_{\ell t}$, are each related individually to the principal tensile strain ε_r, or more accurately the absolute-value sum of the principal strains $(\varepsilon_r - \varepsilon_d)$. These three equations are convenient for the *analysis* of RC 2-D elements.

Equations (5.46)–(5.48) represent the first type of geometric relationship in the Mohr circle, as shown in Figure 5.5(a). The geometric relationship of the half Mohr circle defined by ACD is illustrated in Figure 5.5(b). If CD in Figure 5.5(b) is taken as unity, then $CE = \sin^2 \alpha_r$, $ED = \cos^2 \alpha_r$, and $AE = \sin\alpha_r \cos\alpha_r$. These three trigonometric values are actually the ratios of the three strains $(\varepsilon_\ell - \varepsilon_d)$, $(\varepsilon_t - \varepsilon_d)$ and $\gamma_{\ell t}/2$, respectively, divided by the sum of the principal strains $(\varepsilon_r - \varepsilon_d)$.

Dividing Equation (5.47) by Equation (5.46) we have

$$\tan^2 \alpha_r = \frac{\varepsilon_t - \varepsilon_d}{\varepsilon_\ell - \varepsilon_d} \qquad (5.49)$$

Equation (5.49) is the compatiblity equation to determine the angle α_r. Its derivation signaled the arrival of the Mohr compatibility truss model and the rotating angle softened truss model. These two models will be studied in Sections 5.3 and 5.4, respectively.

Multiplying Equations (5.46) and (5.47) gives

$$(\varepsilon_\ell - \varepsilon_d)(\varepsilon_t - \varepsilon_d) = (\varepsilon_r - \varepsilon_d)^2\, \sin^2 \alpha_r \cos^2 \alpha_r \qquad (5.50)$$

Squaring Equation (5.48) gives

$$\left(\frac{\gamma_{\ell t}}{2}\right)^2 = (\varepsilon_r - \varepsilon_d)^2\, \sin^2 \alpha_r \cos^2 \alpha_r \qquad (5.51)$$

Equating Equations (5.50) and (5.51) and taking square root result in

$$\frac{\gamma_{\ell t}}{2} = \pm\sqrt{(\varepsilon_\ell - \varepsilon_d)(\varepsilon_t - \varepsilon_d)} \qquad (5.52)$$

Equation (5.52) shows that the shear strain $\gamma_{\ell t}/2$ can be calculated without knowing the angle α_r. $\gamma_{\ell t}/2$ is simply the square root of the product of $(\varepsilon_\ell - \varepsilon_d)$ and $(\varepsilon_t - \varepsilon_d)$.

5.2.3 Second Type of Compatibility Equations

The three compatibility equations, (5.46)–(5.48), can be expressed in another form. The three strains ε_ℓ, ε_t and ε_r can each be related individually to the shear strain $\gamma_{\ell t}$. Substituting

$(\varepsilon_r - \varepsilon_d)$ from Equation (5.48) into Equations (5.46) and (5.47) gives

$$(\varepsilon_\ell - \varepsilon_d) = \frac{\gamma_{\ell t}}{2} \cot \alpha_r \tag{5.53}$$

$$(\varepsilon_t - \varepsilon_d) = \frac{\gamma_{\ell t}}{2} \tan \alpha_r \tag{5.54}$$

$$(\varepsilon_r - \varepsilon_d) = \frac{\gamma_{\ell t}}{2} \frac{1}{\sin \alpha_r \cos \alpha_r} \tag{5.55}$$

Equations (5.53)–(5.55) are the second type of expression for the compatibility condition. They represent the second type of geometric relationship in the Mohr circle, as shown in Figure 5.5(c). When AE is taken as unity, then $CE = \tan \alpha_r$, $ED = \cot \alpha_r$, and $CD = 1/\sin \alpha_r \cos \alpha_r$. These three trigonometric values are actually the ratios of the three strains $(\varepsilon_\ell - \varepsilon_d)$, $(\varepsilon_t - \varepsilon_d)$ and $(\varepsilon_r - \varepsilon_d)$, respectively, divided by the shear strain $\gamma_{\ell t}/2$.

The two equations, (5.49) and (5.52) can also be derived from Equations (5.53) and (5.54). Dividing Equation (5.54) by Equation (5.53) gives Equation (5.49), and multiplying Equation (5.53) by Equation (5.54) produces Equation (5.52).

5.2.4 Crack Control

5.2.4.1 Steel Strains for Crack Control

After the cracking of a RC 2-D element, the sizes of the cracks must be controlled, especially at the service load stage. The principal tensile strain ε_r in the cracking direction (r-direction) can be taken as an indicator of the crack widths. In design, we can control the crack widths by specifying an allowable ε_r.

A method of design has been developed on the basis of crack control. In this method, the two principal strains ε_r and ε_d are given, and the problem is to find the steel strains ε_ℓ and ε_t as a function of the angle α_r. To do this, we make Equations (5.46) and (5.47) nondimensional by dividing these two equations by the yield strain of steel ε_y:

$$\frac{\varepsilon_\ell}{\varepsilon_y} = \frac{\varepsilon_d}{\varepsilon_y} + \left(\frac{\varepsilon_r}{\varepsilon_y} - \frac{\varepsilon_d}{\varepsilon_y} \right) \cos^2 \alpha_r \tag{5.56}$$

$$\frac{\varepsilon_t}{\varepsilon_y} = \frac{\varepsilon_d}{\varepsilon_y} + \left(\frac{\varepsilon_r}{\varepsilon_y} - \frac{\varepsilon_d}{\varepsilon_y} \right) \sin^2 \alpha_r \tag{5.57}$$

Equations (5.56) and (5.57) are plotted in Figure 5.6(a) and (b) for $\varepsilon_d/\varepsilon_y = 0$, and -0.25, respectively. In each figure, the two steel strain ratios $\varepsilon_\ell/\varepsilon_y$ and $\varepsilon_t/\varepsilon_y$ are plotted as a function of the angle α_r, using the crack strain ratio $\varepsilon_r/\varepsilon_y$ as a parameter. For 413 MPa (60 ksi) mild steel bars, the yield strain $\varepsilon_y = 0.00207$.

The use of Figures 5.6(a) and (b) can be demonstrated by an example problem: Suppose $\varepsilon_d/\varepsilon_y = 0$ (i.e. very small) and the allowable crack strain ratio, $\varepsilon_r/\varepsilon_y = 1.5$. Find the steel strain ratios, $\varepsilon_\ell/\varepsilon_y$ and $\varepsilon_t/\varepsilon_y$ for an α_r angle of 60°. The solution can be easily found by examining Figure 5.6(a) which is plotted for $\varepsilon_d/\varepsilon_y = 0$. The intersection of the vertical line at $\alpha_r = 60°$ and the $\varepsilon_\ell/\varepsilon_y$-curves for $\varepsilon_r/\varepsilon_y = 1.5$ gives $\varepsilon_\ell = 0.375\varepsilon_y$. The intersection of the same vertical line and the $\varepsilon_t/\varepsilon_y$-curve for $\varepsilon_r/\varepsilon_y = 1.5$ gives $\varepsilon_t = 1.125\varepsilon_y$.

It can be seen from Figure 5.6(a) and (b) that $\alpha_r = 45°$ gives equal strains of ε_ℓ and ε_t, which is lower than ε_ℓ when $\alpha_r < 45°$ and lower than ε_t when $\alpha_r > 45°$. From a serviceability

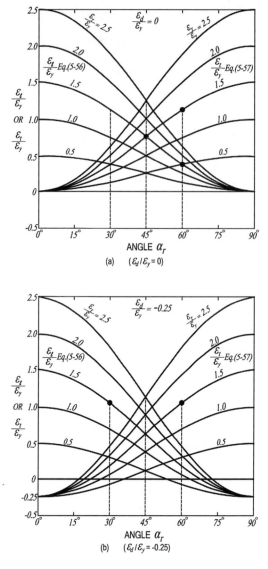

Figure 5.6 Steel strains, ε_ℓ and ε_t, as a function of cracking strain ε_r and angle α_r

point of view, $\alpha_r = 45°$ provides the best design. Substituting $\alpha_r = 45°$ into Equations (5.56) and (5.57) gives the principle of first invariance of strains:

$$\varepsilon_\ell = \varepsilon_t = \frac{1}{2}(\varepsilon_r + \varepsilon_d) \tag{5.58}$$

For $\varepsilon_r = 1.5\varepsilon_y$ and $\varepsilon_d = 0$, both Equation (5.58) and Figure 5.6(a) give the steel strains $\varepsilon_\ell = \varepsilon_t = 0.75\varepsilon_y$.

A method of design based on crack control was pioneered by Gupta (1981, 1984), using the compatibility equations, (5.56) and (5.57), as well as the equilibrium equations, (5.21) and (5.22). Assuming $\varepsilon_d = 0$, $f_\ell = E_s \varepsilon_\ell$, $f_t = E_s \varepsilon_t$, and selecting an α_r angle, we can determine the steel ratios (ρ_ℓ, ρ_t) in the longitudinal and transverse directions.

Gupta's design method produces good approximate solutions. However, this method is not rigorous because the diagonal concrete stress σ_d in the equilibrium equations and the diagonal concrete strain ε_d in the compatibility equations do not have a realistic relationship based on the constitutive law of concrete. A rigorous solution which satisfies Navier's three principles (equilibrium, compatibility and constitutive laws of materials) will be given by the Mohr compatibility truss model in Section 5.3.

5.2.4.2 Cracking Condition at Yielding of Longitudinal Steel

The strain compatibility equations can also be used for the analysis of cracking condition at the yielding of steel. Two cases should be investigated, namely, the yielding of the longitudinal steel and the yielding of the transverse steel.

As shown in Section 5.2.1, the first two compatibility equations, (5.41) and (5.42), involve four normal strains ε_ℓ, ε_t, ε_r and ε_d, plus the angle α_r. In this analysis, the longitudinal steel strain is given as $\varepsilon_\ell = \varepsilon_y$, and the principal compressive strain ε_d is considered a small given value. Our aim is to express the transverse steel strain ε_t and the cracking strain ε_r as a function of ε_ℓ, ε_d and the angle α_r.

The transverse steel strain ε_t is related to the longitudinal steel strain ε_ℓ by Equation (5.49). Rearranging Equation (5.49) to express ε_t gives

$$\varepsilon_t = \varepsilon_d + (\varepsilon_\ell - \varepsilon_d) \tan^2 \alpha_r \qquad (5.59)$$

Substituting ε_t from Equation (5.59) into Equation (5.45) provides the equation for ε_r:

$$\varepsilon_r = \varepsilon_\ell + (\varepsilon_\ell - \varepsilon_d) \tan^2 \alpha_r \qquad (5.60)$$

Dividing Equations (5.59) and (5.60) by ε_y and setting $\varepsilon_\ell = \varepsilon_y$ we obtain the nondimensional equations for ε_t and ε_r as follows:

$$\frac{\varepsilon_t}{\varepsilon_y} = \frac{\varepsilon_d}{\varepsilon_y} + \left(1 - \frac{\varepsilon_d}{\varepsilon_y}\right) \tan^2 \alpha_r \qquad (5.61)$$

$$\frac{\varepsilon_r}{\varepsilon_y} = 1 + \left(1 - \frac{\varepsilon_d}{\varepsilon_y}\right) \tan^2 \alpha_r \qquad (5.62)$$

The transverse steel strain ratio $\varepsilon_t/\varepsilon_y$ and the cracking strain ratio $\varepsilon_r/\varepsilon_y$ are plotted in Figure 5.7 as a function of the angle α_r according to Equations (5.61) and (5.62), respectively. For each equation a range of $\varepsilon_d/\varepsilon_y$ ratios from 0 to -0.25 is given. It can be seen that the effect of the $\varepsilon_d/\varepsilon_y$ ratio is small. When $\alpha_r = 45°$, Figure 5.7 gives $\varepsilon_t = \varepsilon_y$ and $\varepsilon_r = 2\varepsilon_y$ to $2.25\varepsilon_y$. When α_r is increased to $60°$, ε_t increases to the range of $3\varepsilon_y$ to $3.5\varepsilon_y$, and ε_r increases rapidly to the range of $4\varepsilon_y$ to $4.75\varepsilon_y$. These strains increase even faster when α_r is further increased.

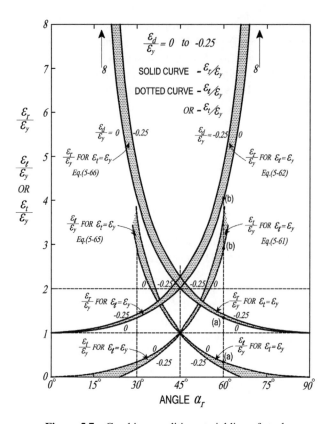

Figure 5.7 Cracking condition at yielding of steel

5.2.4.3 Cracking Condition at Yielding of Transverse Steel

In this case of analysis, the transverse steel strain is given as $\varepsilon_t = \varepsilon_y$, and ε_d is considered a small given value. Our purpose is to express the longitudinal steel strain ε_ℓ and the cracking strain ε_r as a function of ε_t, ε_d and the angle α_r.

The longitudinal steel strain ε_ℓ is related to the transverse steel strain ε_t by Equation (5.49). Rearranging Equation.(5.49) to express ε_ℓ gives

$$\varepsilon_\ell = \varepsilon_d + (\varepsilon_t - \varepsilon_d)\cot^2 \alpha_r \tag{5.63}$$

Substituting ε_ℓ from Equation (5.63) into (5.45) provides the equation for ε_r:

$$\varepsilon_r = \varepsilon_t + (\varepsilon_t - \varepsilon_d)\cot^2 \alpha_r \tag{5.64}$$

Dividing Equations (5.63) and (5.64) by ε_y and setting $\varepsilon_t = \varepsilon_y$ we obtain the nondimensional equations for ε_ℓ and ε_r:

$$\frac{\varepsilon_\ell}{\varepsilon_y} = \frac{\varepsilon_d}{\varepsilon_y} + \left(1 - \frac{\varepsilon_d}{\varepsilon_y}\right)\cot^2 \alpha_r \tag{5.65}$$

$$\frac{\varepsilon_r}{\varepsilon_y} = 1 + \left(1 - \frac{\varepsilon_d}{\varepsilon_y}\right)\cot^2 \alpha_r \tag{5.66}$$

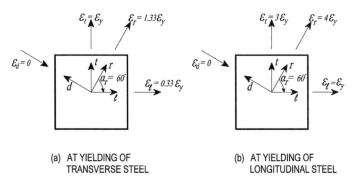

(a) AT YIELDING OF
     TRANSVERSE STEEL

(b) AT YIELDING OF
     LONGITUDINAL STEEL

Figure 5.8   Strain condition when $\alpha_r = 60°$ and $\varepsilon_d = 0$

The longitudinal steel strain ratio $\varepsilon_\ell / \varepsilon_y$ and the cracking strain ratio $\varepsilon_r / \varepsilon_y$ are also plotted in Figure 5.7 as a function of the angle α_r according to Equations (5.65) and (5.66), respectively. For each of these two equations a range of $\varepsilon_d / \varepsilon_y$ ratios from 0 to -0.25 is given. Again, the effect of the $\varepsilon_d / \varepsilon_y$ ratio is shown to be small. When $\alpha_r = 45°$, Figure 5.7 gives $\varepsilon_\ell = \varepsilon_y$, and $\varepsilon_r = 2\varepsilon_y$ to $2.25\varepsilon_y$. When α_r is decreased to $30°$, ε_ℓ increases to the range of $3\varepsilon_y$ to $3.5\varepsilon_y$, and ε_r increases rapidly to the range of $4\varepsilon_y$ to $4.75\varepsilon_y$. These strains increase even faster when α_r is further decreased.

A strain condition is shown in Figure 5.8 when $\alpha_r = 60°$ and $\varepsilon_d = 0$. At the first yielding of the transverse steel (Figure 5.8a) the longitudinal steel strain ε_ℓ is $0.33\varepsilon_y$, and the cracking strain ε_r is $1.33\varepsilon_y$. These values can be obtained from the two points indicated by (a) in Figure 5.7. When straining increases to the stage of yielding in the longitudinal steel (Figure 5.8b) the transverse steel strain ε_t is $3\varepsilon_y$, and the cracking strain ε_r is $4\varepsilon_y$. These values can be obtained from the two points indicated by (b) in Figure 5.7.

The two solid curves for $\varepsilon_r / \varepsilon_y$ in the special case of $\varepsilon_d / \varepsilon_y = 0$ (Figure 5.7) was first presented by Thurlimann (1979). The trend of these two curves shows that the cracking strain ratio $\varepsilon_r / \varepsilon_y$ increases very rapidly after the first yield of steel, when the angle α_r moves away from $45°$. It was, therefore, concluded that the limitation of $30° < \alpha_r < 60°$ is quite valid from the viewpoint of crack control.

5.3 Mohr Compatibility Truss Model (MCTM)

5.3.1 Basic Principles of MCTM

The Bernoulli compatibility truss model has been applied to flexural members in Chapter 3. This bending model satisfies rigorously Navier's three principles of mechanics of materials, namely, the parallel stress equilibrium condition, the Bernoulli linear compatibility condition and the uniaxial constitutive laws. The linear constitutive law (i.e. Hooke's law) is used in Section 3.1, and the nonlinear constitutive law in Section 3.2.

In this chapter we will be dealing with RC 2-D elements. A rigorous analysis of such 2-D elements should also satisfy Navier's three principles of mechanics of materials. In this case of 2-D elements, Navier's principles should consist of the 2-D equilibrium condition, the Mohr circular compatibility condition and the 2-D or biaxial constitutive laws.

However, if the 2-D (or biaxial) constitutive law is replaced by a linear 1-D (or uniaxial) constitutive law, the theory is greatly simplified. This simple model, which is called the *Mohr compatibility truss model*, will be studied in this section. This linear model is applicable up to the service load stage, and could even be used up to the load stage when the steel begins to yield.

If the constitutive laws are based on the nonlinear, softened stress–strain relationships (i.e. 2-D or biaxial), then the theory will be more accurate, but more complex. This nonlinear theory is called the *rotating angle softened truss model* and will be presented in Section 5.4. This nonlinear model should be applicable to the ascending behavioral curve up to the peak load, including both the service load stage and the ultimate load stage.

5.3.2 Summary of Equations

In the Mohr compatibility truss model, the three equilibrium equations used are Equations (5.21)–(5.23), and the three compatibility equations are Equations (5.49), (5.45) and (5.48). They are summarized as follows:

Equilibrium equations

$$\rho_\ell f_\ell = \sigma_\ell + \tau_{\ell t}\ \tan\alpha_r \tag{5.67}$$

$$\rho_t f_t = \sigma_t + \tau_{\ell t}\ \cot\alpha_r \tag{5.68}$$

$$(-\sigma_d) = \tau_{\ell t}\ \frac{1}{\sin\alpha_r\cos\alpha_r} \tag{5.69}$$

Compatibility equations

$$\tan^2\alpha_r = \frac{\varepsilon_t - \varepsilon_d}{\varepsilon_\ell - \varepsilon_d} \tag{5.70}$$

$$\varepsilon_r = \varepsilon_\ell + \varepsilon_t - \varepsilon_d \tag{5.71}$$

$$\frac{\gamma_{\ell t}}{2} = (\varepsilon_r - \varepsilon_d)\ \sin\alpha_r\cos\alpha_r \tag{5.72}$$

The *constitutive laws* used in the Mohr compatibility truss model are Hooke's laws. Hooke's linear stress–strain relationships for steel and concrete are given in 1-D form as follows:

$$\varepsilon_\ell = \frac{f_\ell}{E_s} \tag{5.73}$$

$$\varepsilon_t = \frac{f_t}{E_s} \tag{5.74}$$

$$\varepsilon_d = \frac{\sigma_d}{E_c} \tag{5.75}$$

Similar to flexural members, the tensile stress of concrete is neglected in the design and analysis of cracked 2-D elements, i.e.

$$\sigma_r = 0 \tag{5.76}$$

Equation (5.76) means that the tensile stress–strain relationship of concrete is irrelevant. Then, we have nine equations (5.67)–(5.75) involving fourteen variables. These variables include six stresses (σ_ℓ, σ_t, $\tau_{\ell t}$, σ_d, f_ℓ and f_t), five strains (ε_ℓ, ε_t, $\gamma_{\ell t}$, ε_r and ε_d), two cross-sectional properties (ρ_ℓ and ρ_t) and one geometric parameter α_r. Therefore, five variables must be given before the remaining nine variables can be solved by the nine equations.

In practice, the three stresses σ_ℓ, σ_t, $\tau_{\ell t}$ are obtained from the global analysis of a structure, and are considered given values in the design and analysis of a 2-D element. Therefore, two additional variables must be given before the problem can be solved. Depending on the two additional given variables, the problem is divided into two types:

Types of problem	Given variables	Unknown variables
(a) Analysis:	σ_ℓ σ_t $\tau_{\ell t}$ ρ_ℓ ρ_t	f_ℓ f_t σ_d ε_ℓ ε_t $\gamma_{\ell t}$ ε_r ε_d α_r
(b) Design:	σ_ℓ σ_t $\tau_{\ell t}$ f_ℓ f_t	ρ_ℓ ρ_t σ_d ε_ℓ ε_t $\gamma_{\ell t}$ ε_r ε_d α_r

The problems in the analysis and design of 2-D elements boil down to finding the most efficient way to solve the nine equations. To demonstrate the methodology of the solution process, the nine equations will first be applied to the analysis problem in Sections 5.3.3 and 5.3.4. This will be followed by the application of the nine equations to the design problem in Section 5.3.5.

5.3.3 Solution Algorithm

In order to analyze the stresses and strains in the 2-D element shown in Figure 5.3, the five variables given are the three externally applied stresses σ_ℓ, σ_t and $\tau_{\ell t}$ and the two steel percentages ρ_ℓ and ρ_t. The nine equations (5.67)–(5.75) will then be used to solve the remaining nine unknown variables, including the two steel stresses (f_ℓ and f_t), one concrete stress (ε_d), the five strains (ε_ℓ, ε_t, $\gamma_{\ell t}$, ε_r and ε_d), plus the angle α_r. The problem is summarized below:

Given five variables: σ_ℓ, σ_t, $\tau_{\ell t}$, ρ_ℓ, ρ_t (see Figure 5.3(a))
Find nine unknown variables: f_ℓ, f_t, σ_d, ε_ℓ, ε_t, $\gamma_{\ell t}$, ε_r, ε_d, α_r

From the three equations (5.67)–(5.69), the stresses in the longitudinal steel f_ℓ, transverse steel f_t, and concrete struts σ_d, can be expressed as follows:

$$f_\ell = \frac{\sigma_\ell + \tau_{\ell t} \tan \alpha_r}{\rho_\ell} \tag{5.77}$$

$$f_t = \frac{\sigma_t + \tau_{\ell t} \cot \alpha_r}{\rho_t} \tag{5.78}$$

$$\sigma_d = \frac{-\tau_{\ell t}}{\sin \alpha_r \cos \alpha_r} \tag{5.79}$$

Substituting the stress–strain relationships from Equations (5.73), (5.74) and (5.75) into Equations (5.77), (5.78) and (5.79), respectively, we obtain the strains for the longitudinal steel ε_ℓ, transverse steel ε_t, and concrete element, ε_d:

$$\varepsilon_\ell = \frac{\sigma_\ell + \tau_{\ell t} \tan \alpha_r}{\rho_\ell E_s} \tag{5.80}$$

$$\varepsilon_t = \frac{\sigma_t + \tau_{\ell t} \cot \alpha_r}{\rho_t E_s} \tag{5.81}$$

$$(-\varepsilon_d) = \frac{1}{E_c} \left(\frac{\tau_{\ell t}}{\sin \alpha_r \cos \alpha_r} \right) \tag{5.82}$$

The six equations (5.77)–(5.82) show that all six unknown stresses and strains in the steel reinforcement and concrete struts are expressed in terms of a single unknown variable α_r. This angle α_r can be determined by the compatibility equation (5.70):

$$\tan^2 \alpha_r = \frac{\varepsilon_t - \varepsilon_d}{\varepsilon_\ell - \varepsilon_d} \tag{5.83}$$

Substituting the strains ε_ℓ ε_t and ε_d from Equations (5.80)–(5.82) into Equation (5.83) and simplifying result in:

$$\rho_\ell(1 + \rho_t n)\cot^4 \alpha_r + \frac{\sigma_t}{\tau_{\ell t}}\rho_\ell \cot^3 \alpha_r - \frac{\sigma_\ell}{\tau_{\ell t}}\rho_t \cot \alpha_r - \rho_t(1 + \rho_\ell n) = 0 \tag{5.84}$$

where $n = E_s/E_c$.

The solution of the angle α_r from Equation (5.84) can best be obtained by a trial-and-error procedure. This procedure will be illustrated by an example problem (5.2 in Section 5.3.4).

After the angle α_r is found, the three unknown stresses in the steel bars and concrete struts (f_ℓ, f_t, σ_d) can be found directly from Equations (5.77)–(5.79), and the three corresponding strains (ε_ℓ, ε_t, ε_d) from Hooke's laws (Equations 5.73–5.75). The last two unknown strains (the cracking strain ε_r and the shear strain $\gamma_{\ell t}$) can be calculated from Equations (5.71) and (5.72), respectively.

Now that all the stresses and strains have been calculated, the smeared (or average) strains in RC element and the stresses in the concrete element can be plotted as Mohr circles in Figure 5.9 (b) and (d), respectively. Notice that the rotating angle α_r is the same for these two Mohr circles, because they are both affected by the stresses in the steel grid in Figure 5.9(e). This rotating angle α_r is different from the fixed angle α_1 in the Mohr circle for the RC element (Figure 5.9c). The fixed angle α_1 is calculated directly from the externally applied stresses σ_ℓ, σ_t, $\tau_{\ell t}$.

5.3.4 Example Problem 5.2 using MCTM

5.3.4.1 Statement of Problem

A 2-D RC element has been designed using the equilibrium (plasticity) truss model in Example Problem 5.1, Section 5.1.5 and Figure 5.3. Now we will analyze this element using the Mohr compatibility truss model. This 2-D element is subjected to a set of membrane stresses $\sigma_\ell = 2.13$ MPa (tension), $\sigma_t = -2.13$ MPa (compression) and $\tau_{\ell t} = 3.69$ MPa. For the sake of clarity, these applied stresses are again illustrated in Figure 5.10(a).

Based on the equilibrium (plasticity) truss model, the RC 2-D element is reinforced with 1.03% of steel in both the longitudinal and the transverse directions (i.e. $\rho_\ell = \rho_t = 0.0103$). The steel is designed to have the same grade in both directions $f_{\ell y} = f_{ty} = 413$ MPa, so that the smeared steel stresses $\rho_\ell f_{\ell y}$ and $\rho_t f_{ty} = 4.26$ MPa. In order to use the Mohr compatibility truss model, additional properties of the steel and the concrete are specified as follows: $f_c' = 27.6$ MPa (4000 psi), $E_c = 24\,800$ MPa (3,600 ksi), $E_s = 200\,000$ MPa (29 000 ksi), and $n = 200\,000/24\,800 = 8.06$. Neglect the tensile strength of concrete $\sigma_r = 0$.

Calculate the stress and strain conditions at the first yield of steel. Express all the stresses and strains in terms of Mohr circles.

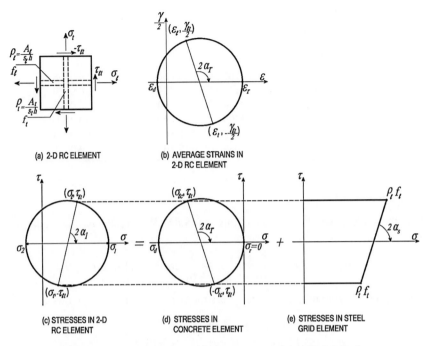

Figure 5.9 Stresses and strains in analysis and design of RC 2-D elements

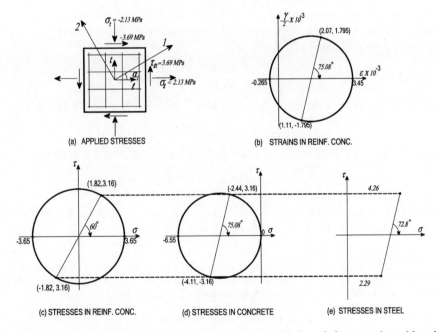

Figure 5.10 Stresses (MPa) in a RC 2-D element at first yield $f_\ell = f_y$ for example problem 5.2

Table 5.1   Trial-and-error method for solution of angle α_r

α_r	$\cot \alpha_r$	$\cot^3 \alpha_r$	$\cot^4 \alpha_r$	Left-hand side of equation
30°	1.7321	5.1961	9.0000	4.6668
35°	1.4281	2.9129	4.1599	0.9170
37.5°	1.3032	2.2134	2.8845	0.0113
37.6°	1.2985	2.1895	2.8431	−0.0170
37.55°	1.3009	2.2014	2.8638	−0.0028
37.54°	1.3013	2.2038	2.8678	−0.0001

5.3.4.2 Solution

First of all, the angle α_r will be solved by Equation (5.84). When $\rho_\ell = \rho_t = \rho$, Equation (5.84) becomes

$$(1 + \rho n) \cot^4 \alpha_r + \frac{\sigma_t}{\tau_{\ell t}} \cot^3 \alpha_r - \frac{\sigma_\ell}{\tau_{\ell t}} \cot \alpha_r - (1 + \rho n) = 0$$

The given property ρn and the external stress ratios $\sigma_\ell / \tau_{\ell t}$ and $\sigma_t / \tau_{\ell t}$ in this equation are:

$$\rho n = 0.0103(8.06) = 0.0833$$

$$\frac{\sigma_\ell}{\tau_{\ell t}} = \frac{2.13}{3.69} = 0.577 \text{ and } \frac{\sigma_t}{\tau_{\ell t}} = \frac{-2.13}{3.69} = -0.577$$

Therefore, $1.0833 \cot^4 \alpha_r - 0.577 \cot^3 \alpha_r - 0.577 \cot \alpha_r - 1.0833 = 0$.
This equation can be easily solved by a trial-and-error procedure using Table 5.1.
For $\alpha_r = 37.54°$, $\tan \alpha_r = 0.7684$, $\cot \alpha_r = 1.3013$, and $\sin \alpha_r \cos \alpha_r = 0.4831$.
The stresses are calculated as follows:

Equation (5.69) $\sigma_d = \dfrac{-\tau_{\ell t}}{\sin \alpha_r \cos \alpha_r} = \dfrac{-3.69}{0.4831} = -7.64 \text{ MPa}$

Equation (5.67) $\rho_\ell f_\ell = \sigma_\ell + \tau_{\ell t} \tan \alpha_r = 2.13 + 3.69(0.7684) = 4.97 \text{ MPa}$

$$f_\ell = \frac{4.97}{0.0103} = 482 \text{ MPa} > 413 \text{ MPa (yield)}$$

Equation (5.68) $\rho_t f_t = \sigma_t + \tau_{\ell t} \cot \alpha_r = -2.13 + 3.69(1.3013) = 2.67 \text{ MPa}$

$$f_t = \frac{2.67}{0.0103} = 259 \text{ MPa} < 413 \text{ MPa (yield)}$$

It can be seen that the longitudinal steel stress has exceeded the yield point while the transverse steel stress is still in the elastic range.

First yield of steel
The first yield of the longitudinal steel should occur when $f_\ell = f_y = 413 \text{ MPa}$. It will occur when the applied stresses are reduced by a factor of $413/482 = 0.857$, i.e.

$$\sigma_\ell = 0.857 (2.13) = 1.82 \text{ MPa}$$
$$\sigma_t = 0.857 (-2.13) = -1.82 \text{ MPa}$$
$$\tau_{\ell t} = 0.857 (3.69) = 3.16 \text{ MPa}$$

These applied stresses at first yield are recorded in terms of the Mohr circle in Figure 5.10(c). Accordingly, the stresses in the steel are:

$$f_\ell = 0.857\,(482) = 413 \text{ MPa}$$
$$f_t = 0.857\,(259) = 222 \text{ MPa}$$
$$\rho_\ell\, f_\ell = 0.857\,(4.97) = 4.26 \text{ MPa}$$
$$\rho_t\, f_t = 0.857\,(2.67) = 2.29 \text{ MPa}$$

These stresses for the steel are recorded in Figure 5.10(e). From the applied stresses and the stresses in the steel we can calculate the stresses in the concrete element at first yield:

$$\sigma_\ell - \rho_\ell\, f_\ell = 1.82 - 4.26 = -2.44 \text{ MPa}$$
$$\sigma_t - \rho_t\, f_t = -1.82 - 2.29 = -4.11 \text{ MPa}$$
$$\sigma_d = 0.857\,(-7.64) = -6.55 \text{ MPa}$$

These stresses for the concrete are used to plot the Mohr circle in Figure 5.10(d). Finally, the strains are calculated as follows:

Equation (5.75) $\quad \varepsilon_d = \dfrac{\sigma_d}{E_c} = \dfrac{-6.55}{24{,}800} = -0.265 \times 10^{-3}$

Equation (5.73) $\quad \varepsilon_\ell = \dfrac{f_\ell}{E_s} = \dfrac{413}{200{,}000} = 2.07 \times 10^{-3}$

Equation (5.74) $\quad \varepsilon_t = \dfrac{f_t}{E_s} = \dfrac{222}{200{,}000} = 1.11 \times 10^{-3}$

Equation (5.71) $\quad \varepsilon_r = \varepsilon_\ell + \varepsilon_t - \varepsilon_d = (2.07 + 1.11 + 0.265) \times 10^{-3} = 3.45 \times 10^{-3}$

Equation (5.72) $\quad \dfrac{\gamma_{\ell t}}{2} = (\varepsilon_r - \varepsilon_d)\, \sin\alpha_r \cos\alpha_r$

$$= (3.45 + 0.265) \times 10^{-3}(0.4831) = 1.795 \times 10^{-3}$$

These strains are given in terms of the Mohr's strain circle in Figure 5.10(b). It should be observed that the α_r value of 37.54° for the strain condition is identical to that for the stress condition in the concrete element, because the same r–d principal coordinate is used for both stresses and strains.

Figure 5.10(b)–(d) shows the stresses and strains of the 2-D element at the first yield of longitudinal steel. The set of externally applied stresses at this load stage is: $\sigma_\ell = 1.82$ MPa (tension), $\sigma_t = -1.82$ MPa (compression) and $\tau_{\ell t} = 3.16$ MPa. This set of applied stresses at first yield is 0.857 times the set of applied stresses shown in Figure 5.10(a). The principal stresses, σ_1 and σ_2, in Figure 5.10(c) are calculated as follows:

$$\sigma_1 = \frac{\sigma_\ell + \sigma_t}{2} + \sqrt{\left(\frac{\sigma_\ell - \sigma_t}{2}\right)^2 + \tau_{\ell t}^2} = \frac{1.82 - 1.82}{2} + \sqrt{\left(\frac{1.82 + 1.82}{2}\right)^2 + 3.16^2}$$
$$= 0 + 3.65 = 3.65 \text{ MPa}$$
$$\sigma_2 = 0 - 3.65 = -3.65 \text{ MPa}$$

Below the first yield load stage the steel and concrete are assumed to obey Hooke's law and the linear behavior of the 2-D element can be predicted by the *Mohr compatibility truss model*. The angle α_r of 37.54° is constant from the cracking load up to the first yield load.

With the application of increasing proportional stresses, however, Hooke's law will no longer be applicable to steel and concrete, and the angle α_r will rotate and decrease. The nonlinear behavior of the 2-D element can then be predicted by the *rotating angle softened truss model* given in Section 5.4. This rotating angle α_r will reach a final value of 30° when both the longitudinal and the transverse steel yield. This plastic load stage at total yield has been treated in Section 5.1.5, Example 5.1, using the *equilibrium (plasticity) truss model*.

5.3.5 Allowable Stress Design of RC 2-D Elements

We will now study the design of reinforcement in the RC 2-D element shown in Figure 5.9. The five variables given in this case are the three applied stresses σ_ℓ, σ_t, $\tau_{\ell t}$, and the two steel stresses f_ℓ and f_t. For design purposes, f_ℓ and f_t will be specified as the allowable stresses $f_{\ell a}$ and f_{ta}. According to the ACI Code Commentary $f_{\ell a}$ and f_{ta} are 138 MPa (20 ksi) for Grade 40 and 50 steel and 165 MPa (24 ksi) for Grade 60 and higher strength steel.

The nine equations (5.67)–(5.75), will then be used to solve the remaining nine unknown variables, including the two steel ratios ρ_ℓ and ρ_t, the one concrete stress σ_d, the five strains ε_ℓ, ε_t, $\gamma_{\ell t}$, ε_r, ε_d, plus the angle α_r. The problem is summarized below:

Given five variables: σ_ℓ, σ_t, $\tau_{\ell t}$, $f_\ell = f_{\ell a}$, $f_t = f_{ta}$
Find nine unknown variables: ρ_ℓ, ρ_t, σ_d, ε_ℓ, ε_t, $\gamma_{\ell t}$, ε_r, ε_d, α_r

Similar to the analysis problem in Sections 5.3.3 and 5.3.4, the solution of the nine equations starts with expressing the three equilibrium equations (5.21)–(5.23) by the second type of equations and setting $\sigma_r = 0$:

$$\rho_\ell = \frac{\sigma_\ell + \tau_{\ell t} \tan \alpha_r}{f_{\ell a}} \tag{5.85}$$

$$\rho_t = \frac{\sigma_t + \tau_{\ell t} \cot \alpha_r}{f_{ta}} \tag{5.86}$$

$$\sigma_d = \frac{-\tau_{\ell t}}{\sin \alpha_r \cos \alpha_r} \tag{5.87}$$

From the linear stress–strain relationships of steel (Equations 5.73 and 5.75), we have

$$\varepsilon_\ell = \frac{f_{\ell a}}{E_s} \tag{5.88}$$

$$\varepsilon_t = \frac{f_{ta}}{E_s} \tag{5.89}$$

$$\varepsilon_d = \frac{\sigma_d}{E_c} \tag{5.90}$$

Substituting Equation (5.87) into the stress–strain equation of compression concrete, (5.90), gives:

$$\varepsilon_d = \frac{1}{E_c} \left(\frac{-\tau_{\ell t}}{\sin \alpha_r \cos \alpha_r} \right) \tag{5.91}$$

It can be seen that the strains ε_ℓ and ε_t are constants from Equations (5.88) and (5.89). The other variables ρ_ℓ ρ_t σ_d and ε_d from Equations (5.85)–(5.87) and (5.91), are all expressed in terms of a single unknown variable α_r. Substituting the strains ε_ℓ, ε_t and ε_d from Equations (5.88), (5.89) and (5.91) into Equation (5.70) results in

$$\cot^2 \alpha_r = \frac{f_{\ell a}\sin\alpha_r\cos\alpha_r + n\tau_{\ell t}}{f_{ta}\sin\alpha_r\cos\alpha_r + n\tau_{\ell t}} \tag{5.92}$$

The angle α_r can be solved by Equation (5.92) using a trial-and-error method.

Examination of Equation (5.92) reveals two observations. First, when the allowable stresses in the two directions are equal ($f_{\ell a} = f_{ta}$), the Mohr compatibility truss model recommends the use of an angle $\alpha_r = 45°$. Second, the second terms in both the numerator and denominator are smaller than the first terms by an order of magnitude. If the two $n\tau_{\ell t}$ terms are neglected in Equation (5.92), then

$$\cot^2 \alpha_r = \frac{f_{\ell a}}{f_{ta}} \tag{5.93}$$

The angle α_r obtained from Equation (5.93) should be a close approximation to that obtained from Equation (5.92) and becomes exact when $f_{\ell a} = f_{ta}$. Once the angle α_r is determined, all the nine unknown variables ρ_ℓ, ρ_t, σ_d, ε_ℓ, ε_t, $\gamma_{\ell t}$, ε_r, ε_d, α_r, can be calculated.

5.4 Rotating Angle Softened Truss Model (RA-STM)

5.4.1 Basic Principles of RA-STM

In Section 5.3 we have studied the Mohr compatibility truss model which utilizes nine equations to analyze the behavior of RC 2-D elements subjected to external membrane stresses. The nine equations include three for equilibrium, three for compatibility and three for the constitutive laws of materials. Since the constitutive equations for both concrete and steel are based on Hooke's law, the predicted behavior of the 2-D element is linear. The Mohr compatibility truss model, therefore, provides a method of linear analysis.

In this section we will introduce a method of nonlinear analysis for RC 2-D elements, which also satisfies Navier's three principles of mechanics of materials. In this model the constitutive equations are based on the actual, observed stress–strain relationships of concrete and steel. The stress–strain curve of concrete must reflect two characteristics. The first is the nonlinear relationship between stress and strain. The second, and perhaps more important, is the softening of concrete in compression, caused by cracking due to tension in the perpendicular direction. Consequently, a softening coefficient will be incorporated in the equation for the compressive stress–strain relationship of concrete.

In view of the crucial importance of the softening effect on the biaxial constitutive laws of reinforced concrete, this model has been named the 'softened truss model'. The word 'softened' implies two characteristics: first, the analysis must be nonlinear and, second, the softening of concrete must be taken into account.

In Section 5.4 we will extend the applicability of the rotating angle softened truss model to prestressed concrete (PC). The three equilibrium equations will include the terms $\rho_{\ell p} f_{\ell p}$ and $\rho_{tp} f_{tp}$ that represent the smeared stresses of the prestressing steel, as shown in

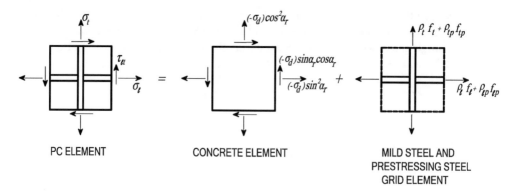

Figure 5.11   Stress condition in prestressed concrete

Figure 5.11. Detailed derivation of these equations can proceed in the same manner as for reinforced concrete. If the prestressing steel is absent, i.e. $\rho_{\ell p} = \rho_{tp} = 0$, these equations, of course, degenerate into those for reinforced concrete.

Similar to the bending theory in Chapter 3, it is very convenient to neglect the tensile stress of concrete, i.e. $\sigma_r = 0$, in the rotating angle shear theory. As a result, the stress–strain relationship of concrete in tension becomes irrelevant.

5.4.2 Summary of Equations

The three stress equilibrium equations for a PC 2-D element and the three strain compatibility equations for the corresponding element are first summarized. The three equilibrium equations are taken from the first type of expression in Section 5.1.2, and the three compatibility equations are taken from the transformation expression in Section 5.2.1.

Equilibrium equations

$$\sigma_\ell - \rho_\ell f_\ell - \rho_{\ell p} f_{\ell p} = \sigma_d \sin^2 \alpha_r \qquad (5.94) \text{ or } \boxed{1}$$

$$\sigma_t - \rho_t f_t - \rho_{tp} f_{tp} = \sigma_d \cos^2 \alpha_r \qquad (5.95) \text{ or } \boxed{2}$$

$$\tau_{\ell t} = (-\sigma_d) \sin \alpha_r \cos \alpha_r \qquad (5.96) \text{ or } \boxed{3}$$

where

$\rho_{\ell p}, \rho_{tp}$ = prestressing steel ratios in the ℓ- and t-directions, respectively,
$f_{\ell p}, f_{tp}$ = stresses in prestressing steel in the ℓ- and t-directions, respectively.

Compatibility equations

$$\varepsilon_\ell = \varepsilon_r \cos^2 \alpha_r + \varepsilon_d \sin^2 \alpha_r \qquad (5.97) \text{ or } \boxed{4}$$

$$\varepsilon_t = \varepsilon_r \sin^2 \alpha_r + \varepsilon_d \cos^2 \alpha_r \qquad (5.98) \text{ or } \boxed{5}$$

$$\frac{\gamma_{\ell t}}{2} = (\varepsilon_r - \varepsilon_d) \sin \alpha_r \cos \alpha_r \qquad (5.99) \text{ or } \boxed{6}$$

It should be pointed out that a new sequence of equation numbers, $\boxed{1}$ to $\boxed{6}$, has been introduced. Each of these equation numbers is simply enclosed in a box without the chapter designation. These fundamental equations will frequently be referred to, especially in the example problems.

The solution of these six equilibrium and compatibility equations require five stress–strain relationships of materials: one relating σ_d to ε_d for the concrete struts, two relating f_ℓ to ε_ℓ and f_t to ε_t for the mild steel in the longitudinal and transverse directions, and two relating $f_{\ell p}$ to $\varepsilon_{\ell p}$ and f_{tp} to ε_{tp} for the prestressed strands in the two directions.

Constitutive law of concrete in compression
Ascending branch

$$\sigma_d = \zeta f_c' \left[2\left(\frac{\varepsilon_d}{\zeta \varepsilon_o}\right) - \left(\frac{\varepsilon_d}{\zeta \varepsilon_o}\right)^2 \right] \quad \varepsilon_d/\zeta \varepsilon_o \leq 1 \qquad (5.100) \text{ or } \boxed{7a}$$

Descending branch

$$\sigma_d = \zeta f_c' \left[1 - \left(\frac{(\varepsilon_d/\zeta \varepsilon_o) - 1}{(2/\zeta) - 1}\right)^2 \right] \quad \varepsilon_d/\zeta \varepsilon_o > 1 \qquad (5.101) \text{ or } \boxed{7b}$$

$$\zeta = \frac{0.9}{\sqrt{1 + 600\varepsilon_r}} \qquad (5.102) \text{ or } \boxed{8}$$

Constitutive law of mild steel

$$f_\ell = E_s \varepsilon_\ell \quad \varepsilon_\ell < \varepsilon_{\ell y} \qquad (5.103) \text{ or } \boxed{9a}$$
$$f_\ell = f_{\ell y} \quad \varepsilon_\ell \geq \varepsilon_{\ell y} \qquad (5.104) \text{ or } \boxed{9b}$$
$$f_t = E_s \varepsilon_t \quad \varepsilon_t < \varepsilon_{ty} \qquad (5.105) \text{ or } \boxed{10a}$$
$$f_t = f_{ty} \quad \varepsilon_t \geq \varepsilon_{ty} \qquad (5.106) \text{ or } \boxed{10b}$$

where $E_s = 200\,000$ MPa (29 000 000 psi).

Constitutive law of prestressing steel

$$f_p \leq 0.7 f_{pu} \quad f_p = E_{ps}(\varepsilon_{dec} + \varepsilon_s) \qquad (5.107) \text{ or } \boxed{11a}\,\boxed{12a}$$

$$f_p > 0.7 f_{pu} \quad f_p = \frac{E_{ps}'(\varepsilon_{dec} + \varepsilon_s)}{\left[1 + \left\{\dfrac{E_{ps}'(\varepsilon_{dec} + \varepsilon_s)}{f_{pu}}\right\}^m\right]^{\frac{1}{m}}} \qquad (5.108) \text{ or } \boxed{11b}\,\boxed{12b}$$

where

  f_p = stress in prestressing steel – f_p becomes $f_{\ell p}$ or f_{tp} when applied to the longitudinal and transverse steel, respectively;

  ε_s = strain in the mild steel – ε_s becomes ε_ℓ or ε_t, when applied to the longitudinal and transverse steel, respectively;

ε_{dec} = strain in prestressing steel at decompression of concrete;
E_{ps} = elastic modulus of prestressed steel, taken as 200 000 MPa (29 000 ksi);
E'_{ps} = tangential modulus of Ramberg–Osgood curve at zero load, taken as 214 000 MPa
(31 060 ksi);
f_{pu} = ultimate strength of prestressing steel;
m = shape parameter (taken as 4).

The strain in prestressing steel at decompression of concrete, ε_{dec}, is considered a known value and can be determined as follows:

$$\varepsilon_{dec} = \varepsilon_{pi} + \varepsilon_i$$

where

ε_{pi} = initial strain in prestressed steel after loss;
ε_i = initial strain in mild steel after loss.

ε_{dec} is approximately equal to 0.005 for grades 1723 MPa (250 ksi) and 1862 MPa (270 ksi) prestressing strands.

The background and application of these constitutive laws are briefly described below:

1. The softened stress–strain curve of concrete in compression, Equations $\boxed{7a}$ and $\boxed{7b}$, are used to relate σ_d to ε_d. This curve is shown in Figure 5.12 and compared with the nonsoftened stress–strain curve. The expressions of the ascending parabolic curve, Equation (5.100) or $\boxed{7a}$, and the descending parabolic curve, Equation (5.101) or $\boxed{7b}$, are explained in Section 6.1.6 (Chapter 6). The softened coefficient ζ which is expressed by Eq. (5-102) or Eq. $\boxed{8}$, is plotted in Figure 5.13 as a function of the tensile strain of concrete ε_r, using a very conservative lower bound of the test data given by Belarbi and Hsu (1995). This ζ coefficient is applicable to normal strength concrete. However, when high strength concrete is used, it is advisable to multiply ζ by the Function of Concrete Strength, $f_2\left(f'_c\right)$, given in Section 6.1.7.2 (Chapter 6).

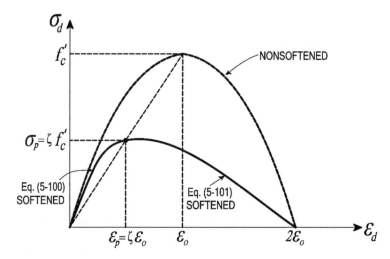

Figure 5.12 Compressive stress–strain curve of concrete

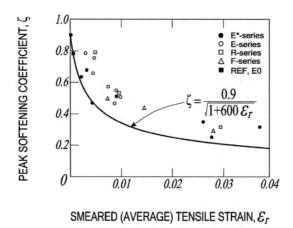

Figure 5.13 Peak softening coefficient of concrete in compression

2. The stress–strain curve of mild steel bars, Equations $\boxed{9a}$ and $\boxed{9b}$ or $\boxed{10a}$ and $\boxed{10b}$, are plotted in Figure 5.14. Each stress–strain curve, made up of a linear portion and a horizontal yield plateau, represents the character of a bare mild steel bar. This simple elastic–perfectly plastic stress–strain curve is accurate in predicting shear strengths, when the tensile strength of concrete is neglected ($\sigma_r = 0$). This combination of steel and concrete material models represents the stress condition at the cracks correctly.
3. The stress–strain curve of prestressing steel strands, Equations $\boxed{11}$ and $\boxed{12}$, is plotted in Figure 5.15. The Ramberg–Osgood curve is suitable to closely approximate an experimental curve that resembles two straight lines connected by a curved knee. The curvature of the knee is controlled by the constant m.

The first almost straight portion of the Ramberg–Osgood curve is asymptotic to the inclined dotted straight lines with a slope of E'_{ps} at the origin. The modulus E'_{ps}, which is slightly greater than E_{ps}, is determined such that the Ramberg–Osgood curve coincides with the elastic straight line (with slope of E_{ps}) at the point of $0.7 f_{pu}$.

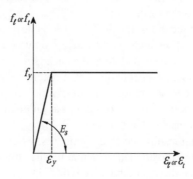

Figure 5.14 Stress–strain relationship of mild steel bars

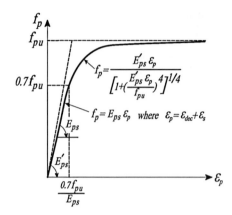

Figure 5.15   Stress–strain relationship of prestressing strands

It is interesting to observe the character of the Ramberg–Osgood curve in two ways. When the first term, 1, in the denominator of Equation $\boxed{11b}$ or $\boxed{12b}$ is omitted, the curve degenerates into the horizontal, asymptotic, straight line $f_p = f_{pu}$. When the second term, $\left\{ \left(E'_{ps}(\varepsilon_{dec} + \varepsilon_s)/f_{pu} \right) \right\}^m$, in the denominator of Equation $\boxed{11b}$ or $\boxed{12b}$ is omitted, the curve degenerates into the asymptotic straight line $f_p = E'_{ps}(\varepsilon_{dec} + \varepsilon_s)$, with a slope of E'_{ps}.

5.4.3 Solution Algorithm

The twelve governing equations for a PC 2-D element, Equations $\boxed{1}$ to $\boxed{12}$, contain fifteen unknown variables. These fifteen unknown variables include eight stresses $(\sigma_\ell, \sigma_t, \tau_{\ell t}, \sigma_d, f_\ell, f_t, f_{\ell p}, f_{tp})$ and five strains $(\varepsilon_\ell, \varepsilon_t, \gamma_{\ell t}, \varepsilon_r, \varepsilon_d)$, as well as the angle α_r and the material coefficient ζ. If three unknown variables are given, then the remaining twelve unknown variables can be solved using the twelve equations.

If the stresses and strains are required throughout the post-cracking loading history, then two variables must be given, and a third variable selected. In general, ε_d is selected as the third variable because it varies monotonically from zero to a maximum. For each given value of ε_d, the remaining twelve unknown variables can be solved. The series of solutions for a sequence of ε_d values allows us to trace the loading history.

In most of the structural applications, the two given variables come from the three externally applied stresses σ_ℓ, σ_t and $\tau_{\ell t}$. They are generally given in two ways. First, the applied normal stresses σ_ℓ and σ_t, are given as constants, while the shear stress $\tau_{\ell t}$ is a variable to be solved. This type of problem occurs in nuclear containment structures. The stresses σ_ℓ and σ_t in a wall element are induced by the internal pressure and are given as constants, while the shear stress $\tau_{\ell t}$ is caused by earthquake motion and is considered as an unknown variable. This first type of problem is treated in Sections 5.4.3 and 5.4.4.

In the second type of relationship, the three applied stresses σ_ℓ, σ_t and $\tau_{\ell t}$ increase proportionally. They are related to the applied principal tensile stress σ_1 by given constants (see Section 5.4.5). In this way, the three variables are reduced to one variable. This type of problem occurs in all elastic structures where the three stresses on a wall element are produced simultaneously by an increasing load. This second type of problem will be treated in Sections 5.4.5–5.4.7.

5.4.3.1 Characteristics of Equations

An efficient algorithm was discovered from a careful observation of the three characteristics of the twelve equations:

1. Equations $\boxed{3}$ and $\boxed{6}$ are independent, because each contains one variable, $\tau_{\ell t}$ and $\gamma_{\ell t}$, respectively, which are not involved in any other equations.
2. Equations $\boxed{7}$ and $\boxed{8}$ for the concrete material law involve only three unknown variables in the d–r coordinate (σ_d, ε_d, and ε_r). If the strains ε_d and ε_r are given, then the stresses σ_d can be calculated from these two equations.
3. The longitudinal steel stresses f_ℓ and $f_{\ell p}$ in equilibrium Equation $\boxed{1}$ are coupled to the longitudinal steel strain ε_ℓ in compatibility Equation $\boxed{4}$ through the longitudinal steel stress–strain relationship of Equations $\boxed{9}$ and $\boxed{11}$. Similarly, the transverse steel stresses f_t and f_{tp} in equilibrium Equation $\boxed{2}$ are coupled to the transverse steel strain ε_t in compatibility Equation $\boxed{5}$ through the transverse steel stress–strain relationships of Equation $\boxed{10}$ and $\boxed{12}$.

5.4.3.2 ε_ℓ as a Function of f_ℓ and $f_{\ell p}$

In view of characteristic 3, the longitudinal steel strain ε_ℓ can be expressed directly as a function of the longitudinal steel stresses f_ℓ and $f_{\ell p}$ by eliminating the angle α_r from Equations $\boxed{1}$ and $\boxed{4}$. Inserting $\varepsilon_r \cos^2 \alpha_r = \varepsilon_r - \varepsilon_r \sin^2 \alpha_r$ into Equation $\boxed{4}$ gives

$$\sin^2 \alpha_r = \frac{\varepsilon_r - \varepsilon_\ell}{\varepsilon_r - \varepsilon_d} \tag{5.109}$$

Substituting Equation (5.109) into Equation $\boxed{1}$ results in

$$\varepsilon_\ell = \varepsilon_r + \frac{\varepsilon_r - \varepsilon_d}{-\sigma_d} \left(\sigma_\ell - \rho_\ell f_\ell - \rho_{\ell p} f_{\ell p} \right) \tag{5.110 or $\boxed{13}$}$$

ε_ℓ, f_ℓ, and $f_{\ell p}$ in Equation $\boxed{13}$ can be solved simultaneously with the two stress–strain relationships of Equations $\boxed{9}$ and $\boxed{11}$ for longitudinal steel.

5.4.3.3 ε_t as a Function of f_t and f_{tp}

Similarly, the transverse steel strain ε_t can be expressed directly as a function of the transverse steel stresses f_t and f_{tp} by eliminating the angle α_r from Equations $\boxed{2}$ and $\boxed{5}$. Inserting $\varepsilon_r \sin^2 \alpha_r = \varepsilon_r - \varepsilon_r \cos^2 \alpha_r$ into Equation $\boxed{5}$ gives

$$\cos^2 \alpha_r = \frac{\varepsilon_r - \varepsilon_t}{\varepsilon_r - \varepsilon_d} \tag{5.111}$$

Substituting Equation (5.111) into Equation $\boxed{2}$ results in

$$\varepsilon_t = \varepsilon_r + \frac{\varepsilon_r - \varepsilon_d}{-\sigma_d} \left(\sigma_t - \rho_t f_t - \rho_{tp} f_{tp} \right) \qquad (5.112) \text{ or } \boxed{14}$$

The variables ε_t, f_t, and f_{tp} in Equation $\boxed{14}$ can be solved simultaneously with the two stress–strain relationships of Equations $\boxed{10}$ and $\boxed{12}$ for transverse steel.

5.4.3.4 ε_r and α_r as Functions of ε_ℓ, ε_t and σ_d

The following two compatibility equations are more convenient to use in the solution procedure described in Section 5.4.3.5. From Equations (5.45) and (5.49) we have:

$$\varepsilon_r = \varepsilon_\ell + \varepsilon_t - \varepsilon_d \qquad (5.113) \text{ or } \boxed{15}$$

$$\tan^2 \alpha_r = \frac{\varepsilon_t - \varepsilon_d}{\varepsilon_\ell - \varepsilon_d} \qquad (5.114) \text{ or } \boxed{16}$$

5.4.3.5 Solution Procedures

A set of solution procedures is proposed, as shown in the flow chart of Figure 5.16. The procedures are described as follows:

Step 1: Select a value of strain in the d-direction ε_d.

Step 2: Assume a value of strain in the r-direction ε_r.

Step 3: Calculate the softened coefficient ζ and the concrete stresses σ_d from Equations $\boxed{8}$ and $\boxed{7}$, respectively.

Step 4: Solve the strains and stresses in the longitudinal steel ε_ℓ, f_ℓ and $f_{\ell p}$ from Equations $\boxed{13}$, $\boxed{9}$ and $\boxed{11}$, and those in the transverse steel ε_t, f_t, and f_{tp} from Equations $\boxed{14}$, $\boxed{10}$, and $\boxed{12}$.

Step 5: Calculate the strain $\varepsilon_r = \varepsilon_\ell + \varepsilon_t - \varepsilon_d$ from Equation $\boxed{15}$. If ε_r is the same as assumed, the values obtained for all the strains are correct. If ε_r is not the same as assumed, then another value of ε_r is assumed, and steps (3) to (5) are repeated.

Step 6: Calculate the angle α_r, the shear stress $\tau_{\ell t}$ and the shear strain $\gamma_{\ell t}$ from Equations $\boxed{16}$, $\boxed{3}$, and $\boxed{6}$, respectively. This will provide one point on the $\tau_{\ell t} - \gamma_{\ell t}$ curve.

Step 7: Select another value of ε_d and repeat steps (2) to (6). Calculation for a series of ε_d values will provide the whole $\tau_{\ell t} - \gamma_{\ell t}$ curve.

The above solution procedures have two distinct advantages. First, the variable angle α_r does not appear in the iteration process from Step 2 to Step 5. Second, the calculation of ε_ℓ and ε_t in Step 4 can easily accommodate the nonlinear stress–strain relationships of reinforcing steel, including those for prestressing strands. These advantages were derived from an understanding of the three characteristics (Section 5.4.3.1) of the twelve governing equations. Steps 1 to 3 are proposed because of characteristic 2. Steps 4 and 5 are the results of characteristic 3, and Step 6 is possible based on characteristic 1.

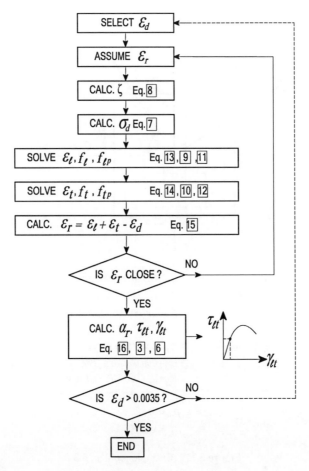

Figure 5.16   Flow chart showing the solution algorithm (Constant Normal Stresses)

5.4.4 Example Problem 5.3 for Sequential Loading

5.4.4.1 Problem statement

A nuclear containment vessel is reinforced with 1.2% of mild steel in both directions and is post-tensioned only in the longitudinal direction by 0.3% of bonded prestressing steel. It is subjected to an internal pressure and a horizontal shear force due to earthquake loading. An element taken from the vessel is shown in Figure 5.17. It is subjected to a longitudinal stress $\sigma_\ell = 3.44$ MPa (500 psi) and a transverse stress $\sigma_t = 1.72$ MPa (250 psi) due to the internal pressure. The earthquake loading induces a shear stress $\tau_{\ell t}$. Analyze the response of the element under the shear stress $\tau_{\ell t}$, i.e. plot the $\tau_{\ell t} - \gamma_{\ell t}$ curve. Use the compressive stress–strain curve of softened concrete expressed by Equations 7 and 8, the stress–strain curve of bare mild steel bars expressed by Equations 9 and 10, and the stress–strain relationships of prestressing steel expressed by 11 and 12.

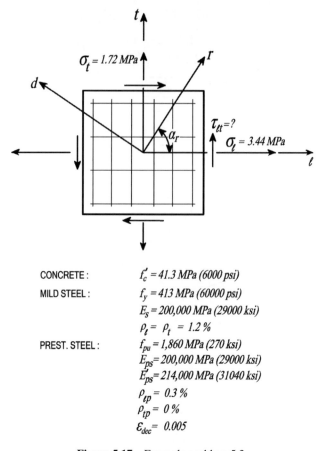

CONCRETE : $f'_c = 41.3\,MPa\,(6000\,psi)$

MILD STEEL : $f_y = 413\,MPa\,(60000\,psi)$

                            $E_s = 200,000\,MPa\,(29000\,ksi)$

                            $\rho_\ell = \rho_t = 1.2\,\%$

PREST. STEEL : $f_{pu} = 1,860\,MPa\,(270\,ksi)$

                            $E_{ps} = 200,000\,MPa\,(29000\,ksi)$

                            $E'_{ps} = 214,000\,MPa\,(31040\,ksi)$

                            $\rho_{\ell p} = 0.3\,\%$

                            $\rho_{tp} = 0\,\%$

                            $\varepsilon_{dec} = 0.005$

Figure 5.17   Example problem 5.3

5.4.4.2 Solution

1. Select $\varepsilon_d = -0.0002$

After several cycles of the trial-and-error process we assume

$$\varepsilon_r = 0.00385$$

Equation $\boxed{8}$ $\zeta = \dfrac{0.9}{\sqrt{1 + 600\varepsilon_r}} = \dfrac{0.9}{\sqrt{1 + 600(0.00385)}} = 0.495$

$\dfrac{\varepsilon_d}{\zeta\varepsilon_o} = \dfrac{-0.0002}{0.495(-0.002)} = 0.202 < 1$ ascending branch

Equation $\boxed{7a}$ $\sigma_d = \zeta f'_c \left[2\left(\dfrac{\varepsilon_d}{\zeta\varepsilon_o}\right) - \left(\dfrac{\varepsilon_d}{\zeta\varepsilon_o}\right)^2 \right]$

$$= 0.495(-41.3)[2(0.202) - (0.202)^2] = -7.43\,MPa$$

Solve the longitudinal steel strain ε_ℓ:

$$\frac{\varepsilon_r - \varepsilon_d}{-\sigma_d} = \frac{0.00385 + 0.0002}{7.43} = 0.545 \times 10^{-3}/\text{MPa}$$

Equation $\boxed{13}$ $\varepsilon_\ell = \varepsilon_r + \dfrac{\varepsilon_r - \varepsilon_d}{-\sigma_d} \left(\sigma_\ell - \rho_\ell f_\ell - \rho_{\ell p} f_{\ell p} \right)$

$$= 0.00385 + 0.545 \times 10^{-3} \, (3.44 - 0.012 f_\ell - 0.003 f_{\ell p})$$

Assume $f_\ell = E_s \varepsilon_\ell = 200 \times 10^3 \varepsilon_\ell$ before yielding

$$f_{\ell p} = E_{ps}(\varepsilon_{dec} + \varepsilon_s) = 200 \times 10^3 (0.005 + \varepsilon_\ell) \text{ before elastic limit}$$

Then $\varepsilon_\ell + 1.308 \ \varepsilon_\ell + 0.327 \ \varepsilon_\ell = 0.00385 + 0.00188 - 0.00164$

$$\varepsilon_\ell = \frac{0.00409}{2.635} = 0.00155 < 0.00207(\varepsilon_y) \text{ OK}$$

$(\varepsilon_{ps} \text{ at } 0.7 \ f_{pu}) = 0.7(1862)/200\ 000 = 0.00652$

$(\varepsilon_{ps} \text{ at } 0.7 f_{pu}) - \varepsilon_{dec} = 0.00652 - 0.005 = 0.00152$

$$\varepsilon_\ell = 0.00155 \approx 0.00152 \text{ OK}$$

Solve the transverse steel strain ε_t:

Equation $\boxed{14}$ $\varepsilon_t = \varepsilon_r + \dfrac{\varepsilon_r - \varepsilon_d}{-\sigma_d} \left(\sigma_t - \rho_t f_t - \rho_{tp} f_{tp} \right)$

$$= 0.00385 + 0.545 \times 10^{-3} \, (1.72 - 0.012 f_t - 0)$$

Assume $f_t = 200 \times 10^3 \varepsilon_t$ before yielding

Then $\varepsilon_t + 1.308 \ \varepsilon_t = 0.00385 + 0.000938$

$$\varepsilon_t = \frac{0.00479}{2.308} = 0.00208 \approx 0.00207 \ (\varepsilon_y) \text{ OK}$$

The tensile strain ε_r can now be checked:

Equation $\boxed{15}$ $\quad \varepsilon_r = \varepsilon_\ell + \varepsilon_t - \varepsilon_d = 0.00155 + 0.00208 + 0.0002$

$$= 0.00383 \approx 0.00385 \text{ (assumed) OK}$$

Now that the strains ε_d, ε_r, ε_ℓ, and ε_t and the stress, σ_d, are calculated from the trial-and-error process, we can determine the angle α_r, the shear stress $\tau_{\ell t}$ and the shear strain $\gamma_{\ell t}$:

Equation $\boxed{16}$ $\tan^2 \alpha_r = \dfrac{\varepsilon_t - \varepsilon_d}{\varepsilon_\ell - \varepsilon_d} = \dfrac{0.00208 + 0.0002}{0.00155 + 0.0002} = 1.303$

$$\tan \alpha_r = 1.1414$$
$$\alpha_r = 48.78° \quad 2\alpha_r = 97.56°$$

Equation $\boxed{3}$ $\quad \tau_{\ell t} = (-\sigma_d) \sin \alpha_r \cos \alpha_r = (7.43)(0.752)(0.659) = 3.68 \text{ MPa}$

Equation $\boxed{6}$ $\quad \dfrac{\gamma_{\ell t}}{2} = (\varepsilon_r - \varepsilon_d) \sin \alpha_r \cos \alpha_r = (0.00383 + 0.0002)(0.752)(0.659)$

$$= 0.00200$$

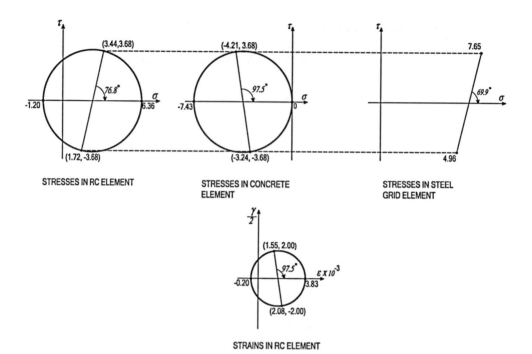

STRESSES IN RC ELEMENT   STRESSES IN CONCRETE   STRESSES IN STEEL
                              ELEMENT   GRID ELEMENT

STRAINS IN RC ELEMENT

Figure 5.18   Stresses (MPa) and strains for $\varepsilon_d = -0.0002$ (close to first yield) in example problem 5.3

It is interesting to note that $\gamma_{\ell t}/2$ can be obtained without knowing the angle α_r by using Equation (5.52), which is directly a simple function of ε_ℓ, ε_t and ε_d,

$$\frac{\gamma_{\ell t}}{2} = \pm\sqrt{(\varepsilon_\ell - \varepsilon_d)(\varepsilon_t - \varepsilon_d)}$$

$$= \pm\sqrt{(0.00155 + 0.0002)(0.00208 + 0.0002)} = 0.00200$$

The stresses in the mild steel and prestressing steel can be calculated from the strains using the stress–strain relationships:

Equation $\boxed{9a}$ $f_\ell = E_s\varepsilon_\ell = 200\,000\,(0.00155) = 310$ MPa

Equation $\boxed{10a}$ $f_t = 413$ MPa (just yielded)

Equation $\boxed{11a}$ $f_{\ell p} = E_{ps}(\varepsilon_{dec} + \varepsilon_\ell) = 200\,000(0.005 + 0.00155) = 1310$ MPa

Equation $\boxed{12a}$ $f_{tp} = 0$

Mohr's circles for stresses on the concrete element, in the steel grid, and on the PC element, as well as Mohr's circle for strains, are all plotted in Figure 5.18 for the selected strain of $\varepsilon_d = -0.0002$. In order to plot these Mohr's circles, the following additional stresses are calculated:

$$\rho_\ell f_\ell + \rho_{\ell p} f_{\ell p} = 0.012\,(310) + 0.003\,(1310) = 7.65 \text{ MPa}$$
$$\rho_t f_t = 0.012\,(413) = 4.96 \text{ MPa}$$
$$\sigma_\ell - \rho_\ell f_\ell - \rho_{\ell p} f_{\ell p} = 3.44 - 7.65 = -4.21 \text{ MPa}$$
$$\sigma_t - \rho_t f_t = 1.72 - 4.96 = -3.24 \text{ MPa}$$

$$\sigma_1 = \frac{\sigma_\ell + \sigma_t}{2} + \sqrt{\left(\frac{\sigma_\ell - \sigma_t}{2}\right)^2 + \tau_{\ell t}^2}$$

$$= \frac{3.44 + 1.72}{2} + \sqrt{\left(\frac{3.44 - 1.72}{2}\right)^2 + 3.68^2}$$

$$= 2.58 + 3.78 = 6.36 \text{ MPa}$$

$$\sigma_2 = 2.58 - 3.78 = -1.20 \text{ MPa}$$

2. Select $\varepsilon_d = -0.0004$

After several cycles of the trial-and-error process, assume

$$\varepsilon_r = 0.0200$$

$$\zeta = 0.250$$

$$\frac{\varepsilon_d}{\zeta \varepsilon_o} = 0.800 < 1 \text{ ascending branch}$$

$$\sigma_d = 0.250(-41.3)[2 (0.800) - (0.800)^2] = -9.92 \text{ psi}$$

Solve the longitudinal steel strain ε_ℓ:

Assume yielding of longitudinal steel, $f_\ell = 413$ MPa, and assume the strain of the longitudinal steel to be $\varepsilon_\ell = 0.00630$

$$E'_{ps}(\varepsilon_{dec} + \varepsilon_\ell) = 214\,000(0.005 + 0.00630) = 2420 \text{ MPa}$$

Equation $\boxed{12b}$ $f_{\ell p} = \dfrac{E'_{ps} (\varepsilon_{dec} + \varepsilon_\ell)}{\left[1 + \left\{\dfrac{E'_{ps} (\varepsilon_{dec} + \varepsilon_\ell)}{f_{pu}}\right\}^4\right]^{\frac{1}{4}}} = \dfrac{2420}{\left[1 + \left\{\dfrac{2420}{1860}\right\}^4\right]^{\frac{1}{4}}} = 1726 \text{ MPa}$

$$\frac{\varepsilon_r - \varepsilon_d}{-\sigma_d} = \frac{0.0200 + 0.0004}{9.92} = 2.056 \times 10^{-3}/\text{MPa}$$

Equation $\boxed{13}$ $\varepsilon_\ell = \varepsilon_r + \dfrac{\varepsilon_r - \varepsilon_d}{-\sigma_d} (\sigma_\ell - \rho_\ell f_\ell - \rho_{\ell p} f_{\ell p})$

$$= 0.0200 + 2.056 \times 10^{-3} [3.44 - 0.012(413) - 0.003(1726)]$$

$$= 0.00623 \approx 0.00630 \text{ OK}$$

Solve the transverse steel strain ε_t:

Assume yielding of transverse steel, $f_t = 413$ MPa

Equation $\boxed{14}$ $\varepsilon_t = \varepsilon_r + \dfrac{\varepsilon_r - \varepsilon_d}{-\sigma_d} (\sigma_t - \rho_t f_t - \rho_{tp} f_{tp})$

$$= 0.0200 + 2.056 \times 10^{-3} [1.72 - 0.012\,(413) - 0]$$

$$= 0.01334 > 0.00207 \text{ OK}$$

The tensile strain ε_r can now be checked:

Equation $\boxed{15}$ $\varepsilon_r = 0.00630 + 0.01334 + 0.0004 = 0.02004 \approx 0.0200$ OK

The angle α_r, the shear stress $\tau_{\ell t}$ and the shear strain $\gamma_{\ell t}$ are:

Equation $\boxed{16}$ $\tan^2 \alpha_r = \dfrac{\varepsilon_t - \varepsilon_d}{\varepsilon_\ell - \varepsilon_d} = \dfrac{0.01335 + 0.0002}{0.00630 + 0.0002} = 2.085$

$$\tan \alpha_r = 1.444$$
$$\alpha_r = 55.2° \quad 2\alpha_r = 110.4°$$

Equation $\boxed{3}$ $\tau_{\ell t} = (-\sigma_d) \sin \alpha_r \cos \alpha_r = (9.92)(0.821)(0.571) = 4.65$ MPa

Equation $\boxed{6}$ $\dfrac{\gamma_{\ell t}}{2} = (\varepsilon_r - \varepsilon_d) \sin \alpha_r \cos \alpha_r = (0.0200 + 0.0004)(0.821)(0.571) = 0.00955$

The stresses in the mild steel and prestressing steel can be calculated from the strains using the stress–strain relationships:

Equation $\boxed{9b}$ $f_l = 413$ MPa
Equation $\boxed{10b}$ $f_t = 413$ MPa
Equation $\boxed{11b}$ $f_{\ell p} = 1726$ MPa
Equation $\boxed{12a}$ $f_{tp} = 0$

The Mohr circles for stresses on the concrete element, in the steel grid, and on the PC element, as well as the Mohr circle for strains, are all plotted in Figure 5.19 for the selected strain of $\varepsilon_d = -0.0004$. In order to plot these Mohr circles, the following additional stresses are calculated:

$$\rho_\ell f_\ell + \rho_{\ell p} f_{\ell p} = 0.012\,(413) + 0.003\,(1726) = 10.13 \text{ MPa}$$
$$\rho_t f_t = 0.012\,(413) = 4.96 \text{ MPa}$$
$$\sigma_\ell - \rho_\ell f_\ell - \rho_{\ell p} f_{\ell p} = 3.44 - 10.13 = -6.69 \text{ MPa}$$
$$\sigma_t - \rho_t f_t = 1.72 - 4.96 = -3.24 \text{ MPa}$$

$$\sigma_1 = \frac{\sigma_\ell + \sigma_t}{2} + \sqrt{\left(\frac{\sigma_\ell - \sigma_t}{2}\right)^2 + \tau_{\ell t}^2}$$

$$= \frac{3.44 + 1.72}{2} + \sqrt{\left(\frac{3.44 - 1.72}{2}\right)^2 + 4.63^2}$$

$$= 2.58 + 4.71 = 7.29 \text{ MPa}$$

$$\sigma_2 = 2.58 - 4.71 = -2.13 \text{ MPa}$$

In addition to the strains $\varepsilon_d = -0.0002$ and -0.0004 selected above, calculations were made for the strains $\varepsilon_d = -0.0003$ and -0.0005. The results of all these calculations are recorded in Table 5.2. Using the $\tau_{\ell t}$ and $\gamma_{\ell t}$ values given in Table 5.2, the shear stress vs. shear strain curve are plotted in Figure 5.20.

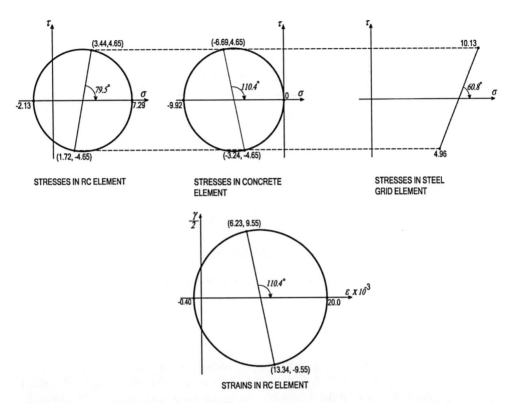

Figure 5.19 Stresses (MPa) and strains for $\varepsilon_d = -0.0004$ (close to ultimate strength) in example problem 5.3

Table 5.2 Results of Calculations for Example Problem 5.3

Variables	Eqs	Calculated values			
ε_d 10^{-3} selected		−0.200	−0.300	−0.400	−0.500
ε_r 10^{-3} last assumed		3.85	11.50	20.00	21.7
ζ	8	0.495	0.320	0.250	0.240
$\varepsilon_d/\zeta\varepsilon_o$		0.202	0.469	0.800	1.042
σ_d MPa	7	−7.43	−9.50	−9.92	−9.92
$(\varepsilon_r - \varepsilon_d)/(-\sigma_d)$ 10^{-3}/MPa		0.545	1.242	2.056	2.238
ε_ℓ 10^{-3}	13	1.55	3.75	6.23	6.63
ε_t 10^{-3}	14	2.08	7.48	13.34	14.45
ε_r 10^{-3} checked	15	3.83	11.53	19.97	21.6
α_r deg.	16	48.78	54.18	55.22	55.37
$\tau_{\ell t}$ MPa		3.68	4.51	4.65	4.64
$\gamma_{\ell t}$ 10^{-3}		4.00	11.2	19.1	20.6
f_ℓ MPa	9	311	413	413	413
f_t MPa	10	413	413	413	413
$f_{\ell p}$ MPa	11	1310	1570	1726	1738
f_{tp} MPa	12	0	0	0	0

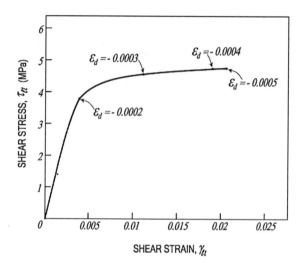

Figure 5.20   Shear stress vs. shear strain curve for example problem 5.3

5.4.4.3 Concluding Remarks

It can be seen from Table 5.2 and Figure 5.20 that the condition under the strain $\varepsilon_d = -0.0002$ represents very closely the *first yield condition*, because the transverse steel strain ε_t of 0.00208 just exceeds the yield stress of 0.00207 while the longitudinal steel strain is still within the elastic range. Under this condition the yield shear stress, $\tau_{\ell t y}$, and the yield shear strain $\gamma_{\ell t y}$ are 3.68 MPa and 0.00400, respectively.

The condition under the strain $\varepsilon_d = -0.0004$, on the other hand, represents very closely the *ultimate condition*. This can be observed as follows: When the strain ε_d is increased negatively from -0.0004 to -0.0005, the concrete struts increase from a level of $\varepsilon_d/\zeta\varepsilon_o = 0.800$ to a level of 1.042, just exceeding the peak stress of unity. In the vicinity of the peak stress, the compressive stress of -9.92 MPa in the concrete levels off and the increases of various strains come to a screeching halt.

It should be mentioned that the rotating angle softened truss model is incapable of predicting the descending branch of the $\tau_{\ell t}-\gamma_{\ell t}$ curve, because neither the Poisson effect nor the shear resistance of concrete struts are taken into account. The prediction of the descending branch will be carefully studied in the softened membrane model in Section 6.1 (Chapter 6).

5.4.5 2-D Elements under Proportional Loading

5.4.5.1 Proportional Membrane Loadings

The stress state described by the three applied stresses σ_ℓ, σ_t and $\tau_{\ell t}$ in the $\ell - t$ coordinate of a 2-D element is shown in Figure 5.21(a). This stress state can also be expressed in terms of three principal stress variables σ_1, S and α_1, in the $1 - 2$ coordinate as shown in

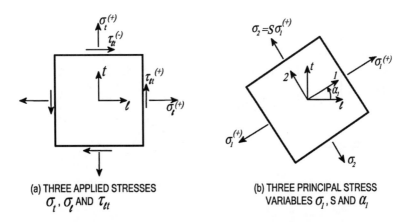

(a) THREE APPLIED STRESSES
σ_t, σ_ℓ AND $\tau_{\ell t}$

(b) THREE PRINCIPAL STRESS
VARIABLES σ_1, S AND α_1

Figure 5.21   Relationship between applied stresses and principal stress variables

Figure 5.21(b). These three principal stress variables are defined as:

σ_1 = larger principal stress, algebraically, in the 1-direction – always positive;
S = σ_2/σ_1, ratio of smaller to larger principal stresses – positive when σ_2 is tension, negative when σ_2 is compression;
α_1 = orientation angle, or angle between the direction of larger principal stress (1-axis) and the longitudinal axis (ℓ-axis).

When an element is subjected to a *proportional 2-D loading*, the larger principal stress σ_1 in the $1-2$ coordinate increases while the other two variables S and α_1 remain constant.

The set of proportional stresses σ_ℓ, σ_t and $\tau_{\ell t}$ in the $\ell - t$ coordinate can also be defined in terms of the principal stress σ_1 as follows:

$$\sigma_\ell = m_\ell \sigma_1 \tag{5.115}$$

$$\sigma_t = m_t \sigma_1 \tag{5.116}$$

$$\tau_{\ell t} = m_{\ell t} \sigma_1 \tag{5.117}$$

The set of three coefficients m_ℓ, m_t and $m_{\ell t}$ in the $\ell - t$ coordinate should remain constant under proportional loading.

Take, for example, a given set of applied proportional stresses ($\sigma_\ell = 1.25$ MPa, $\sigma_t = -0.25$ MPa and $\tau_{\ell t} = 1.30$ MPa) shown in Figure 5.22(a). Then the set of principal stress variables, σ_1, S and α_1, can be calculated as follows:

$$\sigma_1 = \frac{\sigma_\ell + \sigma_t}{2} + \sqrt{\left(\frac{\sigma_\ell - \sigma_t}{2}\right)^2 + \tau_{\ell t}^2} = \frac{1.25 - 0.25}{2} + \sqrt{\left(\frac{1.25 + 0.25}{2}\right)^2 + 1.30^2}$$

$$= 0.50 + 1.50 = 2.00 \text{ MPa}$$

$$\sigma_2 = 0.50 - 1.50 = -1.00 \text{ MPa}$$

$$S = \frac{\sigma_2}{\sigma_1} = \frac{-1.00}{2.00} = -0.5$$

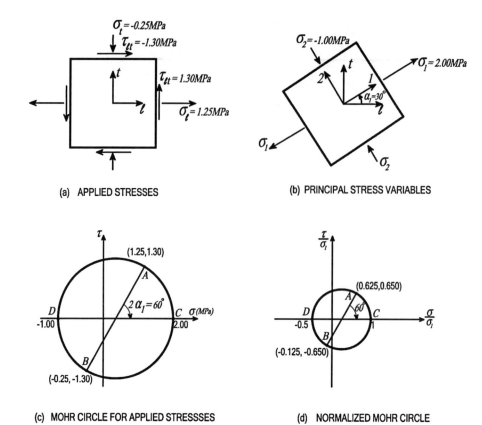

Figure 5.22 m-coefficients and normalized Mohr circle

$$\cot 2\alpha_1 = \frac{\sigma_\ell - \sigma_t}{2\tau_{lt}} = \frac{1.25 + 0.25}{2(1.30)} = 0.577$$
$$2\alpha_1 = 60°, \text{ and } \alpha_1 = 30°$$

These calculated values of σ_1, S and α_1 are recorded in Figure 5.22(b). Both the given set of applied stresses and the calculated set of principal stress variables are shown in Figure 5.22(c) in terms of the Mohr circle. Points A and B give the values of the three applied stresses ($\sigma_\ell = 1.25$ MPa, $\sigma_t = -0.25$ MPa and $\tau_{\ell t} = 1.30$ MPa). Points C and D indicate the principal stresses ($\sigma_1 = 2.00$ MPa and $\sigma_2 = -1.00$ MPa). The angle $2\alpha_1$ is 60°.

The set of m-coefficients are calculated by their definitions:

$$m_\ell = \frac{\sigma_\ell}{\sigma_1} = \frac{1.25}{2.00} = 0.625$$

$$m_t = \frac{\sigma_t}{\sigma_1} = \frac{-0.25}{2.00} = -0.125$$

$$m_{\ell t} = \frac{\tau_{\ell t}}{\sigma_1} = \frac{1.30}{2.00} = 0.650$$

According to these definitions, the coefficients m_ℓ, m_t and $m_{\ell t}$ have a physical meaning. They are simply the applied stresses σ_ℓ, σ_t and $\tau_{\ell t}$ normalized by the principal stress σ_1. These normalized stresses can be represented by a normalized Mohr circle in Figure 5.22(d). The values of the m-coefficients are given by the abscissas and ordinates of points A and B. The normalized principal stresses of σ_1 is, of course, equal to unity, and the normalized principal stress of σ_2 is the ratio $S = -0.5$. These two normalized principal stresses are indicated by the points C and D, respectively. The $2\alpha_1$ angle remains $60°$. In short, the normalized Mohr circle characterizes a given set of proportional loading, either in the $\ell - t$ coordinate (m_ℓ, m_t and $m_{\ell t}$) or in the principal $1 - 2$ coordinate (S and α_1).

The relationship between the set of three coefficients (m_ℓ, m_t and $m_{\ell t}$) in the $\ell - t$ coordinate and the set of two principal stress variables (S and α_1) in the $1 - 2$ coordinate can be obtained through the coordinate transformation relationship. Taking the inverse transformation of Equation (4.7) and setting $\tau_{12} = 0$, the set of stresses (σ_ℓ, σ_t and $\tau_{\ell t}$) in the $\ell - t$ coordinate can be expressed by the set of principal stresses (σ_1 and σ_2) in the $1 - 2$ coordinate as follows:

$$\sigma_\ell = \sigma_1 \cos^2 \alpha_1 + \sigma_2 \sin^2 \alpha_1 \tag{5.118}$$

$$\sigma_t = \sigma_1 \sin^2 \alpha_1 + \sigma_2 \cos^2 \alpha_1 \tag{5.119}$$

$$\tau_{\ell t} = (\sigma_1 - \sigma_2) \sin \alpha_1 \cos \alpha_1 \tag{5.120}$$

Substituting the definition of $\sigma_2 = S \sigma_1$ and the definitions of m-coefficients (Equations 5.115–5.117) into Equations (5.118)–(5.120), and then cancelling the principal stress σ_1 result in:

$$m_\ell = \cos^2 \alpha_1 + S \sin^2 \alpha_1 \tag{5.121}$$

$$m_t = \sin^2 \alpha_1 + S \cos^2 \alpha_1 \tag{5.122}$$

$$m_{\ell t} = (1 - S) \sin \alpha_1 \cos \alpha_1 \tag{5.123}$$

If the two principal stress variables S and α_1 are given, then the three m-coefficients can be calculated from Equations (5.121)–(5.123). Take, for example, the case of $S = 0.5$ and $\alpha_1 = 30°$ as shown in Figure 5.23(a):

$$\sin 30° = 0.50 \quad \cos 30° = 0.866$$
$$\sin^2 30° = 0.25 \quad \cos^2 30° = 0.75 \quad \sin 30° \cos 30° = 0.433$$
$$m_\ell = \cos^2 \alpha_1 + S \sin^2 \alpha_1 = 0.75 + 0.50(0.25) = 0.875$$
$$m_t = \sin^2 \alpha_1 + S \cos^2 \alpha_1 = 0.25 + 0.50(0.75) = 0.625$$
$$m_{\ell t} = (1 - S) \sin \alpha_1 \cos \alpha_1 = (1 - 0.5)(0.433) = 0.2165$$

These calculated m-coefficients are also shown in Figure 5.23(a), together with the normalized Mohr circles.

In addition, Figure 5.23(b) and (c) give two more cases of S and α_1, their corresponding m-coefficients and their normalized Mohr circles. Comparison of the three cases illustrates how the normalized Mohr circles are affected by S and α_1.

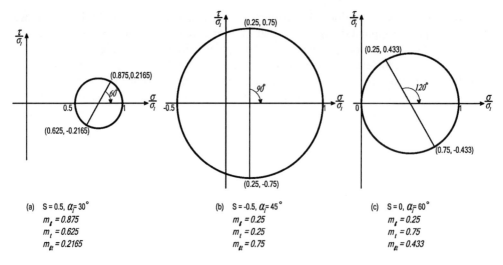

(a) $S = 0.5$, $\alpha_{l} = 30°$
  $m_{t} = 0.875$
  $m_{t} = 0.625$
  $m_{lt} = 0.2165$

(b) $S = -0.5$, $\alpha_{l} = 45°$
  $m_{t} = 0.25$
  $m_{t} = 0.25$
  $m_{lt} = 0.75$

(c) $S = 0$, $\alpha_{l} = 60°$
  $m_{t} = 0.25$
  $m_{t} = 0.75$
  $m_{lt} = 0.433$

Figure 5.23 m-coefficients for three pairs of S and α_{l}

5.4.5.2 Strength of 2-D Elements

The basic equilibrium equations for PC 2-D elements subjected to proportional loadings can be easily derived by substituting the definitions of the m-coefficients (Equations 5.115–5.117), into the equilibrium equations $\boxed{1}$–$\boxed{3}$ of rotating angle theory:

$$m_{\ell}\sigma_1 - \rho_{\ell} f_{\ell} - \rho_{\ell p} f_{\ell p} = \sigma_d \sin^2 \alpha_r \qquad (5.124)$$

$$m_{t}\sigma_1 - \rho_{t} f_{t} - \rho_{tp} f_{tp} = \sigma_d \cos^2 \alpha_r \qquad (5.125)$$

$$m_{\ell t}\sigma_1 = (-\sigma_d) \sin \alpha_r \cos \alpha_r \qquad (5.126)$$

The strength of the element under proportional loadings can be represented by a single stress σ_1. This stress σ_1 can be solved from Equations (5.124)–(5.126) by eliminating the angle α_r. Multiplying Equation (5.124) by Equation (5.125) gives:

$$(m_{\ell}\sigma_1 - \rho_{\ell} f_{\ell} - \rho_{\ell p} f_{\ell p})(m_{t}\sigma_1 - \rho_{t} f_{t} - \rho_{tp} f_{tp}) = (-\sigma_d)^2 \sin^2 \alpha_r \cos^2 \alpha_r \qquad (5.127)$$

Squaring Equation (5.126) gives:

$$(m_{\ell t}\sigma_1)^2 = (-\sigma_d)^2 \sin^2 \alpha_r \cos^2 \alpha_r \qquad (5.128)$$

Equating the left-hand sides of Equations (5.127) and (5.128) and rearranging the terms results in a quadratic equation for σ_1 as follows:

$$(m_{\ell}m_{t} - m_{\ell t}^2)\sigma_1^2 - [m_{\ell}(\rho_{t} f_{t} + \rho_{tp} f_{tp}) + m_{t}(\rho_{\ell} f_{\ell} + \rho_{\ell p} f_{\ell p})]\sigma_1$$
$$+ (\rho_{\ell} f_{\ell} + \rho_{\ell p} f_{\ell p})(\rho_{t} f_{t} + \rho_{tp} f_{tp}) = 0 \qquad (5.129)$$

Define $S = (m_{\ell}m_{t} - m_{\ell t}^2)$ (5.130)

$$B = [m_{\ell}(\rho_{t} f_{t} + \rho_{tp} f_{tp}) + m_{t}(\rho_{\ell} f_{\ell} + \rho_{\ell p} f_{\ell p})] \qquad (5.131)$$

$$C = (\rho_{\ell} f_{\ell} + \rho_{\ell p} f_{\ell p})(\rho_{t} f_{t} + \rho_{tp} f_{tp}) \qquad (5.132)$$

The solution of the quadratic equation is

$$\sigma_1 = \frac{1}{2S}\left(B \pm \sqrt{B^2 - 4SC}\right) \tag{5.133} \text{ or } \boxed{17}$$

It should be noted that the symbol S in Equation (5.130) is the same as the definition of $S = \sigma_2/\sigma_1$. This can be easily proven by inserting Equations (5.121)–(5.123) into Equation (5.130). Equation $\boxed{17}$ will be used in the solution procedure in Section 5.4.5.3, and in the Example Problem 5.4 (Section 5.4.6).

In the case of a nonprestressed element ($\rho_{\ell p} = \rho_{tp} = 0$), and equal yield strengths of mild steel in both directions ($f_\ell = f_t = f_y$), Equation (5.133) can be simplified to the following form:

$$\sigma_1 = \frac{f_y}{2S}\left[(m_\ell\rho_t + m_t\rho_\ell) \pm \sqrt{(m_\ell\rho_t + m_t\rho_\ell)^2 - 4(m_\ell m_t - m_{\ell t}^2)\rho_\ell\rho_t}\right] \tag{5.134}$$

Equation (5.134) was first derived by Han and Mau (1988). It is valid when both the longitudinal and the transverse steel are assumed to yield according to the equilibrium (plasticity) truss model.

5.4.5.3 Method of Solution

The twelve equations, $\boxed{3}$, $\boxed{6}$, and $\boxed{7}$–$\boxed{16}$, which have been successfully used to solve the case of 2-D elements under sequential loading (with constant normal stresses) in Section 5.4.4, can also be applied to the case of elements subjected to proportional loadings. Minor modifications, however, need to be made in Equations $\boxed{14}$ and $\boxed{15}$. Equation $\boxed{14}$ for the longitudinal steel strains ε_ℓ is still valid for proportional loading, except that the applied longitudinal stress σ_ℓ should be replaced by $m_\ell\sigma_1$:

$$\varepsilon_\ell = \varepsilon_r + \frac{\varepsilon_r - \varepsilon_d}{-\sigma_d}\left(m_\ell\sigma_1 - \rho_\ell f_\ell - \rho_{\ell p} f_{\ell p}\right) \tag{5.135} \text{ or } \boxed{13P}$$

The letter P in the equation number $\boxed{13P}$ indicates that it is derived specifically for proportional loading. Similarly, Equation $\boxed{14P}$ for the transverse steel strain ε_t is still valid, except that the applied transverse stresses σ_t should be replaced by $m_\ell\sigma_1$:

$$\varepsilon_t = \varepsilon_r + \frac{\varepsilon_r - \varepsilon_d}{-\sigma_d}\left(m_t\sigma_1 - \rho_t f_t - \rho_{tp} f_{tp}\right) \tag{5.136} \text{ or } \boxed{14P}$$

In Equations $\boxed{13P}$ and $\boxed{14P}$ we notice that a new unknown variable σ_1 has been introduced. Therefore, an additional equation is required. This new equation is furnished by Equation (5.133) or $\boxed{17}$. Based on the thirteen equations, $\boxed{3}$, $\boxed{6}$, and $\boxed{7}$ to $\boxed{17}$, a solution procedure is proposed as shown by the flow chart in Figure 5.24. This procedure is somewhat more complex than the flow chart shown in Figure 5.16 for sequential loading (with constant normal stresses), because of the additional iteration cycles required by the variable σ_1 for the proportional loading. For this trial-and-error procedure with a nested DO-loop for σ_1, it would be convenient to write a computer program and to take advantage of the power and speed of a computer.

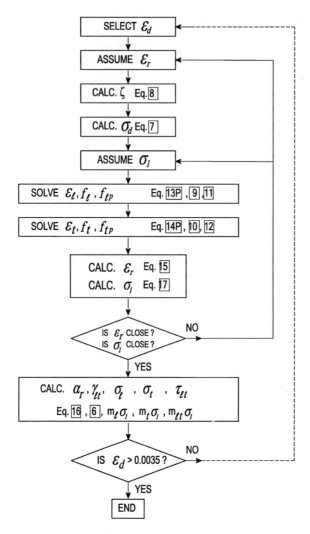

Figure 5.24   Flow chart for proportional loadings

5.4.6 Example Problem 5.4 for Proportional Loading

5.4.6.1 Problem Statement

A nonprestressed 2-D element was designed by the equilibrium (plasticity) truss model in Example Problem 5.1 (see Section 5.1.5 and Figure 5.3), and then analyzed by the Mohr compatibility truss model in Example Problem 5.2 (see Section 5.3.4 and Figure 5.10). This 2-D element is subjected to a set of membrane stresses $\sigma_\ell = 2.13$ MPa (tension), $\sigma_t = -2.13$ MPa (compression) and $\tau_{\ell t} = 3.69$ MPa. Assuming that this set of stresses is applied in a proportional manner, we can now analyze the nonlinear behavior of this 2-D element by the rotating angle softened truss model (RA-STM). Particular interest is placed in the stress and strain conditions at the first yield of steel and at the ultimate load stage.

Based on the equilibrium (plasticity) truss model, the 2-D element is reinforced with 1.03% of steel in both the longitudinal and the transverse directions (Example Problem 5.1). The steel is designed to have the same grade in both directions $f_{\ell y} = f_{ty} = 413$ MPa. In the Mohr compatibility truss model (Example Problem 5.2) the strength of concrete is assumed to be $f'_c = 27.6$ MPa (4000 psi). In order to use the RA-STM, the constitutive laws of concrete and steel are specified as follows: the compression stress–strain curve of softened concrete is expressed by Equations $\boxed{7a}$, $\boxed{7b}$ and $\boxed{8}$. The tensile strength of concrete is neglected (assume $\sigma_r = 0$); and the stress–train curve of bare mild steel bars, Equations $\boxed{9a}$ and $\boxed{9b}$ (or $\boxed{10a}$ and $\boxed{10b}$), is adopted.

5.4.6.2 Solution

For the given set of applied stresses the principal stress variables σ_1, S and α_1 are:

$$\sigma_1 = \frac{\sigma_\ell + \sigma_t}{2} + \sqrt{\left(\frac{\sigma_\ell - \sigma_t}{2}\right)^2 + \tau_{\ell t}^2} = \frac{2.13 - 2.13}{2} + \sqrt{\left(\frac{2.13 + 2.13}{2}\right)^2 + 3.69^2}$$

$$= 0 + 4.26 = 4.26 \text{ MPa}$$

$$\sigma_2 = 0 - 4.26 = -4.26 \text{ MPa}$$

$$S = \frac{\sigma_2}{\sigma_1} = \frac{-4.26}{4.26} = -1$$

$$\cot 2\alpha_1 = \frac{(\sigma_\ell - \sigma_t)}{2\tau_{\ell t}} = \frac{2.13 - (-2.13)}{2(3.69)} = 0.577$$

$$2\alpha_1 = 60°, \quad \text{and } \alpha_1 = 30°.$$

The three m-coefficients are:

$$m_\ell = \frac{\sigma_\ell}{\sigma_1} = \frac{2.13}{4.26} = 0.5$$

$$m_t = \frac{\sigma_t}{\sigma_1} = \frac{-2.13}{4.26} = -0.5$$

$$m_{\ell t} = \frac{\tau_{\ell t}}{\sigma_1} = \frac{3.69}{4.26} = 0.866$$

The normalized Mohr circle for this case of proportional loading is shown in Figure 5.25. This normalized Mohr circle is characterized by $S = -1$ and $\alpha_1 = 30°$ in the $1 - 2$ coordinate, and $m_\ell = 0.5$, $m_t = -0.5$ and $m_{\ell t} = 0.886$ in the $\ell - t$ coordinate.

1. Select $\varepsilon_d = -0.000275$ (for the first yield condition)

The stress and strain conditions at first yield have been given in Figure 5.10 (see Example Problem 5.2, Section 5.3.4) based on the Mohr compatibility truss model. At this load stage $\varepsilon_d = -0.000265$, $\varepsilon_r = 0.00345$ and $\sigma_1 = 3.65$ MPa. In this softened truss model, however, ε_d should be slightly greater than -0.000265 (in an absolute sense), because of the slightly nonlinear stress–strain relationship of concrete at low level of loading. After several cycles of trial-and-error process we select the values of $\varepsilon_d = -0.000275$, and assume

$$\varepsilon_r = 0.00345 \quad \text{and} \quad \sigma_1 = 3.65 \text{ MPa}$$

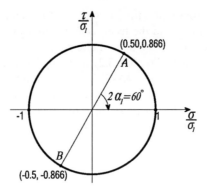

PRINCIPAL STRESS VARIABLES :

$S = -1,\qquad \alpha_1 = 30°$

m-COEFFICIENT :

$m_\ell = 0.5,\ m_t = -0.5,\ m_{\ell t} = 0.866$

Figure 5.25   Normalized Mohr circle for proportional loading in example problem 5.4

Equation $\boxed{8}$ $\quad \zeta = \dfrac{0.9}{\sqrt{1+600\varepsilon_r}} = \dfrac{0.9}{\sqrt{1+600(0.00345)}} = 0.514$

$\dfrac{\varepsilon_d}{\zeta\varepsilon_o} = \dfrac{-0.000275}{0.514(-0.002)} = 0.268 < 1$ ascending branch

Equation $\boxed{7a}$ $\quad \sigma_d = \zeta f'_c\left[2\left(\dfrac{\varepsilon_d}{\zeta\varepsilon_o}\right) - \left(\dfrac{\varepsilon_d}{\zeta\varepsilon_o}\right)^2\right]$

$\qquad\qquad = 0.514(-27.6)\,[2(0.268) - (0.268)^2] = -6.57 \text{ MPa}$

Solve the longitudinal steel strain ε_ℓ:

$$\frac{\varepsilon_r - \varepsilon_d}{-\sigma_d} = \frac{0.00345 + 0.000275}{6.57} = 0.567\times 10^{-3}/\text{MPa}$$

Equation $\boxed{13P}$ $\quad \varepsilon_\ell = \varepsilon_r + \dfrac{\varepsilon_r - \varepsilon_d}{-\sigma_d}\left(m_\ell\sigma_1 - \rho_\ell f_\ell - \rho_{\ell p}f_{\ell p}\right)$

$\qquad\qquad = 0.00345 + 0.567\times 10^{-3}\,(0.5(3.65) - 0.0103\,f_\ell - 0)$

Assume $f_\ell = 200\times 10^3 \varepsilon_\ell$ before yielding
Then $\varepsilon_\ell + 1.168\,\varepsilon_\ell = 0.00345 + 0.001035$

$$\varepsilon_\ell = \frac{0.004485}{2.168} = 0.00207 = 0.00207\ (\varepsilon_y)\ \text{OK}$$

Solve the transverse steel strain ε_t:

Equation $\boxed{14P}$ $\quad \varepsilon_t = \varepsilon_r + \dfrac{\varepsilon_r - \varepsilon_d}{-\sigma_d}\left(m_t\sigma_1 - \rho_t f_t - \rho_{tp} f_{tp}\right)$

$$= 0.00345 + 0.567 \times 10^{-3}\,(-0.5(3.65) - 0.0103\,f_t - 0)$$

Assume $f_t = 200 \times 10^3 \varepsilon_t$ before yielding
Then $\varepsilon_t + 1.168\,\varepsilon_t = 0.00345 - 0.001035$

$$\varepsilon_t = \frac{0.002415}{2.168} = 0.00111 < 0.00207(\varepsilon_y)\ \text{OK}$$

The tensile strain ε_r can now be checked:

Equation $\boxed{15}$ $\quad \varepsilon_r = \varepsilon_\ell + \varepsilon_t - \varepsilon_d = 0.00207 + 0.00111 + 0.000275$
$$= 0.003455 \approx 0.00345\ \text{(assumed)}\ \text{OK}$$

Check σ_1:

Equation $\boxed{9a}$ $\quad f_\ell = E_s\,\varepsilon_\ell = (200\ 000)(0.00207) = 413\ \text{MPa (just yielded)}$
Equation $\boxed{10a}$ $\quad f_t = E_s\,\varepsilon_t = (200\ 000)(0.00111) = 222\ \text{MPa (elastic range)}$
$\qquad \rho_\ell f_\ell = 0.0103(413) = 4.26\ \text{MPa}$
$\qquad \rho_t f_t = 0.0103(222) = 2.29\ \text{MPa}$
$\qquad B = m_\ell(\rho_t f_t) + m_t(\rho_\ell f_\ell) = 0.5(2.29) - 0.5(4.26) = -0.985\ \text{MPa}$
$\qquad C = (\rho_\ell f_\ell)(\rho_t f_t) = (4.26)(2.29) = 9.755\ \text{MPa}$

Equation $\boxed{17}$ $\quad \sigma_1 = \dfrac{1}{2S}\left(B \pm \sqrt{B^2 - 4SC}\right)$

$$= \frac{1}{2(-1)}\left(-0.985 \pm \sqrt{(-0.985)^2 - 4(-1)(9.755)}\right)$$

$$= -0.5(-0.985 - 6.32) = 3.65\ \text{MPa} = 3.65\ \text{MPa (assumed)}\ \text{OK}$$

Now that the strains ε_d, ε_r, ε_ℓ, and ε_t and the stresses, σ_d and σ_1, are calculated from the trial-and-error process, we can determine the angle α_r, the shear stress $\tau_{\ell t}$ and the shear strain $\gamma_{\ell t}$:

Equation $\boxed{16}$ $\tan^2\alpha_r = \dfrac{\varepsilon_t - \varepsilon_d}{\varepsilon_\ell - \varepsilon_d} = \dfrac{0.00111 + 0.000275}{0.00207 + 0.000275} = 0.5906$

$\qquad \tan\alpha_r = 0.7685, \quad \sin\alpha_r = 0.609, \quad \cos\alpha_r = 0.793$

$\qquad \alpha_r = 37.54° \quad 2\alpha_r = 75.08°$

Equation $\boxed{3}$ $\quad \tau_{\ell t} = (-\sigma_d)\sin\alpha_r\cos\alpha_r = (6.57)(0.609)(0.793)$
$$= 3.17\ \text{MPa}$$

Equation $\boxed{6}$ $\quad \gamma_{\ell t}/2 = (\varepsilon_r - \varepsilon_d)\sin\alpha_r\cos\alpha_r$
$$= (0.00345 + 0.000275)(0.609)(0.793) = 0.00180$$

The applied stresses at the level of first yield are:

$$\sigma_\ell = m_\ell \sigma_1 = 0.5(3.65) = 1.82 \text{ MPa}$$
$$\sigma_t = m_t \sigma_1 = -0.5(3.65) = -1.82 \text{ MPa}$$
$$\tau_{\ell t} = m_{\ell t} \sigma_1 = 0.866(3.65) = 3.16 \text{ MPa} \approx 3.17 \text{ MPa OK.}$$

These calculations show that the stress and strain conditions analyzed by the softened truss model are almost identical to those obtained by the Mohr compatibility truss model in Figure 5.10, Example Problem 5.2, Section 5.3.4. Therefore, the Mohr circles for the stresses and strains calculated by the RA-STM in this problem will not be given here.

2. Select $\varepsilon_d = -0.0005$ (after yielding of steel in both directions)
This is a point after the yielding of both the longitudinal and the transverse steel. The stress condition of the 2-D element is shown in Figure 5.3(a)–(d). The applied principal tensile stress must be $\sigma_1 = 4.26$ MPa. The strain condition, however, will be determined here. After several cycles of the trial-and-error process we assume:

$$\varepsilon_r = 0.0108 \quad \text{and} \quad \sigma_1 = 4.26 \text{ MPa}$$

Equation $\boxed{8}$ $\zeta = \dfrac{0.9}{\sqrt{1 + 600\varepsilon_r}} = \dfrac{0.9}{\sqrt{1 + 600(0.0108)}} = 0.329$

$\dfrac{\varepsilon_d}{\zeta\varepsilon_o} = \dfrac{-0.0005}{0.329(-0.002)} = 0.760 < 1$ Ascending branch

Equation $\boxed{7a}$ $\sigma_d = \zeta f_c' \left[2\left(\dfrac{\varepsilon_d}{\zeta\varepsilon_o}\right) - \left(\dfrac{\varepsilon_d}{\zeta\varepsilon_o}\right)^2 \right]$

$$= 0.329(-27.6)[2(0.760) - (0.760)^2] = -8.54 \text{ MPa}$$

Solve the longitudinal steel strain ε_ℓ (assume yielding of steel)

$$\dfrac{\varepsilon_r - \varepsilon_d}{-\sigma_d} = \dfrac{0.0108 + 0.0005}{8.54} = 1.323 \times 10^{-3}/\text{MPa}$$

Equation $\boxed{13P}$ $\varepsilon_\ell = \varepsilon_r + \dfrac{\varepsilon_r - \varepsilon_d}{-\sigma_d}\left(m_\ell \sigma_1 - \rho_\ell f_\ell - \rho_{\ell p} f_{\ell p}\right)$

$$= 0.0108 + 1.323 \times 10^{-3} (0.5(4.26) - 4.26 - 0)$$
$$= 0.00798 > 0.00207 \ (\varepsilon_y) \quad \text{OK}$$

Solve the transverse steel strain ε_t (assume yielding of steel)

Equation $\boxed{14P}$ $\varepsilon_t = \varepsilon_r + \dfrac{\varepsilon_r - \varepsilon_d}{-\sigma_d}\left(m_t \sigma_1 - \rho_t f_t - \rho_{tp} f_{tp}\right)$

$$= 0.0108 + 1.323 \times 10^{-3} (-0.5(4.26) - 4.26 - 0)$$
$$= 0.00234 > 0.00207 \ (\varepsilon_y), \quad \text{OK, shortly after yielding}$$

The tensile strain ε_r can now be checked:

Equation $\boxed{15}$ $\qquad \varepsilon_r = \varepsilon_\ell + \varepsilon_t - \varepsilon_d = 0.00798 + 0.00234 + 0.0005$
$$= 0.01082 \approx 0.0108 \text{ (assumed) OK}$$

Check σ_1:

Equation $\boxed{9a}$ $\qquad f_\ell = 413$ MPa (extensively yielded since $\varepsilon_\ell = 0.0079 \gg 0.00207$)

Equation $\boxed{10a}$ $\qquad f_t = 413$ MPa (shortly after yielding since $\varepsilon_t = 0.00234$)
$$\rho_\ell f_\ell = 0.0103(413) = 4.26 \text{ MPa}$$
$$\rho_t f_t = 0.0103(413) = 4.26 \text{ MPa}$$
$$B = m_\ell(\rho_t f_t) + m_t(\rho_\ell f_\ell) = 0.5(4.26) - 0.5(4.26) = 0$$
$$C = (\rho_\ell f_\ell)(\rho_t f_t) = (4.26)(4.26) = 18.15 \text{ MPa}$$

Equation $\boxed{17}$ $\qquad \sigma_1 = \dfrac{1}{2S}\left(B \pm \sqrt{B^2 - 4SC}\right) = \dfrac{1}{2(-1)}\left(-0 \pm \sqrt{(0)^2 - 4(-1)(18.15)}\right)$
$$= -0.5\,(-8.52) = 4.26 \text{ MPa} = 4.26 \text{ MPa (assumed)} \quad \text{OK}$$

Now that the strains ε_d, ε_r, ε_ℓ, and ε_t and the stresses, σ_d and σ_1, are calculated from the trial-and-error process, we can determine the angle α_r, the shear stress $\tau_{\ell t}$ and the shear strain $\gamma_{\ell t}$:

Equation $\boxed{16}$ $\tan^2 \alpha_r = \dfrac{\varepsilon_t - \varepsilon_d}{\varepsilon_\ell - \varepsilon_d} = \dfrac{0.00234 + 0.0005}{0.00798 + 0.0005} = 0.334$
$$\tan \alpha_r = 0.577$$
$$\alpha_r = 30° \quad 2\alpha_r = 60°$$

Equation $\boxed{3}$ $\qquad \tau_{\ell t} = (-\sigma_d)\sin \alpha_r \cos \alpha_r = (8.54)(0.500)(0.866)$
$$= 3.69 \text{ MPa}$$

Equation $\boxed{6}$ $\qquad \gamma_{\ell t}/2 = (\varepsilon_r - \varepsilon_d)\sin \alpha_r \cos \alpha_r$
$$= (0.0108 + 0.0005)(0.500)(0.866) = 0.00489$$

The applied stresses at the level near ultimate where both steel have yielded are:

$$\sigma_\ell = m_\ell \sigma_1 = 0.5(4.26) = 2.13 \text{ MPa}$$
$$\sigma_t = m_t \sigma_1 = -0.5(4.26) = -2.13 \text{ MPa}$$
$$\tau_{\ell t} = m_{\ell t}\sigma_1 = 0.866(4.26) = 3.69 \text{ MPa}$$

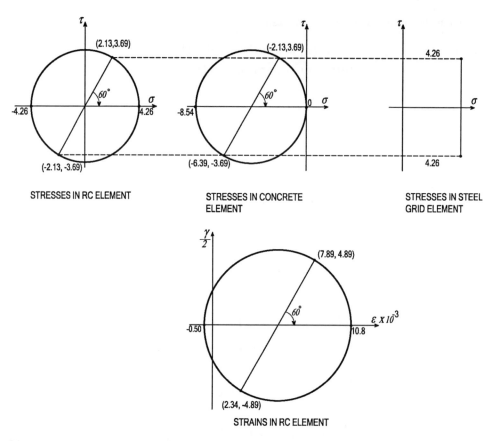

Figure 5.26 Stresses (MPa) and strains for $\varepsilon_d = -0.0005$ (at peak strength in example problem 5.4)

The Mohr circles for stresses on the concrete element, in the steel grid, and on the PC element, as well as the Mohr circle for strains, are all plotted in Figure 5.26 for the selected strain of $\varepsilon_d = -0.0005$. In order to plot these Mohr circles, the following additional stresses are calculated:

$$\sigma_\ell - \rho_\ell f_\ell = 2.13 - 4.26 = -2.13 \text{ MPa}$$
$$\sigma_t - \rho_t f_t = -2.13 - 4.26 = -6.39 \text{ MPa}$$
$$\sigma_1 = \frac{\sigma_\ell + \sigma_t}{2} + \sqrt{\left(\frac{\sigma_\ell - \sigma_t}{2}\right)^2 + \tau_{\ell t}^2}$$
$$= \frac{2.13 - 2.13}{2} + \sqrt{\left(\frac{2.13 + 2.13}{2}\right)^2 + 3.69^2}$$
$$= 0 + 4.26 = 4.26 \text{ MPa}$$
$$\sigma_2 = 0 - 4.26 = -4.26 \text{ MPa}$$

Table 5.3 Results of calculations for Example Problem 5.4.

Variables	Eqs	Calculated values				
$\varepsilon_d\ 10^{-3}$ selected		−0.275	−0.400	−0.500	−0.600	−0.620
$\varepsilon_r\ 10^{-3}$ last assumed		3.45	7.90	10.80	12.35	12.40
σ_1 MPa last assumed		3.65	4.16	4.26	4.26	4.26
ζ	[8]	0.514	0.376	0.329	0.3103	0.3098
$\varepsilon_d/\zeta\varepsilon_o$		0.268	0.532	0.760	0.967	1.000
σ_d MPa	[7]	−6.57	−8.09	−8.54	−8.54	−8.53
$(\varepsilon_r-\varepsilon_d)/(-\sigma_d)\ 10^{-3}/MPa$		0.567	0.706	1.323	1.515	1.525
$\varepsilon_\ell\ 10^{-3}$	[13P]	2.07	5.65	7.98	9.11	9.14
$\varepsilon_t\ 10^{-3}$	[14P]	1.11	1.85	2.33	2.64	2.63
$\varepsilon_r\ 10^{-3}$ checked	[15]	3.455	7.90	10.82	12.35	12.39
f_ℓ MPa	[9]	413	413	413	413	413
f_t MPa	[10]	222	369	413	413	413
$f_{\ell p}$ MPa	[11]	0	0	0	0	0
f_{tp} MPa	[12]	0	0	0	0	0
σ_1 MPa checked	[17]	3.65	4.16	4.26	4.26	4.26
α_r degrees.	[16]	37.54	31.38	30.0	30.0	30.0
$\tau_{\ell t}$ MPa	[3]	3.17	3.59	3.69	3.69	3.69
$\gamma_{\ell t}\ 10^{-3}$	[6]	3.60	7.38	9.78	11.22	11.28
σ_ℓ MPa		1.82	2.08	2.13	2.13	2.13
σ_t MPa		−1.82	−2.08	−2.13	−2.13	−2.13
$\tau_{\ell t}$ MPa		3.16	3.59	3.69	3.69	3.69

5.4.6.3 Discussion

The results of calculations for $\varepsilon_d = -0.000275$, -0.0004 and -0.0005, -0.0006 and -0.00062 are summarized in Table 5.3. Comparison of these five cases illustrates clearly the trends of all the variables. It should be kept in mind that $\varepsilon_d = -0.000275$ is the point of first yield of the longitudinal steel. $\varepsilon_d = -0.0004$ represents a point after the yielding of longitudinal steel, but before the yielding of transverse steel. $\varepsilon_d = -0.0005$ gives a point after the yielding of both the longitudinal and the transverse steel. Computation could not proceed after $\varepsilon_d = -0.00062$.

Let us compare the Mohr stress circles for concrete and steel at peak load stage in Figure 5.26 with those at first yield in Figure 5.10. The $2\alpha_r = 75.08°$ at first yield (Figure 5.10) has gradually reduced to $2\alpha_r = 60°$ at peak load stage (Figure 5.26). The transverse steel stress $\rho_t f_t = 2.29$ MPa at first yield (Figure 5.10) has gradually increased to the yield level of $\rho_t f_t = 4.26$ MPa at peak load stage. This increase of transverse steel stress is responsible for the corresponding increase of the applied stresses (σ_ℓ, σ_t, $\tau_{\ell t}$).

Let us also compare the Mohr stress circles for applied stresses, concrete stresses and steel stresses at peak load stage in Figure 5.26 and those in Figure 5.3. It can be seen that they are identical. In other words, when the steel in both the longitudinal and transverse directions reaches the yield point, the predictions of RA-STM is the same as the prediction of the equilibrium (plasticity) truss model. Of course, RA-STM is more powerful theory, because the incorporation of the Mohr compatibility condition allows RA-STM to produce the Mohr

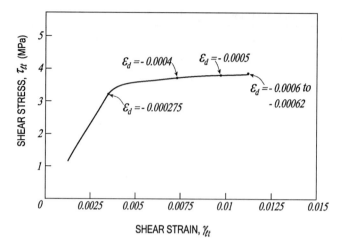

Figure 5.27   Shear stress–shear strain curve for example problem 5.4

strain circle shown in Figure 5.26. The nonlinear constitutive laws of concrete and steel also allow RA-STM to predict the nonlinear behavior of RC 2-D elements throughout the loading history up to the peak point.

The relationship between the shear stress $\tau_{\ell t}$ and shear strain $\gamma_{\ell t}$ is shown in Figure 5.27. The shear strains at first yield and at ultimate are 0.00360 and 0.01128, respectively. It is interesting to point out that RA-STM model cannot predict the descending branch of the shear stress–shear strain curve ($\tau_{\ell t} - \gamma_{\ell t}$ curve), because both the Poisson effect and the shear resistance of concrete struts are not taken into account. The descending branches of various behavioral curves of RC 2-D elements will be studied in Section 6.1 (Chapter 6) in connection with the Softened Membrane Theory (SMM).

5.4.7 Failure Modes of RC 2-D Elements

A 2-D element subjected to a given set of shear and normal stresses may be designed to fail in different modes. Depending on the selected thickness of the 2-D element, the steel may yield before the crushing of concrete or the concrete may crush before the yielding of steel. Depending on the percentage ratios of steel in the two directions, yielding of the steel may first occur in the longitudinal bars or in the transverse bars. A 2-D element designed in different ways will behave very differently.

Depending on the longitudinal steel percentages (ρ_ℓ) and the transverse steel percentages (ρ_t), a 2-D element may fail in four modes:

1. *Under-reinforced 2-D element.* Both the longitudinal steel and the transverse steel yield before the crushing of concrete.
2. *2-D element partially under-reinforced in ℓ-direction.* Longitudinal steel yields before the crushing of concrete. Transverse steel does not yield.

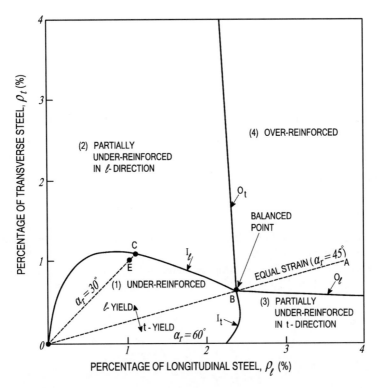

Figure 5.28   Failure modes diagram for example problem 5.4 (proportional loading, $S = -1, \alpha_r = 30°$)

3. *2-D element partially under-reinforced in t-direction.* Transverse steel yields before the crushing of concrete. Longitudinal steel does not yield.
4. *Over-reinforced 2-D element.* Concrete crushes before the yielding of steel in both directions.

These four modes of failure can be divided graphically by *failure modes diagrams*, Figure 5.28. In this diagram the percentage of longitudinal steel, ρ_ℓ, and the percentage of transverse steel ρ_t are represented by the horizontal axis and the vertical axis, respectively. Figure 5.28 is constructed for Example Problem 5.4, in which a 2-D element is subjected to proportional loading defined by the normalized Mohr circle in Figure 5.25. This proportional loading has a pair of principal stress variables of $S = -1$ and $\alpha_1 = 30°$ and a set of *m*-coefficients: $m_\ell = 0.5$, $m_t = -0.5$ and $m_{\ell t} = 0.866$. Such failure modes diagrams for proportional loading were first constructed by Han and Mau (1988).

To understand Figure 5.28, we must first study the *equal strain condition* for steel in Section 5.4.7.1 and the *balanced condition* of steel and concrete in Section 5.4.7.2. After the equal strain line OA and the balanced point B are located, the boundaries for the four failure modes are constructed in Section 5.4.7.3.

5.4.7.1 Equal Strain Condition

Suppose that the longitudinal steel and transverse steel have the same stress $f_\ell = f_t$ throughout the loading history, then the strains in both directions must be equal, i.e.

$$\varepsilon_\ell = \varepsilon_t \tag{5.137}$$

Inserting the compatibility equations, (5.41) and (5.42), into Equation (5.137) we have

$$\varepsilon_r \cos^2 \alpha_r + \varepsilon_d \sin^2 \alpha_r = \varepsilon_r \sin^2 \alpha_r + \varepsilon_d \cos^2 \alpha_r \tag{5.138}$$

Regrouping the terms results in

$$(\varepsilon_r - \varepsilon_d)(\cos^2 \alpha_r - \sin^2 \alpha_r) = 0 \tag{5.139}$$

Equation (5.139) is satisfied only if $\alpha_r = \pm 45°$, except when $\varepsilon_r = \varepsilon_d$, the special case of uniform tension in biaxial stresses.

Let us now limit our study to non-prestressed elements $(\rho_{\ell p} = \rho_{tp} = 0)$. The three equilibrium equations, (5.124)–(5.126), can be further simplified by taking $\alpha_r = \pm 45°$:

$$m_\ell \sigma_1 = \frac{\sigma_d}{2} + \rho_\ell f_\ell \tag{5.140}$$

$$m_t \sigma_1 = \frac{\sigma_d}{2} + \rho_t f_t \tag{5.141}$$

$$m_{\ell t} \sigma_1 = -\frac{\sigma_d}{2} \tag{5.142}$$

Multiply Equation (5.140) by $m_{\ell t}$ and Equation (5.142) by m_ℓ. Then, we can subtract the latter from the former to eliminate the variable σ_1 and obtain

$$\rho_\ell = \frac{-\sigma_d}{2 f_\ell} \frac{m_\ell + m_{\ell t}}{m_{\ell t}} \tag{5.143}$$

Similarly, multiplying Equation (5.141) by $m_{\ell t}$ and Equation (5.142) by m_t, and then subtracting the latter from the former give:

$$\rho_t = \frac{-\sigma_d}{2 f_t} \frac{m_t + m_{\ell t}}{m_{\ell t}} \tag{5.144}$$

Dividing Equation (5.143) by Equation (5.144) and taking $f_\ell = f_t$ result in the equal strain condition:

$$\frac{\rho_\ell}{\rho_t} = \frac{m_\ell + m_{\ell t}}{m_t + m_{\ell t}} \tag{5.145}$$

The equal strain condition in Eq. (5-145) was first derived by Fialkow (1985). This equation expresses the ratio of the percentage of longitudinal steel to the percentage of transverse steel in terms of the three m-coefficients for proportional loadings. If a 2-D element is designed according to this equal strain condition, the yielding of the steel is expected to occur simultaneously in the longitudinal and the transverse steel.

In Example Problem 5.4, Section 5.4.6, the three m-coefficients were found to be $m_\ell = 0.5$, $m_t = -0.5$ and $m_{\ell t} = 0.866$. The ρ_ℓ / ρ_t ratio under the equal strain condition is

$$\frac{\rho_\ell}{\rho_t} = \frac{0.5 + 0.866}{-0.5 + 0.866} = 3.73$$

The ρ_ℓ / ρ_t ratio of 3.73 is represented by the dotted straight line OA in Figure 5.28. Because the 2-D element in Example Problem 5.4 was designed to have a ρ_ℓ / ρ_t ratio of unity, which is much less than 3.73, the yielding of the steel is expected to occur much earlier in the longitudinal bars than in the transverse bars. The ρ_ℓ / ρ_t ratio of unity is represented by the dotted straight line OC in Figure 5.28.

5.4.7.2 Balanced Condition

Now that the steel in a 2-D element can be designed to yield simultaneously under the equal strain condition, we can proceed to determine the balanced condition between the steel and the concrete. The balanced condition defines a mode of failure where both the longitudinal and the transverse steel yield simultaneouly with the crushing of concrete. The balanced percentage of steel, therefore, divides the under-reinforced 2-D element from the over-renforced 2-D element under the equal strain condition.

Adding Equations (5.143) and (5.144), and taking $f_\ell = f_t = f_s$ result in

$$\rho_\ell + \rho_t = \frac{-\sigma_d}{f_s}\left(\frac{m_\ell + 2m_{\ell t} + m_t}{2m_{\ell t}}\right) \tag{5.146}$$

Equation (5.146) is a parametric representation of a point on the equal strain line with the ratio σ_d / f_s as an unknown parameter. The relationship between the concrete stress σ_d and the steel stress f_s will have to be determined from the strain compatibility condition and the stress–strain relationships of concrete and steel. The balanced condition is obtained when the normalized concrete stress, σ_d / f_c', peaks at the same time the steel stress f_s reaches the yield point.

Assuming the yielding of steel under the equal strain condition, i.e. $\varepsilon_\ell = \varepsilon_t = \varepsilon_y$ and $\alpha_r = 45°$, both the strain compatibility equations, (5.97) or $\boxed{4}$ and (5.98) or $\boxed{5}$, are reduced to

$$\varepsilon_r = 2\varepsilon_y - \varepsilon_d \tag{5.147}$$

Inserting ε_r from Equation (5.147) into Equation (5.102) or $\boxed{8}$ for the softening coefficient of concrete:

$$\zeta = \frac{0.9}{\sqrt{1 + 600(2\varepsilon_y - \varepsilon_d)}} \tag{5.148}$$

The yield strain ε_y in Equation (5.148) is equal to 0.00207 for a mild steel with $f_y = 413$ MPa (60 000 psi).

Using the softened stress–strain relationship of concrete represented by Eqs. $\boxed{7a}$ and $\boxed{7b}$, and the softening coefficient expressed by Eq. (5.148), we can now trace the normalized concrete stress, σ_d / f_c', with the increase of the concrete strain ε_d (in an absolute sense) as shown in Table 5.4.

Table 5.4   Tracing the peak of σ_d/f_c' ($\alpha_r = 45°$)

ε_d	ζ Eq. (5-148)	$\dfrac{\varepsilon_d}{\zeta\varepsilon_o}$	σ_d/f_c' Eq. $\boxed{7a}$	σ_d/f_c' Eq. $\boxed{7b}$
−0.00050	0.46266	0.54035 < 1	0.36493	
−0.00060	0.45904	0.65354 < 1	0.40392	
−0.00070	0.45550	0.76839 < 1	0.43106	
−0.00080	0.45204	0.88488 < 1	0.44605	
−0.00086	0.45000	0.95556 < 1	0.44911	
−0.00087	0.44966	0.96739 < 1	0.44918 peak	
−0.00088	0.44933	0.97924 < 1	0.44914	
−0.00090	0.44866	1.00298 > 1		0.44866
−0.00100	0.44535	1.12271 > 1		0.44480

In Table 5.4 the peak strain ε_o of nonsoftened concrete is taken as −0.002 in the calculation of the ratio $\varepsilon_d/\zeta\varepsilon_o$. It should be recalled that the concrete strain ε_d is in the ascending branch when $\varepsilon_d/\zeta\varepsilon_o < 1$ and is in the descending branch when $\varepsilon_d/\zeta\varepsilon_o > 1$. Table 5.4 illustrates that the normalized concrete stress σ_d/f_c' peaks at a value of $0.44918 \approx 0.4492$ at a concrete strain ε_d of 0.00087. The corresponding $\varepsilon_d/\zeta\varepsilon_o$ value of 0.96739 is in the ascending branch but is very close to the peak stress of $\zeta f_c'$.

Substituting $\sigma_d = 0.4492 f_c'$ and $f_s = f_y$ into Equation (5.146) we obtain the total percentage of steel for the balanced condition:

$$\rho_\ell + \rho_t = \frac{-0.4492 f_c'}{f_y}\left(\frac{m_\ell + 2m_{\ell t} + m_t}{2m_{\ell t}}\right) \tag{5.149}$$

In the case of Example Problem 5.4, Section 5.4.6, where $m_\ell = 0.5$, $m_t = -0.5$, $m_{\ell t} = 0.866$, $f_c' = 27.56$ MPa (4000 psi) and $f_y = 413$ MPa (60 000 psi), the total percentage of steel $\rho_\ell + \rho_t$ is

$$\rho_\ell + \rho_t = \frac{-0.4492(-27.56)}{413}\left(\frac{0.5 + 2(0.866) - 0.5}{2(0.866)}\right) = 0.0300 \text{ or } 3.00\%$$

Solving this equation with the equal strain condition of $\rho_\ell/\rho_t = 3.73$, the percentages of the steel in the longitudinal and the transverse directions for the balanced condition are:

$$\rho_\ell = \frac{3.73}{1+3.73}(0.0300) = 2.36\%$$

$$\rho_t = \frac{1}{1+3.73}(0.0300) = 0.633\%.$$

The balanced point B, having a coordinate of $\rho_\ell = 2.36\%$ and $\rho_t = 0.633\%$, lies on the dotted line OA in Figure 5.28.

5.4.7.3 Failure Regions and Boundaries

In Figure 5.28, the inclined dotted line OA represents the equal strain condition with a ρ_ℓ/ρ_t ratio of 3.73. The point indicated by the letter B represents the balanced point with a coordinate of $\rho_\ell = 2.36\%$ and $\rho_t = 0.633\%$ as calculated above.

The failure modes diagram is divided into four regions by the four solid curves radiating from the balanced point. These four regions give the four modes of failure as indicated in the diagram. They are: (1) under-reinforced; (2) partially under-reinforced in the ℓ-direction; (3) partially under-reinforced in the t-direction; and (4) over-reinforced. These four regions are defined as follows:

1. Under-reinforced: $\varepsilon_t > \varepsilon_y$, $\varepsilon_\ell > \varepsilon_y$ and σ_d/f_c' peaks
2. Partially under-reinforced in ℓ-direction: $\varepsilon_t < \varepsilon_y$, $\varepsilon_\ell > \varepsilon_y$ and σ_d/f_c' peaks
3. Partially under-reinforced in t-direction: $\varepsilon_t > \varepsilon_y$, $\varepsilon_\ell < \varepsilon_y$ and σ_d/f_c' peaks
4. Over-reinforced: $\varepsilon_t < \varepsilon_y$, $\varepsilon_\ell < \varepsilon_y$ and σ_d/f_c' peaks

The boundaries between the regions are defined as follows:

Inner boundary I_ℓ: $\varepsilon_t = \varepsilon_y$, $\varepsilon_\ell > \varepsilon_y$ and σ_d/f_c' peaks.   Between regions (1) and (2)
Inner boundary I_t: $\varepsilon_t > \varepsilon_y$, $\varepsilon_\ell = \varepsilon_y$ and σ_d/f_c' peaks.   Between regions (1) and (3)
Outer boundary O_t: $\varepsilon_t < \varepsilon_y$, $\varepsilon_\ell = \varepsilon_y$ and σ_d/f_c' peaks.   Between regions (2) and (4)
Outer boundary O_ℓ : $\varepsilon_t = \varepsilon_y$, $\varepsilon_\ell < \varepsilon_y$ and σ_d/f_c' peaks.   Between regions (3) and (4)

The boundaries of the four failure modes can be constructed by the same procedures used in Sections 5.4.7.1 and 5.4.7.2, except that the angle α_r is no longer $45°$ as required by the equal strain condition. Therefore, the equilibrium and compatibility equations will have to be generalized to include the angle α_r.

In the case of nonprestressed elements ($\rho_{\ell p} = \rho_{tp} = 0$), the three equilibrium equations for proportional loadings, (5.124)–(5.126), are reduced to:

$$m_\ell \sigma_1 = \sigma_d \sin^2 \alpha_r + \rho_\ell f_\ell \tag{5.150}$$

$$m_t \sigma_1 = \sigma_d \cos^2 \alpha_r + \rho_t f_t \tag{5.151}$$

$$m_{\ell t} \sigma_1 = -\sigma_d \sin \alpha_r \cos \alpha_r \tag{5.152}$$

Multiply Equation (5.150) by $m_{\ell t}$ and Equation (5.152) by m_ℓ. Then, we can subtract the latter from the former to eliminate the variable σ_1 and obtain

$$\rho_\ell = \frac{-\sigma_d}{f_\ell} \left(\frac{m_\ell}{m_{\ell t}} \sin \alpha_r \cos \alpha_r + \sin^2 \alpha_r \right) \tag{5.153}$$

Similarly, multiplying Equation (5.151) by $m_{\ell t}$ and Equation (5.152) by m_t, and then subtracting the latter from the former give:

$$\rho_t = \frac{-\sigma_d}{f_t} \left(\frac{m_t}{m_{\ell t}} \sin \alpha_r \cos \alpha_r + \cos^2 \alpha_r \right) \tag{5.154}$$

The relationship between the concrete stress σ_d and the steel stresses, f_ℓ and f_t, in Equations (5.153) and (5.154) will have to be determined from the strain compatibility conditions and the stress–strain relationships of concrete and steel.

In the case of $\varepsilon_\ell = \varepsilon_y$ for mild steel, the compatibility equation for ε_ℓ, Equation (5.97) or $\boxed{4}$, becomes:

$$\varepsilon_r = -\varepsilon_d \tan^2 \alpha_r + \varepsilon_y \frac{1}{\cos^2 \alpha_r} \tag{5.155}$$

Similarly, in the case of $\varepsilon_t = \varepsilon_y$, the compatibility equation for ε_t, Equation (5.98) or $\boxed{5}$, give:

$$\varepsilon_r = -\varepsilon_d \cot^2 \alpha_r + \varepsilon_y \frac{1}{\sin^2 \alpha_r} \tag{5.156}$$

The tensile strain ε_r in Equations (5.155) or (5.156) can be used in conjunction with Equation $\boxed{8}$ to calculate the softening coefficient of concrete ζ. Then the softened stress-strain relationship of Eqs. $\boxed{7a}$ and $\boxed{7b}$ will be used to trace and to locate the peak of the concrete stress. The procedures to plot the inner boundary curves, I_ℓ and I_t, are summarized as follows:

Step 1: Select an angle α_r and insert it into Equations (5.153)–(5-156).
Step 2: By increasing the strain ε_d incrementally and calculating the softening coefficient ζ by Equation $\boxed{8}$ in conjunction with the strain ε_r from Equation (5.155) or (5.156), we can trace and locate the peak of the normalized concrete stress σ_d/f_c' by the stress–strain relationship of Equations $\boxed{7a}$ or $\boxed{7b}$, similar to the process in Table 5.4.
Step 3: Substituting the peak value of σ_d and $f_\ell = f_t = f_y$ into Equations (5.153) and (5.154) we obtain the percentages of the longitudinal and the transverse steel ρ_ℓ and ρ_t respectively, to locate one point on the inner boundary curves for the selected angle α_r.
Step 4: Select the angle α_r incrementally and repeat steps 1–3; we can generate a series of points to plot the inner boundary curves.

Take $\alpha_r = 30°$, for example, and locate a point on the inner boundary curve I_ℓ, where $\varepsilon_t = \varepsilon_y$ and $\varepsilon_\ell > \varepsilon_y$:

Equation (5.156) $\varepsilon_r = -\varepsilon_d \cot^2 30° + \varepsilon_y \dfrac{1}{\sin^2 30°} = -\varepsilon_d(3) + 0.00207(4)$

$$= 3(-\varepsilon_d) + 0.00828$$

Now we can find the peak of the concrete stress as shown in the following Table 5.5.
Since the peak of concrete stress occurs at $\sigma_d/f_c' = 0.33644$, the longitudinal steel and the transverse steel can be calculated from Equations (5.153) and (5.154) as:

$$\rho_\ell = \frac{-0.3364 f_c'}{f_y} \left(\frac{m_\ell}{m_{\ell t}} \sin 30° \cos 30° + \sin^2 30° \right)$$

$$= \frac{-0.3364(27.56)}{413} \left(\frac{0.5}{0.866}(0.5)(0.866) + (0.5)^2 \right) = 0.0112$$

$$\rho_t = \frac{-0.3364 f_c'}{f_y} \left(\frac{m_t}{m_{\ell t}} \sin 30° \cos 30° + \cos^2 30° \right)$$

$$= \frac{-0.3364(27.56)}{413} \left(\frac{-0.5}{0.866}(0.5)(0.866) + (0.866)^2 \right) = 0.0112$$

Table 5.5   Tracing the peak of σ_d/f_c' ($\alpha_r = 30°$ and $\varepsilon_t = \varepsilon_y$)

	ε_r	ζ		σ_d/f_c'	
ε_d	Eq. (5-156)	Eq. $\boxed{8}$	$\dfrac{\varepsilon_d}{\zeta \varepsilon_o}$	Eq. $\boxed{7a}$	Eq. $\boxed{7b}$
−0.00050	0.00978	0.34342	0.72797 < 1	0.31801	
−0.00060	0.01008	0.33901	0.88493 < 1	0.33452	
−0.00064	0.01020	0.33729	0.94874 < 1	0.33640	
−0.00065	0.01023	0.33686	0.96478 < 1	0.33644 peak	
−0.00066	0.01026	0.33644	0.98086 < 1	0.33632	
−0.00070	0.01038	0.33476	1.04552 > 1		0.33473
−0.00080	0.01068	0.33067	1.20966 > 1		0.33007

The coordinate of $\rho_\ell = \rho_t = 0.0112$ are given as point C in Figure 5.28. It is a point on the inner boundary I_ℓ with $\alpha_r = 30°$. By varying the angle α_r from 0° to 45° and repeating the same procedures we have the whole inner boundary curve I_ℓ. Similarly, the inner boundary curve I_t can be obtained using Equation (5.155) and varying the angle α_r from 45° to 60°. On this boundary curve $\varepsilon_\ell = \varepsilon_y$ and $\varepsilon_t > \varepsilon_y$.

It is interesting to point out that the design of reinforcement using the equilibrium (plasticity) truss model in Example Problem 5.1, Section 5.1.5, resulted in the steel percentage of $\rho_\ell = \rho_t = 0.0103$. This pair of ρ_ℓ and ρ_t is plotted as point E in Figure 5.28 for a concrete strength of $f_c' = 27.56$ MPa (4000 psi) and a steel yield strength of $f_y = 413$ MPa (60 000 psi). Point E is located in region (1), just within the inner boundary I_ℓ, meaning that the 27.56 MPa (4000 psi) concrete did not crush when both the longitudinal and the transverse steel yielded. If a lower strength of concrete (say, 20 MPa or 3000 psi) is used, point E would be located in region (2) outside the inner boundary I_ℓ. This situation means that the concrete would crush before the yielding of the transverse steel, and the assumption of the yielding of all the steel ($f_\ell = f_t = f_y$) could not be ensured.

The outer boundary curves, O_ℓ and O_t, in Figure 5.28 can be obtained by the preceding four-step procedures with one exception in step 3. In the case of O_t, the longitudinal steel stress will yield, $f_\ell = f_y$, but the transverse steel stress will not, $f_t < f_y$. Hence, before the calculation of transverse steel percentage ρ_t by Equation (5.154) the pre-yield transverse steel stress f_t has to be calculated from the transverse steel strain ε_t by the strain compatibility condition, $\varepsilon_t = \varepsilon_r - \varepsilon_\ell + \varepsilon_d$. The strain ε_ℓ is, of course, equal to the yield stress ε_y, and the strains ε_d and ε_r are obtained in step 2 after the peak of the normalized concrete stress is located.

5.5 Concluding Remarks

The rotating angle softened truss model (RA-STM) for shear has the following characteristics when compared with the fixed angle theory discussed in Chapter 6.

1. The direction of cracks is assumed to follow the principal $r - d$ coordinate of the concrete element. Consequently, the concrete shear stress τ_{rd} in the $r - d$ coordinate must vanish and the 'contribution of concrete' (V_c) must be zero. Chapter 5 clearly shows the elegance and the purity of the rotating angle theory.

2. The elegance and purity of the rotating angle theory is further enhanced by two simplifying assumptions: (a) neglecting the tensile strength of concrete by taking $\sigma_r = 0$; and (b) using the elastic–perfectly plastic stress–strain curve of bare mild steel bars. Since the first measure is conservative and the second is nonnconservative, the errors induced by these two measures should cancel each other in terms of strength. This cancellation of strength errors is conceptually correct because these two assumptions actually occur at the crack sections. The theory, however, will seriously overestimate the deformations because the stiffening effect of the steel bars due to concrete is neglected. The smeared (or average) stress–strain curves of steel bars and the smeared (or average) tensile stress of concrete will be discussed in Chapter 6.

3. The rotating angle softened truss model (RA-STM) in this chapter could not predict the descending portion of the shear stress–shear strain curve ($\tau_{\ell t} - \gamma_{\ell t}$ curve). It is also incapable of predicting the shear ductility. This limitation stems from two sources: First, rotating angle shear theory in Chapter 5 does not take into account the shear resistance of concrete struts. The fixed angle shear theory to be studied in Chapter 6, however, will incorporate the stress–strain relationship of concrete in shear in Section 6.1.10. Second, the rotating angle shear theory does not take into account the 2-D Poisson effect. The Poisson effect in RC 2-D elements will be discussed in Section 6.1.3 and 6.1.4, and this becomes an essential component of the softened membrane model (SMM).

4. The softened coefficient, ζ, in the concrete stress–strain curve of RA-STM is made, on purpose, to be simple and conservative. It would not be wise to make the softened co-efficient too complicated, when we are striving to make the theory more elegant. The ζ coefficient given in Chapter 5 is reasonable for normal-strength concrete, and when the steel percentages in the longitudinal and transverse directions do not differ greatly. A more comprehensive and accurate softened coefficient, but more complicated, will be studied in Chapter 6. The coefficient ζ in Chapter 6 should be used when applied to high-strength concrete and when the steel percentage in one direction is much smaller than that in the other direction.

6

Fixed Angle Shear Theories

6.1 Softened Membrane Model (SMM)

6.1.1 Basic Principles of SMM

In this chapter we will study three models in the fixed angle theory: the fixed angle softened truss model (FA-STM), the softened membrane model (SMM), and the cyclic softened membrane model (CSMM). The second and third models, SMM and CSMM are also based on the fixed angle concept, even without the special fixed angle (FA) label in the title. Since the FA-STM is a special case of SMM, we will first study SMM in this section before reducing it to FA-STM in Section 6.2. Then, the CSMM will be treated in Section 6.3.

From Section 4.3.2, the SMM satisfies Navier's three principles as follows:

6.1.1.1 Stress equilibrium equations

$$\sigma_\ell = \sigma_1^c \cos^2 \alpha_1 + \sigma_2^c \sin^2 \alpha_1 - \tau_{12}^c 2 \sin \alpha_1 \cos \alpha_1 + \rho_\ell f_\ell \tag{6.1}$$

$$\sigma_t = \sigma_1^c \sin^2 \alpha_1 + \sigma_2^c \cos^2 \alpha_1 + \tau_{12}^c 2 \sin \alpha_1 \cos \alpha_1 + \rho_t f_t \tag{6.2}$$

$$\tau_{\ell t} = (\sigma_1^c - \sigma_2^c) \sin \alpha_1 \cos \alpha_1 + \tau_{12}^c (\cos^2 \alpha_1 - \sin^2 \alpha_1) \tag{6.3}$$

It should be emphasized that these three equilibrium equations for fixed angle theory exhibit the following characteristics. The equations are established on the basis of the principal 1–2 coordinate of the applied stresses on the RC element, in which the shear stress $\tau_{12} = 0$. However, the concrete shear stress τ_{12}^c, is not equal to zero (see Figure 4.14a). These τ_{12}^c terms are the sources of the 'contribution of concrete' (V_c).

Strain compatibility equations

$$\varepsilon_\ell = \varepsilon_1 \cos^2 \alpha_1 + \varepsilon_2 \sin^2 \alpha_1 - \frac{\gamma_{12}}{2} 2 \sin \alpha_1 \cos \alpha_1 \tag{6.4}$$

$$\varepsilon_t = \varepsilon_1 \sin^2 \alpha_1 + \varepsilon_2 \cos^2 \alpha_1 + \frac{\gamma_{12}}{2} 2 \sin \alpha_1 \cos \alpha_1 \tag{6.5}$$

$$\frac{\gamma_{\ell t}}{2} = (\varepsilon_1 - \varepsilon_2) \sin \alpha_1 \cos \alpha_1 + \frac{\gamma_{12}}{2} \left(\cos^2 \alpha_1 - \sin^2 \alpha_1 \right) \tag{6.6}$$

Unified Theory of Concrete Structures   Thomas Hsu and Yi-Lung Mo
© 2010 John Wiley & Sons, Ltd

Notice that the three compatibility equations for fixed angle theory contain the terms with concrete shear strain, γ_{12}. In all the fixed angle models, it is necessary to have a constitutive relationship that relates γ_{12} in the compatibility equations to τ_{12}^c in the equilibrium equations.

Constitutive relationships

The solution of the three stress equilibrium equations and the three strain compatibility equations requires two constitutive matrices, one for concrete and one for steel. Assuming reinforced concrete to be a *continuous orthotropic* material, the concrete constitutive matrix is:

$$\begin{bmatrix} \sigma_1^c \\ \sigma_2^c \\ \tau_{12}^c \end{bmatrix} = \begin{bmatrix} E_1^c & \nu_{12} E_1^c & 0 \\ \nu_{21} E_2^c & E_2^c & 0 \\ 0 & 0 & G_{12}^c \end{bmatrix} = \begin{bmatrix} \varepsilon_1 \\ \varepsilon_2 \\ \dfrac{\gamma_{12}}{2} \end{bmatrix} \tag{6.7}$$

The 3×3 matrix in Equation (6.7) contains three diagonal elements E_1^c, E_2^c and G_{12}^c. Element E_1^c is a nonlinear modulus, representing the tensile stress–strain curve of concrete ($\sigma_1^c - \varepsilon_1$ curve). Modulus E_2^c represents the compressive stress–strain curve of concrete ($\sigma_2^c - \varepsilon_2$ curve); and G_{12}^c is the shear stress–strain curve of concrete ($\tau_{12}^c - \gamma_{12}/2$ curve).

The first two diagonal elements E_1^c and E_2^c in the matrix represent curves obtained from experiments, and the third diagonal element G_{12}^c is the shear modulus in the 1–2 coordinate. In the past, G_{12}^c was thought to be an independent material property, which must also be established from experiments. The resulting experimental expressions for G_{12}^c were often very complicated and made the analytical method very difficult to apply. One great advantage of using the smeared crack concept is that the cracked concrete can be treated as a *continuous* material, which requires only two independent moduli, rather than three. A theoretical derivation of G_{12}^c as a function of the two independent moduli E_1^c and E_2^c will be derived in Section 6.1.10.

The two off-diagonal elements $\nu_{12} E_1^c$ and $\nu_{21} E_2^c$ in Equation (6.7), represent the Poisson effect, i.e. the mutual effect of normal strains in the 1–2 coordinate. The symbol ν_{12} is the ratio of the resulting strain increment in 1-direction to the source strain increment in 2-direction, and the symbol ν_{21} is the ratio of the resulting strain increment in 2-direction to the source strain increment in 1-direction.

The ratios ν_{12} and ν_{21} are the well-known Poisson ratios for continuous isotropic materials, and vary from zero to 0.5. In a linear, isotropic unit cube subjected to triaxial compressive stresses, the volume will expand if the Poisson ratio is larger than 0.5 (Boresi, *et al.*, 1993). For cracked reinforced concrete, however, Equation (6.7) is assumed to be a constitutive matrix for *continuous orthotropic* materials, so that the smeared (or average) behavior of this cracked composite can be evaluated by continuum mechanics. In this case, ν_{12} and ν_{21} are the Hsu/Zhu ratios, and ν_{12} is allowed to exceed 0.5 because of the smeared cracks.

The steel constitutive matrix is:

$$\begin{bmatrix} \rho_\ell f_\ell \\ \rho_t f_t \\ 0 \end{bmatrix} = \begin{bmatrix} \rho_\ell E_\ell^s & 0 & 0 \\ 0 & \rho_t E_t^s & 0 \\ 0 & 0 & 0 \end{bmatrix} \begin{bmatrix} \varepsilon_\ell \\ \varepsilon_t \\ 0 \end{bmatrix} \tag{6.8}$$

The two diagonal elements, $\rho_\ell E_\ell^s$ and $\rho_t E_t^s$, are moduli for steel bars in the ℓ- and t-directions, respectively. ρ_ℓ and ρ_t are the steel ratios. E_ℓ^s and E_t^s represent the smeared stress–strain curves of mild steel bars embedded in concrete ($f_\ell - \varepsilon_\ell$ curve or $f_t - \varepsilon_t$ curve).

The constitutive matrices of concrete and steel shown in Equations (6.7) and (6.8) will be carefully studied in Sections 6.1.2–6.1.10. Section 6.1.2 provides a historical overview of research in RC 2-D elements. Section 6.1.3 studies the Poisson effect in RC 2-D elements, including the concept of biaxial strains versus uniaxial strains, and the constitutive matrices of smeared concrete and smeared steel expressed in terms of Hsu/Zhu ratios. Section 6.1.4 carefully studies the Hsu/Zhu ratios ν_{12} and ν_{21}, including the test methods, the experimental formulas, the effect of Hsu/Zhu ratios on the post-peak behavior, and the strain conversion matrix $[V]$.

The stress–strain curves of concrete and steel in Equations (6.7) and (6.8) must be 'smeared' or 'averaged', because the stress equilibrium equations (6.1)–(6.3), and the strain compatibility equations (6.4)–(6.6), are derived for *continuous* materials. The smeared stress–strain curves of concrete are studied in Sections 6.1.5–6.1.8. The smeared stress–strain curves of mild steel bars are studied in Section 6.1.9.

6.1.2 Research in RC 2-D Elements

The truss model had been applied to treat shear (Ritter, 1899; Morsch, 1902) and torsion (Rausch, 1929) of reinforced concrete since the turn of the 20th century. However, the prediction based on truss model consistently overestimated the shear and torsional strengths of tested specimens. The overestimation might exceed 50% in the case of low-rise shear walls and 30% in the case of torsional members. This nagging mystery had plagued the researchers for over half a century. The source of this difficulty was first understood by Robinson and Demorieux (1972). They realized that a reinforced concrete membrane element subjected to shear stresses is actually subjected to biaxial stresses (principal compression and principal tension). Viewing the shear action as a two-dimensional problem, they discovered that the compressive strength in the direction of principal compression was reduced by cracking due to principal tension in the perpendicular direction. Applying this softened effect of concrete struts to the thin webs of eight I-section test beams, they were able to explain the equilibrium of stresses in the webs according to the truss model. Apparently, the mistake in applying the truss model theory before 1972 was the use of the compressive stress–strain relationship of concrete obtained from the uniaxial tests of standard cylinders without considering this two-dimensional softening effect.

The tests of Robinson and Demorieux, unfortunately, could not delineate the variables that govern the softening coefficient, because of the technical difficulties in the biaxial testing of 2-D elements. The quantification of the softening phenomenon, therefore, had to wait for a decade until a unique 'shear rig' test facility was built by Vecchio and Collins (1981) at the University of Toronto. Based on their tests of 19 panels, 0.89 m square and 70 mm thick, they proposed a softening coefficient that was a function of the tensile principal strains.

The softening coefficient was significantly improved by Hsu and his colleagues at the University of Houston from 1988 to 2009 using the Universal Panel Tester (UPT) (Hsu, Belarbi and Pang, 1995). This large test facility is shown in Figure 6.1. It stands 5 m tall, weighs nearly 40 tons and contains more than a mile of pipes to transport oil pressure to its 40 jacks. Because each jack has a high force capacity of 100 tons, the UPT is capable of testing full-size RC 2-D elements of 1.4 m (55 in.) square and 178 mm (7 in.) thick, as shown in Figure 6.2. Tests of 39 such large specimens by Belarbi and Hsu (1994, 1995) and Pang and

(a) South View

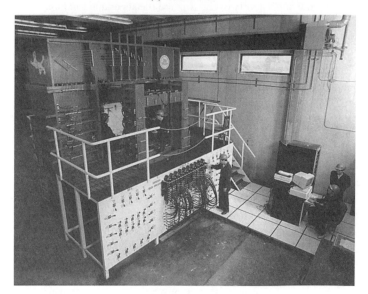

(b) North View

Figure 6.1   Universal panel tester at the University of Houston

Hsu (1995, 1996) resulted in an improved softened coefficient as a function of principal tensile strain $\bar{\varepsilon}_1$ (see Section 6.1.7.1).

In 1995, a servo-control system was installed on the UPT (Hsu, Zhang and Gomez, 1995) so that strain control tests could be performed. Using this equipment, Zhang and Hsu (1998) experimented with RC 2-D elements made of high-strength concrete up to 100 MPa. They

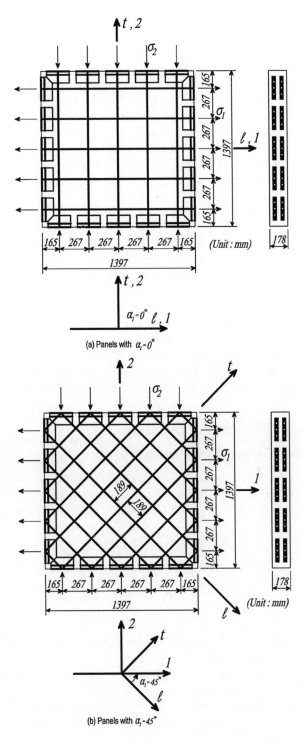

Figure 6.2   Dimension and coordinates of test panels for RC 2-D elements

found experimentally that the softening coefficient was not only a function of the perpendicular tensile strain $\bar{\varepsilon}_1$, but also a function of the concrete compressive strength f'_c. The softening coefficient was then improved to become a function of both $\bar{\varepsilon}_1$ and f'_c (see Section 6.1.7.2).

In addition to the variables $\bar{\varepsilon}_1$ and f'_c, Chintrakarn (2001) and Wang (2006) showed that the softening coefficient is also a function of the deviation angle β (see Section 6.1.7.3) when the fixed angle shear theory is used (including SMM). A new function $f(\beta)$ was established to relate the softening of concrete struts to the deviation angle (β) and, in turn, to the ratio of longitudinal to transverse steel (ρ_ℓ/ρ_t). By treating the softening coefficient (see Section 6.1.7) as a function of all three variables (ε_1, f'_c and β), the fixed angle shear theory (including SMM) becomes very powerful, applicable to 2-D elements with any ρ_ℓ/ρ_t ratio from unity to infinity, and any concrete strength up to 100 MPa (see Section 6.1.7.3).

Using the strain-control feature of UPT, Zhu and Hsu (2002) quantified the Poisson effect of RC 2-D elements (see section 6.1.3) and characterized this property by two Hsu/Zhu ratios (see Section 6.1.4). Taking into account the Poisson effect, Hsu and Zhu (2002) developed the softened membrane model (SMM). SMM can satisfactorily predict the entire monotonic response of the RC 2-D elements, including both the ascending and the descending branches, as well as both the pre-cracking and post-cracking responses.

The servo-control system of UPT also allows the researchers at UH to conduct reversed cyclic shear tests of RC 2-D elements (Mansour and Hsu, 2005a,b). The constitutive relationships of concrete and steel under unloading and reloading were established from these tests, and a cyclic softened membrane model (CSMM) was developed. This model allows us to predict the shear stiffness, the shape of the hysteresis loop, the shear ductility, and the energy dissipation capacity of a 2-D element. CSMM will be studied in Section 6.3.

6.1.3 Poisson Effect in Reinforced Concrete

6.1.3.1 Biaxial Strains versus Uniaxial Strains

A 2-D concrete element defined in the 1–2 coordinate is subjected to two stresses σ_1^c and σ_2^c as shown in Figure 6.3(a). The side lengths of the 2-D element are taken as unity. When the Hsu/Zhu ratios (ν_{12} and ν_{21}) are considered, the strains ε_1 and ε_2 are expressed as follows:

$$\varepsilon_1 = \frac{\sigma_1^c}{\bar{E}_1^c} - \nu_{12}\frac{\sigma_2^c}{\bar{E}_2^c} \tag{6.9}$$

$$\varepsilon_2 = \frac{\sigma_2^c}{\bar{E}_2^c} - \nu_{21}\frac{\sigma_1^c}{\bar{E}_1^c} \tag{6.10}$$

where:

ν_{12} = ratio of the resulting strain increment in the 1-direction to the source strain increment in the 2-direction;

ν_{21} = ratio of the resulting strain increment in the 2-direction to the source strain increment in the 1-direction;

$\bar{E}_1^c, \bar{E}_2^c$ = moduli of concrete in the 1- and 2-directions, respectively, when a panel is subjected to uniaxial loading; or subjected to biaxial loading, but assuming the Hsu/Zhu ratios to be zero.

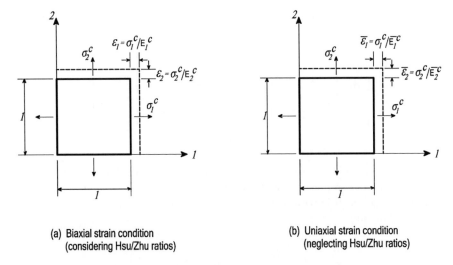

(a) Biaxial strain condition
(considering Hsu/Zhu ratios)

(b) Uniaxial strain condition
(neglecting Hsu/Zhu ratios)

Figure 6.3 Biaxial strains versus uniaxial strains

Let's define $\bar{\varepsilon}_1 = \sigma_1^c/\bar{E}_1^c$ and $\bar{\varepsilon}_2 = \sigma_2^c/\bar{E}_2^c$ in Equations (6.9) and (6.10) as the uniaxial strains; these two equations can now be written as

$$\varepsilon_1 = \bar{\varepsilon}_1 - \nu_{12}\bar{\varepsilon}_2 \qquad (6.11)$$

$$\varepsilon_2 = \bar{\varepsilon}_2 - \nu_{21}\bar{\varepsilon}_1 \qquad (6.12)$$

where:

$\varepsilon_1, \varepsilon_2$ = smeared (average) strains in the 1- and 2-directions, respectively, when a panel is subjected to *biaxial* loading and taking into account the Hsu/Zhu ratios;

$\bar{\varepsilon}_1, \bar{\varepsilon}_2$ = smeared (average) strains in the 1- and 2-directions, respectively, when a panel is subjected to *uniaxial* loading; or subjected to biaxial loading, but assuming the Hsu/Zhu ratios to be zero.

When the Hsu/Zhu ratios are assumed to be zero in Equations (6.11) and (6.12), the biaxial strains are the same as the unixial strains ($\varepsilon_1 = \bar{\varepsilon}_1$ and $\varepsilon_2 = \bar{\varepsilon}_2$). This condition is illustrated in Figure 6.3(b), where the stresses σ_1^c and σ_2^c are simply related to the uniaxial strains $\bar{\varepsilon}_1$ and $\bar{\varepsilon}_2$ by the uniaxial moduli $\bar{E}_1^c$ and $\bar{E}_2^c$, respectively.

6.1.3.2 Constitutive Matrices in Terms of Hsu/Zhu Ratios

Constitutive matrix of smeared concrete
Solving Equations (6.11) and (6.12) gives

$$\bar{\varepsilon}_1 = \frac{1}{1 - \nu_{12}\nu_{21}}\varepsilon_1 + \frac{\nu_{12}}{1 - \nu_{12}\nu_{21}}\varepsilon_2 \qquad (6.13)$$

$$\bar{\varepsilon}_2 = \frac{\nu_{21}}{1 - \nu_{12}\nu_{21}}\varepsilon_1 + \frac{1}{1 - \nu_{12}\nu_{21}}\varepsilon_2 \qquad (6.14)$$

Substituting $\bar{\varepsilon}_1 = \sigma_1^c/\bar{E}_1^c$ and $\bar{\varepsilon}_2 = \sigma_2^c/\bar{E}_2^c$ into Equations (6.13) and (6.14), respectively, gives:

$$
\begin{bmatrix} \sigma_1^c \\ \sigma_2^c \\ \tau_{12}^c \end{bmatrix} = \begin{bmatrix} \dfrac{\bar{E}_1^c}{1-v_{12}v_{21}} & \dfrac{v_{12}\bar{E}_1^c}{1-v_{12}v_{21}} & 0 \\ \dfrac{v_{21}\bar{E}_2^c}{1-v_{12}v_{21}} & \dfrac{\bar{E}_2^c}{1-v_{12}v_{21}} & 0 \\ 0 & 0 & G_{12}^c \end{bmatrix} \begin{bmatrix} \varepsilon_1 \\ \varepsilon_2 \\ \dfrac{\gamma_{12}}{2} \end{bmatrix}
\tag{6.15}
$$

where σ_1^c, σ_2^c = smeared (average) stresses of concrete in the 1- and 2-directions, respectively. The symbols, σ_1^c and σ_2^c, represent both the *biaxial* stresses and the *uniaxial* stresses.

Comparing Equation (6.7) with (6.15) gives the relationships between the biaxial moduli and the uniaxial moduli:

$$
E_1^c = \frac{\bar{E}_1^c}{(1-v_{12}v_{21})}
\tag{6.16}
$$

$$
E_2^c = \frac{\bar{E}_2^c}{(1-v_{12}v_{21})}
\tag{6.17}
$$

Equations (6.16) and (6.17) state that the biaxial moduli of concrete are simply the uniaxial moduli of concrete divided by $(1 - v_{12}v_{21})$. This is applicable in both directions of principal compression and principal tension.

Constitutive matrix of smeared steel bars

In the SMM, the constitutive matrix of steel bars is derived from the compatibility equations (6.4) and (6.5) for biaxial strains ε_1, ε_2, ε_ℓ and ε_t. Substituting Equations (6.11) and (6.12) into Equations (6.4) and (6.5) results in a set of two long equations, each with five terms on the right-hand side:

$$
\varepsilon_\ell = \bar{\varepsilon}_1 \cos^2\alpha_1 + \bar{\varepsilon}_2 \sin^2\alpha_1 - \frac{\gamma_{12}}{2}2\sin\alpha_1\cos\alpha_1 - v_{12}\bar{\varepsilon}_2\cos^2\alpha_1 - v_{21}\bar{\varepsilon}_1\sin^2\alpha_1 \tag{6.18}
$$

$$
\varepsilon_t = \bar{\varepsilon}_1 \sin^2\alpha_1 + \bar{\varepsilon}_2 \cos^2\alpha_1 + \frac{\gamma_{12}}{2}2\sin\alpha_1\cos\alpha_1 - v_{12}\bar{\varepsilon}_2\sin^2\alpha_1 - v_{21}\bar{\varepsilon}_1\cos^2\alpha_1 \tag{6.19}
$$

Notice that the first three terms on the right-hand side of Equations (6.18) and (6.19) are the transformation forms of unaxial strains $\bar{\varepsilon}_\ell$ and $\bar{\varepsilon}_t$:

$$
\bar{\varepsilon}_\ell = \bar{\varepsilon}_1 \cos^2\alpha_1 + \bar{\varepsilon}_2 \sin^2\alpha_1 - \frac{\gamma_{12}}{2}2\sin\alpha_1\cos\alpha_1 \tag{6.20}
$$

$$
\bar{\varepsilon}_t = \bar{\varepsilon}_1 \sin^2\alpha_1 + \bar{\varepsilon}_2 \cos^2\alpha_1 + \frac{\gamma_{12}}{2}2\sin\alpha_1\cos\alpha_1 \tag{6.21}
$$

Substituting Equations (6.20) and (6.21) into Equations (6.18) and (6.19), respectively, gives:

$$
\varepsilon_\ell = \bar{\varepsilon}_\ell - v_{12}\bar{\varepsilon}_2\cos^2\alpha_1 - v_{21}\bar{\varepsilon}_1\sin^2\alpha_1 \tag{6.22}
$$

$$
\varepsilon_t = \bar{\varepsilon}_t - v_{12}\bar{\varepsilon}_2\sin^2\alpha_1 - v_{21}\bar{\varepsilon}_1\cos^2\alpha_1 \tag{6.23}
$$

Equations (6.22) and (6.23) show that the biaxial steel strains (ε_ℓ, ε_t) are the uniaxial steel strains ($\bar{\varepsilon}_\ell$, $\bar{\varepsilon}_t$) minus two strain terms. These two additional steel strains are the components of the products of Hsu/Zhu ratios and concrete strains ($\nu_{12}\bar{\varepsilon}_2$ and $\nu_{21}\bar{\varepsilon}_1$), and represent Poisson effect on the stresses and strains of smeared steel bars.

Substituting $\bar{\varepsilon}_\ell = f_\ell/\bar{E}_\ell^s$ and $\bar{\varepsilon}_t = f_t/\bar{E}_t^s$ into Equations (6.22) and (6.23) results in:

$$
\begin{bmatrix} \rho_\ell f_\ell \\ \rho_t f_t \\ 0 \end{bmatrix} = \begin{bmatrix} \rho_\ell \bar{E}_\ell^s & 0 & 0 \\ 0 & \rho_t \bar{E}_t^s & 0 \\ 0 & 0 & 0 \end{bmatrix} \begin{bmatrix} \varepsilon_\ell \\ \varepsilon_t \\ 0 \end{bmatrix} + \begin{bmatrix} \rho_\ell \bar{E}_\ell^s(\nu_{12}\bar{\varepsilon}_2 \cos^2 \alpha_1) + \rho_\ell \bar{E}_\ell^s(\nu_{21}\bar{\varepsilon}_1 \sin^2 \alpha_1) \\ \rho_t \bar{E}_t^s(\nu_{12}\bar{\varepsilon}_2 \sin^2 \alpha_1) + \rho_t \bar{E}_t^s(\nu_{21}\bar{\varepsilon}_1 \cos^2 \alpha_1) \\ 0 \end{bmatrix}
$$

$$(6.24)$$

where:

f_ℓ, f_t = smeared steel stress of steel bars embedded in concrete in the ℓ- and t-directions, respectively. The symbols f_ℓ and f_t represent the *biaxial* steel stresses, as well as the *uniaxial* steel stresses;

$\bar{E}_\ell^s, \bar{E}_t^s$ = secant modulus of steel bars embedded in concrete in the ℓ and t directions, respectively, calculated from the constitutive law of embedded steel bars (smeared stress versus smeared strain) obtained under uniaxial loading.

Equation (6.24) shows that, under biaxial stress conditions, the smeared steel stresses in the longitudinal direction $\rho_\ell f_\ell$ and in the transverse direction $\rho_t f_t$, are each made up of two parts. The first part is the uniaxial steel stresses using the uniaxial steel moduli without taking into account the Hsu/Zhu ratios. The second part is the steel stresses produced by the two Hsu/Zhu ratios and the two uniaxial strains. If the Hsu/Zhu ratios are assumed to be zero, then the second part will disappear and Equation (6.24) is reduced to the familiar expression of smeared steel stress under uniaxial condition (note that ε_ℓ becomes $\bar{\varepsilon}_\ell$ and ε_t becomes $\bar{\varepsilon}_t$). Comparison of Equations (6.24) and (6.8) gives the relationships between the biaxial steel moduli (E_ℓ^s, E_t^s) and the uniaxial steel moduli ($\bar{E}_\ell^s, \bar{E}_t^s$) as:

$$E_\ell^s = \bar{E}_\ell^s \left[1 + \nu_{12}(\bar{\varepsilon}_2/\varepsilon_\ell) \cos^2 \alpha_1 + \nu_{21}(\bar{\varepsilon}_1/\varepsilon_\ell) \sin^2 \alpha_1 \right] \qquad (6.25)$$

$$E_t^s = \bar{E}_t^s \left[1 + \nu_{12}(\bar{\varepsilon}_2/\varepsilon_t) \sin^2 \alpha_1 + \nu_{21}(\bar{\varepsilon}_1/\varepsilon_t) \cos^2 \alpha_1 \right] \qquad (6.26)$$

Equations (6.25) and (6.26) state that the biaxial moduli of steel is the uniaxial moduli of steel plus two additional term caused by Poisson effect. This concept is applicable in both directions of longitudinal steel and transverse steel.

6.1.4 Hsu/Zhu Ratios ν_{12} and ν_{21}

6.1.4.1 Test Methods

Reinforced concrete 2-D elements subjected to shear behave very differently before and after cracking. The Poisson effect before cracking is characterized by the well-known Poisson ratio, but after cracking is characterized by two Hsu/Zhu ratios.

Twelve panels as shown in Figure 6.2(a) and (b) were tested to establish the formulas for Hsu/Zhu ratios ν_{12} and ν_{21}. In the 1–2 coordinate system, the panels were subjected to a horizontal tensile stress σ_1 and a vertical compressive stress σ_2 in a proportional manner. The

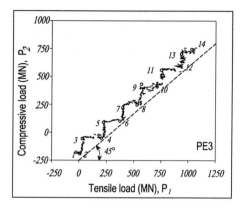

(a) Loading procedure using load-control mode before yielding

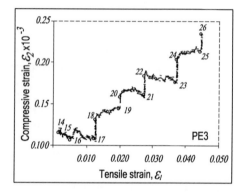

(b) Loading procedure using strain-control mode after yielding

Figure 6.4   Loading procedure

proportional loading was simulated by small, step-wise increments so that the Hsu/Zhu ratios, v_{12} and v_{21}, could be measured at each load step. A load-control procedure was used to apply the external stresses σ_1 and σ_2 before the steel yielded, while a strain-control procedure was used after the yielding of steel.

Load-control procedure before yielding

Figure 6.4(a) shows a step-wise proportional loading using load-control procedure before yielding. When the compressive stress σ_2 was increased and the tensile stress σ_1 was maintained constant, the source strain increment $\Delta\varepsilon_2$ and the resulting strain increment $\Delta\varepsilon_1$ were measured at this loading step. The Hsu/Zhu ratio v_{12} is then calculated from the measured strain increments as

$$v_{12} = -\left(\frac{\Delta\varepsilon_1}{\Delta\varepsilon_2}\right) \tag{6.27}$$

Similarly, when the tensile stress σ_1 was increased with a constant compressive stress σ_2, the source strain increment $\Delta\varepsilon_1$ and the resulting strain increment $\Delta\varepsilon_2$ were measured. The Hsu/Zhu ratio ν_{21} was then calculated by

$$\nu_{21} = -\left(\frac{\Delta\varepsilon_2}{\Delta\varepsilon_1}\right) \tag{6.28}$$

Strain-control procedure after yielding

Figure 6.4(b) shows a step-wise proportional loading using strain-control procedure after yielding. When the load approached the first yielding of steel bars, a mode switch was made from load-control to strain-control. When the compressive strain ε_2 was increased and the tensile strain ε_1 was maintained constant, the strain increment $\Delta\varepsilon_2$ and the stress increment $\Delta\sigma_1$ were measured. The Hsu/Zhu ratio ν_{12} was then calculated as

$$\nu_{12} = -\left(\frac{\Delta\sigma_1}{\bar{E}_1 \Delta\varepsilon_2}\right) \tag{6.29}$$

where $\bar{E}_1$ (always positive) was the unloading modulus of the RC 2-D element because the tensile stress σ_1 always decreased (i.e. $\Delta\sigma_1$ was negative) under constant strain ε_1. $\bar{E}_1$ was calculated from the next reloading modulus, because the unloading modulus had been observed experimentally to be equal to the initial linear portion of the reloading σ_1–ε_1 curve. $\bar{E}_1$ was then calculated by

$$\bar{E}_1 = \frac{(\Delta\sigma_1)_{linear}}{(\Delta\varepsilon_1)_{linear}} \tag{6.30}$$

where $(\Delta\sigma_1)_{linear}$ and $(\Delta\varepsilon_1)_{linear}$ were the stress increment and strain increment, respectively, in the linear portion of the reloading σ_1– ε_1 curve.

In the step where the tensile strain ε_1 was increased and the compressive strain ε_2 was maintained constant, the strain increment $\Delta\varepsilon_1$ and the stress increment $\Delta\sigma_2$ were measured. The ratio ν_{21} was then calculated by

$$\nu_{21} = -\left(\frac{\Delta\sigma_2}{\bar{E}_2 \Delta\varepsilon_1}\right) \tag{6.31}$$

where $\bar{E}_2$ (always positive) was either the unloading or the loading modulus of the concrete element, depending on whether $\Delta\sigma_2$ is positive or negative. For example, in the end level 22 to 23, the compression stress increased ($\Delta\sigma_2$ is positive) and ν_{21} became negative.

6.1.4.2 Formulas for Hsu/Zhu Ratios

The measured Hsu/Zhu ratios, ν_{12} and ν_{21}, are plotted in Figure 6.5(a) and (b), respectively, against the steel strain ε_{sf}. The symbol ε_{sf} is defined as the strain in the steel bars that yield first. In panels with $\alpha_1 = 0°$ (Figure 6.2a), ε_ℓ is always in tension and ε_t is always in compression. Therefore, $\varepsilon_{sf} = \varepsilon_\ell$. In panels with $\alpha_1 = 45°$, Fig. 6.2 (b), ε_{sf} could be ε_ℓ or ε_t, whichever yields first.

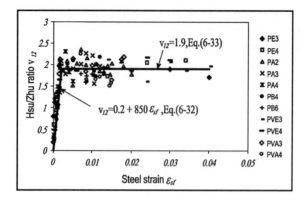

(a) Hsu/Zhu ratio v $_{12}$

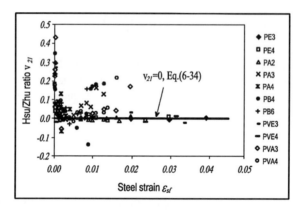

(b) Hsu/Zhu ratio v $_{21}$

Figure 6.5   Hsu/Zhu ratio ν_{12} and ν_{21}

Based on the test data shown in Figure 6.5(a), the following equation is suggested for the Hsu/Zhu ratio ν_{12}:

$$\nu_{12} = 0.2 + 850\varepsilon_{sf} \qquad \varepsilon_{sf} \leq \varepsilon_y \qquad (6.32)$$

$$\nu_{12} = 1.9 \qquad \varepsilon_{sf} > \varepsilon_y \qquad (6.33)$$

Figure 6.5(a) clearly shows that the Hsu/Zhu ratio ν_{12} lies outside the valid range of 0–0.5 for the Poisson ratio of continuous materials.

The measured data for Hsu/Zhu ratio ν_{21} are plotted in Figure 6.5(b). It can be seen that ν_{21} is about 0.2–0.25 before cracking took place in the test panels. After cracking, however, ν_{21} reduces rapidly to about 0.05 and then gradually approaches zero at yielding. Near the yielding stage, negative ν_{21} begins to occur. It is interesting to note that the negative and positive values of ν_{21} occur alternatively in some specimens. For simplicity in an analytical model, ν_{21} is assumed to be zero for the whole post-cracking range as

$$\nu_{21} = 0 \tag{6.34}$$

It is interesting to note that twelve full-size RC 2-D elements (panels) have been tested to establish the two Hsu/Zhu ratios. These panels include four variables: the percentage of steel ($\rho = 0.77$–3.04%), the steel bar orientation ($\alpha_1 = 0°$ and $45°$), the ratio of steel percentages in the transverse and longitudinal direction ($\eta = 0.24$–1.0), and the strength of concrete (50 and 90 MPa). Test results show that the Hsu/Zhu ratios are not a function of any of these four variables within the usable ranges.

6.1.4.3 Effect of Hsu/Zhu Ratios on Post-peak Behavior

Now let us demonstrate the important effect of Hsu/Zhu ratios on the post-peak behavior of RC 2-D elements by examining the equilibrium equation (6.1). In order to simplify Equation (6.1) we take the following three measures. (a) Under pure shear loading, the applied stresses, σ_ℓ, on the left-hand side of Equation (6.1) should be zero. (b) For a specimen with the same steel ratios in both the longitudinal and the transverse directions, the smeared shear stress of concrete τ_{12}^c is zero, because the deviation angle $\beta = 0$. (c) The smeared tensile stress of concrete σ_1^c can be neglected, because its magnitude is very small when compared with the smeared compressive stress of concrete (σ_2^c) and the smeared steel stresses ($\rho_\ell f_\ell$ and $\rho_t f_t$).

Setting $\sigma_\ell = 0$, $\tau_{12}^c = 0$, and $\sigma_1^c = 0$, Equation (6.1) is simplified to:

$$0 = \sigma_2^c \sin^2 \alpha_1 + \rho_\ell f_\ell \tag{6.35}$$

If $\rho_\ell f_\ell$ in Equation (6.35) is treated as a uniaxial steel stress, Equation (6.35) represents the basic 'truss-concept' of the internal balance between the compressive stress of concrete struts σ_2^c and the tensile stresses of steel bars $\rho_\ell f_\ell$. In the ascending branch of the load–deformation curves, both the concrete stress σ_2^c and the steel stress $\rho_\ell f_\ell$ increase, and the internal equilibrium is maintained. The peak point of a load–deformation curve represents physically the concrete compressive stress σ_2^c reaching its peak. Beyond the peak point, the concrete stress σ_2^c begins to decrease, while the steel stresses $\rho_\ell f_\ell$ continue to increase. As a result, the equilibrium equation (6.35) cannot be satisfied and the computer operation comes to a halt at the peak point.

If $\rho_\ell f_\ell$ in Equation (6.35) is treated as a biaxial steel stress subjected to Poisson effect, however, $\rho_\ell f_\ell$ in Equation (6.35) should be replaced by the expression in Equation (6.24) to give:

$$0 = \sigma_2^c \sin^2 \alpha_1 + \rho_\ell \bar{E}_\ell^s \varepsilon_\ell + \rho_\ell \bar{E}_\ell^s \left(\nu_{12} \bar{\varepsilon}_2 \cos^2 \alpha_1 \right) + \rho_\ell \bar{E}_\ell^s \left(\nu_{21} \bar{\varepsilon}_1 \sin^2 \alpha_1 \right) \tag{6.36}$$

Equation (6.36) includes two additional terms with Hsu/Zhu ratios. The last term is zero because $\nu_{21} = 0$. The remaining term with Hsu/Zhu ratio ν_{12} is caused by compressive strain $\bar{\varepsilon}_2$ and should be negative. This term, which is induced by Poisson effect, reduces the uniaxial steel stress, $\rho_\ell \bar{E}_\ell^s \varepsilon_\ell$, so that equilibrium can be restored in the post-peak range.

6.1.4.4 Strain Conversion Matrices

To solve the equilibrium and compatibility equations, (6.1)–(6.6), the constitutive relationships of concrete and steel have to be introduced. As pointed out previously, the compatibility equations (6.4)–(6.6) are expressed in terms of biaxial strains (ε_1, ε_2, ε_ℓ, ε_t). These biaxial strains could not be related directly to the stresses (σ_1^c, σ_2^c, f_ℓ, f_t) in the equilibrium equations (6.1)–(6.3) by experimental constitutive laws, because the Poisson effect could not be ignored in the tests. Therefore, the biaxial strains (ε_1, ε_2, ε_ℓ, ε_t) must first be converted to the uniaxial strains ($\bar\varepsilon_1$, $\bar\varepsilon_2$, $\bar\varepsilon_\ell$, $\bar\varepsilon_t$).

Uniaxial strains ($\bar\varepsilon_1$, $\bar\varepsilon_2$) of smeared concrete
The uniaxial strains in the 1–2 coordinate ($\bar\varepsilon_1$, $\bar\varepsilon_2$) of smeared concrete can be calculated from the biaxial strains in the 1–2 coordinate (ε_1, ε_2) using Equations (6.13) and (6.14). These strain conversion relationships can be generalized into in a 3×3 matrix equation as follows:

$$
\begin{bmatrix} \bar\varepsilon_1 \\ \bar\varepsilon_2 \\ \dfrac{\gamma_{12}}{2} \end{bmatrix} = \begin{bmatrix} \dfrac{1}{1-\nu_{12}\nu_{21}} & \dfrac{\nu_{12}}{1-\nu_{12}\nu_{21}} & 0 \\[2ex] \dfrac{\nu_{21}}{1-\nu_{12}\nu_{21}} & \dfrac{1}{1-\nu_{12}\nu_{21}} & 0 \\[2ex] 0 & 0 & 1 \end{bmatrix} \begin{bmatrix} \varepsilon_1 \\ \varepsilon_2 \\ \dfrac{\gamma_{12}}{2} \end{bmatrix} \tag{6.37}
$$

The 3×3 matrix in Equation (6.37) is called the strain conversion matrix $[V]$ for smeared concrete.

In Equations (6.37), or (6.13) and (6.14), the two Hsu/Zhu ratios ν_{12} and ν_{21} have been determined experimentally in Section 6.1.4.2. When $\nu_{21} = 0$ after cracking, Equation (6.37) can be reduced to:

$$\bar\varepsilon_1 = \varepsilon_1 + \nu_{12}\varepsilon_2 \tag{6.38}$$

$$\bar\varepsilon_2 = \varepsilon_2 \tag{6.39}$$

It is interesting to observe in Equation (6.39) that the uniaxial strain $\bar\varepsilon_2$ is identical to the biaxial strain ε_2 in principal compression. In Equation (6.38), however, the term $\nu_{12}\varepsilon_2$ represents the Poisson effect. Since ε_2 is always negative and $\nu_{12} = 1.9$ after the yielding of the steel, the uniaxial strain $\bar\varepsilon_1$ is equal to the biaxial strain ε_1 in principal tension minus 1.9 times the biaxial strain of principal compression ε_2. This $\nu_{12}\varepsilon_2$ term can be quite large in the post-peak range of the load–deformation curves.

The uniaxial strains ($\bar\varepsilon_1$ and $\bar\varepsilon_2$) obtained using the conversion matrix $[V]$ in Equation (6.37), or from Equations (6.13) and (6.14), can be used to calculate the stresses in the concrete (σ_1^c and σ_2^c) in the equilibrium equations (6.1)–(6.3). The constitutive relationships between stresses (σ_1^c, σ_2^c) and the uniaxial strains ($\bar\varepsilon_1$, $\bar\varepsilon_2$) are obtained directly from tests, and are given in the following Sections 6.1.6–6.1.8.

Uniaxial strains ($\bar\varepsilon_\ell$ and $\bar\varepsilon_t$) of smeared steel bars
The uniaxial strains in the $\ell - t$ coordinate ($\bar\varepsilon_\ell$ and $\bar\varepsilon_t$) of the smeared steel bars can be calculated from the uniaxial strains in the 1–2 coordinate ($\bar\varepsilon_1$, $\bar\varepsilon_2$) using the transformation equations (6.20) and (6.21). These strain transformation relationships can be generalized into

a 3 × 3 matrix equation as follows:

$$
\begin{bmatrix} \bar{\varepsilon}_\ell \\ \bar{\varepsilon}_t \\ \dfrac{\gamma_{\ell t}}{2} \end{bmatrix} = [T]^{-1} \begin{bmatrix} \bar{\varepsilon}_1 \\ \bar{\varepsilon}_2 \\ \dfrac{\gamma_{12}}{2} \end{bmatrix}
\tag{6.40}
$$

where $[T]^{-1}$ is the inverse transformation matrix given in Equation (4.49), Chapter 4.

The two uniaxial strains $(\bar{\varepsilon}_\ell, \bar{\varepsilon}_t)$ obtained from Equation (6.40), or the two equations (6.20) and (6.21), can then be used to calculate the stresses in the smeared steel bars (f_ℓ and f_t) in the equilibrium equations (6.1)–(6.3). The constitutive relationships connecting the stresses (f_ℓ, f_t) and the uniaxial strains $(\bar{\varepsilon}_\ell, \bar{\varepsilon}_t)$ are obtained directly from tests, and are given in Section 6.1.9.

6.1.5 Experimental Stress–Strain Curves

The experimental stress-strain curves are obtained from testing RC 2-D elements, as shown in Figure 6.2, in the universal panel tester (UPT) shown in Figure 6.1. All the test specimens are subjected to a tensile stress σ_1 in the horizontal direction, and to a compressive stress σ_2 in the vertical direction. As such, the horizontal and vertical axes constitute the principal 1–2 coordinate. The mild steel bars in a 2-D element, which define the $\ell - t$ coordinate, could be oriented in any directions. Figure 6.2(a) and (b) show two typical cases where the angle α_1 is equal to $0°$ and $45°$, respectively. The special case of pure shear is achieved by applying the stresses σ_1 and σ_2 in equal magnitude on a 2-D element with $\alpha_1 = 45°$ (Figure 6.2b). A pure shear stress $\tau_{\ell t}$ in the $\ell - t$ coordinate is created in the $45°$ direction. This shear stress $\tau_{\ell t}$ could be increased in load stages until the failure of the test panel.

The strains of a test panel were measured by two sets of LVDT (linear voltage differential transformer) rosettes, one on each face of a panel. Each set of rosette consists of six LVDTs: two in the horizontal direction to measure the principal tensile strains ε_1, two in the vertical direction to measure the principal compressive strains ε_2, and two oriented at $45°$ to measure the steel bar strains, ε_ℓ and ε_t, in the case of pure shear panels, Figure 6.2(b). In order to increase the accuracy of the measured compressive strains ε_2 (which is much smaller than the other three strains ε_1, ε_ℓ and ε_t), four additional LVDTs (two on each face) were added in the vertical directions. In short, a total of 16 LVDTs were installed.

The six LVDTs of a rosette are anchored to the panel at the four corners of a 0.8 m (31.5 in.) square, so that the four horizontal and vertical LVDTs measure the displacements over a length of 0.8 m and the two diagonal LVDTs over a length of 1.13 m. The principal strains ε_1 and ε_2 are calculated from the measured displacements divided by a length of 0.8 m, while the diagonal strains ε_ℓ and ε_t are divided by a length of 1.13 m. Since the measured lengths of 0.8 and 1.13 m span over several cracks, the measured strains are 'smeared strains', or 'average strains', that include the gaps of the crack widths.

The pure shear test described above could produce a softened compressive stress–strain $(\sigma_2^c - \bar{\varepsilon}_2)$ curve for concrete. The plotting of this $\sigma_2^c - \bar{\varepsilon}_2$ curve from the test results was based on the following methodology:

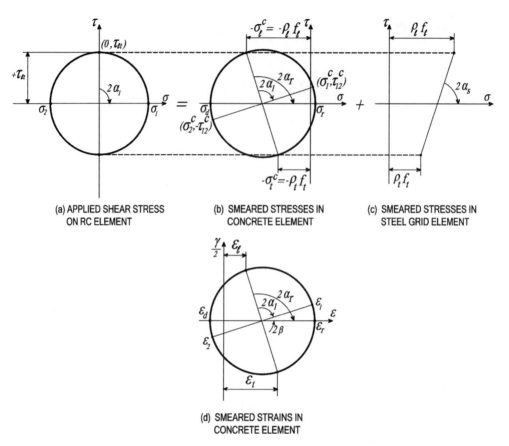

(a) APPLIED SHEAR STRESS
ON RC ELEMENT

(b) SMEARED STRESSES IN
CONCRETE ELEMENT

(c) SMEARED STRESSES IN
STEEL GRID ELEMENT

(d) SMEARED STRAINS IN
CONCRETE ELEMENT

Figure 6.6   Mohr circles for smeared stresses and smeared strains in 2-D test elements under pure shear

1. At each load stage in the test panel, the stress condition due to the applied shear stress $\tau_{\ell t}$ was shown by a Mohr circle in Figure 6.6(a).
2. At each load stage, the smeared strains were measured on the faces of a panel in all four directions (ε_1, ε_2, ε_ℓ, ε_t). From these four measured strains the Mohr circle for strains was established in Figure 6.6(d) by minimizing the measurement errors (Actually only three strains are required to establish a Mohr circle, the fourth being an added redundant strain to increase the accuracy). From this Mohr strain circle, the strain in the $r - d$ coordinate, ε_r and ε_d, and the rotating angle α_r could be determined.
3. From the two measured steel strains, ε_ℓ and ε_t, the smeared steel stresses $\rho_\ell f_\ell$ and $\rho_t f_t$ were calculated from the constitutive laws of embedded steel and were recorded in Figure 6.6(c). From the equilibrium of stresses in the longitudinal and transverse directions the concrete stresses were computed: $-\sigma_\ell^c = -\rho_\ell f_\ell$, and $-\sigma_t^c = -\rho_t f_t$. Using the concrete stresses σ_ℓ^c, σ_t^c and $\tau_{\ell t}^c$ ($\tau_{\ell t}^c$ is equal to $\tau_{\ell t}$), the Mohr circle for concrete stresses was completed in Figure 6.6(b). From this Mohr circle, the concrete stresses in the $r - d$ coordinate σ_r and σ_d and the rotating-angle α_r could be obtained. By repeating this procedure for all the load

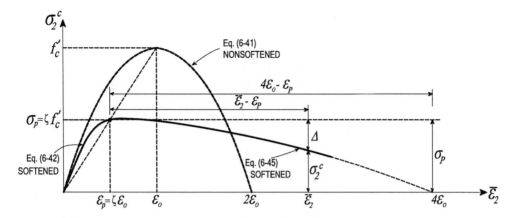

Figure 6.7 Compressive stress–strain curve of concrete

stages, a complete σ_d–ε_d curve can be established for the rotating angle theory as given in Figures 5.12 and 5.13 (Chapter 5).

4. In the case of fixed angle theory, the measured biaxial strain ε_2 shown in Figure 6.6(d) is equal to the uniaxial strain $\bar{\varepsilon}_2$ (see Equation 6.39). From the Mohr stress circle in Figure 6.6(b), the concrete stresses in the 1–2 coordinate (σ_1^c, σ_2^c and τ_{12}^c) could be determined. The concrete stress σ_2^c and the strain $\bar{\varepsilon}_2$ give a point on the softened compressive stress–strain curve of concrete, as shown in Figure 6.7. By repeating this procedure for all the load stages, a complete σ_2^c–$\bar{\varepsilon}_2$ curve can be established.

The analytical stress–strain ($\sigma_2^c - \bar{\varepsilon}_2$) curve for the softened membrane model (SMM), however, is not obtained simply from one panel according to step 4. This is because the characteristic of the curve would be affected by the shear stress τ_{12}^c on the concrete struts. Thus, the stress–strain curve must be a function of the deviation angle β, which depends on the steel percentages in the longitudinal and transverse directions. The softened coefficient ζ in the stress–strain curve must be a function of β, and must be obtained from tests of many panels with various steel percentages in the two directions. This function of β will be derived in Section 6.1.7.3.

6.1.6 Softened Stress–Strain Relationship of Concrete in Compression

The stress–strain curve of a standard concrete cylinder subjected to a uniaxial compression is usually expressed mathematically by a parabolic curve:

$$\sigma_2^c = f_c' \left[2 \left(\frac{\bar{\varepsilon}_2}{\varepsilon_o} \right) - \left(\frac{\bar{\varepsilon}_2}{\varepsilon_o} \right)^2 \right] \tag{6.41}$$

where ε_o is the strain at the peak stress f_c', and is usually taken as 0.002. Equation (6.41) is shown as a dotted curve in Figure 6.7. It is labeled as a nonsoftened stress–strain curve, because the uniaxial tests are performed without perpendicular tensile stress.

The compressive stress–strain curve of concrete in a 2-D element subjected to shear exhibits three characteristics, as shown in Figure 6.7. First, the peak point is reduced or 'softened' in both stress and strain. The locus of the peak point follows a straight line from the nonsoftened peak to the origin. Second, the pre-peak ascending curve is found to be parabolic. Third, the post-peak descending curve is also a parabolic curve, but the gently sloping curve intersects the horizontal axis at a large strain of $4\varepsilon_o$.

The *ascending branch* of the softened stress–strain curve can be expressed as:

$$\sigma_2^c = \zeta f_c' \left[2 \left(\frac{\bar{\varepsilon}_2}{\zeta \varepsilon_o} \right) - \left(\frac{\bar{\varepsilon}_2}{\zeta \varepsilon_o} \right)^2 \right] \qquad \bar{\varepsilon}_2/\zeta \varepsilon_o \le 1 \tag{6.42}$$

where ζ is the softened coefficient. Notice that peak concrete stress f_c' is multiplied by ζ to achieve the effect of stress softening, and the strain ε_o at peak point is multiplied by ζ to achieve the effect of strain softening. Proportional softening is accomplished by using the same ζ for both stress and strain softening. For a family of such curves with decreasing softening coefficient ζ, the locus of the peak points traces a straight line passing through the origin.

The *descending branch* of the softened stress–strain curve is derived as follows. The descending portion of the solid curve in Figure 6.7 is assumed to be a parabolic curve from the peak point to the point of $4\varepsilon_o$ on the horizontal axis. The vertical distance from the parabolic curve to the peak stress level is designated as Δ. This vertical distance Δ is located at a horizontal distance $\bar{\varepsilon}_2 - \varepsilon_p$ from the peak point. At a horizontal distance of $4\varepsilon_o - \varepsilon_p$ from the peak point, however, the vertical distance from the parabolic curve to the peak stress level is σ_p. The ratio of Δ/σ_p can be obtained from the geometry of parabolic shape:

$$\frac{\Delta}{\sigma_p} = \left(\frac{\bar{\varepsilon}_2 - \varepsilon_p}{4\varepsilon_o - \varepsilon_p} \right)^2 \tag{6.43}$$

Then, the stress σ_2^c at the location of $\bar{\varepsilon}_2$ is

$$\sigma_2^c = \sigma_p - \Delta = \sigma_p \left[1 - \left(\frac{\bar{\varepsilon}_2 - \varepsilon_p}{4\varepsilon_o - \varepsilon_p} \right)^2 \right] \tag{6.44}$$

or more convenient for calculation:

$$\sigma_2^c = \zeta f_c' \left[1 - \left(\frac{\bar{\varepsilon}_2/\zeta \varepsilon_o - 1}{4/\zeta - 1} \right)^2 \right] \qquad \bar{\varepsilon}_2/\zeta \varepsilon_o \ge 1 \tag{6.45}$$

The compressive stress σ_2^c calculated by Equation (6.45) should be limited to $0.5\zeta f_c'$, based on test results.

6.1.7 Softening Coefficient ζ

The softening coefficient ζ in Equations (6.42) and (6.45) is the most important parameter affecting the compressive stress–strain relationship of cracked reinforced concrete. ζ is a function of three variables: the uniaxial tensile strain $\bar{\varepsilon}_1$ in the perpendicular direction, the

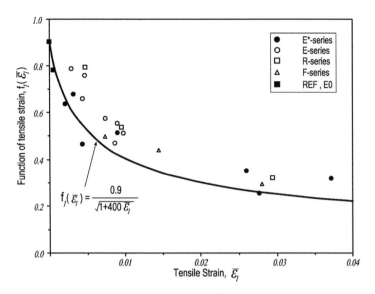

Figure 6.8   Function of tensile strain, $f_1(\bar{\varepsilon}_1)$

concrete compressive strength f_c', and the deviation angle β as follows:

$$\zeta = f_1(\bar{\varepsilon}_1)\, f_2\left(f_c'\right) f_3(\beta) \tag{6.46}$$

where

$$f_1(\bar{\varepsilon}_1) = \frac{1}{\sqrt{1 + 400\bar{\varepsilon}_1}} \tag{6.47}$$

$$f_2\left(f_c'\right) = \frac{5.8}{\sqrt{f_c'}} \le 0.9 \;\left(f_c' \text{ and } \sqrt{f_c'} \text{ in MPa}\right), \tag{6.48}$$

$$f_3(\beta) = 1 - \frac{|\beta|}{24^\circ} \tag{6.49}$$

These three functions, $f_1(\bar{\varepsilon}_1)$, $f_2\left(f_c'\right)$ and $f_3(\beta)$, will be elaborated below.

6.1.7.1 Function of Tensile Strain $f_1(\bar{\varepsilon}_1)$

Belarbi and Hsu (1995) studied the effect of five variables in 2-D elements (Figure 6.2a) on the softening coefficient ζ: (1) the principal tensile strain $\bar{\varepsilon}_1$; (2) the presence of tensile stress at failure σ_1; (3) the loading path (sequential and proportional); (4) spacing of reinforcing bars; and (5) the amount of main longitudinal reinforcement ρ. They confirmed that the principal tensile strain is the dominant variable, as shown in Figure 6.8. The presence of tensile stress at failure has no effect. The load path, the spacing of reinforcing bars and the amount of longitudinal reinforcement have some small effect, but can be neglected for simplicity.

The fact that the softening coefficient ζ is primarily a function of the tensile strain ε_1 makes physical sense. ζ must decrease with an increase of a parameter that measures the severity

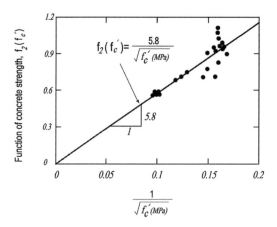

Figure 6.9   Function of concrete strength, $f_2(f'_c)$

of cracking. The most important parameter to measure the severity of cracking is the tensile strain $\bar{\varepsilon}_1$.

Pang and Hsu (1996) studied the behavior of 2-D elements subjected to pure shear (Figure 6.2b). The test specimens were reinforced with various amount of steel bars oriented in the 45° directions ($\ell - t$ coordinate). The test results of 13 panel specimens confirmed the following equation given by Belarbi and Hsu (1995) for the softened coefficient:

$$\zeta = \frac{0.9}{\sqrt{1 + 400\bar{\varepsilon}_1}} \tag{6.50}$$

Equation (6.50) is valid for normal strength concrete of 42 MPa (6000 psi) or less. When the tensile strain $\bar{\varepsilon}_1 = 0$, $\zeta = 0.9$. The constant of 0.9 was determined by the tests of two reference panels subjected to uniaxial compression. It takes into account the size effect, the shape effect and the loading rate effect between the testing of standard cylinders and the testing of panels.

6.1.7.2   Function of Concrete Strength $f_2\left(f'_c\right)$

Utilizing the servo-control system on the UPT, Zhang and Hsu (1998) experimented with RC 2-D elements made of high-strength concrete of 70 and 100 MPa. They found experimentally that the softening coefficient was not only a function of the perpendicular tensile strain $\bar{\varepsilon}_1$, but also a function of the concrete compressive strength f'_c. Figure 6.9 shows that the function of concrete strength $f_2\left(f'_c\right)$ is inversely proportional to $\sqrt{f'_c}$ and can be expressed as $5.8/\sqrt{f'_c(MPa)}$. The groups of test data include panels with f'_c of 42, 70 and 100 MPa.

Thus, for concrete strength up to 100 MPa, the softening coefficient ζ can be expressed as a function of both ε_1 and f'_c as follows:

$$\zeta = \left(\frac{5.8}{\sqrt{f'_c(MPa)}} \leq 0.9\right)\left(\frac{1}{\sqrt{1 + 400\bar{\varepsilon}_1}}\right) \tag{6.51}$$

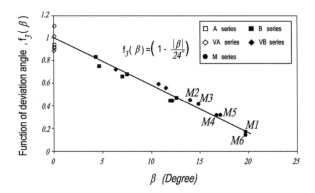

Figure 6.10 Function of deviation angle, $f_3(\beta)$

When concrete strength f'_c is less than 42 MPa, $5.8/\sqrt{f'_c(MPa)}$ is greater than 0.9. The constant of 0.9 governs and Eq. (6.51) degenerates into Eq. (6.50). When f'_c is greater than 42 MPa, however, $5.8/\sqrt{f'_c(MPa)}$ is less than 0.9 and this term will dominate. ζ calculated by Equation (6.51) will be less than that calculated by Equation (6.50). Equation (6.51) provides a simple concept: the higher the concrete strength f'_c, the lower the softened coefficient ζ.

6.1.7.3 Function of Deviation Angle $f_3(\beta)$

In the fixed angle theory, the softening coefficient ζ is also a function of the deviation angle β, in addition to the variables ε_1 and f'_c. The angle β is defined as: $\beta = \alpha_1 - \alpha_r$, the difference between the angle α_1 of the applied principal stresses in the 1–2 coordinate and the angle α_r of the principal concrete stresses in the $r - d$ coordinate. Angle β has been discussed in Sections 4.3.1 and 4.3.5, and in Figure 4.14.

Figure 6.10 shows that $f_3(\beta)$ is a linear function of the deviation angle β, and can be expressed by a simple formula, $f_3(\beta) = 1 - |\beta|/24°$. This straight-line expression was determined by regression analysis (Wang, 2006) based on three sets of test data:

1. Pang and Hsu (1995) tested ten panels (A series and B series) made of normal-strength concrete of approximately 42 MPa and reinforced with steel percentages from 0.6 to 3.0% in the longitudinal and transverse directions. The ratio ρ_ℓ/ρ_t ranged from 1 to 5.
2. Zhang and Hsu (1998) tested nine panels (VA series and VB series) made of high-strength concrete of approximately 100 MPa, and reinforced with steel percentages in the range 0.42–6.0%. The ratio ρ_ℓ/ρ_t also varied from 1 to 5.
3. Chintrakarn (2001) tested six panels (M series) with concrete strength of approximately 45 MPa. The ratios ρ_ℓ/ρ_t in panels M1–M6 were much larger than those tested by Pang and Hsu (1995) and Zhang and Hsu (1998). The four panels M2, M3, M4 and M5 had ρ_ℓ/ρ_t ratios of 4.0, 9.0, 7.2, and 15.6, respectively. No steel was placed in the transverse direction in panels M1 and M6, giving a ρ_ℓ/ρ_t ratio of infinity.

The straight-line function $f_3(\beta)$ has an intercept of $f_3(\beta) = 1$, when $\beta = 0$. When a 2-D shear element is reinforced with the same amounts of steel bars in the two directions

($\rho_\ell = \rho_t$), both the angle β and the concrete shear stress τ_{12}^c are equal to zero. With increasing ρ_ℓ/ρ_t ratio, both β and τ_{12}^c are increased. An increase of τ_{12}^c reduces the compressive capacity of the concrete struts through the reduction of $f_3(\beta)$ and, in turn, the softening coefficient ζ.

Substituting $f_3(\beta)$ into Equation (6.46), the softened coefficient ζ becomes:

$$\zeta = \left(\frac{5.8}{\sqrt{f_c'}} \leq 0.9\right)\left(\frac{1}{\sqrt{1+400\bar{\varepsilon}_1}}\right)\left(1 - \frac{|\beta|}{24°}\right) \tag{6.52}$$

In the fixed angle theory, β can be calculated from the three strains ε_1, ε_2 and γ_{21} using the compatibility equation:

$$\beta = \frac{1}{2}\tan^{-1}\left[\frac{\gamma_{12}}{(\varepsilon_1 - \varepsilon_2)}\right] \tag{6.53}$$

Equation (6.53) can also be derived from the Mohr strain circle, as will be shown later in Figure 6.22(b).

6.1.8 Smeared Stress–Strain Relationship of Concrete in Tension

6.1.8.1 Smeared Stress σ_1^c and Smeared Strain $\bar{\varepsilon}_1$ of Concrete

In the measured Mohr circle for concrete stresses, as shown in Figure. 6.6, it can be seen that the tensile stress of concrete σ_1^c is small compared with the compressive stress σ_2^c, but not zero. This stress σ_1^c is an uniform tensile stress of concrete, representing the stiffening of the steel bars by concrete in tension.

Figure 6.11 shows a typical tensile stress–strain curve of concrete. The curve consists of two distinct branches. Before cracking the stress–strain relationship is essentially linear. After cracking, however, a drastic drop of strength occurs and the descending branch of the curve becomes concave. In the descending branch, the concrete is cracked and the concept of concrete tensile stress σ_1^c and concrete tensile strain $\bar{\varepsilon}_1$ are quite different from those before cracking. σ_1^c is defined as the *smeared* (or *average*) concrete tensile stress and $\bar{\varepsilon}_1$ is the *smeared* (or *average*) concrete tensile strain. These terms will be elaborated in Section 6.1.8.2.

Based on the tests of 35 full-size panels, Belarbi and Hsu (1994) and Pang and Hsu (1995) proposed the following analytical expressions for the σ_1^c–$\bar{\varepsilon}_1$ curve:

Ascending branch ($\bar{\varepsilon}_1 \leq \varepsilon_{cr}$)

$$\sigma_1^c = E_c\bar{\varepsilon}_1 \tag{6.54}$$

where:

E_c = modulus of elasticity of concrete, taken as $3875\sqrt{f_c'(\text{MPa})}$, where f_c' and $\sqrt{f_c'}$ are in MPa;

ε_{cr} = cracking strain of concrete, taken as 0.00008 mm/mm,

Descending branch ($\bar{\varepsilon}_1 > \varepsilon_{cr}$)

$$\sigma_1^c = f_{cr}\left(\frac{\varepsilon_{cr}}{\bar{\varepsilon}_1}\right)^{0.4} \tag{6.55}$$

where f_{cr} = cracking stress of concrete, taken as $0.31\sqrt{f_c'(\text{MPa})}$.

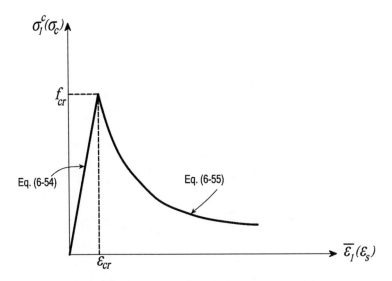

Figure 6.11   Tensile stress–strain curve of concrete

Equation (6.55) was first proposed by Tamai *et al.* (1987) based on tension of mild steel bars embedded in concrete. This equation was also found to be applicable to 2-D elements as shown in Section 6.1.8.2.

6.1.8.2 Validation of Equation (6.55)

In the panel with $\alpha_1 = 0°$, as shown in Figure 6.2(a), steel bars surrounded by concrete are subjected to horizontal tensile stress σ_1 in the principal 1-direction, as well as in the longitudinal ℓ-direction. This condition is simulated in Figure 6.12 by one steel bar surrounded by concrete and subjected to a tensile force P. After cracking, a portion of the tensile member is isolated which includes two cracks with a spacing of L. At the two cracks indicated, the steel stress will be designated f_{so}, as indicated in Figure 6.12(b). At any section a distance x from the first left crack, however, the steel stress $f_s(x)$ will be less than f_{so}, and the difference will be carried by the concrete in tension $\sigma_c(x)$. The longitudinal equilibrium of forces at this section (Figure 6.12a), gives:

$$P = P_s(x) + P_c(x) \tag{6.56}$$

Expressing the forces in Equation (6.56) in terms of stresses:

$$A_s f_{so} = A_s f_s(x) + A_c \sigma_c(x) \tag{6.57}$$

where:

A_s = cross-sectional area of steel bars;
A_c = area of net concrete section, not including A_s.

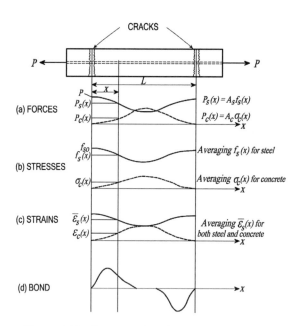

Figure 6.12 Stresses and strains between two cracks

Rearranging Equation (6.57) gives:

$$f_{so} = f_s(x) + \frac{1}{\rho}\sigma_c(x) \tag{6.58}$$

where $\rho = A_s/A_c$, the percentage of steel based on the net concrete section. Equation (6.58) states that, at any section between the two cracks, the sum of the steel stress $f_s(x)$ and the concrete stress $\sigma_c(x)$ divided by ρ must be equal to the steel stress at the cracked section f_{so}.

The steel strain $\bar{\varepsilon}_s(x)$ and the concrete strain $\varepsilon_c(x)$ are also sketched in Figure 6.12(c). The steel strain $\bar{\varepsilon}_s(x)$ decreases from a maximum at the crack to a minimum at the midpoint between the two cracks. In contrast, the concrete strain $\varepsilon_c(x)$ should be zero at the crack and increases to a maximum at the midpoint. The difference between $\bar{\varepsilon}_s(x)$ and $\varepsilon_c(x)$ is caused by the slip between the steel bar and the surrounding concrete. The slip results in the bond stresses sketched in Figure 6.12(d) and the gaps that constitute the cracks.

Let us recall that the solution of the equilibrium and compatibility equations, (6.1)–(6.6), requires the constitutive relationship between the *smeared (or average) tensile stress of concrete* σ_1^c and the *smeared (or average) tensile strain of concrete* $\bar{\varepsilon}_1$, in the principal 1-direction. The smeared strain of concrete $\bar{\varepsilon}_1$ should be measured along a length that traverses several cracks. This smeared strain $\bar{\varepsilon}_1$ is not the average value of the concrete strain $\varepsilon_c(x)$ shown in Figure 6.12(c), because the smeared strain $\bar{\varepsilon}_1$ includes not only the strain of the concrete itself, but also the strain contributed by the crack widths. Hence, the smeared strain $\bar{\varepsilon}_1$ must be obtained from averaging the steel strain $\bar{\varepsilon}_s(x)$ along the steel bar, not $\varepsilon_c(x)$. In other words, the smeared strain $\bar{\varepsilon}_1$ represents both the smeared strain of steel $\bar{\varepsilon}_s$ and the smeared strain of concrete including the crack widths. With this understanding in mind the *smeared tensile strain of concrete*, $\bar{\varepsilon}_1 = \bar{\varepsilon}_s$, is obtained by averaging the strain $\bar{\varepsilon}_s(x)$ from

$x = 0$ to $x = L$ as follows:

$$\bar{\varepsilon}_1 = \bar{\varepsilon}_s = \frac{1}{L}\int_0^L \bar{\varepsilon}_s(x)\mathrm{d}x \tag{6.59}$$

Before the first yielding of the steel, the linear relationship $\bar{\varepsilon}_s(x) = f_s(x)/E_s$ is valid at any cross-section x. Substituting $\bar{\varepsilon}_s(x)$ into Equation (6.59) gives

$$\bar{\varepsilon}_1 = \bar{\varepsilon}_s = \frac{1}{E_s}\left(\frac{1}{L}\int_0^L f_s(x)\mathrm{d}x\right) \tag{6.60}$$

The term in the parenthesis of Equation (6.60) is defined as the *smeared stress of steel* f_s, i.e.

$$f_s = \left(\frac{1}{L}\int_0^L f_s(x)\mathrm{d}x\right) \tag{6.61}$$

Substituting Equation (6.61) into (6.60) gives a linear relationship between the smeared stress of steel f_s and the smeared strain of steel $\bar{\varepsilon}_1$ or $\bar{\varepsilon}_s$:

$$\bar{\varepsilon}_1 = \bar{\varepsilon}_s = \frac{1}{E_s}f_s \tag{6.62}$$

Using this concept of averaging, we can now average the steel stresses $f_s(x)$ and the concrete stresses $\sigma_c(x)$ in Equation (6.58) by integrating these stresses from $x = 0$ to $x = L$ and divided by the length L:

$$f_{so} = \left(\frac{1}{L}\int_0^L f_s(x)\mathrm{d}x\right) + \frac{1}{\rho}\left(\frac{1}{L}\int_0^L \sigma_c(x)\mathrm{d}x\right) \tag{6.63}$$

The first term on the right-hand side of Equation (6.63) is obviously the smeared stress of steel f_s, as defined in Equation (6.61). The quantity enclosed by the parenthesis in the second term is the smeared tensile stress of concrete σ_1^c or σ_c:

$$\sigma_1^c = \sigma_c = \left(\frac{1}{L}\int_0^L \sigma_c(x)\mathrm{d}x\right) \tag{6.64}$$

Then Equation (6.63) becomes

$$f_{so} = f_s + \frac{1}{\rho}\sigma_c \tag{6.65}$$

Substituting $f_s = E_s\bar{\varepsilon}_s$ from Equation (6.62) into (6.65), we derive the relationship between the smeared tensile stress of concrete $\sigma_1^c(\sigma_c)$ and the smeared tensile strain of concrete $\bar{\varepsilon}_1(\bar{\varepsilon}_s)$:

$$\sigma_1^c = \rho(f_{so} - E_s\bar{\varepsilon}_1) = \rho\left(\frac{P}{A_s} - E_s\bar{\varepsilon}_1\right) = \frac{P}{A_c} - \rho E_s\bar{\varepsilon}_1 \tag{6.66}$$

For each load stage in a panel test, the load P was measured by load cells, and the smeared tensile strain $\bar{\varepsilon}_1(\bar{\varepsilon}_s)$ was measured by LVDTs over a length of 0.8 m (31.5 in.). A smeared stress of concrete $\sigma_1^c(\sigma_c)$ can then be calculated by Equation (6.66). Plotting σ_1^c/f_{cr} against $\bar{\varepsilon}_1(\bar{\varepsilon}_s)$ consecutively for all load stages will give an experimental tensile stress–strain curve for

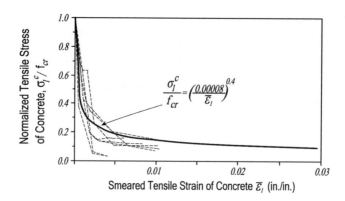

Figure 6.13 Desending branch of smeared stress–strain curves of concrete in tension obtained from test panel

concrete in the principal 1-direction. Experimental curves of 17 panels tested by Belarbi and Hsu (1995) were plotted in Figure 6.13. They showed that the best mathematical expression to fit the descending branch of the experimental stress–strain curves was given by Equation (6.55).

6.1.9 Smeared Stress–Strain Relationship of Mild Steel Bars in Concrete

6.1.9.1 Bare Steel Bars

The stress–strain curve of bare mild steel bars is usually assumed to be elastic–perfectly plastic, i.e.

$$f_s = E_s \bar{\varepsilon}_s \qquad \text{when} \quad \bar{\varepsilon}_s \leq \varepsilon_{sy}, \tag{6.67}$$

$$f_s = f_{sy} \qquad \text{when} \quad \bar{\varepsilon}_s > \varepsilon_{sy}, \tag{6.68}$$

where

E_s = modulus of elasticity of mild steel bars;
f_s = stresses in mild steel bars;
f_{sy} = yield stress in mild steel bars;
$\bar{\varepsilon}_s$ = uniaxial strain in mild steel bars;
ε_{sy} = yield strain in mild steel bars.

In Equations (6.67) and (6.68), subscript s in the symbols is replaced by ℓ for the longitudinal steel, and by t for the transverse steel.

Equations (6.67) and (6.68), obtained from the testing of bare, mild steel bars, are plotted in Figure 6.14. The smeared stress–strain curve of steel bars embedded in concrete is also shown in the figure. It can be seen that these two stress–strain curves are quite different. The stress–strain curve of a bare bar relates the stress to the strain at a local point, while the smeared stress–strain curve of a steel bar in concrete relates the *smeared (or average) stress* f_s to the *smeared (or average) strain* $\bar{\varepsilon}_s$. Figure 6.14 also shows the 'smeared' bilinear model (using dashed lines), which will be discussed in Section 6.1.9.4.

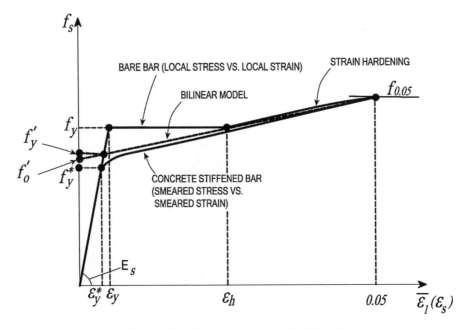

Figure 6.14   Stress–strain curve of mild steel

6.1.9.2 Smeared Yield Stress of Mild Steel f_y^*

Inserting $\bar{\varepsilon}_1 = f_s/E_s$ from Equation (6.62) into (6.55) and, in turn, substituting $\sigma_1^c = \sigma_c$ from Equation (6.55) into Equation (6.65) gives:

$$f_{so} - f_s = \frac{f_{cr}}{\rho} \left(\frac{E_s \varepsilon_{cr}}{f_s} \right)^{0.4} \tag{6.69}$$

Equation (6.69) states that the difference between the steel stress at the crack f_{so} and the smeared steel stress f_s, i.e. $f_{so} - f_s$, depends on the parameter on the right-hand side of the equation.

Yielding of a reinforced concrete panel occurs when the steel stress at the cracked section reaches the yield plateau, i.e. $f_{so} = f_y$. At the same time, the smeared steel stress reaches a level which we shall call the 'smeared yield stress of steel' (f_y^*), i.e. $f_s = f_y^*$. Substituting $f_{so} = f_y$ and $f_s = f_y^*$ into Equation (6.69), dividing by f_y, then multiplying both sides of the equation by $(f_y^*/f_y)^{0.4}$, result in:

$$\left(\frac{f_y^*}{f_y} \right)^{0.4} - \left(\frac{f_y^*}{f_y} \right)^{1.4} = \frac{f_{cr}}{\rho f_y} \left(\frac{E_s \varepsilon_{cr}}{f_y} \right)^{0.4} \tag{6.70}$$

Using the relationships $\varepsilon_{cr} = f_{cr}/E_c$ and the modulus ratio $n = E_s/E_c$, Equation (6.70) becomes

$$\left(\frac{f_y^*}{f_y} \right)^{0.4} - \left(\frac{f_y^*}{f_y} \right)^{1.4} = \frac{n^{0.4}}{\rho} \left(\frac{f_{cr}}{f_y} \right)^{1.4} \tag{6.71}$$

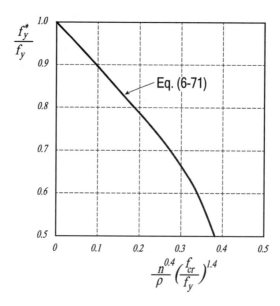

Figure 6.15   Graphical solution of Equation (6.71)

The solution of Equation (6.71) determines the smeared yield stress ratio f_y^*/f_y.

Equation (6.71) can be solved by a graphical method, as shown in Figure 6.15. When the term $(n^{0.4}/\rho)(f_{cr}/f_y)^{1.4}$ on the right-hand side of Equation (6.71) is given, locate this value on the horizontal axis in Figure 6.15. Follow this value vertically up to the curve designated Equation (6.71), then horizontally to the left to locate the value of f_y^*/f_y.

In normal application, f_y^*/f_y varies from 1.0 to 0.6. Figure 6.15 shows that in this range f_y^*/f_y is nearly a linear function of the parameter $(n^{0.4}/\rho)(f_{cr}/f_y)^{1.4}$. According to Belarbi and Hsu (1995), this nearly linear relationship can be closely approximated by the following expression:

$$\frac{f_y^*}{f_y} = 1 - \frac{4}{\rho}\left(\frac{f_{cr}}{f_y}\right)^{1.5} \tag{6.72}$$

If $(1/\rho)(f_{cr}/f_y)^{1.5}$ in Equation (6.72) is defined as a parameter B, then the expression $f_y^*/f_y = 1 - 4B$ represents a straight line. The parameter B is the product of two nondimensional variables $(1/\rho)$ and $(f_{cr}/f_y)^{1.5}$. The constant 4 is the slope of the straight line.

In Figure 6.16, f_y^*/f_y is plotted as a function of ρ and f_{cr}/f_y. Figure 6.16 reveals two trends: First, the lower the percentage of steel ρ, the lower the yield stress ratio f_y^*/f_y. This ratio f_y^*/f_y is particularly sensitive to ρ at low percentages of steel. Second, the higher the value of f_{cr}/f_y, the lower the smeared yield stress ratio f_y^*/f_y. Physically, both trends express the fact that the steel bar in Figure 6.12 is receiving more help from the concrete to resist the tensile force P.

The calculation of f_y^*/f_y can be illustrated by the following example:

Given: $\rho = 0.75\%$; $f_{cr} = 2.07$ MPa (300 psi); $f_y = 413$ MPa (60 000 psi); $n = 7$.

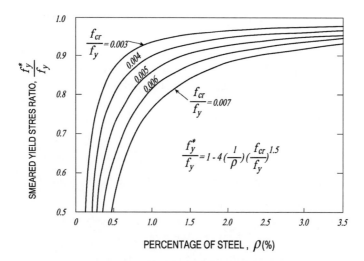

Figure 6.16 f_y^*/f_y as a function of ρ and f_{cr}/f_y

Exact solution: $f_{cr}/f_y = 2.07/413 = 0.005$

$$\frac{n^{0.4}}{\rho}\left(\frac{f_{cr}}{f_y}\right)^{1.4} = \frac{7^{0.4}}{0.0075}\left(0.005^{1.4}\right) = 0.290 \times 10^3 \left(0.601 \times 10^{-3}\right) = 0.174$$

From Figure 6.15 or Equation (6.71)

$$\frac{f_y^*}{f_y} = 0.811$$

Approximate solution according to Belarbi and Hsu's Equation (6.72):

$$\frac{f_y^*}{f_y} = 1 - 4\left(\frac{1}{0.0075}\right)(0.005^{1.5}) = 1 - 4(0.133 \times 10^3)(0.353 \times 10^{-3})$$
$$= 1 - 0.188 = 0.812$$

very close to the exact value of 0.811.

Answer: $f_y^* = 0.812 f_y = 0.812(413\text{MPa}) = 335$ MPa.

6.1.9.3 Smeared Stress–Strain Curve of Mild Steel After Yield

The smeared stress–strain relationship of mild steel bars embedded in concrete is more difficult to determine after yielding, because the steel strain at the cracked sections increases rapidly to reach the strain hardening region of the stress–strain curve. The averaging of the steel strains and the corresponding steel stresses along the length L becomes mathematically more complex. The integration process involved in the averaging requires numerical integration and the use of an electronic computer.

In order to simplify the averaging process, two assumptions are made by Tamai *et al.* (1987).

(a) The stress distribution in the steel between two adjacent cracks is assumed to follow a full cosine curve.
(b) The smeared stress–strain relationship of concrete in tension (Equation 6.55), is valid both before and after yielding. That is to say, Equation (6.55), which had been calibrated to fit the test results before yielding, remains valid after yielding.

From the first assumption we can write

$$f_{so} = f_s + a_s \cos \frac{2\pi x}{L} \tag{6.73}$$

where a_s is the amplitude of the cosine curve. At the cracked sections, $f_s(x) = f_{so}$ and $\cos \dfrac{2\pi x}{L} = 1$ (i.e. x equals to 0 or L). Therefore

$$a_s = f_{so} - f_s \tag{6.74}$$

Substituting $(f_{so} - f_s)$ from Equation (6.65) into (6.74) gives

$$a_s = \frac{\sigma_c}{\rho} \tag{6.75}$$

Substituting a_s from Equation (6.75) into (6.73) gives

$$f_s(x) = f_s + \frac{\sigma_c}{\rho} \cos \frac{2\pi x}{L} \tag{6.76}$$

Now, using the second assumption and substituting $\sigma_c = \sigma_1^c$ and $\bar{\varepsilon}_1 = \bar{\varepsilon}_s$ from Equation (6.55) into (6.76) gives:

$$f_s(x) = f_s + \frac{1}{\rho} f_{cr} \left(\frac{\varepsilon_{cr}}{\bar{\varepsilon}_s}\right)^{0.4} \cos \frac{2\pi x}{L} \tag{6.77}$$

With these two assumptions, the averaging process is summarized as follows:

1. Select a value of the smeared steel stress f_s.
2. Assume a smeared steel strain $\bar{\varepsilon}_s$.
3. Calculate the distribution of steel stress $f_s(x)$ from Equation (6.77).
4. Determine the corresponding distribution of steel strain $\bar{\varepsilon}_s(x)$ according to the stress–strain curve of bare bars (Equations 6.67–6.68).
5. Calculate the smeared steel strain $\bar{\varepsilon}_s$ by numerical integration of the following integral:

$$\bar{\varepsilon}_s = \frac{1}{L} \int_0^L \bar{\varepsilon}_s(x) \mathrm{d}x \tag{6.78}$$

6. If $\bar{\varepsilon}_s$ calculated from Equation (6.78) is not the same as that assumed, repeat steps 2–5 until the calculated $\bar{\varepsilon}_s$ is sufficiently close to the assumed value. The calculated $\bar{\varepsilon}_s$ and the selected value f_s provide one point on the smeared stress–strain curve in the post-yield range.
7. By selecting a series of f_s values and find their corresponding $\bar{\varepsilon}_s$ values from steps 2–6, the whole smeared stress–strain curve in the post-yielding range can be plotted.

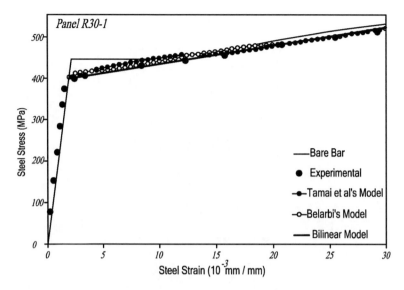

Figure 6.17   Smeared stress–strain curves of mild steel-theories and tests

A smeared stress–strain ($f_s - \bar{\varepsilon}_s$) curve of steel obtained using the model of Tamai *et al.* is compared with an experimental stress–strain curve of steel in Figure 6.17. The agreement is acceptable.

The prediction of this theoretical method can be improved if the stress distribution of the steel bar is expressed by a function closer to the actual condition. Belarbi (1991) modified the cosine function in Eq. (6.73) by adding more terms in the following form:

$$f_{so} = f_s + a_s \cos \frac{2\pi x}{L} + b_s \left(\sin \frac{3\pi x}{L} - 0.6 \sin \frac{5\pi x}{L} - 0.1358 \right) \qquad (6.79)$$

where a_s is given by Equation (6.75). In the added terms b_s, 0.6 and 0.1358 were chosen to satisfy the boundary conditions at the cracks and at the midpoint between the cracks. The additional sinusoidal and constant terms and the resulting stress distribution are given in Figure 6.18. The post-yield $f_s - \bar{\varepsilon}_s$ curve can then be calculated using the seven-step iteration procedure used for solving Equation (6.77).

Belarbi's improved $f_s - \bar{\varepsilon}_s$ curve based on the stress distribution of Equation (6.79) is also shown in Figure 6.17. It can be seen that the improved method is closer to the test points. Notice also that the post-yield theoretical curve is very close to the experimental curve of the bare bar in the strain hardening region. This is because the smeared tensile stress of concrete becomes negligible in this region of very large strains.

A parametric study using Belarbi's method was carried out to examine the effect of the percentage of steel (ρ) and the cracking strength of concrete (f_{cr}) on the post-yielding stress–strain curves of steel. Figure 6.19(a) gives a family of stress–strain curves using the percentage of steel (ρ) as parameter. Figure 6.19(b) gives a family of stress–strain curves using the cracking strength of concrete (f_{cr}) as parameter. It can be seen that the smeared stress–strain curves move downward when ρ is decreased and when f_{cr} is increased.

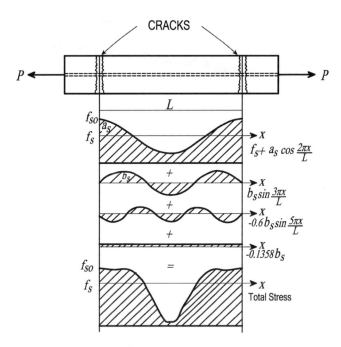

Figure 6.18 Belarbi's stress distribution along a reinforcing bar after cracking

6.1.9.4 Bilinear Model for Smeared Stress–Strain Curve

As shown in Figure 6.19(a) and (b), the shape of the average stress–strain curve of mild steel resembles two straight lines. These two straight lines will have a slope of E_s before yielding and a slope of E_p' after yielding, as illustrated in Figure 6.20. The plastic modulus E_p' after yielding is only a small fraction of the elastic modulus E_s before yielding. The stress level at which the two straight lines intersect is designated as the smeared yield stress f_y'. The equations of these two lines are then given as follows:

$$f_s = E_s \bar{\varepsilon}_s \qquad \text{when } f_s \leq f_y' \qquad (6.80)$$
$$f_s = f_o' + E_p' \bar{\varepsilon}_s \qquad \text{when } f_s > f_y' \qquad (6.81)$$

where f_o' is the vertical intercept of the post-yield straight line. This vertical intercept f_o' can be calculated by

$$f_o' = \frac{E_s - E_p'}{E_s} f_y' \qquad (6.82)$$

It should be noted that f_y' is quite different from f_y^*. The symbol f_y' is the smeared yield stress of the bilinear model shown in Figure 6.20, while the symbol f_y^* is the smeared yield stress derived from the theoretical model explained in Section 6.1.9.2. The smeared yield stress f_y' and the plastic modulus E_p' are determined to best approximate the smeared yield stress f_y^* (Section 6.1.9.2) as well as the post-yield smeared stress–strain curve (Section 6.1.9.3). The lower the f_y^*, the lower the f_y', and the higher the E_p'. The nondimensional ratios f_y'/f_y and

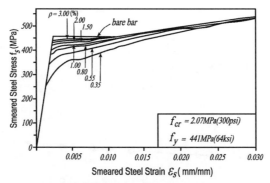

(a) PERCENTAGE OF STEEL (ρ) AS A PARAMETER

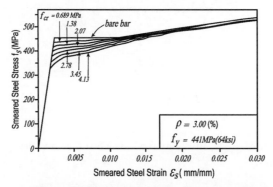

(b) CRACKING STRENGTH OF CONCRETE (f_{cr}) AS A PARAMETER

Figure 6.19   Smeared stress–strain curves of mild steel

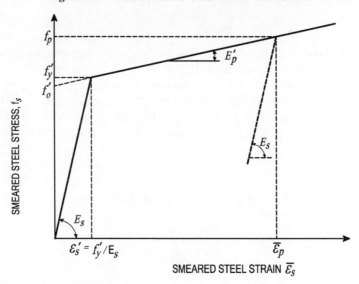

Figure 6.20   Stress–strain relationship for mild steel using Bilinear model

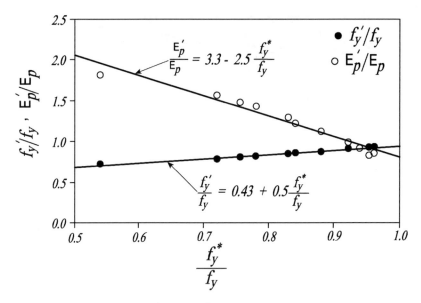

Figure 6.21 f_y'/f_y and E_p'/E_p as a function of f_y^*/f_y

E_p'/E_p are plotted against f_y^*/f_y in Figure 6.21. The test data are closely represented by two straight lines as follows:

$$\frac{f_y'}{f_y} = 0.43 + 0.5\frac{f_y^*}{f_y} \tag{6.83}$$

$$\frac{E_p'}{E_p} = 3.3 - 2.5\frac{f_y^*}{f_y} \tag{6.84}$$

The plastic modulus E_p in Equation (6.84) is the slope of the strain-hardening region of the bare mild steel bars, and is assumed to be $0.025E_s$.

A simple bilinear model of the smeared stress–strain relationship of mild steel embedded in concrete can now be derived. First, substituting f_y^*/f_y from Equation (6.72) into (6.83) and (6.84); and second, inserting f_o', f_y', and E_p' from Equations (6.82)–(6.84) into (6.80) and (6.81) and rounding up the term f_o'. These two steps result in two simple straight lines given by:

$$f_s = E_s\bar{\varepsilon}_s \qquad\qquad\qquad\qquad\qquad \text{when } \bar{\varepsilon}_s \leq \varepsilon_y' \tag{6.85}$$
$$f_s = (0.91 - 2B)f_y + (0.02 + 0.25B)E_s\bar{\varepsilon}_s \quad \text{when } \bar{\varepsilon}_s > \varepsilon_y' \tag{6.86}$$

where

$$\varepsilon_y' = f_y'/E_s \qquad\qquad f_y' = (0.93 - 2B)f_y \tag{6.87}$$

$$B = \frac{1}{\rho}\left(\frac{f_{cr}}{f_y}\right)^{1.5} \qquad f_{cr} = 0.31\sqrt{f_c'(\text{MPa})} \text{ and } \rho \geq 0.15\% \tag{6.88}$$

In calculating the parameter B in Equation (6.88), the percentage of steel ρ should not be less than 0.15%. This $\rho_{min} = 0.15\%$ should not impose any difficulty, because it is the minimum specified by the ACI Code for deformed bars in walls.

Equations (6.85) and (6.86) are quite simple to use, because they are functions of only one parameter B defined in Equation. (6.88). These two equations are applicable to both longitudinal and transverse steel as follows:

$f_s = f_\ell$ or f_t when applied to longitudinal steel or transverse steel, respectively,
$\bar{\varepsilon}_s = \bar{\varepsilon}_\ell$ or $\bar{\varepsilon}_t$ when applied to longitudinal steel or transverse steel, respectively,

The application of Equation (6.86) is illustrated by the following example:

Given: $\rho = 0.75\%$, $f_{cr} = 2.07\%$ MPa (300 psi); $f_y = 413$ MPa (60 ksi);
$\quad\quad E_s = 200\,000$ MPa (29 000 ksi)

Find: the smeared steel stress f_s at a smeared tensile strain $\bar{\varepsilon}_s = 0.01$

$$B = \frac{1}{\rho}\left(\frac{f_{cr}}{f_y}\right)^{1.5} = \left(\frac{1}{0.0075}\right)\left(\frac{2.07}{413}\right)^{1.5} = (0.133 \times 10^3)(0.353 \times 10^{-3}) = 0.0469$$

$f_y' = (0.93 - 2B)f_y = (0.93 - 2 \times 0.0469)f_y = 0.836\,(413) = 345$ MPa
$\varepsilon_y' = f_y'/E_s = 345/200,000 = 0.00173$

Since $\bar{\varepsilon}_s = 0.01 > 0.00173$ (ε_y'), use Equation (6.86):

$$\begin{aligned} f_s &= (0.91 - 2B)f_y + (0.02 + 0.25B)E_s\bar{\varepsilon}_s \\ &= (0.91 - 2 \times 0.0469)f_y + (0.02 + 0.25 \times 0.0469)E_s\bar{\varepsilon}_s \\ &= 0.816(413) + (0.0317)(200\,000)(0.01) = 337 + 63 = 400 \text{ MPa.} \end{aligned}$$

When the steel stress f_s reaches a peak stress f_p and starts to unload, the unloading stress–strain curve is assumed to be a straight line with a slope of E_s. This unloading straight line can be expressed as follows:

$$f_s = f_p - E_s(\bar{\varepsilon}_p - \bar{\varepsilon}_s) \quad\quad \bar{\varepsilon}_s < \bar{\varepsilon}_p \quad\quad\quad\quad (6.89)$$

Equation (6.89) is also shown in Figure 6.20 as a bold dotted straight line.

6.1.10 Smeared Stress–Strain Relationship of Concrete in Shear

A rational and simple shear modulus has been derived theoretically by Zhu, Hsu, and Lee (2001) for a stress and strain analysis based on the smeared-crack concept. Referring to the Mohr circles for stresses and strains in Figure 6.22(a) and (b), and assuming that the direction of the principal stress of concrete coincides with the principal strain, the following two equations can be obtained:

$$\tau_{12}^c = \frac{\sigma_1^c - \sigma_2^c}{2} \tan 2\beta \quad\quad\quad\quad (6.90)$$

$$\gamma_{12} = (\varepsilon_1 - \varepsilon_2) \tan 2\beta \quad\quad\quad\quad (6.91)$$

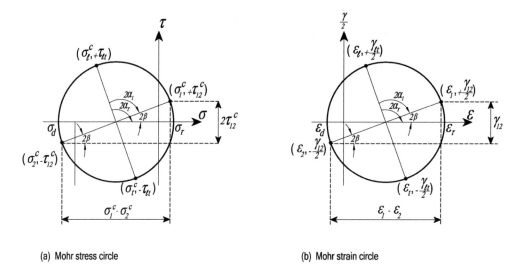

(a) Mohr stress circle   (b) Mohr strain circle

Figure 6.22   Mohr Circle for stresses and strains

Dividing Equation (6.90) by Equation (6.91) gives the new constitutive law of concrete in shear as:

$$G_{12}^c = \frac{\tau_{12}^c}{(\gamma_{12}/2)} = \frac{\sigma_1^c - \sigma_2^c}{(\varepsilon_1 - \varepsilon_2)} \tag{6.92}$$

where ε_1 and ε_2 are biaxial strains.

6.1.11 Solution Algorithm

6.1.11.1 Summary of Governing Equations

The eighteen governing equations for RC 2-D elements are summarized below:

Stress equilibrium equations

$$\sigma_\ell = \sigma_1^c \cos^2 \alpha_1 + \sigma_2^c \sin^2 \alpha_1 - \tau_{12}^c 2 \sin \alpha_1 \cos \alpha_1 + \rho_\ell f_\ell \qquad \boxed{1}$$

$$\sigma_t = \sigma_1^c \sin^2 \alpha_1 + \sigma_2^c \cos^2 \alpha_1 + \tau_{12}^c 2 \sin \alpha_1 \cos \alpha_1 + \rho_t f_t \qquad \boxed{2}$$

$$\tau_{\ell t} = (\sigma_1^c - \sigma_2^c) \sin \alpha_1 \cos \alpha_1 + \tau_{12}^c (\cos^2 \alpha_1 - \sin^2 \alpha_1) \qquad \boxed{3}$$

Strain compatibility equations

$$\varepsilon_\ell = \varepsilon_1 \cos^2 \alpha_1 + \varepsilon_2 \sin^2 \alpha_1 - \frac{\gamma_{12}}{2} 2 \sin \alpha_1 \cos \alpha_1 \qquad \boxed{4}$$

$$\varepsilon_t = \varepsilon_1 \sin^2 \alpha_1 + \varepsilon_2 \cos^2 \alpha_1 + \frac{\gamma_{12}}{2} 2 \sin \alpha_1 \cos \alpha_1 \qquad \boxed{5}$$

$$\frac{\gamma_{\ell t}}{2} = (\varepsilon_1 - \varepsilon_2) \sin \alpha_1 \cos \alpha_1 + \frac{\gamma_{12}}{2} (\cos^2 \alpha_1 - \sin^2 \alpha_1) \qquad \boxed{6}$$

Poisson effect

$$v_{12} = 0.2 + 850\varepsilon_{sf} \qquad \varepsilon_{sf} \le \varepsilon_y \qquad \boxed{7a}$$

$$v_{12} = 1.9 \qquad \varepsilon_{sf} > \varepsilon_y \qquad \boxed{7b}$$

$$v_{21} = 0 \text{ after cracking}, \quad v_{21} = 0.2 \text{ before cracking} \qquad \boxed{8}$$

$$\bar{\varepsilon}_1 = \frac{1}{1 - v_{12}v_{21}}\varepsilon_1 + \frac{v_{12}}{1 - v_{12}v_{21}}\varepsilon_2 \qquad \boxed{9}$$

$$\bar{\varepsilon}_2 = \frac{v_{21}}{1 - v_{12}v_{21}}\varepsilon_1 + \frac{1}{1 - v_{12}v_{21}}\varepsilon_2 \qquad \boxed{10}$$

$$\bar{\varepsilon}_\ell = \bar{\varepsilon}_1 \cos^2 \alpha_1 + \bar{\varepsilon}_2 \sin^2 \alpha_1 - \frac{\gamma_{12}}{2} 2 \sin \alpha_1 \cos \alpha_1 \qquad \boxed{11}$$

$$\bar{\varepsilon}_t = \bar{\varepsilon}_1 \sin^2 \alpha_1 + \bar{\varepsilon}_2 \cos^2 \alpha_1 + \frac{\gamma_{12}}{2} 2 \sin \alpha_1 \cos \alpha_1 \qquad \boxed{12}$$

Constitutive laws of materials
Concrete in compression

$$\sigma_2^c = \zeta f_c' \left[2 \left(\frac{\bar{\varepsilon}_2}{\zeta \varepsilon_o} \right) - \left(\frac{\bar{\varepsilon}_2}{\zeta \varepsilon_o} \right)^2 \right] \qquad \bar{\varepsilon}_2 / \zeta \varepsilon_o \le 1 \qquad \boxed{13a}$$

$$\sigma_2^c = \zeta f_c' \left[1 - \left(\frac{(\bar{\varepsilon}_2 / \zeta \varepsilon_o) - 1}{(4/\zeta) - 1} \right)^2 \right] \qquad \bar{\varepsilon}_2 / \zeta \varepsilon_o \ge 1 \qquad \boxed{13b}$$

$$\zeta = \left(\frac{5.8}{\sqrt{f_c'}} \le 0.9 \right) \left(\frac{1}{\sqrt{1 + 400\bar{\varepsilon}_1}} \right) \left(1 - \frac{|\beta|}{24°} \right) \qquad \boxed{14}$$

$$\beta = \frac{1}{2} \tan^{-1} \left[\frac{\gamma_{12}}{(\varepsilon_1 - \varepsilon_2)} \right] \qquad \boxed{15}$$

Concrete in tension

$$\sigma_1^c = E_c \bar{\varepsilon}_1 \qquad \bar{\varepsilon}_1 \le \varepsilon_{cr} \quad \text{and} \quad \varepsilon_{cr} = 0.00008 \text{ mm/mm}, \qquad \boxed{16a}$$

$$\sigma_1^c = f_{cr} \left(\frac{\varepsilon_{cr}}{\bar{\varepsilon}_1} \right)^{0.4} \qquad \bar{\varepsilon}_1 > \varepsilon_{cr} \quad \text{and} \quad f_{cr} = 0.31\sqrt{f_c'(MPa)} \qquad \boxed{16b}$$

Concrete in shear

$$\tau_{12}^c = \frac{\sigma_1^c - \sigma_2^c}{2(\varepsilon_1 - \varepsilon_2)}\gamma_{12} \qquad \boxed{17}$$

Mild steel

$$f_s = E_s \bar{\varepsilon}_s \qquad\qquad\qquad\qquad\qquad \bar{\varepsilon}_s \le \varepsilon'_y \qquad\qquad \boxed{18a}, \boxed{19a}$$

$$f_s = (0.91 - 2B)f_y + (0.02 + 0.25B)E_s\bar{\varepsilon}_s \quad \bar{\varepsilon}_s > \varepsilon'_y \qquad\qquad \boxed{18b}, \boxed{19b}$$

$$f_s = f_p - E_s(\bar{\varepsilon}_p - \bar{\varepsilon}_s) \qquad\qquad\qquad \bar{\varepsilon}_s < \bar{\varepsilon}_p \qquad\qquad \boxed{18c}, \boxed{19c}$$

$$\varepsilon'_y = f'_y/E_s \qquad\qquad f'_y = (0.93 - 2B)f_y \qquad\qquad\qquad\quad \boxed{18d}, \boxed{19d}$$

$$B = \frac{1}{\rho}\left(\frac{f_{cr}}{f_y}\right)^{1.5} \quad f_{cr} = 0.31\sqrt{f'_c(MPa)} \text{ and } \rho \ge 0.15\% \qquad \boxed{18e}, \boxed{19e}$$

Equations $\boxed{18a}$ to $\boxed{18e}$ are intended for the longitudinal steel when the subscripts s are replaced by the subscript ℓ, and Eqs. $\boxed{19a}$ to $\boxed{19e}$ are intended for the transverse steel when the subscripts s are replaced by the subscript t.

6.1.11.2 Solution Algorithm

The 19 equations, Eqs. $\boxed{1}$ to $\boxed{19}$ will be used to solve 19 unknown variables. These 19 equations involve 22 variables, namely, 8 stresses (σ_ℓ, σ_t, $\tau_{\ell t}$, σ^c_1, σ^c_2, τ^c_{12}, f_ℓ, f_t), 10 strains (ε_ℓ, ε_t, $\gamma_{\ell t}$, ε_1, ε_2, γ_{12}, $\bar{\varepsilon}_1$, $\bar{\varepsilon}_2$, $\bar{\varepsilon}_\ell$, $\bar{\varepsilon}_t$), the deviation angle β, the softening coefficients ζ, and the two Hsu/Zhu ratios ν_{12} and ν_{21}. In the case of pure shear, the two applied normal stresses $\sigma_\ell = 0$ and $\sigma_t = 0$. When a value of ε_2 is selected, the remaining 19 unknown variables can be solved by the 19 equations.

To develop an efficient solution algorithm, the first two equilibrium equations $\boxed{1}$ and $\boxed{2}$ are summed and subtracted to obtain the following two equations, which are used as the convergence criteria for the solution procedure (Hsu and Zhu, 2002):

$$\rho_\ell f_\ell + \rho_t f_t = (\sigma_\ell + \sigma_t) - (\sigma^c_1 + \sigma^c_2) \qquad\qquad\qquad\qquad \boxed{20}$$

$$\rho_\ell f_\ell - \rho_t f_t = (\sigma_\ell - \sigma_t) - (\sigma^c_1 - \sigma^c_2)\cos 2\alpha_1 + 2\tau^c_{12}\sin 2\alpha_1 \qquad \boxed{21}$$

The iterative procedure of SMM is illustrated by the flow chart in Figure 6.23. First select a ε_2 value and assume two values for ε_1 and γ_{12}. After completing the DO-loops in the flow chart, Figure 6.23, using Equations $\boxed{4}$, $\boxed{5}$, and $\boxed{7}$–$\boxed{21}$, the shear stress $\tau_{\ell t}$ and the shear strain $\gamma_{\ell t}$ for a selected value of ε_2 are calculated from Equations $\boxed{3}$ and $\boxed{6}$, respectively. By selecting a series of ε_2 values, it is possible to plot the $\tau_{\ell t} - \gamma_{\ell t}$ curves. Curves relating any two variables can similarly be plotted.

6.1.12 Example Problem 6.1

6.1.12.1 Problem Statement

A RC 2-D element M3, as shown in Figure 6.2(b) with $\alpha_1 = 45°$, has been tested by Chintrakarn (2001). This 2-D element was reinforced with 2#6 mild steel bars at 189 mm spacing in the longitudinal direction and 2#2 mild steel bars at 189 mm spacing in the transverse direction. The properties of the longitudinal steel are: $\rho_\ell = 1.70\%$, $f_{\ell y} = 425.4$ MPa, and $E_s = 212\,700$ MPa. For transverse steel: $\rho_t = 0.19\%$, $f_{ty} = 457.9$ MPa, and $E_s = 184\,600$ MPa. The

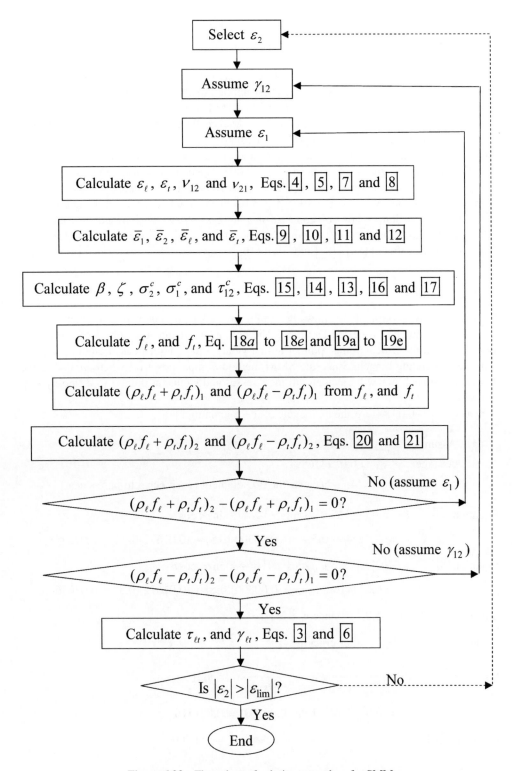

Figure 6.23 Flow chart of solution procedure for SMM

properties of concrete are: $f'_c = 48.1$ MPa, $\varepsilon_o = 0.0024$, and $E_c = 3875\sqrt{f'_c} = 26\,875$ MPa. Since panel M3 has a very large ρ_ℓ / ρ_t ratio of 9.0, and thus a very large β value of 16.7° at peak load, it is a good example to illustrate the power of SMM.

Panel M3 was subjected to a principal tensile stress σ_1, and a principal compressive stress σ_2 of equal magnitude in the 1–2 coordinate. These applied stresses resulted in a pure shear stress in the $\ell - t$ coordinate and during its loading history: i.e. $\sigma_\ell = 0$ MPa, $\sigma_t = 0$ MPa and an applied shear stress $\tau_{\ell t}$ that increased until failure. Analyze the behavior of this 2-D element M3 by the softened membrane model (SMM). Particular interest is placed in the stress and strain conditions at the first yield of steel, at the peak load stage, and in the descending branch when the selected concrete compressive strain $\varepsilon_2 = -0.003$. The computer program was terminated at $|\varepsilon_2| > |\varepsilon_{\lim}|$, where the limiting concrete compressive strain $\varepsilon_{\lim}$ is taken as -0.008.

In order to use the SMM, the Poisson effect must be taken into account by using Equations $\boxed{7}$–$\boxed{12}$. The constitutive laws of concrete in compression, in tension and in shear were calculated by Equations $\boxed{13}$–$\boxed{17}$, while the constitutive laws of smeared mild steel were calculated by Equations $\boxed{18a}$–$\boxed{18e}$ and $\boxed{19a}$–$\boxed{19e}$.

6.1.12.2 Solution

The calculations according to the flow chart in Figure 6.23 are best done by computer, because the procedure involves a nested DO-loop. The computer-calculated results are recorded in Table 6.1 for three critical points: point 1 at first yield; point 2 at the peak strength; and point 3 in the descending branch. Since the SMM method is most powerful when applied to the descending branch, we will calculate all the stresses and strains at point 3 in a step-by-step manner.

Select $\varepsilon_2 = -0.00300$. At the final cycle of the calculation:

$$\text{Assume} \qquad \gamma_{12} = 0.01672$$
$$\text{Assume} \qquad \varepsilon_1 = 0.027416$$
$$\alpha_1 = 45°, \text{ because the 2-D element is subjected to pure shear.}$$

Equation $\boxed{4}$ $\qquad \varepsilon_\ell = \varepsilon_1 \cos^2 \alpha_1 + \varepsilon_2 \sin^2 \alpha_1 - \frac{\gamma_{12}}{2} 2 \sin \alpha_1 \cos \alpha_1$

$$= 0.027416(0.5) + (-0.00300)(0.5) - (0.01672)(0.5) = 0.003848$$

Equation $\boxed{5}$ $\qquad \varepsilon_t = \varepsilon_1 \sin^2 \alpha_1 + \varepsilon_2 \cos^2 \alpha_1 + \frac{\gamma_{12}}{2} 2 \sin \alpha_1 \cos \alpha_1$

$$= 0.027416(0.5) + (-0.00300)(0.5) + (0.01672)(0.5) = 0.020568$$

Equation $\boxed{7b}$ $\quad \nu_{12} = 1.9 \qquad$ since $\varepsilon_t > \varepsilon_y$

Equation $\boxed{8}$ $\quad \nu_{21} = 0 \quad$ after cracking

Equation $\boxed{9}$ $\qquad \bar{\varepsilon}_1 = \dfrac{1}{1 - \nu_{12}\nu_{21}} \varepsilon_1 + \dfrac{\nu_{12}}{1 - \nu_{12}\nu_{21}} \varepsilon_2 = \varepsilon_1 + \nu_{12}\varepsilon_2$

$$= 0.027416 + 1.9(-0.00300) = 0.021716$$

Equation $\boxed{10}$ $\quad \bar{\varepsilon}_2 = \frac{\nu_{21}}{1 - \nu_{12}\nu_{21}} \varepsilon_1 + \frac{1}{1 - \nu_{12}\nu_{21}} \varepsilon_2 = \varepsilon_2 = -0.00300$

Table 6.1   Calculation results of panel M3

Variables	Calculated values		
	Point 1 (first yielding)	Point 2 (peak point)	Point 3 (descending)
ε_2 selected	−0.000082	−0.00027	−0.00300
ε_1 last assumed	0.002254	0.010181	0.027416
γ_{12} last assumed	0.00094	0.00689	0.01672
ε_ℓ	0.000618	0.00151	0.003848
ε_t	0.001554	0.00835	0.020568
ν_{12}	1.52	1.90	1.90
$\bar{\varepsilon}_1$	0.002129	0.009668	0.021716
$\bar{\varepsilon}_2$	−0.000082	−0.00027	−0.00300
$\bar{\varepsilon}_\ell$	0.000556	0.001253	0.000998
$\bar{\varepsilon}_t$	0.001492	0.008144	0.017718
β (degrees)	10.9	16.7	14.4
ζ	0.335	0.115	0.107
σ_2^c (MPa)	−3.12	−5.54	−4.73
σ_1^c (MPa)	0.579	0.316	0.229
τ_{12}^c (MPa)	0.741	1.932	1.361
f_ℓ (MPa)	118.2	266.6	212.2
f_t (MPa)	278.8	355.3	465.5
$(\rho_\ell f_\ell + \rho_t f_t)_1$	2.540	5.207	4.492
$(\rho_\ell f_\ell - \rho_t f_t)_1$	1.480	3.857	2.723
$(\rho_\ell f_\ell + \rho_t f_t)_2$	2.541	5.227	4.496
$(\rho_\ell f_\ell - \rho_t f_t)_2$	1.481	3.863	2.723
$\tau_{\ell t}$ (MPa)	1.849	2.929	2.477
$\gamma_{\ell t}$	0.002336	0.010451	0.030416

Equation 11 $\bar{\varepsilon}_\ell = \bar{\varepsilon}_1 \cos^2 \alpha_1 + \bar{\varepsilon}_2 \sin^2 \alpha_1 - \frac{\gamma_{12}}{2} 2 \sin \alpha_1 \cos \alpha_1$

$$= 0.021716(0.5) + (-0.00300)(0.5) - (0.01672)(0.5) = 0.000998$$

Notice that the uniaxial strain $\bar{\varepsilon}_\ell = 0.000998$ is much smaller than the biaxial strain $\varepsilon_\ell = 0.003848$ due to Poisson effect.

Equation 12 $\bar{\varepsilon}_t = \bar{\varepsilon}_1 \sin^2 \alpha_1 + \bar{\varepsilon}_2 \cos^2 \alpha_1 + \frac{\gamma_{12}}{2} 2 \sin \alpha_1 \cos \alpha_1$

$$= 0.021716(0.5) + (-0.00300)(0.5) + (0.01672)(0.5) = 0.017718$$

Equation 15 $\beta = \dfrac{1}{2} \tan^{-1} \left[\dfrac{\gamma_{12}}{(\varepsilon_1 - \varepsilon_2)} \right] = \dfrac{1}{2} \tan^{-1} \left[\dfrac{0.01672}{(0.027416 + 0.00300)} \right] = 14.40°$

Equation 14 $\zeta = \left(\dfrac{5.8}{\sqrt{f_c'}} \le 0.9 \right) \left(\dfrac{1}{\sqrt{1 + 400\bar{\varepsilon}_1}} \right) \left(1 - \dfrac{|\beta|}{24°} \right)$

$$= \left(\frac{5.8}{\sqrt{48.1}} \le 0.9 \right) \left(\frac{1}{\sqrt{1 + 400(0.021716)}} \right) \left(1 - \frac{14.40°}{24°} \right)$$

$$= (0.8363)\ (0.3213)\ (0.400) = 0.1075$$

Equation $\boxed{13b}$ $\dfrac{\bar{\varepsilon}_2}{\zeta\varepsilon_o} = \dfrac{-0.00300}{0.1075(-0.0024)} = 11.63 > 1$ descending branch

$$\sigma_2^c = \zeta f_c' \left[1 - \left(\frac{(\bar{\varepsilon}_2/\zeta\varepsilon_o) - 1}{(4/\zeta) - 1} \right)^2 \right]$$

$$= 0.1075\,(-48.1) \left[1 - \left(\frac{11.63 - 1}{(4/0.1075) - 1} \right)^2 \right]$$

$$= 0.1075\,(-48.1)\,(0.9138) = -4.725 \text{ MPa}$$

Equation $\boxed{16b}$ $\bar{\varepsilon}_1 = 0.21716 > \varepsilon_{cr} = 0.00008$ descending branch

$$f_{cr} = 0.31\sqrt{f_c'(MPa)} = 0.31\sqrt{48.1} = 2.150 \text{ MPa}$$

$$\sigma_1^c = f_{cr} \left(\frac{\varepsilon_{cr}}{\bar{\varepsilon}_1} \right)^{0.4} = 2.150 \left(\frac{0.00008}{0.021716} \right)^{0.4} = 0.2285 \text{ MPa}$$

Equation $\boxed{17}$ $\tau_{12}^c = \dfrac{\sigma_1^c - \sigma_2^c}{2(\varepsilon_1 - \varepsilon_2)}\gamma_{12} = \dfrac{0.2285 + 4.725}{2(0.027416 + 0.00300)}(0.01672)$

$$= 81.42(0.01672) = 1.361 \text{ MPa}$$

Calculate longitudinal steel stress f_ℓ:

Equation $\boxed{18e}$ $B = \dfrac{1}{\rho_\ell} \left(\dfrac{f_{cr}}{f_{\ell y}} \right)^{1.5} = \dfrac{1}{0.0170} \left(\dfrac{2.150}{425.4} \right)^{1.5} = 0.0211$

Equation $\boxed{18d}$ $f_y' = (0.93 - 2B)f_{\ell y} = (0.93 - 2 \times 0.0211)(425.4) = 377.7 \text{ MPa}$

$$\varepsilon_y' = f_y'/E_s = 377.7/212,700 = 0.00178 > \bar{\varepsilon}_\ell = 0.000998$$

Notice that longitudinal steel did not yield.

Equation $\boxed{18a}$ $f_\ell = E_s\bar{\varepsilon}_\ell = 212\,700\,(0.000998) = 212.3 \text{ MPa}$

Calculate transverse steel stress f_t:

Equation $\boxed{19e}$ $B = \dfrac{1}{\rho_t} \left(\dfrac{f_{cr}}{f_{ty}} \right)^{1.5} = \dfrac{1}{0.0019} \left(\dfrac{2.150}{457.9} \right)^{1.5} = 0.1693$

Equation $\boxed{19d}$ $f_y' = (0.93 - 2B)f_{ty} = (0.93 - 2 \times 0.1693)(457.9) = 270.8 \text{ MPa}$

$$\varepsilon_y' = f_y'/E_s = 270.8/184,600 = 0.00147 < \bar{\varepsilon}_t = 0.01772$$

Notice that the transverse steel reached into the strain-hardening range.

Equation $\boxed{19b}$ $f_t = (0.91 - 2B)f_{ty} + (0.02 + 0.25B)E_s\bar{\varepsilon}_t$

$$= (0.91 - 2 \times 0.1693)\,(457.9)$$

$$+ (0.02 + 0.25 \times 0.1693)\,(184\,600)\,(0.01772)$$

$$= 261.6 + 203.9 = 465.5 \text{ MPa.}$$

Check equilibrium equations $\boxed{20}$ and $\boxed{21}$:

$$(\rho_\ell f_\ell + \rho_t f_t)_1 = 0.0170(212.3) + 0.0019(465.5)$$
$$= 3.609 + 0.884 = 4.493 \text{ MPa}$$
$$(\rho_\ell f_\ell - \rho_t f_t)_1 = 0.0170(212.3) - 0.0019(465.5)$$
$$= 3.609 - 0.884 = 2.725 \text{ MPa}$$

Equation $\boxed{20}$ $\quad (\rho_\ell f_\ell + \rho_t f_t)_2 = (\sigma_\ell + \sigma_t) - (\sigma_1^c + \sigma_2^c)$
$$= (0 + 0) - (0.2285 - 4.725)$$
$$= 4.496 \text{ MPa} \approx 4.493 \text{ MPa OK}$$

Equation $\boxed{21}$ $\quad (\rho_\ell f_\ell - \rho_t f_t)_2 = (\sigma_\ell - \sigma_t) - (\sigma_1^c - \sigma_2^c) \cos 2\alpha_1 + 2\tau_{12}^c \sin 2\alpha_1$
$$= (0 - 0) - (0.2285 + 4.725)(0) + 2(1.361)(1)$$
$$= 2.722 \text{ MPa} \approx 2.725 \text{ MPa OK}$$

Equation $\boxed{3}$ $\quad \tau_{\ell t} = (\sigma_1^c - \sigma_2^c) \sin \alpha_1 \cos \alpha_1 + \tau_{12}^c(\cos^2 \alpha_1 - \sin^2 \alpha_1)$
$$= (0.2285 + 4.725)(0.5) + 1.361(0.5 - 0.5) = 2.477 \text{ MPa}$$

Equation $\boxed{6}$ $\quad \dfrac{\gamma_{\ell t}}{2} = (\varepsilon_1 - \varepsilon_2) \sin \alpha_1 \cos \alpha_1 + \dfrac{\gamma_{12}}{2}(\cos^2 \alpha_1 - \sin^2 \alpha_1)$
$$= (0.027416 + 0.00300)(0.5) + \dfrac{0.01672}{2}(0.5 - 0.5)$$
$$= 0.015208.$$

$$\gamma_{\ell t} = 2(0.015208) = 0.030416.$$

6.1.12.3 Characteristics of Panels with Large ρ_ℓ/ρ_t Ratios

Figure 6.24 shows seven behavioral curves for panel M3, including: (a) applied shear stress versus shear strain relationship; (b) concrete compressive stress–strain relationship; (c) concrete shear stress–strain relationship; (d) steel stress versus biaxial strain in the longitudinal direction; (e) steel stress versus uniaxial strain in the longitudinal direction; (f) steel stress versus biaxial strain in the transverse direction; and (g) steel stress versus uniaxial strain in the transverse direction.

Examination of Figure 6.24 reveals five characteristics of panel M3 because of its large ρ_ℓ/ρ_t ratio of 9.0:

1. In a panel with transverse steel ($\rho_t = 0.19\%$) much less than the longitudinal steel ($\rho_\ell = 1.70\%$), the first yield point in the shear stress versus shear strain curve, designated by point 1, is due to the yielding of the transverse steel as shown in Figure 6.24(f) and (g). The steel stresses in the longitudinal direction were much less than the yield strength as shown in Figure 6.24(d) and (e).
2. In Table 6.1, the Hsu/Zhu ratio ν_{12} at first yielding (point 1) is 1.52, much less than 1.9. This is because the transverse steel ratio is very small ($\rho_t = 0.0019$), resulting in a large B value of 0.129 and a small smeared yield strain ($\varepsilon_y' = 0.001554$). Substituting $\varepsilon_{sf} = \varepsilon_t = 0.001554$ into Equation $\boxed{7a}$ gives a ν_{12} value of 1.52.
3. In panel M3, the peak point 2 of the applied shear stress (Figure 6.24a) is accompanied by the peaks of concrete properties in Figure 6.24 (b) and (c), indicating that the compressive

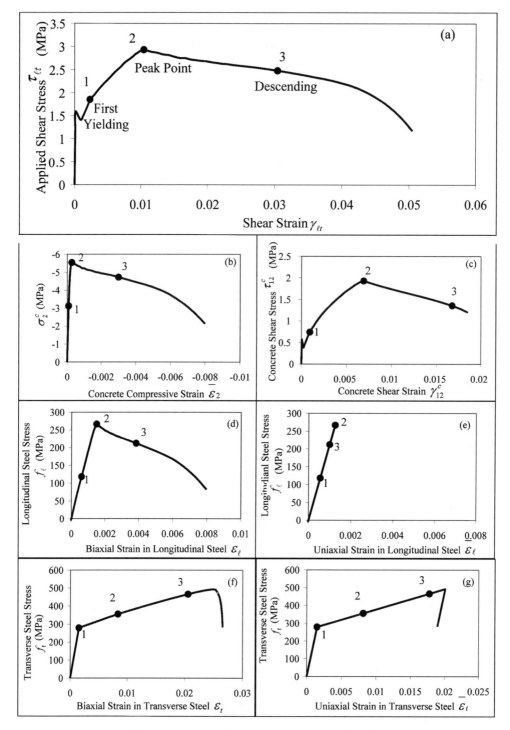

Figure 6.24   Calculated results of panel M3 using SMM

strength (Figure 6.24b) and the shear strength (Figure 6.24c) of the concrete are exhausted. Looking at the steel bars, the peak point is accompanied by large strains of the transverse steel bars in Figure 6.24(f) and (g), but not accompanied by the yielding of the longitudinal steel bars in Figure 6.24(d) and (e). In other words, failure is caused by the crushing of concrete and the yielding of transverse steel, but without the yielding of the longitudinal steel.

4. The strains in the longitudinal steel behave in a very interesting way after reaching point 2 (less than the yield point). While the uniaxial strain decreases elastically after point 2, Figure 6.24(e), the biaxial strain increases significantly along a descending branch (Figure 6.24d) due to the Poisson effect. The incorporation of Poisson effect allows SMM to predict the descending branches shown in Figure 6.24 (a), (b) and (c).

5. While the loading history of shear stress versus shear strain curve in Figure 6.24(a) moves from point 2 to a typical point 3 in the descending branch, both the stresses and the strains in the transverse steel increase rapidly into the strain-hardening region, as shown in Figure 6.24(f) and (g). Beyond point 3, the transverse steel stresses eventually decrease. However, this decrease of stresses is accompanied by an elastic decrease of uniaxial strain in Figure 6.24 (g), but is also accompanied by an increase of biaxial strain in Figure 6.24 (f).

6.1.12.4 Comparison of Predicted and Experimental $\tau_{\ell t}$–$\gamma_{\ell t}$ Curves

Figure 6.25 compares the SMM predicted shear stress versus shear strain ($\tau_{\ell t}$–$\gamma_{\ell t}$) curves of 16 panels with the experimental curves. The agreement is very good. The 16 panels include 6 panels from series B of Pang and Hsu (1995), 4 panels from series VB of Zhang and Hsu (1998) and 6 panels from series M of Chintrakarn (2001). Series B has a concrete strength of 42 MPa and ρ_ℓ/ρ_t ratios ranging from 1.5 to 5. Series VB has a concrete strength of 100 MPa and ρ_ℓ/ρ_t ratios ranging from 2 to 5. Series M has a concrete strength of 42 MPa and ρ_ℓ/ρ_t ratios ranging from 4 to infinity. It can be concluded that SMM is applicable to RC 2-D elements with concrete strengths up to 100 MPa and ρ_ℓ/ρ_t ratios up to infinity.

6.2 Fixed Angle Softened Truss Model (FA-STM)

6.2.1 Basic Principles of FA-STM

In Section 6.2, we will study the fixed angle softened truss model (FA-STM). FA-STM was developed before the Poisson effect was understood and before the establishment of the softened membrane model (SMM), as described in Section 6.1.

The essence of FA-STM is to neglect the Poisson effect in SMM. This has two implications. First, both the Hsu/Zhu ratios ν_{12} and ν_{21} are assumed to be zero in SMM and second, the biaxial strains (ε_1, ε_2, ε_ℓ, ε_t) are identical to the uniaxial strains ($\bar{\varepsilon}_1$, $\bar{\varepsilon}_2$, $\bar{\varepsilon}_\ell$, $\bar{\varepsilon}_t$). As a result, FA-STM can be considered a special case of SMM, and is simpler than SMM. However, FA-STM can not correctly predict the descending branch of the load–deformation curves.

FA-STM is useful when we are interested only in the peak shear strength and the behavior before the peak, and when we would like to have a more accurate prediction than RA-STM given in Section 5.4.

In FA-STM, the three equilibrium equations are the same as those in SMM, i.e Equations (6.1)–(6.3). The three compatibility equations, however, are different from those in SMM, i.e.

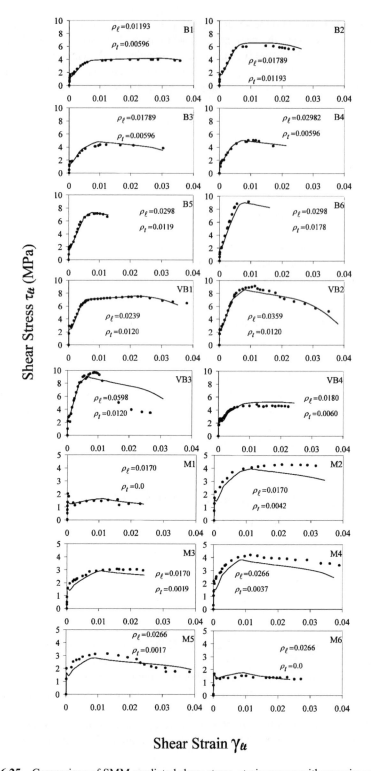

Figure 6.25 Comparison of SMM predicted shear stress–strain curves with experimental data

Strain compatibility equations

$$\bar{\varepsilon}_\ell = \bar{\varepsilon}_1 \cos^2 \alpha_1 + \bar{\varepsilon}_2 \sin^2 \alpha_1 - \frac{\gamma_{12}}{2} 2 \sin \alpha_1 \cos \alpha_1 \tag{6.93}$$

$$\bar{\varepsilon}_t = \bar{\varepsilon}_1 \sin^2 \alpha_1 + \bar{\varepsilon}_2 \cos^2 \alpha_1 + \frac{\gamma_{12}}{2} 2 \sin \alpha_1 \cos \alpha_1 \tag{6.94}$$

$$\frac{\gamma_{\ell t}}{2} = (\bar{\varepsilon}_1 - \bar{\varepsilon}_2) \sin \alpha_1 \cos \alpha_1 + \frac{\gamma_{12}}{2} (\cos^2 \alpha_1 - \sin^2 \alpha_1) \tag{6.95}$$

Notice in Equations (6.93)–(6.95) that the symbols $\bar{\varepsilon}_1$, $\bar{\varepsilon}_2$, $\bar{\varepsilon}_\ell$ and $\bar{\varepsilon}_t$ are the *uniaxial* strains, not the biaxial strains (ε_1, ε_2, ε_ℓ and ε_t) in SMM. The advantage of using uniaxial strains ($\bar{\varepsilon}_1$, $\bar{\varepsilon}_2$, $\bar{\varepsilon}_\ell$ and $\bar{\varepsilon}_t$) in FA-STM is that they can be directly related to the stresses (σ_1^c, σ_2^c, τ_{12}^c, f_ℓ and f_t) in the equilibrium equations by uniaxial stress–strain relationships obtained directly from the uniaxial tests of RC 2-D elements. The uniaxial stress–strain relationships of smeared concrete and smeared steel bars are given in Sections 6.1.6–6.1.10 and are summarized as follows.

Uniaxial constitutive relationships

The solution of the three stress equilibrium equations and the three strain compatibility equations requires two constitutive matrices, one for concrete and one for steel. The concrete constitutive matrix is:

$$\begin{bmatrix} \sigma_1^c \\ \sigma_2^c \\ \tau_{12}^c \end{bmatrix} = \begin{bmatrix} \bar{E}_1^c & 0 & 0 \\ 0 & \bar{E}_2^c & 0 \\ 0 & 0 & G_{12}^c \end{bmatrix} \begin{bmatrix} \bar{\varepsilon}_1 \\ \bar{\varepsilon}_2 \\ \dfrac{\gamma_{12}}{2} \end{bmatrix} \tag{6.96}$$

The 3×3 matrix in Equation (6.96) contains three diagonal nonlinear moduli $\bar{E}_1^c$, $\bar{E}_2^c$ and G_{12}^c. $\bar{E}_1^c$ represents the $\sigma_1^c - \bar{\varepsilon}_1$ curve given in Section 6.1.8. $\bar{E}_2^c$ represents the $\sigma_2^c - \bar{\varepsilon}_2$ curve given in Section 6.1.6 and 6.1.7; and G_{12}^c gives the $\tau_{12}^c - \gamma_{12}$ curve in section 6.1.10.

The steel constitutive matrix is:

$$\begin{bmatrix} \rho_\ell f_\ell \\ \rho_t f_t \\ 0 \end{bmatrix} = \begin{bmatrix} \rho_\ell \bar{E}_\ell^s & 0 & 0 \\ 0 & \rho_t \bar{E}_t^s & 0 \\ 0 & 0 & 0 \end{bmatrix} \begin{bmatrix} \bar{\varepsilon}_\ell \\ \bar{\varepsilon}_t \\ 0 \end{bmatrix} \tag{6.97}$$

The two diagonal elements in the 3×3 matrix $\rho_\ell \bar{E}_\ell^s$ and $\rho_t \bar{E}_t^s$ are for mild steel bars in the ℓ- and t-directions, respectively. ρ_ℓ and ρ_t are the steel ratios. $\bar{E}_\ell^s$ and $\bar{E}_t^s$ represent the $f_\ell - \bar{\varepsilon}_\ell$ and $f_t - \bar{\varepsilon}_t$ curves, respectively, for the smeared stress–strain curves of mild steel bars embedded in concrete. These curves are given in Section 6.1.9.

6.2.2 Solution Algorithm

6.2.2.1 Summary of Governing Equations

The twelve governing equations for RC 2-D elements are summarized below:

Stress equilibrium equations

$$\sigma_\ell = \sigma_1^c \cos^2 \alpha_1 + \sigma_2^c \sin^2 \alpha_1 - \tau_{12}^c 2 \sin \alpha_1 \cos \alpha_1 + \rho_\ell f_\ell \qquad \boxed{1}$$

$$\sigma_t = \sigma_1^c \sin^2 \alpha_1 + \sigma_2^c \cos^2 \alpha_1 + \tau_{12}^c 2 \sin \alpha_1 \cos \alpha_1 + \rho_t f_t \qquad \boxed{2}$$

$$\tau_{\ell t} = (\sigma_1^c - \sigma_2^c) \sin \alpha_1 \cos \alpha_1 + \tau_{12}^c (\cos^2 \alpha_1 - \sin^2 \alpha_1) \qquad \boxed{3}$$

Strain compatibility equations

$$\bar{\varepsilon}_\ell = \bar{\varepsilon}_1 \cos^2 \alpha_1 + \bar{\varepsilon}_2 \sin^2 \alpha_1 - \frac{\gamma_{12}}{2} 2 \sin \alpha_1 \cos \alpha_1 \qquad \boxed{11}$$

$$\bar{\varepsilon}_t = \bar{\varepsilon}_1 \sin^2 \alpha_1 + \bar{\varepsilon}_2 \cos^2 \alpha_1 + \frac{\gamma_{12}}{2} 2 \sin \alpha_1 \cos \alpha_1 \qquad \boxed{12}$$

$$\frac{\gamma_{\ell t}}{2} = (\bar{\varepsilon}_1 - \bar{\varepsilon}_2) \sin \alpha_1 \cos \alpha_1 + \frac{\gamma_{12}}{2}(\cos^2 \alpha_1 - \sin^2 \alpha_1) \qquad \boxed{6a}$$

Constitutive laws of materials
Concrete in compression

$$\sigma_2^c = \zeta f_c' \left[2 \left(\frac{\bar{\varepsilon}_2}{\zeta \varepsilon_o} \right) - \left(\frac{\bar{\varepsilon}_2}{\zeta \varepsilon_o} \right)^2 \right] \qquad \bar{\varepsilon}_2/\zeta \varepsilon_o \leq 1 \qquad \boxed{13a}$$

$$\sigma_2^c = \zeta f_c' \left[1 - \left(\frac{\bar{\varepsilon}_2/\zeta \varepsilon_o - 1}{4/\zeta - 1} \right)^2 \right] \qquad \bar{\varepsilon}_2/\zeta \varepsilon_o \geq 1 \qquad \boxed{13b}$$

$$\zeta = \left(\frac{5.8}{\sqrt{f_c'}} \leq 0.9 \right) \left(\frac{1}{\sqrt{1 + 400 \bar{\varepsilon}_1}} \right) \left(1 - \frac{|\beta|}{24°} \right) \qquad \boxed{14}$$

$$\beta = \frac{1}{2} \tan^{-1} \left[\frac{\gamma_{12}}{(\bar{\varepsilon}_1 - \bar{\varepsilon}_2)} \right] \qquad \boxed{15}$$

Concrete in tension

$$\sigma_1^c = E_c \bar{\varepsilon}_1 \qquad \bar{\varepsilon}_1 \leq \varepsilon_{cr} \quad \text{and} \quad \varepsilon_{cr} = 0.00008 \text{ mm/mm}, \qquad \boxed{16a}$$

$$\sigma_1^c = f_{cr} \left(\frac{\varepsilon_{cr}}{\bar{\varepsilon}_1} \right)^{0.4} \qquad \bar{\varepsilon}_1 > \varepsilon_{cr} \quad \text{and} \quad f_{cr} = 0.31 \sqrt{f_c'(MPa)} \qquad \boxed{16b}$$

Concrete in shear

$$\tau_{12}^c = \frac{\sigma_1^c - \sigma_2^c}{2(\bar{\varepsilon}_1 - \bar{\varepsilon}_2)} \gamma_{12} \qquad \boxed{17}$$

Mild steel

$$f_s = E_s \bar{\varepsilon}_s \qquad\qquad\qquad\qquad\qquad \bar{\varepsilon}_s \leq \varepsilon_y' \qquad \boxed{18a}, \boxed{19a}$$

$$f_s = (0.91 - 2B)f_y + (0.02 + 0.25B)E_s \bar{\varepsilon}_s \qquad \bar{\varepsilon}_s > \varepsilon_y' \qquad \boxed{18b}, \boxed{19b}$$

$$f_s = f_p - E_s(\bar{\varepsilon}_p - \bar{\varepsilon}_s) \qquad\qquad\qquad \bar{\varepsilon}_s < \bar{\varepsilon}_p \qquad \boxed{18c}, \boxed{19c}$$

$$\varepsilon_y' = f_y'/E_s \qquad f_y' = (0.93 - 2B)f_y \qquad\qquad\qquad \boxed{18d}, \boxed{19d}$$

$$B = \frac{1}{\rho} \left(\frac{f_{cr}}{f_y} \right)^{1.5} \qquad f_{cr} = 0.31 \sqrt{f_c'(MPa)} \text{ and } \rho \geq 0.15\% \qquad \boxed{18e}, \boxed{19e}$$

6.2.2.2 Solution Algorithm

To develop an efficient solution algorithm, the first two equilibrium equations [1] and [2] are summed and subtracted to obtain the following two equations, which are used as the convergence criteria for the solution procedure:

$$\rho_\ell f_\ell + \rho_t f_t = (\sigma_\ell + \sigma_t) - \left(\sigma_1^c + \sigma_2^c\right) \tag{20}$$

$$\rho_\ell f_\ell - \rho_t f_t = (\sigma_\ell - \sigma_t) - \left(\sigma_1^c - \sigma_2^c\right) \cos 2\alpha_1 + 2\tau_{12}^c \sin 2\alpha_1 \tag{21}$$

The iterative procedure of FA-STM is illustrated by the flow chart in Figure 6.26. First select a $\bar{\varepsilon}_2$ value and assume two values for $\bar{\varepsilon}_1$ and γ_{12}. After completing the DO-loops in the flow chart using Equations [20]–[21], the shear stress $\tau_{\ell t}$ and the shear strain $\gamma_{\ell t}$ for a selected value of ε_2 are calculated from equations [3] and [6a], respectively. By selecting a series of $\bar{\varepsilon}_2$ values, we can plot the $\tau_{\ell t} - \gamma_{\ell t}$ curves up to the peak point. Curves relating any two variables can similarly be plotted.

6.2.3 Example Problem 6.2

6.2.3.1 Problem Statement

A RC 2-D element B2, as shown in Figure 6.2(b) with $\alpha_1 = 45°$, has been tested by Pang and Hsu (1995). This 2-D element will be used to illustrate the step-by-step calculation procedure of FA-STM. The properties of concrete and steel and the loadings are as follows:

Longitudinal steel: $\rho_\ell = 1.789\%$, $f_{\ell y} = 446.5$ MPa, and $E_s = 200\,000$ MPa.
Transverse steel: $\rho_t = 1.193\%$, $f_{ty} = 462.6$ MPa, and $E_s = 192\,400$ MPa.
Properties of concrete: $f_c' = 44.1$ MPa and $\varepsilon_o = 0.00235$.
Applied stresses: $\sigma_\ell = 0$ MPa, $\sigma_t = 0$ MPa, and $\tau_{\ell t}$ increased until failure.
Notice that the ρ_ℓ / ρ_t ratio ($= 1.5$) in this panel is not far from unity, Therefore, the contribution of concrete shear stress, τ_{12}^c, will not be large.

Analyze the behavior of this 2-D element B2 by the fixed angle softened truss model (FA-STM). Plot the $\tau_{\ell t} - \gamma_{\ell t}$ curve, the $\sigma_2^c - \bar{\varepsilon}_2$ curve, the $\tau_{12}^c - \gamma_{12}$ curve, the $f_\ell - \varepsilon_\ell$ curve, the $f_\ell - \bar{\varepsilon}_\ell$ curve, the $f_t - \varepsilon_t$ curve, and the $f_t - \bar{\varepsilon}_t$ curve. Compare these curves with those predicted by SMM.

6.2.3.2 Solution

The calculations according to the flow chart in Figure 6.26 are best done by computer, because the procedure involves a nested DO-loop. The calculations were terminated when the error is less than 0.1% and when the concrete compressive strain ε_2 reaches a limiting value $\varepsilon_{\lim}$ of -0.005.

The computer-calculated results are recorded in Table 6.2 for three critical points: point 1 at first yield; point 2 at second yield; point 3 at the peak strength of $\tau_{\ell t}$. For comparison, the SMM-predicted results are also given in Table 6.2. The three critical points are labeled as point 1^* at first yield, point 2^* at second yield, point 3^* at the peak strength of $\tau_{\ell t}$.

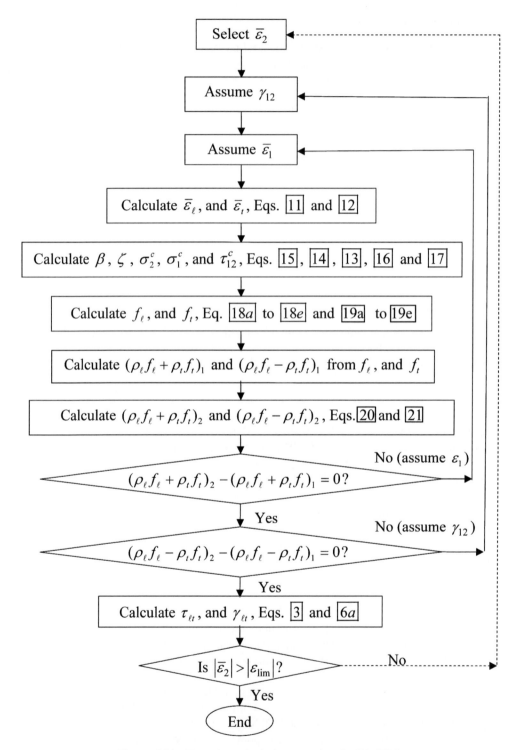

Figure 6.26 Flow chart of solution procedure for FA-STM

Table 6.2 Calculated results of panel B2

Variables	Calculated values (FA-STM)			Calculated values (SMM)		
	Point 1 transverse steel yielding	Point 2 longitudinal steel yielding	Point 3 peak point	Point 1* transverse steel yielding	Point 2* longitudinal steel yielding	Point 3* peak point
ε_2 Selected	-0.00036	-0.00044	-0.00067	-0.00036	-0.00044	-0.00067
ε_1 last assumed	0.0041695	0.0056824	0.0118800	0.0049156	0.0065430	0.0131990
γ_{12} last assumed	0.0004425	0.0010643	0.0021431	0.0005100	0.0012085	0.0023529
ε_ℓ	0.001684	0.002089	0.004534	0.002023	0.002447	0.005088
ε_t	0.002126	0.003153	0.006677	0.002533	0.003656	0.007441
ν_{12}	0	0	0	1.9	1.9	1.9
$\bar{\varepsilon}_1$	0.004170	0.005682	0.011880	0.004232	0.005707	0.011926
$\bar{\varepsilon}_2$	-0.00036	-0.00044	-0.00067	-0.00036	-0.00044	-0.00067
$\bar{\varepsilon}_\ell$	0.001684	0.002089	0.004534	0.001681	0.002029	0.004452
$\bar{\varepsilon}_t$	0.002126	0.003153	0.006677	0.002191	0.003238	0.006804
β (degrees)	2.79	4.93	4.85	2.76	4.91	4.81
ζ	0.4726	0.3836	0.2906	0.4710	0.3834	0.2907
σ_2^c (MPa)	-11.32	-12.48	-12.81	-11.31	-12.48	-12.81
σ_1^c (MPa)	0.4234	0.3741	0.2785	0.4209	0.3735	0.2781
τ_{12}^c (MPa)	0.5737	1.1176	1.1177	0.5673	1.1124	1.1105
f_ℓ (MPa)	336.7	400.9	412.8	336.2	400.6	412.4
f_t (MPa)	408.7	413.9	431.6	409.0	414.3	432.3
$(\rho_\ell f_\ell + \rho_t f_t)_1$	10.90	12.11	12.53	10.89	12.11	12.54
$(\rho_\ell f_\ell - \rho_t f_t)_1$	1.148	2.234	2.236	1.135	2.224	2.221
$(\rho_\ell f_\ell + \rho_t f_t)_2$	10.90	12.11	12.53	10.89	12.11	12.54
$(\rho_\ell f_\ell - \rho_t f_t)_2$	1.147	2.235	2.235	1.135	2.225	2.221
$\tau_{\ell t}$ (MPa)	5.873	6.429	6.546	5.868	6.428	6.546
$\gamma_{\ell t}$	0.004530	0.006122	0.012550	0.005276	0.006983	0.013869

Calculation at the peak point 3 using FA-STM:

> Select $\bar{\varepsilon}_2 = -0.00067$

> Assume $\bar{\varepsilon}_1 = 0.01188$

> Assume $\gamma_{12} = 0.002143$

> $\alpha_1 = 45°$, because the 2-D element is subjected to pure shear.

Equation $\boxed{11}$ $\bar{\varepsilon}_\ell = \bar{\varepsilon}_1 \cos^2 \alpha_1 + \bar{\varepsilon}_2 \sin^2 \alpha_1 - \frac{\gamma_{12}}{2} 2 \sin \alpha_1 \cos \alpha_1$

$$= 0.01188(0.5) + (-0.00067)(0.5) - (0.002143)(0.5)$$

$$= 0.004534$$

Equation $\boxed{12}$ $\bar{\varepsilon}_t = \bar{\varepsilon}_1 \sin^2 \alpha_1 + \bar{\varepsilon}_2 \cos^2 \alpha_1 + \frac{\gamma_{12}}{2} 2 \sin \alpha_1 \cos \alpha_1$

$$= 0.01188(0.5) + (-0.00067)(0.5) + (0.002143)(0.5)$$

$$= 0.006677$$

Equation $\boxed{15}$ $\beta = \frac{1}{2} \tan^{-1} \left[\frac{\gamma_{12}}{(\bar{\varepsilon}_1 - \bar{\varepsilon}_2)} \right] = \frac{1}{2} \tan^{-1} \left[\frac{0.002143}{(0.01188 + 0.00067)} \right] = 4.845°$

Equation $\boxed{14}$ $\zeta = \left(\frac{5.8}{\sqrt{f'_c}} \le 0.9 \right) \left(\frac{1}{\sqrt{1 + 400\bar{\varepsilon}_1}} \right) \left(1 - \frac{|\beta|}{24°} \right)$

$$= \left(\frac{5.8}{\sqrt{44.1}} \le 0.9 \right) \left(\frac{1}{\sqrt{1 + 400(0.01188)}} \right) \left(1 - \frac{4.845°}{24°} \right)$$

$$= (0.8734)(0.4170)(0.7981) = 0.2906$$

Equation $\boxed{13b}$ $\dfrac{\bar{\varepsilon}_2}{\zeta \varepsilon_o} = \dfrac{-0.00067}{0.2906(-0.00235)} = 0.981 < 1$ ascending branch

$$\sigma_2^c = \zeta f'_c \left[2 \left(\frac{\bar{\varepsilon}_2}{\zeta \varepsilon_o} \right) - \left(\frac{\bar{\varepsilon}_2}{\zeta \varepsilon_o} \right)^2 \right]$$

$$= 0.2906(-44.1) \left[2(0.981) - (0.981)^2 \right]$$

$$= 0.2906(-44.1)(0.9996) = -12.81 \text{ MPa}$$

Equation $\boxed{16b}$ $\bar{\varepsilon}_1 = 0.01188 > \varepsilon_{cr} = 0.00008$ descending branch

$$f_{cr} = 0.31\sqrt{f'_c(MPa)} = 0.31\sqrt{44.1} = 2.059 \text{ MPa}$$

$$\sigma_1^c = f_{cr} \left(\frac{\varepsilon_{cr}}{\bar{\varepsilon}_1} \right)^{0.4} = 2.059 \left(\frac{0.00008}{0.01188} \right)^{0.4} = 0.2785 \text{ MPa}$$

Equation $\boxed{17}$ $\tau_{12}^c = \dfrac{\sigma_1^c - \sigma_2^c}{2(\bar{\varepsilon}_1 - \bar{\varepsilon}_2)} \gamma_{12} = \dfrac{0.2785 + 12.81}{2(0.01188 + 0.00067)}(0.002143)$

$$= 521.58(0.002143) = 1.1177 \text{ MPa}$$

Calculate the longitudinal steel stress f_ℓ:

Equation 18e $\quad B = \dfrac{1}{\rho_\ell}\left(\dfrac{f_{cr}}{f_{\ell y}}\right)^{1.5} = \dfrac{1}{0.01789}\left(\dfrac{2.059}{446.5}\right)^{1.5} = 0.0175$

Equation 18d $\quad f_y' = (0.93 - 2B)f_{\ell y} = (0.93 - 2 \times 0.0175)\,446.5 = 399.6$ MPa

$\qquad\qquad \varepsilon_y' = f_y'/E_s = 399.6/200\,000 = 0.00200 < \bar{\varepsilon}_\ell = 0.004534$

Notice that the longitudinal steel has yielded.

Equation 18b $\quad f_\ell = (0.91 - 2B)f_{\ell y} + (0.02 + 0.25B)E_\ell \bar{\varepsilon}_\ell$

$\qquad\qquad = (0.91 - 2 \times 0.0175)\,446.5$

$\qquad\qquad\quad +(0.02 + 0.25 \times 0.0175)\,200\,000\,(0.004534)$

$\qquad\qquad = 390.7 + 22.10 = 412.8$ MPa

Calculate the transverse steel stress f_t:

Equation 19e $\quad B = \dfrac{1}{\rho_t}\left(\dfrac{f_{cr}}{f_{ty}}\right)^{1.5} = \dfrac{1}{0.01193}\left(\dfrac{2.059}{462.6}\right)^{1.5} = 0.0249$

Equation 19d $\quad f_y' = (0.93 - 2B)f_{ty} = (0.93 - 2 \times 0.0249)\,462.6 = 407.2$ MPa

$\qquad\qquad \varepsilon_y' = f_y'/E_s = 407.2/192\,400 = 0.00212 < \bar{\varepsilon}_t = 0.006677$

Notice that the transverse steel strain has also yielded and its strain $(\bar{\varepsilon}_t)$ is greater than the longitudinal steel strain $(\bar{\varepsilon}_\ell)$.

Equation 19b $\quad f_t = (0.91 - 2B)f_{ty} + (0.02 + 0.25B)E_t\bar{\varepsilon}_t$

$\qquad\qquad = (0.91 - 2 \times 0.0249)\,462.6$

$\qquad\qquad\quad +(0.02 + 0.25 \times 0.0249)\,192\,400\,(0.006677)$

$\qquad\qquad = 397.9 + 33.70 = 431.6$ MPa

Check equilibrium equations 20 and 21 :

$\qquad\qquad (\rho_\ell f_\ell + \rho_t f_t)_1 = 0.01789(412.5) + 0.01193(431.6)$

$\qquad\qquad\qquad\qquad = 7.385 + 5.149 = 12.53$ MPa

$\qquad\qquad (\rho_\ell f_\ell - \rho_t f_t)_1 = 0.01789(412.5) - 0.01193(431.6)$

$\qquad\qquad\qquad\qquad = 7.385 - 5.149 = 2.236$ MPa

Equation 20 $\quad (\rho_\ell f_\ell + \rho_t f_t)_2 = (\sigma_\ell + \sigma_t) - (\sigma_1^c + \sigma_2^c)$

$\qquad\qquad\qquad = (0 + 0) - (0.2785 - 12.81)$

$\qquad\qquad\qquad = 12.53$ MPa ≈ 12.53 MPa OK

Equation $\boxed{21}$ $(\rho_\ell f_\ell - \rho_t f_t)_2 = (\sigma_\ell - \sigma_t) - (\sigma_1^c - \sigma_2^c) \cos 2\alpha_1 + 2\tau_{12}^c \sin 2\alpha_1$

$$= (0 + 0) - (0.2785 + 12.81)(0) + 2(1.1177)(1)$$

$$= 2.235 \text{ MPa} \approx 2.236 \text{ MPa OK}$$

Equation $\boxed{3}$ $\tau_{\ell t} = (\sigma_1^c - \sigma_2^c) \sin \alpha_1 \cos \alpha_1 + \tau_{12}^c (\cos^2 \alpha_1 - \sin^2 \alpha_1)$

$$= (0.2785 + 12.81)(0.5) + 1.1177(0.5 - 0.5) = 6.546 \text{ MPa}$$

Equation $\boxed{6}$ $\dfrac{\gamma_{\ell t}}{2} = (\bar\varepsilon_1 - \bar\varepsilon_2) \sin \alpha_1 \cos \alpha_1 + \dfrac{\gamma_{12}}{2}(\cos^2 \alpha_1 - \sin^2 \alpha_1)$

$$= (0.01188 + 0.00067)(0.5) + \frac{0.002143}{2}(0.5 - 0.5) = 0.006275$$

$$\gamma_{\ell t} = 2(0.006275) = 0.01255$$

6.2.3.3 Comparison of Curves Predicted by FA-STM and SMM

The 2-D element B2 is analyzed by the fixed angle softened truss model (FA-STM) and the softened membrane model (SMM). The calculated results are recorded in Figure 6.27(a)–(g) in terms of the following seven curves: (a) applied shear stress versus shear strain relationship ($\tau_{\ell t} - \gamma_{\ell t}$ curve); (b) concrete compressive stress versus uniaxial strain relationship ($\sigma_2^c - \bar\varepsilon_2$ curve); (c) concrete shear stress–strain relationship ($\tau_{12}^c - \gamma_{12}$ curve); (d) steel stress versus biaxial strain in the longitudinal direction ($f_\ell - \varepsilon_\ell$ curve), (e) steel stress versus uniaxial strain in the longitudinal direction ($f_\ell - \bar\varepsilon_\ell$ curve); (f) steel stress versus biaxial strain in the transverse direction ($f_t - \varepsilon_t$ curve); and (g) steel stress versus uniaxial strain in the transverse direction ($f_t - \bar\varepsilon_t$ curve). The following observations are noted:

1. Figure 6.27(a) shows that the peak strength $\tau_{\ell t}$ predicted by FA-STM and SMM are essentially the same. In other words, the absence of Poisson effect (i.e. assuming the Hsu/Zhu ratios to be zero) in FA-STM has very little influence on its prediction of peak strength.
2. In Figure 6.27(a), the shear strains, $\gamma_{\ell t}$, are noticeably influenced by the Poisson effect at the first yield point 1. The difference between FA-STM and SMM increases at second yield point 2, at the peak strength point 3 and particularly in the descending branch. In the descending branch, $\gamma_{\ell t}$ terminates at a strain of 0.020 for FA-STM, much less than the termination strain of 0.030 for SMM.
3. In panel B2 with transverse steel ($\rho_t = 1.193\%$) less than the longitudinal steel ($\rho_\ell = 1.789\%$), the first yield at point 1 in Figure 6.27(a) is due to yielding of transverse steel as shown in Figure 6.27(f) and (g). The second yield at point 2 is due to yielding of longitudinal steel as shown in Figure 6.27(d) and (e). The yielding of transverse steel (point 1) is followed shortly by the yielding of the longitudinal steel (point 2), because $\rho_\ell / \rho_t = 1.5$ is not far from unity.
4. The peak point 3 in Figure 6.27(a) is accompanied by the peak point of compressive stress–strain ($\sigma_2^c - \bar\varepsilon_2$) curve of concrete in Figure 6.27(b) and by the peak point of the shear stress–strain ($\tau_{12}^c - \gamma_{12}$) curve of concrete in Figure 6.27(c). In other words, the peak shear strength $\tau_{\ell t}$ corresponds to the failure of concrete struts due to combined compression and shear.

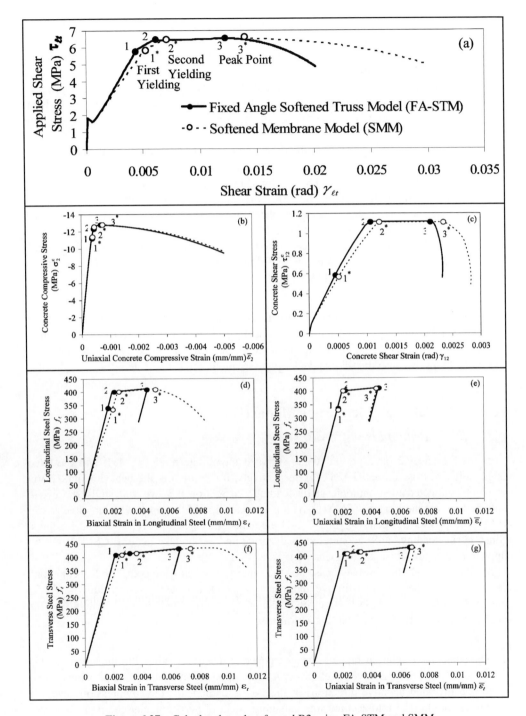

Figure 6.27   Calculated results of panel B2 using FA-STM and SMM

5. Beyond the peak point 3, the long descending branch predicted by SMM in Figure 6.27(a) is accompanied by the long descending branches of biaxial steel strains (ε_ℓ, ε_t) predicted by SMM in Figure 6.27(d) and (f). These large biaxial steel strains in the longitudinal and transverse steel are the direct results of considering the large Poisson effect (Hsu/Zhu ratio $\nu_{12} = 1.9$). It should be noted, however, that the concrete compressive strain, $\bar\varepsilon_2 = \varepsilon_2$, is not affected by Poisson effect, because Hsu/Zhu ratio $\nu_{21} = 0$. Even so, the uniaxial concrete compressive strain, $\bar\varepsilon_2$, in Figure 6.27(b) shows long descending branches.

6. The influence of Poisson effect in the descending branches of longitudinal steel can also be observed by comparing the biaxial steel strains ε_ℓ and ε_t in Figure 6.27(d) and (f), respectively, calculated by FA-STM and by SMM. The biaxial steel strains calculated by SMM, which considers the Hsu/Zhu ratios, provides a long and gently descending curve. In contrast, the biaxial steel strains calculated by FA-STM, which neglects the Hsu/Zhu ratios, follows an elastic, unloading straight line.

7. The uniaxial steel strains $\bar\varepsilon_\ell$ and $\bar\varepsilon_t$ calculated by FA-STM and SMM and shown in Figure 6.27(e) and (g), respectively, are essentially the same. The descending branches always exhibit elastic, unloading straight line.

In summary, in the design of concrete structures subjected to monotonic, static loading, the FA-STM can be very useful in calculating the peak shear strength. However, in the design of concrete structures subjected to earthquake loading, the SMM must be used to calculate the hysteretic loops, because ductility and energy dissipation play a vital role.

6.3 Cyclic Softened Membrane Model (CSMM)

6.3.1 Basic Principles of CSMM

Sections 6.1 and 6.2 have presented the softened membrane model (SMM) and the fixed angle softened truss model (FA-STM), respectively, for predicting the behavior of RC 2-D elements subjected to monotonic static loading. In Section 6.3, we will study the cyclic softened membrane model (CSMM) which is an extension of the softened membrane model (SMM) to predict the behavior of RC 2-D elements subjected to reversed cyclic loadings. Since cyclic loadings are encountered in dynamic and earthquake actions, CSMM will be applied to structures discussed in Chapters 9 and 10.

Similar to SMM, the CSMM is based on the same equilibrium and compatibility equations (6.1)–(6-6), and the same constitutive matrices for concrete and steel given by Equations (6.7) and (6.8). However, the constitutive relationships of concrete and steel are generalized for application to reversed cyclic loading as follows:

1. The constitutive relationships of smeared concrete must be modified in three aspects: (a) the tensile and compressive envelope curves are expressed by equations applicable to both the positive and negative directions of the cyclic loading; (b) the compression envelope curves should take into account the damaging effect of perpendicular compressive stress in previous loading cycles; and (c) the unloading and reloading curves in both the tension

and compression regions are defined by a series of straight lines. These modifications are studied in Section 6.3.2.

2. The constitutive relationships of smeared mild steel bars must also include two modifications: (a) the envelope curves of mild steel are identical to the monotonic curves, except that a compressive yielding stage should be added; and (b) the unloading and reloading curves should be defined to express the Bauschinger effect. These modifications are studied in Section 6.3.3.

3. The Hsu/Zhu ratios need to be adjusted for application to cyclic loading. This subject is studied in Section 6.3.4.

Once the constitutive relationships of materials under cyclic loading are derived, we can then discuss the solution procedure in Section 6.3.5.

The results of the solution are presented graphically as hysteretic loops. Section 6.3.6 presents the interesting and crucial problem of a 'pinched shape' in the hysteretic loops. This is followed by a discussion of the mechanism of pinching and failure under cyclic shear in Section 6.3.7.

Finally, the hysteretic loops of eight demonstrative panels are generated by CSMM in Section 6.3.8, in order to study the effect of the two variables (steel bar angle and steel percentage) on the three properties of RC 2-D elements under cyclic shear. The three cyclic properties, which are shear stiffness, shear ductility, and shear energy dissipation, are discussed in Sections 6.3.9, 6.3.10, and 6.3.11, respectively.

6.3.2 Cyclic Stress–Strain Curves of Concrete

When Equations (6.1)–6.8) are applied to monotonic static loading in SMM, the 1–2 coordinate is defined as follows: 1-axis is the direction of the principal tensile stress and strain, and 2-axis is the direction of the principal compressive stress and strain. When Equations (6.1)–(6.8) are applied to reversed cyclic loading in CSMM, however, the 1–2 coordinate is defined in a more general manner. The 1–2 coordinate is still the coordinate of principal stress and strain, but both 1-axis and 2-axis are alternately the directions of principal tension and the principal compression. In other words, when the cyclic load is in the positive direction, the 1-axis is the principal tension direction and the 2-axis is the principal compression direction (as in the monotonic loading). When the cyclic load is in the *negative* direction, however, the 1-axis is the principal *compression* direction and the 2-axis is the principal *tension* direction.

The cyclic uniaxial constitutive relationships of cracked concrete in compression and tension are summarized in Figure 6.28. In the graph, the vertical axis represents the cyclic stress σ^c, with positive tensile stress above the origin and negative compressive stress below the origin. The horizontal axis represents the cyclic uniaxial strain $\bar{\varepsilon}$, with positive tensile strain to the right of origin and negative compressive strain to the left of origin.

The upper right quadrant gives the tensile envelope stress–strain curves T1 and T2. In the lower left quadrant is the compression envelope stress–strain curves C1 and C2. The unloading and reloading curves are represented by the series of straight lines C3–C7 in the compressive strain regions, and T3, T4 in the tensile strain region. Each straight line connects two points with their coordinates specified in the lower right quadrant.

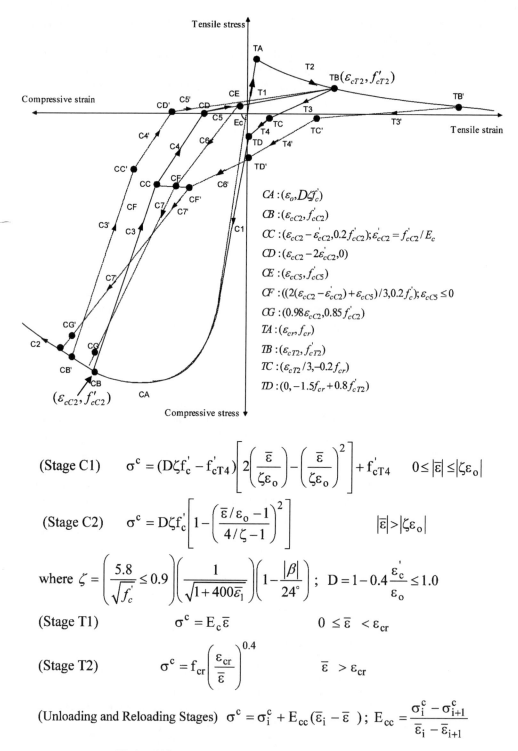

Figure 6.28 Cyclic smeared stress–strain curve of concrete

The figure contains the following labeled points:

$$CA : (\varepsilon_o, D\mathcal{J}_c')$$
$$CB : (\varepsilon_{cC2}, f_{cC2}')$$
$$CC : (\varepsilon_{cC2} - \varepsilon_{cC2}', 0.2f_{cC2}'); \varepsilon_{cC2}' = f_{cC2}'/E_c$$
$$CD : (\varepsilon_{cC2} - 2\varepsilon_{cC2}', 0)$$
$$CE : (\varepsilon_{cC5}, f_{cC5}')$$
$$CF : ((2(\varepsilon_{cC2} - \varepsilon_{cC2}') + \varepsilon_{cC5})/3, 0.2f_c'); \varepsilon_{cC5} \le 0$$
$$CG : (0.98\varepsilon_{cC2}, 0.85f_{cC2}')$$
$$TA : (\varepsilon_{cr}, f_{cr})$$
$$TB : (\varepsilon_{cT2}, f_{cT2}')$$
$$TC : (\varepsilon_{cT2}/3, -0.2f_{cr})$$
$$TD : (0, -1.5f_{cr} + 0.8f_{cT2}')$$

$$\text{(Stage C1)} \qquad \sigma^c = (D\zeta f_c' - f_{cT4}')\left[2\left(\frac{\bar{\varepsilon}}{\zeta\varepsilon_o}\right) - \left(\frac{\bar{\varepsilon}}{\zeta\varepsilon_o}\right)^2\right] + f_{cT4}' \qquad 0 \le |\bar{\varepsilon}| \le |\zeta\varepsilon_o|$$

$$\text{(Stage C2)} \qquad \sigma^c = D\zeta f_c'\left[1 - \left(\frac{\bar{\varepsilon}/\varepsilon_o - 1}{4/\zeta - 1}\right)^2\right] \qquad\qquad |\bar{\varepsilon}| > |\zeta\varepsilon_o|$$

$$\text{where } \zeta = \left(\frac{5.8}{\sqrt{f_c'}} \le 0.9\right)\left(\frac{1}{\sqrt{1 + 400\bar{\varepsilon}_1}}\right)\left(1 - \frac{|\beta|}{24°}\right); \quad D = 1 - 0.4\frac{\varepsilon_c'}{\varepsilon_o} \le 1.0$$

$$\text{(Stage T1)} \qquad\qquad \sigma^c = E_c\bar{\varepsilon} \qquad\qquad 0 \le \bar{\varepsilon} < \varepsilon_{cr}$$

$$\text{(Stage T2)} \qquad\qquad \sigma^c = f_{cr}\left(\frac{\varepsilon_{cr}}{\bar{\varepsilon}}\right)^{0.4} \qquad\qquad \bar{\varepsilon} > \varepsilon_{cr}$$

$$\text{(Unloading and Reloading Stages)} \quad \sigma^c = \sigma_i^c + E_{cc}(\bar{\varepsilon}_i - \bar{\varepsilon}); \quad E_{cc} = \frac{\sigma_i^c - \sigma_{i+1}^c}{\bar{\varepsilon}_i - \bar{\varepsilon}_{i+1}}$$

6.3.2.1 Envelope Curves of Concrete (C1, C2, T1 and T2)

Tensile envelope curves T1 and T2

In theory, the tensile stress–strain curve of concrete subjected to biaxial stresses should be different from those subjected to uniaxial stresses. However, the tensile envelope curves, T1 and T2 in the biaxial condition, were found to be close to the monotonic curves given in Section 6.1.8. Since the tensile stress is very small compared with the compressive stress, the tensile envelope curves T1 and T2 are taken, for simplicity, to be the same as the monotonic curves expressed by Equations (6.54) and (6.55):

$$\text{Stage T1} \qquad \sigma^c = E_c \bar{\varepsilon} \qquad\qquad 0 \leq \bar{\varepsilon} \leq \varepsilon_{cr} \qquad\qquad (6.98)$$

$$\text{Stage T2} \qquad \sigma^c = f_{cr} \left(\frac{\varepsilon_{cr}}{\bar{\varepsilon}} \right)^{0.4} \qquad \bar{\varepsilon} > \varepsilon_{cr} \qquad\qquad (6.99)$$

Notice in Equations (6.98) and (6.99) that the stress and strain symbols σ^c and $\bar{\varepsilon}$ of concrete in compression do not have a subscript. This is because they can be applied to either the horizontal 1-direction or the vertical 2-direction. When the cyclic load is in the positive direction, $\sigma^c = \sigma_1^c$ and $\bar{\varepsilon} = \bar{\varepsilon}_1$. When the cyclic load is in the negative direction, however, $\sigma^c = \sigma_2^c$ and $\bar{\varepsilon} = \bar{\varepsilon}_2$.

Compressive envelope curves C1 and C2

In the case of concrete cylinders under uniaxial cyclic compression, Karsan and Jirsa (1968) showed that the hysteretic loops of the compression stress–strain curves produced an envelope curve that was virtually identical to the curve for monotonic compressive loading. In a more general case of RC 2-D elements subjected to a cyclic normal stress in one principal direction and a contant tensile strain in the other principal direction, Mansour *et al.* (2001b) showed that the *envelope* compression curves of the hysteretic loops were similar to the *monotonic* compression curves (Equation 6.42 and 6.45), proposed by Belarbi and Hsu (1994, 1995) in Section 6.1.6.

The difference between the above two experiments is the fact that the strain normal to the cyclic compression direction was zero in Karsan and Jirsa's tests, while a constant tensile strain was applied normal to the cyclic compression direction in the 2-D elements of Mansour *et al.* (2001b). This constant tensile strain in the orthogonal direction caused a 'softening' of the concrete compressive strength. The tests carried out by Mansour *et al.* showed that the 'softening coefficient' ζ given by Equations (6.46)–(6.49) in Section 6.1.7 was valid not only for the monotonic loading curves (Zhang and Hsu, 1998), but also for the envelope curves of cyclic loading.

6.3.2.2 Damage Coefficient D for Compression Envelope

When a 2-D element is subjected to *cyclic shear* loading, however, an additional phenomenon needs to be considered when modeling the constitutive relationships of concrete in compression. Since the horizontal and vertical principal applied stresses are subjected to out-of-phase compression–tension stresses, a damage coefficient D needs to be incorporated in the envelope compression stress–strain curves of concrete to take into account the damage caused by the history of tensile and compressive stress reversal normal to the compression direction being considered. As such, the *compressive envelope curves* for the cyclic stress–strain curves are taken from the monotonic curves of Equations (6.42) and (6.45) with two modifications:

Stage C1 $\quad \sigma^c = (D\zeta f_c' - f_{cT4}')\left[2\left(\dfrac{\bar{\varepsilon}}{\zeta\varepsilon_o}\right) - \left(\dfrac{\bar{\varepsilon}}{\zeta\varepsilon_o}\right)^2\right] + f_{cT4}' \qquad \varepsilon_o \le \bar{\varepsilon} < 0$ $\quad$ (6.100)

Stage C2 $\quad \sigma^c = D\zeta f_c'\left[1 - \left(\dfrac{\bar{\varepsilon}/\varepsilon_o - 1}{4/\zeta - 1}\right)^2\right] \qquad\qquad\qquad\qquad \bar{\varepsilon} < \varepsilon_o$ $\quad$ (6.101)

where ζ is the softening coefficient given by Equation (6.52). Also, when the cyclic load is in the positive direction, $\sigma^c = \sigma_2^c$ and $\bar{\varepsilon} = \bar{\varepsilon}_2$. When the cyclic load is in the negative direction, however, $\sigma^c = \sigma_1^c$ and $\bar{\varepsilon} = \bar{\varepsilon}_1$.

The first modification in Equations (6.100) and (6.101) is the incorporation of a damage coefficient D (Mansour, 2001; Mansour and Hsu, 2005b). This damage coefficient D takes into account the effect of *cyclic shear* loading, where cyclic compression and tension occur in both principal directions.

To be consistent with the concept of a softening coefficient due to tensile strain, the damage coefficient D is taken as a linear function of the compression strain ε_c':

$$D = 1 - \psi\frac{\varepsilon_c'}{\varepsilon_o} \le 1.0 \qquad\qquad (6.102)$$

The strain ε_c' (always negative) in Equation (6.102) is the maximum compression strain normal to the compression direction under consideration, and occurred in the previous loading cycles. The compressive strain ε_o (always negative) is the concrete cylinder compressive strain at the peak cylinder stress f_c'. The symbol ψ is a constant taken as 0.4. The value of $\psi = 0.4$ was chosen to best fit the test results of the cyclic shear stress–strain curves of the test panels (Mansour, 2001). Because the damaging effect of the perpendicular tensile strain ε_t' (or the uniaxial $\bar{\varepsilon}_t'$) is taken care of by the softening coefficient ζ, the damage coefficient D in Equation (6.102) cannot be greater than unity, and the strain ε_c' cannot be positive.

The second modification is the incorporation of a stress f_{cT4}' in Equation (6.100). f_{cT4}' is the concrete stress of point TD on the vertical axis at the end of stage T4. Because the envelope curve C1 starts from the point TD, rather than the origin, Equation (6.100) needs to be adjusted accordingly.

6.3.2.3 Unloading and Reloading Curves of Concrete (C3–C7 and T3–T4)

In Figure 6.28, stage C3–C7 and stage T3–T4 represent the stress–strain curves of unloading and reloading. Starting with the cyclic load in tension from the origin, the response is elastic as long as the tensile stress is less than the cracking stress at point TA (f_{cr}). The loading and unloading behavior follows the straight line T1 given by Equation (6.98). If the load is reversed from tension to compression before cracking, the compressive response follows the dotted line with a slope equal to the initial modulus of concrete E_c. Once the cracking stress of concrete at point TA is exceeded, the tensile response follows a concave curve T2 given by Equation (6.99). As the load is reversed at point TB from the tensile direction to the compressive direction, there is initially a region that represents the closure of the cracks (stage T3) up to point TC. As the crack continues to close, the second region, stage T4, will be stiffer. This stage T4 represents the increase in concrete stiffness before the complete closing of the crack. Stage T4 ends at point TD $(0, f'_{cT4})$ where the stress $f'_{cT4} = -1.5 f_{cr} + 0.8 f'_{cT2}$ is taken to be a function of the cracking stress, f_{cr}, and the maximum tensile stress, f'_{cT2} attained at point TB.

As the load progresses in the compression region, the first compressive virgin loading, after cracking, follows the envelope compression curves, stage C1 and stage C2, given by Equations (6.100) and (6.101), respectively. As the load is reversed at point CB from the compressive direction to the tensile direction, the response follows two straight lines, Stage C3 (from CB to CC) and stage C4 (from CC to CD). The slope of stage C3 is 80% of E_c, the initial modulus of elasticity of concrete, while the slope of stage C4 is 20% of E_c. A straight line is also assumed in stage C5, when the strain continues to increase in the positive direction from point CD to point TB.

If in stage C5 the direction of loading is reversed at point CE, the compressive reloading response will follow stage C6 up to point CF. The slope of stage C6 was chosen to decrease as the horizontal distance between points CC and point CE increases. The strain at point CF is assumed to be $\varepsilon_{cC6} = \varepsilon_{cCC} - (\varepsilon_{cCC} - \varepsilon_{cC5})/3$, where ε_{cCC} is the compressive strain at point CC and ε_{cC5} is the compressive strain at point CE. The strain at point CC is equal to $\varepsilon_{cCC} = \varepsilon_{cC2} - \varepsilon'_{cC2}$, where ε_{cC2} is the strain at point CB and $\varepsilon'_{cC2} = f'_{cC2}/E_c$. Hence, the strain at point CF becomes $\varepsilon_{cC6} = \left[2\left(\varepsilon_{cC2} - \varepsilon'_{cC2}\right) + \varepsilon_{cC5}\right]/3$. It should be noted that the strain at point CE (ε_{cC5}) should be taken as zero when the strain is in tension. The resulting strain would be equal to $\varepsilon'_{cC6} = 2\left(\varepsilon_{cC2} - \varepsilon'_{cC2}\right)/3$ which defines the strain at point CF'. As the compression load continues to increase from CF, the response follows stage C7, passing through point CG $(0.85 f'_{cC2}, 0.98\varepsilon_{cC2})$ until reaching the envelope curve designated by C1 and C2.

Unloading for the second time from the envelope compression curve at point CB', the response will follow the two straight lines defined by stage C3' and stage C4', respectively. As the cyclic loading progresses from compression to tension, the response follows stage C5' from point CD' in the compression region to the previous point TB in the tension region.

Further increase in the tensile load will cause the concrete response to follow the tensile envelope curve (stage T2) given by Equation (6.99). As the load is reversed at point TB' from the tensile direction to the compressive direction, the concrete response follows stage T3' and stage T4' until point TD' is reached. As the load continues to progress in the compression region, the response will now follow stage C6' and C7', passing by point CF' and CG', respectively, until reaching the envelope compression curve designated by C1 and C2. After that, the process of loading, unloading and reloading continues as described earlier.

The unloading and reloading curves are constructed by connecting consecutively a set of points given in Figure 6.28. The linear expression between two points is given as:

$$\sigma^c = \sigma_i^c + E_{cc}(\bar{\varepsilon} - \bar{\varepsilon}_i) \tag{6.103}$$

where σ_i^c and $\bar{\varepsilon}_i$ are concrete stress and strain at the load reversal point 'i' or at the point where the stages change; E_{cc} is the slope of the linear expression and is taken to be

$$E_{cc} = \frac{\sigma_i^c - \sigma_{i+1}^c}{\bar{\varepsilon}_i - \bar{\varepsilon}_{i+1}} \tag{6.104}$$

where σ_{i+1}^c and $\bar{\varepsilon}_{i+1}$ are the concrete stress and strain at the end of the stage under consideration.

6.3.3 Cyclic Stress–Strain Curves of Mild Steel

The cyclic constitutive relationships of reinforcing steel bars embedded in concrete and subjected to uniaxial strains (Mansour *et al.*, 2001b) are summarized in Figure 6.29. The solid curves represent the smeared stress–strain curves of steel bars, while the dotted curves are the monotonic stress–strain relationship of a bare bar. In the smeared stress–strain curves, stage 1, 2T and 2C are the envelope curves and stages 3 and 4 are the unloading and reloading curves.

6.3.3.1 Envelope Curves of Mild Steel (Stages 1, 2T and 2C)

When the envelope curves of embedded steel bars under cyclic loading are compared with the monotonic curves, Belarbi and Hsu (1994, 1995) and Hsu and Zhang (1996) found that the envelope curves for the cyclic stress–strain curves closely resemble the monotonic stress–strain curves given in Section 6.1.9. In other words, stages 1 and 2T can be described by Equations (6.85) and (6-86) in Section 6.1.9.4 for a bilinear model:

$$(\text{Stage 1}) \quad f_s = E_s \varepsilon_s \qquad\qquad\qquad\qquad (\bar\varepsilon_s \leq \varepsilon'_y) \qquad (6.105)$$

$$(\text{Stage 2T}) \quad f_s = (0.91 - 2B)f_y + (0.02 + 0.25B)E_s \bar\varepsilon_s \qquad (\bar\varepsilon_s > \varepsilon'_y) \qquad (6.106)$$

If the steel stress progresses in the compression region, the smeared maximum stress f_s is limited to the compressive yield stress $-f_y$, as indicated in stage 2C:

$$(\text{Stage 2C}) \quad f_s = -f_y \qquad (f_s \leq -f_y) \qquad\qquad\qquad\qquad (6.107)$$

6.3.3.2 Unloading and Reloading Curves of Mild Steel (Stages 3 and 4)

The unloading and reloading stress vs. strain curves of embedded steel bars, stages 3 and 4, take into account the Bauschinger effect, as shown in Figure 6.29. When the unloading branch starts from a point beyond the smeared yield point, the unloading stress vs. strain curve of a steel bar follows essentially a straight line in the tension region, and then becomes curved in the compression region. This curve was found by Mansour *et al.* (2001b) to be well represented by the Ramberg–Osgood type of expression first used by Yokoo and Nakamura (1977):

$$(\text{Stages 3 and 4}) \quad \bar\varepsilon_s - \bar\varepsilon_{si} = \frac{f_s - f_i}{E_s}\left[1 + A^{-R}\left|\frac{f_s - f_i}{f_y}\right|^{R-1}\right] \qquad (6.108)$$

where f_s and $\bar\varepsilon_s$ are the smeared stress and smeared uniaxial strain of an embedded steel bar; f_i and $\bar\varepsilon_{si}$ are the smeared stress and smeared uniaxial strain of steel bars at the initial load reversal point.

The coefficients A and R in Equation (6.108) were determined from the reversed cyclic loading tests to best fit the test results: $A = 1.9k_p^{-0.1}$, $R = 10k_p^{-0.2}$. The parameter in the coefficients A and R is the plastic strain ratio k_p which is defined as the ratio $\bar\varepsilon_p/\varepsilon'_y = (\bar\varepsilon_{si} - \varepsilon'_y)/\varepsilon'_y$. In this expression $\bar\varepsilon_p$ is the smeared plastic strain, and ε'_y is the smeared yield strain.

The coefficients A and R and the ratio k_p control the shape of the stress–strain curves of an embedded steel bar to best fit the test data. The coefficient A controls the strains at which

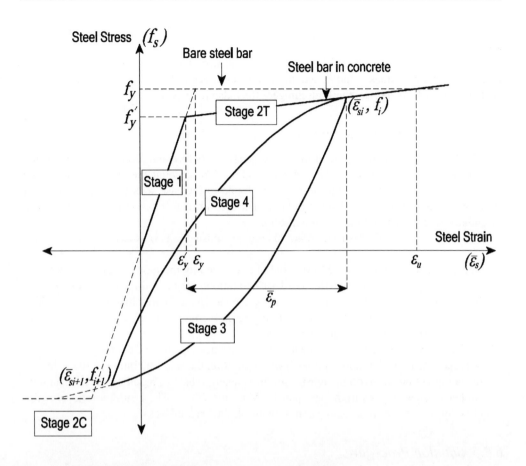

(Stage 1) $f_s = E_s \bar{\varepsilon}_s$ $(\bar{\varepsilon}_s \le \varepsilon'_y)$

(Stage 2T) $f_s = (0.91 - 2B) f_y + (0.02 + 0.25B) E_s \bar{\varepsilon}_s$ $(\bar{\varepsilon}_s > \varepsilon'_y)$

where $B = \dfrac{1}{\rho} \left(\dfrac{f_{cr}}{f_y} \right)^{1.5}$ and $\rho \ge 0.15\%$; $\varepsilon'_y = (0.93 - 2B)\varepsilon_y$

(Stage 2C) $f_s = -f_y$ $(f_s \le -f_y)$

(Stage 3 and Stage 4) $\bar{\varepsilon}_s - \bar{\varepsilon}_{si} = \dfrac{f_s - f_i}{E_s} \left[1 + A^{-R} \left| \dfrac{f_s - f_i}{f_y} \right|^{R-1} \right]$

where $A = 1.9 k_p^{-0.1}$, $R = 10 k_p^{-0.2}$ and $k_p = \dfrac{\bar{\varepsilon}_p}{\varepsilon'_y}$

Figure 6.29   Cyclic smeared stress–strain curve of mild steel bars

stage 3 changes to stage 2C and stage 4 changes to stage 2T. The coefficient R controls the knee curvature of the stress–strain curves in stage 3 and stage 4. As the value of R increases, the curve in stage 3 and stage 4 becomes more concave. The plastic strain ratio k_p is the parameter that determines A and R, and, therefore, controls the Bauschinger effect as a whole. As the value of k_p increases, the Bauschinger effect also increases.

6.3.4 Hsu/Zhu Ratios υ_{TC} and υ_{CT}

The SMM was developed to predict the entire behavior (before and after peak point) of RC 2-D elements under *monotonic* shear stresses, taking into account the Hsu/Zhu ratios (see Section 6.1.4). The first Hsu/Zhu ratio v_{12} was found to increase with the increase of steel strain and became a constant 1.9 after yielding. The second Hsu/Zhu ratio v_{21} was found to be essentially zero throughout the loading history.

In CSMM, however, the Hsu/Zhu ratios should be adjusted in two aspects:

(a) The symbols v_{12} and v_{21} for SMM should be replaced by the more general symbols v_{TC} and v_{CT}, respectively, for CSMM. The Hsu/Zhu ratio v_{TC} is the ratio of the resulting strain increment in principal *tension* to the source strain increment in principal *compression*. The Hsu/Zhu ratio v_{CT} is the ratio of the resulting strain increment in principal *compression* to the source strain increment in principal *tension*.
(b) Under reversed cyclic loading, repeated cycles of crack closing and opening are expected to reduce the Hsu/Zhu ratio v_{TC}. In CSMM, therefore, the Hsu/Zhu ratio v_{TC} (the effect of the compression strain on the tensile strain) is taken to be 1.0, based on the comparative studies of shear strains in the test panels (Mansour, 2001). The Hsu/Zhu ratio v_{CT} (the effect of the tensile strain on the compression strain) remains zero.

6.3.5 Solution Procedure

The solution procedure for the cyclic softened membrane model (CSMM) is shown by the flow chart in Figure 6.30. It can be seen that this flow chart is very similar to the one for the softened membrane model (SMM) in Figure 6.23, except in the following two aspects:

6.3.5.1 Modification of Coordinates and Installation of Strain History

(1) In CSMM of a RC 2-D element subjected to reversed cyclic shear, the 1–2 coordinate is defined in a general manner. Both the 1-axis and the 2-axis are alternately the directions of principal tension and the principal compression. When the cyclic load is in the positive direction, the horizontal strain ε_1 is in the principal tension direction and the vertical strain ε_2 is in the principal compression direction (as in the SMM for monotonic loading). When the cyclic load is in the *negative* direction, however, the horizontal strain ε_1 is in the principal *compression* direction and the vertical strain ε_2 is in the principal *tension* direction.
(2) A strain history must be specified to generate the hysteretic loops. The solution procedure starts with the selection of a horizontal strain ε_1 from the input history. Then assume the shear strain γ_{12} and the vertical strain ε_2 to calculate all the strains and stresses according to the iterative procedure of the flow chart to arrive at a pair of shear stress $\tau_{\ell t}$ and shear

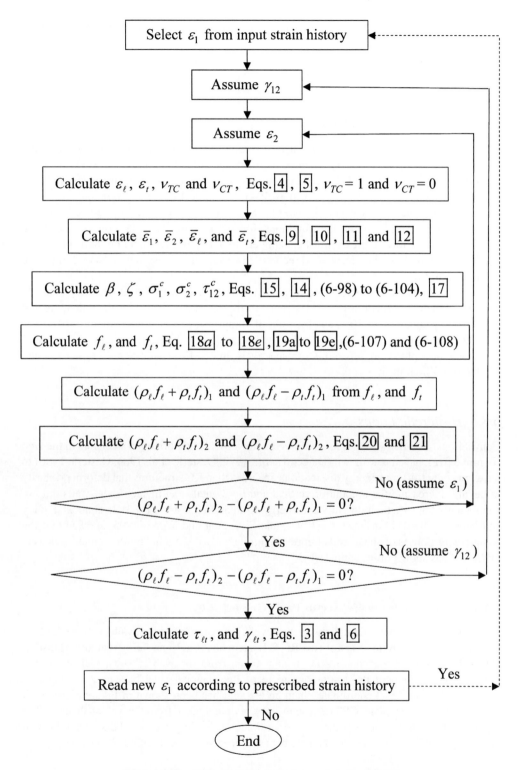

Figure 6.30 Flow chart of solution procedure for CSMM

strain $\gamma_{\ell t}$. By reading a new strain ε_1 according to the prescribed strain history and starting a new cycle, we can obtain the next pair of $\tau_{\ell t}$ and $\gamma_{\ell t}$. By repeating this process, a hysteretic loop of $\tau_{\ell t}$ and $\gamma_{\ell t}$ can be plotted. It is clear that the calculation is very tedious and must be performed by computer.

6.3.5.2 Modifications of Material Laws

(1) In SMM Equations $\boxed{13}$–$\boxed{17}$ are used to calculate the constitutive law of concrete, including β, ζ, σ_1^c, σ_2^c, τ_{12}^c. Three adjustments should be made in CSMM. First, Equations $\boxed{13a}$ and $\boxed{13b}$ for calculating concrete compressive stresses σ_2^c should be replaced by Equations (6.100)–(6.104). Second, Equation $\boxed{16a}$ and $\boxed{16b}$ for calculating the concrete tensile stress σ_1^c should be replaced by Equations (6.98) and (6.99). Third, in all the equations for σ_1^c and σ_2^c, the subscripts 1 or 2 are for tension and compression when applied to the positive direction of cyclic load. When applied to the negative direction, however, the subscripts 1 and 2 must be interchanged.
(2) In CSMM the equations required for the calculation of smeared steel stresses f_ℓ, and f_t, should include Equations $\boxed{18a}$–$\boxed{18e}$ and $\boxed{19a}$–$\boxed{19e}$ (for SMM), plus the unloading and reloading equation, (6.108), and the limit of compressive steel stress (Equation 107).
(3) The Hsu/Zhu ratios ν_{12} and ν_{21} calculated according to Equations $\boxed{7}$ and $\boxed{8}$ in SMM should be replaced by ν_{TC} and ν_{CT}, respectively, in CSMM. Hsu/Zhu ratio $\nu_{TC} = 1.0$, and the Hsu/Zhu ratio $\nu_{CT} = 0$.

6.3.6 Hysteretic Loops

Conventional low-rise shear walls reinforced with vertical and horizontal steel bars in the web (Figure 6.31a), have been shown (Derecho et al., 1979, Oesterle et al. 1984, Oesterle 1986) to have lower ductility and energy dissipation capacities than RC structures that deform primarily in flexure. Increasing the amount of vertical and horizontal steel in such shear walls did not significantly improve their ductility and energy dissipation (Oesterle 1986). In contrast, Paulay and Binney (1974), Mansur et al. (1992), and Sittippunt and Wood (1995) showed that the hysteretic responses of shear walls improved significantly if diagonal reinforcement was used in the web, as shown in Figure 6.31(b).

6.3.6.1 Panel CA3 ($\alpha_1 = 45°$) versus Panel CE3 ($\alpha_1 = 0°$)

The effect of steel bar orientation on the cyclic response of shear walls was studied by comparing the hysteretic loops of two RC 2-D elements, CA3 and CE3 (Hsu and Mansour, 2002). Panel CA3 represents a 2-D element taken from the web of shear wall of Figure 6.31(a). In this element the $\ell - t$ coordinate of the diagonal steel bars is at an angle of 45° to the principal 1–2 coordinate. Such an element has been shown in Figure 6.2(b) with $\alpha_1 = 45°$. In contrast, panel CE3 represents a 2-D element taken from the web of shear wall of Figure 6.31(b). In this element the $\ell - t$ coordinate of the diagonal steel bars coincides with the principal 1–2 coordinate. Such an element has been shown in Figure 6.2(a) with $\alpha_1 = 0°$.

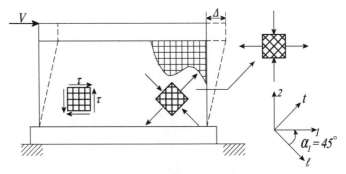

(a) Low-rise shear wall with conventional horizontal and vertical steel bars

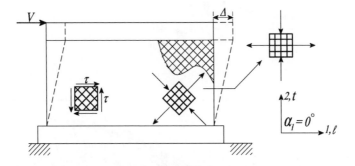

(b) Low-rise shear wall with diagonal steel bars

Figure 6.31   Shear dominant structures

Both panels CA3 and CE3 are 1.4 m (55 in.) square and 178 mm (7 in.) thick. They are reinforced with two layers of steel grids made of #6 deformed bars (bar area = 284 mm²), but with different spacing. The material properties of these panels are given in Table 6.3

Panels CA3 and CE3 were subjected to reversed cyclic shear strains with increasing magnitude using the strain-control feature of the universal panel tester (Figure 6.1). The experimental hysteretic loops of panels CA3 and CE3 are recorded by solid curves in Figure 6.32(a) and (b), respectively. In these diagrams, the shear stress $\tau_{45°}$ is positive, as shown in Figure 6.33(a), which corresponds to a tensile horizontal stress σ_H and a compressive vertical stress σ_V. The shear stress $\tau_{45°}$ is negative as shown in Figure 6.33(b), which corresponds to a compressive σ_H and a tensile σ_V.

Table 6.3   Material properties of panels CA3 and CE3

	Concrete		Steel in ℓ-direction			Steel in t-irection		
Panel	f_c'(MPa)	ε_o	Spacing of #6 bars (mm)	ρ_ℓ (%)	$f_{\ell y}$(MPa)	Spacing of #6 bars (mm)	ρ_t (%)	f_{ty}(MPa)
CA3	45	0.0025	189	1.7	428	189	1.7	428
CE3	50	0.0023	267	1.2	428	267	1.2	428

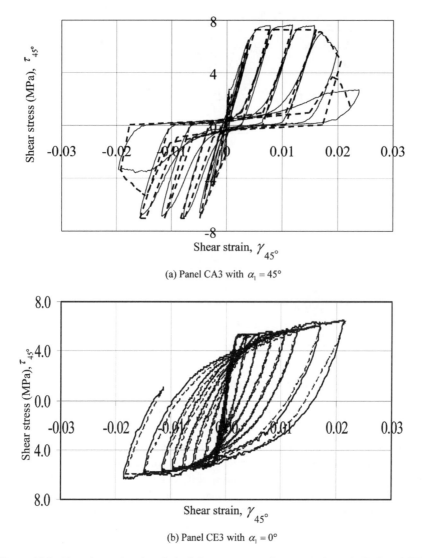

Figure 6.32 Experimental and analytical shear stress–strain curves of panels CA3 and CE3

The hysteretic loops predicted by CSMM are also given in Figure 6.32(a) and (b) by dotted curves. It can be seen that the experimental and analytical shear stress–strain curves of panels CA3 and CE3 agree very well. In other words, CSMM can correctly predict the shape of the hysteretic loops.

6.3.6.2 Presence and Absence of Pinching Effect

The most important difference between the shapes of the two sets of hysteretic loops in Figure 6.32(a) and (b) is the so-called pinching effect. The hysteretic loops of panel CA3

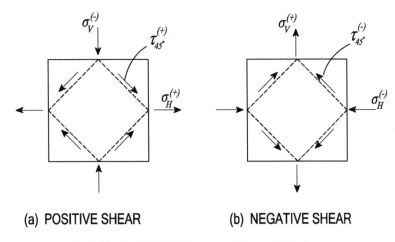

(a) POSITIVE SHEAR   (b) NEGATIVE SHEAR

Figure 6.33   2-D elements under cyclic shear

($\alpha_1 = 45°$) are severely 'pinched' near the origin, while the hysteretic loops of panel CE3 ($\alpha_1 = 0°$) are fully rounded and robust. The 'pinched' hysteretic loops of panel CA3 reduce its shear ductility and energy dissipation capacity by an order of magnitude as compared to the 'fully rounded' hysteretic loops of panel CE3.

The behavioral comparison of panels CA3 and CE3 in Figure 6.32 clearly shows that the magnitude of shear ductility and energy dissipation capacity are strongly effected by the orientation of steel bars with respect to the directions of principal stresses (1–2 coordinate). When the steel bars are oriented in the principal 1–2 coordinate, the hysteretic loops are fully rounded and the behavior is ductile. When the steel bars are oriented at 45° to the principal 1–2 coordinate, the hysteretic loops are severely pinched and the behavior is much less ductile. The behavior of panels CA3 and CE3 clearly explains the observations of previous researchers that the shear walls with vertical and horizontal steel bars (Figure 6.31a), are not ductile, while the shear walls with diagonal steel bars (Figure 6.31b), are ductile.

Figure 6.32 also confirmed the validity of CSMM. This analytical model is capable of predicting the pinched shape of the hysteretic loops of panels CA3 as well as the fully rounded hysteretic loops of panel CE3. In Section 6.3.7, CSMM is also used to explain the presence and absence of the pinching mechanism in RC 2-D elements under cyclic shear loading, as well as the failure mechanisms of such elements (Mansour *et al.*, 2001a; Hsu and Mansour, 2002; Mansour and Hsu, 2005b).

6.3.6.3 Deformation Characteristics

Figure 6.34(a) and (b) plots the horizontal strain ε_H versus vertical strain ε_V under cyclic shear for panels CA3 ($\alpha_1 = 45°$) and CE3 ($\alpha_1 = 0°$), respectively. The predicted and experimental curves are given side by side, the experimental solid curves to the right and the predicted dotted curves to the left. Figure 6.33(a) and (b) shows that under positive shear stress, the horizontal strain ε_H increases, while the vertical strains ε_V decreases. Under negative shear stress, however, the horizontal strain ε_H decreases, while the vertical strain ε_V increases.

Comparison of Figures 6.34(a) and (b) shows an important difference in the deformation characteristics of panels CA3 ($\alpha_1 = 45°$) and CE3 ($\alpha_1 = 0°$). In panels CA3 ($\alpha_1 = 45°$) Figure 6.34(a) shows that under positive shear stress (Figure 6.33a), the vertical strain ε_V reaches the compression region in every cycle, meaning that the horizontal cracks are fully closed and vertical concrete struts are formed to resist the vertical compression. Under negative shear stress (Figure 6.33b), the horizontal strain ε_H reaches the compression region in every cycle, meaning that the vertical cracks are fully closed and horizontal concrete struts are formed to resist the horizontal compression.

In contrast, in panels CE3 ($\alpha_1 = 0°$) Figure 6.34(b) shows that under positive shear stress the vertical strains ε_V do not reach the compression region and the end strains are further away from the compression region with every cycle. Physically, this means that the horizontal cracks remain open, and widen with each cycle. The vertical compressive stresses are resisted by the vertical steel bars in compression, rather than concrete. Under negative shear stress, the horizontal strains ε_H do not reach the compression region and the end strains are further away from the compression region with every cycle. Physically, this means that the vertical cracks remain open and widen with each cycle. The horizontal compressive stresses are resisted by the horizontal steel bars in compression, rather than concrete.

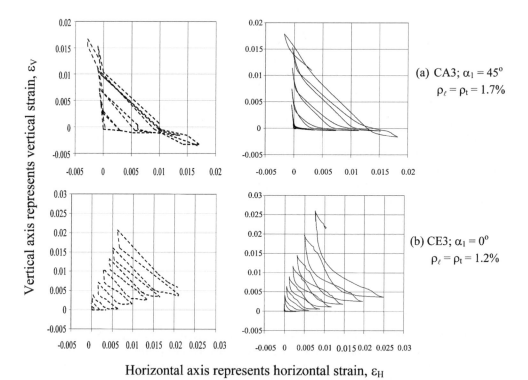

Figure 6.34 Predicted and experimental horizontal-strain versus vertical-strain curves of panels CA3 and CE3 (dotted curves show predicted results, solid curves show experimental results.)

Figure 6.34 shows a good agreement between the predicted curves (dotted curves) and the experimental curves (solid curves), and thus illustrates the success of CSMM in predicting the deformation history.

6.3.7 Mechanism of Pinching and Failure under Cyclic Shear

6.3.7.1 Pinching Mechanism

Panel CA3 ($\alpha_1 = 45°$)
The first cycle of the hysteretic loops beyond yielding for panel CA3 is plotted in Figure 6.35(a). Four points A, B, C, and D are chosen in Figure 6.35(a) to illustrate the *presence* of the pinched shape under negative shear stress (Figure 6.33b). Point A is at the maximum positive shear strain of the first cycle beyond yielding. Point B is at the stage where the shear stress is zero after unloading. Point C, with a very low shear stress, is taken in the negative shear strain region at the end of the low-stress pinching zone, just before the sudden increase in stiffness. Point D is at the maximum negative shear strain of the first cycle beyond yielding. The three segments of curves under negative shear from point A to point D in Figure 6.35(a) clearly define the pinched shape of the hysteretic loops.

Panel CE3 ($\alpha_1 = 0°$)
The first cycle of the hysteretic loops beyond yielding for panel CE3 is plotted in Figure 6.35(b). Four points A, B, C, and D are chosen in Figure 6.35(b) to illustrate the *absence* of the pinched shape under negative shear stress. Points A, B and D correspond to the same three points in Figure 6.35(a). However, point C with a high shear stress is taken in the negative shear strain region when the horizontal compression steel reaches yielding. The three segments of curves under negative shear stress from point A to point D in Figure 6.35(b) clearly show the absence of pinching.

6.3.7.2 Physical Visualization

Panel CA3 ($\alpha_1 = 45°$)
The presence of the pinching mechanism in panel CA3 can be explained by examining a cracked 2-D element with 45° steel bars as shown in the drawing on the left-hand side of Figure 6.36(a). The 2-D element has both vertical and horizontal cracks induced by previous cycles of positive and negative shear stresses, respectively.

The reverse loading stage under negative shear from point B to point C (Figure 6.35a) defines the region where pinching occurs. Since both the vertical and the horizontal cracks are open, the applied compressive stress σ_H and the tensile stress σ_V must be resisted by the two 45° steel bars. The effect of σ_H and σ_V are separated and shown in the second and third diagrams of Figure 6.36(a). The horizontal compressive stress σ_H induces a compressive stress in the two 45° steel bars, while the vertical tensile stress σ_V induces a tensile stress of equal magnitude in the same two 45° bars. The two stresses induced by σ_H and σ_V in the two 45° steel bars cancel each other out. As a result, the element offers no shear resistance to the applied negative shear stress in the 45° direction, while the negative shear strain in the 45° direction increases rapidly due to the closing of the vertical cracks and the opening of the horizontal cracks. Therefore, the shear stiffness in the BC regions becomes close to zero.

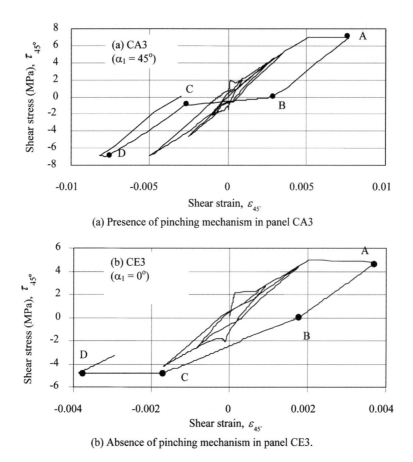

(a) Presence of pinching mechanism in panel CA3

(b) Absence of pinching mechanism in panel CE3.

Figure 6.35 Presence and absence of pinching mechanism.

At point C the vertical cracks close, while the horizontal cracks remain wide open. This condition results in the formation of a series of horizontal concrete struts to resist the horizontal compressive stress σ_H. At the same time, the vertical tensile stress σ_V is resisted by the two 45° steel bars. Conceptually, a truss is formed to resist the applied negative shear stress in the 45° direction and the shear stiffness increases dramatically. The two segments of BC and CD in Figure 6.35(a) create the pinched shape in the hysteretic loops of the shear stress–shear strain curves.

Panel CE3 ($\alpha_1 = 0°$).
The absence of pinching mechanism in panel CE3 can be visualized by considering a cracked element with horizontal and vertical steel bars, as shown in Figure 6.36(b). The drawing on the left shows a 2-D element subjected to a negative shear stress, which is equivalent to applying a horizontal compressive stress σ_H and vertical tensile stress σ_V of equal magnitude. This 2-D element also has vertical and horizontal cracks induced by previous cycles of shear stresses.

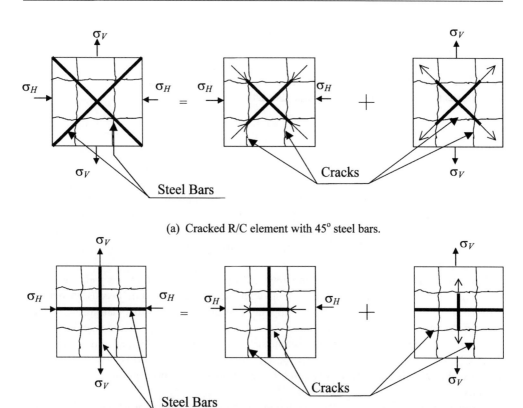

(a) Cracked R/C element with 45° steel bars.

(b) Cracked R/C element with 0° steel bars.

Figure 6.36   Effect of 45° and 0° steel bars on pinching mechanism in cracked RC 2-D element (in negative direction of a pure shear cycle)

In the reverse loading stage under negative shear stress from point B to point C (Figure 6.35b) both the vertical and horizontal cracks are open. The second drawing in Figure 6.36(b) shows that the horizontal compressive stress σ_H is resisted by the compressive stress in the horizontal steel bars, while the third drawing shows the vertical tensile stress σ_V to be resisted by a tensile stress in the vertical bar. Both the compressive stress in the horizontal steel bar and the tensile stress in the vertical steel bar contribute to the negative shear stress in the 45° direction. As a result, the shear stress increases proportionally to the shear strain, and the large shear stiffness in the BC segment in Figure 6.35(b) is sustained until the yielding of steel bars at point C. The large stiffness of the BC segment and the horizontal CD segment creates a fully rounded and robust hysteretic loop without the pinched shape.

Mohr circles
The smeared strains, the applied stresses, the smeared concrete stresses, and the smeared steel stresses at the four points (A, B, C and D) chosen in Figure 6.35(a) and (b) have been calculated by CSMM and are represented by Mohr circles for panels CA3 and CE3 (Mansour

and Hsu, 2005b). Using the Mohr circles has two advantages: First, Mohr circles represent the entire stress or strain state in an element, i.e. stresses or strains in all directions. Second, Mohr circles are accurate means to illustrate the mechanism behind the pinching effect and the failure mechanism of reinforced concrete elements under cyclic loading. Interested readers are referred to the paper by Mansour and Hsu (2005b).

6.3.7.3 Failure Mechanism

When a reinforced concrete structure is subjected to static loading, the principal compression stresses in the structure can be resisted by concrete struts while the principal tensile stresses are resisted by the reinforcing bars. This struts-and-ties concept can be used to design all reinforced concrete structures under static loads. However, comparison of the deformation behavior of panel CA3 ($\alpha_1 = 45°$) and panel CE3 ($\alpha_1 = 0°$) in Figure 6.34(a) and (b), as well as Figures 6.35 and 6.36, clearly shows that the struts-and-ties model is no longer valid for application to reversed cyclic loading.

When RC 2-D elements are resisting reversed cyclic shear beyond the yielding of the steel, crack widths increase in both directions with each cycle of loading because panels are expanding. Cracks will not close after unloading and the concrete struts can not be formed to resist the reversed loading while the cracks are still open. Therefore, the steel bars must be designed to resist principal compression stresses as well as principal tension stresses. Panel CE3 ($\alpha_1 = 0°$), which is designed based on this concept, performs very well. It does not have the pinching problem, and exhibits ductile behavior and high energy dissipation capacity.

However, when the steel bars are oriented at an angle of 45° to the principal applied stress coordinate as in the panel CA3 ($\alpha_1 = 45°$), the longitudinal and transverse steel bars in the diagonal directions can not form a truss in the absence of concrete compression struts when the cracks are open. In order to form the concrete compression struts, the cracks must be forced to close under the negative shear stress. This closing of cracks, which is associated with large shear strains and small shear stress, is the reason why the undesirable pinched shape of hysteretic loop is produced, as explained in Section 6.3.7. Furthermore, this closing of cracks and the subsequent reopening of cracks in each cycle of loading represent a very destructive failure mechanism, leading to the rapid deterioration of bond and the ultimate disintegration of concrete. This undesirable failure mechanism is responsible for the early arrival of the descending branch, the low ductility, and the low capacity in energy dissipations.

6.3.8 Eight Demonstration Panels

The CSMM is used to generate the hysteretic loops of eight RC 2-D elements (panels) subjected to reversed cyclic shear stresses. These eight panels were designed to isolate the two variables (steel bar orientation and the steel percentage) and to demonstrate their separate effects on the stiffness, ductility and energy dissipation capacity. The panels used in the investigation can each be visualized as 2-D elements isolated from larger shear-dominated structures, such as the web of low-rise shear walls in Figure 6.31. Understanding the cyclic characteristics of such RC 2-D elements will allow structural engineers to design ductile shear-dominated RC

Table 6.4   Steel grid orientations and steel percentages in panels

Panel	Steel grid orientation α_1 (degrees)	Steel percentage in ℓ or t-direction (ρ_l, or ρ_t)
G-45-0.54	45	0.54
G-45-1.2	45	1.2
G-45-2.7	45	2.7
G-21.8-1.2	21.8	1.2
G-10.2-1.2	10.2	1.2
G-0-0.54	0	0.54
G-0-1.2	0	1.2
G-0-2.7	0	2.7

structures with large stiffness and high energy dissipation capacity, just like flexure-dominant RC structures.

6.3.8.1 Two Variables (α_1 and ρ)

The steel bar angle (α_1) and the steel percentages (ρ_ℓ, ρ_t) of all eight panels are given in Table 6.4. Each panel label starts out with the letter G" to identify that its cyclic response was 'generated' by the CSMM. The first and second numbers after the G represent the steel bar angle α_1 and the steel percentage (ρ_ℓ, ρ_t), respectively. For example, panel G-21.8-1.2 means that it belongs to the G group of panels, has a steel bar angle of $21.8°$ and a steel percentage of 1.2%.

Steel bar angle
The demonstration panels are reinforced with two layers of steel grids oriented at an angle α_1 from $0°$ to $45°$, as shown in Figure 6.2(a) and (b), where α_1 is the angle between the 1–2 principal coordinate system and the $\ell - t$ coordinate of the steel bars. Four values of α_1 are considered in this investigation: $45°$, $21.8°$, $10.2°$, and $0°$. When the steel bar angle α_1 is $45°$, the steel grid is set in the direction of the pure shear stresses. When the steel bar angle α_1 is $0°$, the steel grid is oriented in the direction of the principal applied stresses.

Steel percentage
The eight demonstration panels were also designed to isolate the effect of the steel percentage on the cyclic shear response of RC 2-D elements. The steel percentages in the longitudinal and transverse directions (ρ_ℓ, ρ_t) were kept equal within a panel. The three steel percentages (in each direction) were 0.54, 1.2 and 2.7%, the values used in the panels tested by Mansour and Hsu (2005a).

6.3.8.2 Three Series of Panels

The eight demonstration panels given in Table 6.4 provide three series of panels:

Series P

The four panels G-45-1.2, G-21.8-1.2, G-10.2-1.2 and G-0-1.2 form a series of panels with a constant steel percentage (ρ_ℓ, ρ_t) of 1.2%. Since the only variable in this series of panels is the steel bar angle (α_1), the generated hysteretic loops of panels in series P can be used to assess the effect of steel bar angle on the ductility, the stiffness and the energy dissipation capacity.

Series A45

Panels G-45-0.54, G-45-1.2 and G-45-2.7 form a series of panels with constant steel bar angle of 45°. Since the only variable in this series of panels is the steel percentage, the generated hysteretic loops of panels in series A45 can be used to assess the effect of steel percentage on the ductility, the stiffness and the energy dissipation capacity, when $\alpha_1 = 45°$.

Series A0

Panels G-0-0.54, G-0-1.2 and G-0-2.7 form a series of panels with a constant steel bar angle of 0°. Since the only variable in this series of panels is the steel percentage, the generated hysteretic loops of panels in series A0 can be used to assess the effect of steel percentage on the cyclic responses of the panels when $\alpha_1 = 0°$.

6.3.8.3 Generated Hysteretic Loops

Figure 6.37 shows the shear strain history used to generate the hysteretic loops of all the eight demonstration panels. This shear strain history was chosen to be a multiple of 0.004, a value that represents approximately the yield shear strain of panels reinforced with steel bars oriented at 45° to the applied principal stresses. An α_1 angle of 45° is used in conventional steel layouts of RC shear walls and is taken as a reference.

Figures 6.38 and 6.39 show the CSMM generated hysteretic loops of the shear stress–strain curves of all eight panels. The hysteretic loops in these two figures will be used to assess the effect of varying the steel bar angle and the steel percentage on the ductilities, stiffnesses and energy dissipation capacities of RC panels under cyclic shear.

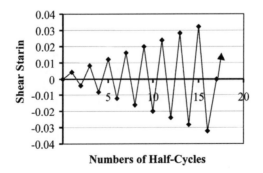

Figure 6.37   Shear strain history used for all panels

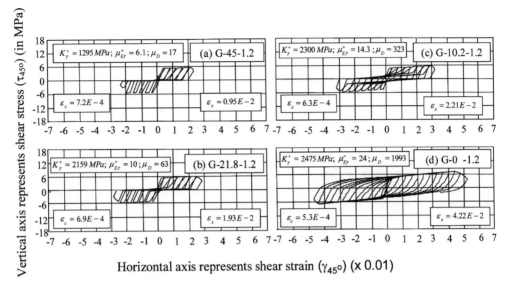

Horizontal axis represents shear strain (γ_{45^o}) (x 0.01)

Figure 6.38 Shear properties of panels in series P.

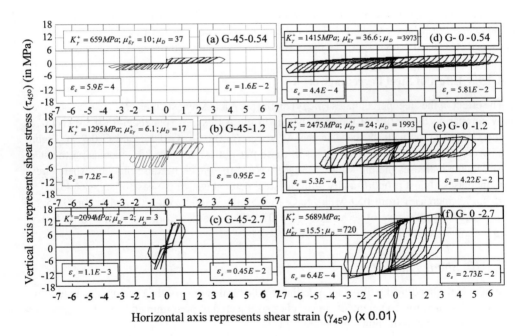

Horizontal axis represents shear strain (γ_{45^o}) (x 0.01)

Figure 6.39 Shear properties of panels in series 45 and A0

6.3.9 Shear Stiffness

6.3.9.1 Definition of Pre-yield Shear Stiffness (K_y)

The pre-yield shear stiffness (K_y) of a panel subjected to cyclic shear is defined as:

$$K_y = \frac{\tau_y}{\gamma_y} \tag{6.109}$$

where τ_y and γ_y are the yield shear stress and the yield shear strain, respectively, at 45° to the principal 1–2 coordinate. In the post-cracking stage, but still below the yield shear stress, the shear stiffness of a typical panel is found to decrease as load intensity increases and as more cracks occur. In this pre-yield stage, the pre-yield shear stiffness K_y approaches an asymptotic value defined by Equation (6.109), because the steel bars are still in their linear, elastic stage. The positive and negative pre-yield shear stiffnesses (K_y^+ and K_y^-), correspond to the positive and negative loading directions, respectively, of cyclic shear.

6.3.9.2 Effect of Steel Bar Angle on K_y

Series P with panels G-45-1.2, G-21.8-1.2, G-10.2-1.2 and G-0-1.2 is designed to study the effect of the steel bar angle (α_1) on the pre-yield shear stiffness of the panels, while keeping the percentage of steel (ρ_ℓ, ρ_t) constant at 1.2%. Figure 6.38 records the values of positive pre-yield shear stiffnesses K_y^+ of these four panels. It can be seen that K_y^+ increases as the steel bar angle (α_1) decreases. The K_y^+ value of 2475 MPa for G-0-1.2 is about twice the K_y^+ value of 1295 MPa for G-45-1.2. This increase of pre-yield shear stiffness from $\alpha_1 = 45°$ to $\alpha_1 = 0°$ is caused primarily by the decrease of the yield shear strain γ_y.

6.3.9.3 Effect of Steel Percentage on K_y

The effect of the steel percentages (ρ_ℓ, ρ_t) on the pre-yield shear stiffness, K_y^+, can be observed by comparing the results of the three panels in the A45 series (panels G-45-0.54, G-45-1.2 and G-45-2.7), as well as those in the A0 series (G-0-0.54, G-0-1.2 and G-0-2.7). Figure 6.39 records the K_y^+ values for panels in these two series. As expected, increasing the steel percentage increases the pre-yield shear stiffness of the panel, because the yield shear stress τ_y increases almost proportionally with the steel percentage. In the case of series A45 ($\alpha_1 = 45°$), K_y^+ increases from 659 MPa for G-45-0.54 ($\rho = 0.54\%$) to 2094 MPa for G-45-2.7 ($\rho = 2.7\%$). In the case of series A0 ($\alpha_1 = 0°$), K_y^+ increases from 1415 MPa for G-0-0.54 ($\rho = 0.54\%$) to 5689 MPa for G-0-2.7 ($\rho = 2.7\%$).

6.3.10 Shear Ductility

Similar to the bending ductility factor defined for flexural members, the envelope shear ductility factor is defined for 2-D elements in structures subjected to predominant shear force, such as the low-rise shear wall shown in Figure 6.31.

6.3.10.1 Definition of Envelope Shear Ductility Factor ($\mu_{E\gamma}$)

The envelope shear ductility factor ($\mu_{E\gamma}$) in cyclic loading is based on the envelope curve that houses the shear stress–shear strain hysteretic loops of a RC 2-D element (or panel). It is defined as the ratio of the envelope shear strain at ultimate γ_u to that at the onset of yielding

γ_y as follows:

$$\mu_{E\gamma} = \frac{\gamma_u}{\gamma_y} \tag{6.110}$$

In the case of an ideal elastic perfectly plastic behavior, γ_u and γ_y are easily defined. However, for RC 2-D elements the definition of γ_u is somewhat subjective. The ultimate shear strain, γ_u, is taken as the shear strain that corresponds to 80% of the shear stress capacity in the post-peak descending branch of the envelope curve. The yielding shear strain γ_y is the shear strain at the onset of yielding of the reinforcing bars. Since the positive and negative envelope shear ductility factors $\mu_{E\gamma}^+$ and $\mu_{E\gamma}^-$ of a given panel are almost identical, only the positive envelope shear ductility factors $\mu_{E\gamma}^+$ are given in Figures 6.38 and 6.39.

6.3.10.2 Effect of Steel Bar Angle on $\mu_{E\gamma}$

Figure 6.38 shows the positive envelope shear ductility factors $\mu_{E\gamma}^+$ as a function of steel bar angle (α_1) for series P (panels G-45-1.2, G-21.8-1.2, G-10.2-1.2, and G-0-1.2). The figure indicates that $\mu_{E\gamma}^+$ are 6.1 and 24 for panels G-45-1.2 and G-0-1.2, respectively. In other words, $\mu_{E\gamma}^+$ increases by a factor of almost four as α_1 decreases from 45° to 0°.

6.3.10.3 Effect of Steel Percentage on $\mu_{E\gamma}$

Figure 6.39 shows the positive envelope shear ductility factors $\mu_{E\gamma}^+$ as functions of percentage of steel (ρ_ℓ or ρ_t) for panels in series A45 and series A0. When the steel ratio in a shear panel increases from 0.54 to 2.7% in series A45, $\mu_{E\gamma}^+$ reduces from 10 to 2, a reduction of 80%. When the steel ratio in a shear panel increases from 0.54 to 2.7% in series A0, $\mu_{E\gamma}^+$ reduces from 36.6 to 15.5, a reduction of about 60%. In short, increasing the steel percentage in a shear panel leads to a decrease in the envelope shear ductility factor, i.e. the hysteretic loops exhibit a shorter envelope curve and the descending branch arrives earlier.

6.3.11 Shear Energy Dissipation

The shear resistance of a reinforced concrete element is contributed by the concrete in compression and the steel bars in tension. The concrete component experiences significant degradation of strength and stiffness when subjected to large cyclic displacements. Therefore, the primary source of energy dissipation must be provided by the inelastic behavior of reinforcing steel. Improvement in the energy dissipation capacity can then be achieved by designing the reinforced concrete structural elements to: (1) promote large strains after yielding of reinforcement bars; and (2) reduce the degradation of concrete in terms of strength and stiffness.

6.3.11.1 Definition of Shear Energy Dissipation

The total amount of hysteretic energy (i.e. the cumulative area of the hysteretic loops) when the structure fails under many cycles of reversed loading should reflect three characteristics of the hysteretic loops: (1) the pinched shape of each loop cycle; (2) the number of cycles before failure, and (3) the ultimate shear strain as measured by the envelope shear ductility factor ($\mu_{E\gamma}$).

The shear energy dissipation in each RC 2-D element is calculated by integrating the area under the shear stress–strain curves at every cycle after the first yield. The integration was performed numerically by using the trapezoidal rule, in which the area under the shear stress-strain curve was approximated by a series of trapezoids. Good accuracy can be obtained by dividing the area into small intervals. The shear energy dissipation of one cycle of loading at a specified shear strain is defined as the shear energy dissipation ratio (R_A), and the shear energy dissipation to resist the cumulative cycles is defined as the shear energy dissipation factor (μ_D)

6.3.11.2 Shear Energy Dissipation Ratio (R_A)

The calculated area, denoted by A_{pi}, under the shear stress–strain curve is defined in Figure 6.40(b) at a shear strain γ_i, where $i = 2, 3, 4$, etc. To make A_{pi} nondimensional, we will define the shear energy dissipation ratio R_A at a certain shear strain γ_i, as

$$R_A = \frac{A_{pi}}{A_y} \tag{6.111}$$

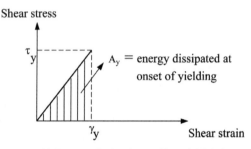

(a) Energy dissipation at first yield (A_y)

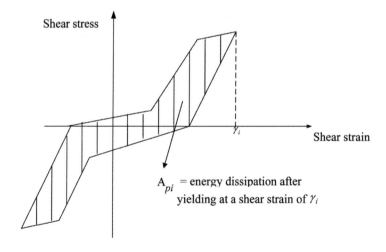

(b) Energy dissipation area (A_{pi}) at shear strain (γ_i).

Figure 6.40 Definition of shear energy dissipation ratio R_A

where A_{pi} is the area under the shear stress–strain curve in one reversed cycle at a given shear strain γ_i and A_y is the area under the shear stress–strain curve at the onset of yielding (Figure 6.40a).

To normalize the shear strain γ_i, we will define a shear strain ratio R_γ as follows:

$$R_\gamma = \frac{\gamma_i}{\gamma_y} \qquad (6.112)$$

where γ_y is the shear strain at the onset of yielding. The shear energy dissipation ratio (R_A) at a certain shear strain ratio (R_γ) is a good measure of the pinching of the hysteretic loop. The lower the R_A at a certain R_γ, the more severe is the pinching effect.

6.3.11.3 Shear Energy Dissipation Factor (μ_D)

The shear energy dissipation factor μ_D is defined as the area A_{RR} under the $R_A - R_\gamma$ curve as shown in Figure 6.41 for series P. This factor μ_D is a broader measure than the shear energy dissipation ratio R_A, because it takes into account not only the 'pinching effect', measured by the ratio R_A at every cycle, but also the number of cycles or the ultimate shear strain as measured by the envelope shear ductility factor $\mu_{E\gamma}$. The factors μ_D of all eight panels are given in Figures 6.38 and 6.39.

6.3.11.4 Effects of Steel Bar Angle on Ratio R_A and Factor μ_D

The effect of the steel bar angle (α_1) on the shear energy dissipation ratio (R_A) and the shear energy dissipation factor (μ_D) can be observed using series P (panels G-45-1.2, G-21.8-1.2,

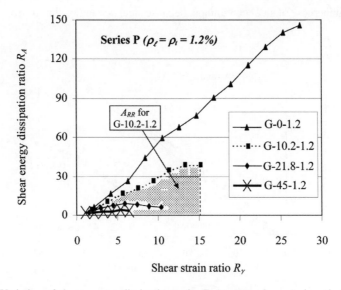

Figure 6.41 Variation of shear energy dissipation ratio (R_A) versus shear strain ratio (R_γ) for panels with different steel orientation but same amount of steel percentages (1.2%).

G-10.2-1.2, and G-0-1.2). All four panels in this series have the same amount of steel percentage (1.2%) in the transverse and longitudinal steel directions. Figure 6.41 shows the variation of the shear energy dissipation ratio (R_A) versus the shear strain ratio (R_γ) of these four panels.

Figure 6.41 clearly indicates that the shear energy dissipation ratios R_A of the four panels decrease tremendously when the steel bar orientation α_1 deviates from 0°. For example, when the shear strain ratio $R_\gamma = 5$, the ratio R_A are equal to 23, 16, 7.0 and 3.9 for panels G-0-1.2, G-10.2-1.2, G-21.8-1.2, and G-45-1.2, respectively. The decrease in the slope of the curves stems from the increasing pinching of the hysteretic loops. Figure 6.41 also indicates that the curves tend to terminate at lower shear strain ratios as the steel angle α_1 deviates from the 0° angle. For example, the maximum shear strain ratios R_γ are equal to 27.4, 15.2, 10.4, and 6.5 of panels G-0-1.2, G-10.2-1.2, G-21.8-1.2, and G-45-1.2, respectively. These differences in the maximum R_γ stem from the differences in ultimate shear strains.

Figure 6.38 gives the shear energy dissipation factors μ_D as 17, 63, 323 and 1993 for panels G-45-1.2, G-21.8-1.2, G-10.2-1.2, and G-0-1.2, respectively. It can be seen that μ_D increases very rapidly by a factor of $1993/17 \approx 120$ times, when α_1 decreases from 45° to 0°. This large difference is caused by the combined effects of pinching and ultimate shear strain. Apparently, the steel bar orientation in an RC 2-D element plays a dramatic role in determining the shear energy dissipation.

6.3.11.5 Effects of Steel Percentage on Ratio R_A and Factor μ_D

Figure 6.42 shows the shear energy dissipation ratio R_A as a function of the shear strain ratio R_γ for series A0 (G-0-0.54, G-0-1.2, and G-0-2.7) and series A45 (panels G-45-0.54, G-45-1.2, and G-45-2.7). For both series of panels, Figure 6.42 shows that the slope of the $R_A - R_\gamma$ curves is not a function of the steel percentage, because the shape of the hysteretic loops is similar in each series of panels. However, the maximum shear strain ratio decreases rapidly with an increase of steel percentage, because steel percentage has a strong effect on the

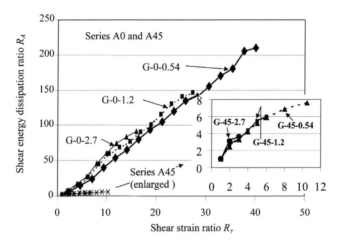

Figure 6.42   Variation of shear energy dissipation ratio versus shear strain ratio for panels with steel grid angle of 45° and 0°

ultimate shear strain. For example, the maximum shear strain ratios (R_y) are approximately equal to 40, 27 and 16 for panels G-0-0.54, G-0-1.2, and G-0-2.7, respectively.

Figure 6.39 gives the values of shear energy dissipation factor μ_D as a function of steel percentages (ρ_ℓ or ρ_t) for panels in the A0 and A45 series. Two trends are evident: (a) the factor μ_D for panels in A0 series is about two orders of magnitude larger than that in A45 series; and (b) the factor μ_D decreases with an increase of steel percentage from 0.54 to 2.7% for panels in both the A0 series and the A45° series.

7

Torsion

7.1 Analysis of Torsion

In Chapter 5, we have applied the rotating angle theory to 2-D elements subjected to shear and normal stresses. In this chapter we will apply the rotating angle softened truss model (RA-STM), presented in Section 5.4, to torsion (Hsu, 1984; Hsu and Mo, 1985a, 1985b, 1985c; Hsu, 1988, 1990, 1991a, 1991b, 1993). Section 7.1 will be devoted to the analysis of torsion, while Section 7.2 will include several important topics in the design for torsion.

The softened membrane model (SMM) studied in Chapter 6, Section 6.1, can also be applied to torsion (Jeng and Hsu, 2009). However, this analytical model will not be presented in this book.

7.1.1 Equilibrium Equations

7.1.1.1 Shear Elements in Shear Flow Zone

A reinforced concrete prismatic member is subjected to an external torque T as shown in Figure 7.1(a). This external torque is resisted by an internal torque formed by the circulatory shear flow q along the periphery of the cross-section. This shear flow q occupies a zone, called the shear flow zone, which has a thickness denoted as t_d. This thickness t_d is a variable determined from the equilibrium and compatibility conditions. It is not the same as the given wall thickness h of a hollow member.

A 2-D element A in the shear flow zone (Figure 7.1a), is subjected to a shear stress $\tau_{\ell t} = q/t_d$, as shown in Figure 7.1(b). The in-plane equilibrium of this element should satisfy the three equations, (5.94)–(5.96), in Section 5.4.2:

$$\sigma_\ell - \rho_\ell f_\ell - \rho_{\ell p} f_{\ell p} = \sigma_d \sin^2 \alpha_r \qquad \text{(7.1) or [1]}$$

$$\sigma_t - \rho_t f_t - \rho_{tp} f_{tp} = \sigma_d \cos^2 \alpha_r \qquad \text{(7.2) or [2]}$$

$$\tau_{\ell t} = (-\sigma_d) \sin \alpha_r \cos \alpha_r \qquad \text{(7.3) or [3]}$$

Unified Theory of Concrete Structures   Thomas Hsu and Yi-Lung Mo
© 2010 John Wiley & Sons, Ltd

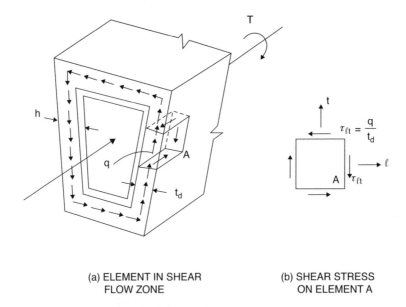

(a) ELEMENT IN SHEAR   (b) SHEAR STRESS
    FLOW ZONE   ON ELEMENT A

Figure 7.1   Hollow box subjected to torsion

where:

ρ_ℓ, ρ_t = mild steel ratio in the ℓ- and t-directions, respectively,
f_ℓ, f_t = stresses in mild steel in the ℓ- and t-directions, respectively.
$\rho_{\ell p}$, ρ_{tp} = prestressing steel ratio in the ℓ- and t-directions, respectively,
$f_{\ell p}$, f_{tp} = stresses in prestressing steel in the ℓ- and t-directions, respectively.

In Equations [1]–[3] the steel ratios, ρ_ℓ, ρ_t, $\rho_{\ell p}$, ρ_{tp}, should be taken with respect to the thickness of the shear flow zone t_d as follows:

$$\rho_\ell = \frac{A_\ell}{p_o t_d} \tag{7.4}$$

$$\rho_t = \frac{A_t}{s t_d} \tag{7.5}$$

$$\rho_{\ell p} = \frac{A_{\ell p}}{p_o t_d} \tag{7.6}$$

$$\rho_{tp} = \frac{A_{tp}}{s t_d} \tag{7.7}$$

where:

A_ℓ, $A_{\ell p}$ = total cross-sectional areas of mild steel and prestressing steel, respectively, in the longitudinal direction;
A_t, A_{tp} = cross-sectional areas of one mild steel bar and one prestressing steel strand, respectively, in the transverse or hoop direction;
p_o = perimeter of centerline of shear flow around a cross-section;
s = spacing of hoop steel along a member.

Since Equations [1] and [2] involve the steel ratios ρ_ℓ, ρ_t, $\rho_{\ell p}$, ρ_{tp}, these two equilibrium equations are coupled to the compatibility equations through the variable t_d. The thickness t_d is a geometric variable which has to be determined not only by the equilibrium conditions, but also by the compatibility conditions. This is similar to the determination of the neutral axis in the bending theory (see Chapter 3), which requires the plane section compatibility condition.

7.1.1.2 Bredt's Equilibrium Equation

In Section 7.1.1.1 three equations, [1]–[3], are derived from the equilibrium of a membrane element in the shear flow zone. To maintain equilibrium of the whole cross-section, however, a fourth equation must be satisfied. The derivation of this additional equilibrium equation has been given in Section 2.1.4.1. The resulting equation (2.46) is expressed in terms of the shear flow q. For a shear flow zone thickness of t_d the shear stress, $\tau_{\ell t}$, is:

$$\tau_{\ell t} = \frac{T}{2A_o t_d} \qquad (7.8) \text{ or } [4]$$

This additional equation [4] also introduces an additional variable, the torque T.

The thickness of the shear flow zone t_d, is strongly involved in Equation [4] not only explicitly through t_d, but also implicitly through A_o, which is a function of t_d. Consequently, the three equilibrium equations, [1], [2] and [4], are now coupled, through the variable t_d, to the compatibility equations. These compatibility equations will be derived in Section 7.1.2.

7.1.2 Compatibility Equations

7.1.2.1 2-D elements in shear

As shown in Figure 7.1(a) and (b), the 2-D element A in the shear flow zone is subjected to a shear stress. The in-plane deformation of this element should satisfy the three compatibility equations (5.97)–(5.99), in Section 5.4.2:

$$\varepsilon_\ell = \varepsilon_r \cos^2 \alpha_r + \varepsilon_d \sin^2 \alpha_r \qquad (7.9) \text{ or } [5]$$

$$\varepsilon_t = \varepsilon_r \sin^2 \alpha_r + \varepsilon_d \cos^2 \alpha_r \qquad (7.10) \text{ or } [6]$$

$$\frac{\gamma_{\ell t}}{2} = (\varepsilon_r - \varepsilon_d) \sin \alpha_r \cos \alpha_r \qquad (7.11) \text{ or } [7]$$

This 2-D element A will also be subjected to out-of-plane deformation. To study this out-of-plane deformation, we will first relate the shear strain $\gamma_{\ell t}$ in Equation [7] to the angle of twist θ by the geometric relationship presented in Section 7.1.2.2.

7.1.2.2 Shear Strain due to Twisting

When a tube is subjected to torsion, the relationship between the shear strain $\gamma_{\ell t}$ in the wall of the tube and the angle of twist θ of the member, can be derived from the compatibility condition of warping deformation. In Figure 7.2, a longitudinal cut is made in an infinitesimal length $d\ell$ of a tube. Progressing from one side of the cut along the circumference to the other side of the cut, the differential warping displacement (in the ℓ-direction) must be zero when integrating throughout the whole perimeter.

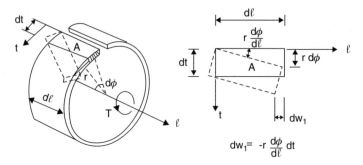

(a) WARPING DISPLACEMENT (IN ℓ-DIRECTION)
    DUE TO ANGLE OF TWIST

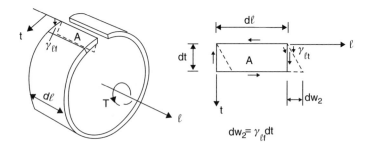

(b) WARPING DISPLACEMENT (IN ℓ-DIRECTION)
    DUE TO SHEAR DEFORMATION

Figure 7.2   Warping displacement in a tube

The warping displacement of a differential 2-D element A, shown in Figure 7.2, is composed of two parts. The first part is induced by the rigid rotation $d\phi$, as shown in Figure 7.2(a). The second part is caused by the shear deformation $\gamma_{\ell t}$, as shown in Figure 7.2(b). Under torsion, the 2-D element A, $d\ell$ by dt, in Figure 7.2(a) rotates through an angle $rd\phi/d\ell$ in the $\ell - t$ plane. The symbol r is the distance from the center of twist to the centerline of the element. The differential warping displacement dw_1 is therefore:

$$dw_1 = -r\frac{d\phi}{d\ell}dt = -r\theta dt \qquad (7.12)$$

The differential warping displacement dw_2, of 2-D element A due to shear deformation in Figure 7.2(b) is:

$$dw_2 = \gamma_{\ell t} dt \qquad (7.13)$$

Adding Equations (7.12) and (7.13) gives the total differential warping displacement dw due to both the rotation and the shear deformation:

$$dw = dw_1 + dw_2 = -r\theta dt + \gamma_{\ell t} dt \qquad (7.14)$$

For a closed section, the total differential warping displacement integrating around the whole perimeter must be equal to zero, giving

$$\oint dw = -\theta \oint r\,dt + \oint \gamma_{\ell t}\,dt = 0 \tag{7.15}$$

Recalling $\oint r\,dt = 2A_o$ from Equation (2.45) gives

$$\oint \gamma_{\ell t}\,dt = 2A_o\theta \tag{7.16}$$

When the wall thickness of the tube is uniform, the shear stress $\tau_{\ell t}$ is a constant, resulting in a uniform shear strain of $\gamma_{\ell t}$. Then $\gamma_{\ell t}$ could be taken out of the integral in Equation (7.16), giving

$$\gamma_{\ell t} \oint dt = 2A_o\theta \tag{7.17}$$

Since $\oint dt$ is the perimeter of the centerline of the shear flow, p_o, we have

$$\theta = \frac{p_o}{2A_o}\gamma_{\ell t} \tag{7.18) or [8]}$$

It is clear from Equation [8] that the angle of twist θ will produce a shear strain $\gamma_{\ell t}$ in the 2-D elements of the shear flow zone. This shear strain $\gamma_{\ell t}$ will induce the steel strains ε_ℓ and ε_t in the $\ell - t$ coordinate and the concrete strains ε_r and ε_d in the $r - d$ coordinate. The relationship between the strains in the $\ell - t$ coordinate (ε_ℓ, ε_t, and $\gamma_{\ell t}$) and the strains in the $r - d$ coordinate (ε_r and ε_d) is described by the three compatibility equations, [5]–[7].

7.1.2.3 Bending of Diagonal Concrete Struts

In a torsional member, the angle of twist θ also produces warping in the wall of the member, which, in turn, causes bending in the concrete struts. In other words, the concrete struts are not only subjected to compression due to the circulatory shear, but also subjected to bending due to the warping of the wall. The relationship between the angle of twist, θ and the bending curvature of concrete struts ψ will now be studied.

A box member with four walls of thickness t_d and subjected to a torsional moment T is shown in Figure 7.3(a). The length of the member is taken to be L. The centerline of shear flow q along the top wall has a width of $L \cot \alpha_r$, so that the diagonal line along the center plane of top wall OABC has an angle of α_r with respect to the shear flow q. When this member receives an angle of twist θ, this center plane will become a hyperbolic paraboloid surface OADC, as shown in Figure 7.3(b). The edge CB of the plane rotates to the position CD through an angle θL. The curve OD will have a curvature of ψ to be determined.

A 3-D coordinate system with axes, x, y and w, is imposed on the center plane OABC, as shown in Figure 7.3(b). The axes x and y are along the edges OC and OA, while the axis w is normal to the center plane. The hyperbolic paraboloid surface OADC can then be expressed by:

$$w = \theta xy \tag{7.19}$$

where w is the displacement perpendicular to the $x - y$ plane.

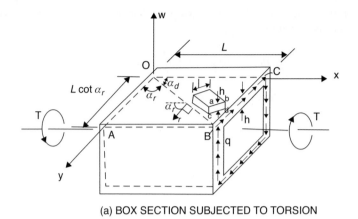

(a) BOX SECTION SUBJECTED TO TORSION

(b) DEFORMATION OF TOP WALL

Figure 7.3   Bending of a concrete strut in the wall of a box section subjected to torsion

Imposing an axis s through OB along the direction of the diagonal concrete struts, the slope of the curve OD can be obtained by differentiating w with respect to s. Utilizing the chain rule, the slope dw/ds is:

$$\frac{dw}{ds} = \frac{\partial w}{\partial x}\frac{dx}{ds} + \frac{\partial w}{\partial y}\frac{dy}{ds} = (\theta y)\sin\alpha_r + (\theta x)\cos\alpha_r \qquad (7.20)$$

The curvature of the concrete struts ψ is the second derivative of w with respect to s, resulting in

$$\frac{d^2 w}{d^2 s} = \frac{\partial \left(\frac{dw}{ds}\right)}{\partial x}\frac{dx}{ds} + \frac{\partial \left(\frac{dw}{ds}\right)}{\partial y}\frac{dy}{ds} = (\theta\cos\alpha_r)\sin\alpha_r + (\theta\sin\alpha_r)\cos\alpha_r \qquad (7.21)$$

Hence,

$$\psi = \theta\sin 2\alpha_r \qquad \text{(7.22) or [9]}$$

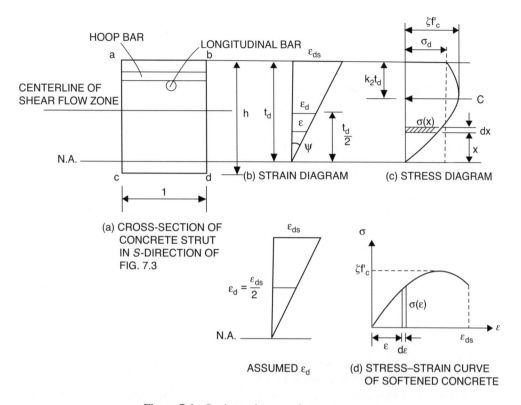

Figure 7.4   Strains and stresses in concrete struts

The derivation of Equation [9] has been illustrated by a rectangular box section in Figure 7.3, because the imposed curvature is easy to visualize in such a section. In actuality, this equation is applicable to any arbitrary bulky sections with multiple walls.

7.1.2.4 Strain Distribution in Concrete Struts

The curvature ψ derived in Equation [9] produces a nonuniform strain distribution in the concrete struts. Figure 7.4(a) shows a unit width of a concrete strut in a hollow section with a wall thickness h. The tension area in the inner portion of the cross-section is neglected. The area in the outer portion that is in compression is considered to be effective to resist the shear flow. The depth of the compression zone from the neutral axis (N.A.) to the extreme compression fiber is defined as the thickness of the shear flow zone t_d. Within this thickness t_d, the strain distribution is assumed to be linear, as shown in Figure 7.4(b). This assumption is identical to Bernoulli's plane section hypothesis used in the bending theory of Chapter 3. The thickness t_d can, therefore, be related to the curvature ψ and the maximum strain at the surface ε_{ds} by the simple relationship:

$$t_d = \frac{\varepsilon_{ds}}{\psi}$$

(7.23) or [10]

The average strain ε_d can be simply related to the maximum strain ε_{ds} by:

$$\varepsilon_d = \frac{\varepsilon_{ds}}{2} \qquad\qquad \text{(7.24) or [11]}$$

Equations [8], [9], [10] and [11] are the four additional compatibility equations. They introduce four additional variables, θ, ψ, t_d and ε_{ds}.

7.1.3 Constitutive Relationships of Concrete

7.1.3.1 Softened Compression Stress Block

When the strain distribution in the concrete struts is assumed to be linear, the stress distribution is represented by a curve as shown in Figure 7.4(c). This softened compression stress block has a peak stress σ_p defined by

$$\sigma_p = \zeta f_c' \qquad\qquad (7.25)$$

where the softening of the concrete struts is taken into account by the coefficient ζ. The average stress of the concrete struts σ_d is defined as

$$\sigma_d = k_1 \sigma_p = k_1 \zeta f_c' \qquad\qquad \text{(7.26) or [12]}$$

where the nondimensional coefficient k_1 is defined as the ratio of the average stress to the peak stress. The symbol σ_d in Equation [12] has been generalized to represent the average compression stress in a concrete strut subjected to bending and compression, rather than the compression stress of axially loaded concrete struts in a membrane element as defined for the symbol σ_d in the equilibrium equations [1]–[3]. The generalization of the symbol σ_d implies two assumptions. First, the relationship between the average stress σ_d defined by Equation [12] and average strain ε_d defined by Equation [11] is identical to the stress–strain relationship of an axially loaded concrete strut. Second, the softening coefficient ζ, which has been determined from the tests of 2-D elements with concrete struts under axial compression, is assumed to be applicable to the concrete struts under combined bending and compression.

The resultant C of the softened compression stress block has a magnitude of

$$C = \sigma_d t_d = k_1 \sigma_p t_d = k_1 \zeta f_c' t_d \qquad\qquad (7.27)$$

This resultant C is located at a distance $k_2 t_d$ from the extreme compression fiber, Figure 7.4(c). The nondimensional coefficient k_2 is defined as the ratio of the distance between the resultant C and the extreme compression fiber to the depth of the compression zone t_d. The compression stress block is statically defined, when the two coefficients k_1 and k_2 are determined.

The coefficient k_1 can be determined from the equilibrium of forces by integrating the compression stress block. Designating the stress at a distance x from the neutral axis as $\sigma(x)$ (Figure 7.4c), gives

$$C = k_1 \sigma_p t_d = \int_o^{t_d} \sigma(x)\mathrm{d}x \qquad\qquad (7.28)$$

The geometric shape of the compression stress block in Figure 7.4(c) will be identical to that of the stress–strain curve of softened concrete (Figure 7.4d), if two assumptions are made. First, the strain distribution in Figure 7.4(b) is assumed to be linear, and the strain ε at a distance x from the neutral axis is related to x by similar triangles as follows:

$$x = \frac{t_d}{\varepsilon_{ds}}\varepsilon \tag{7.29}$$

or

$$dx = \frac{t_d}{\varepsilon_{ds}}d\varepsilon \tag{7.30}$$

Second, the strain gradient in the stress block is assumed to have no effect on the stress–strain curve. Substituting dx from Equation (7.30) into (7.28) and changing the integration limit from the distance t_d to the strain ε_{ds}, we have

$$C = k_1\sigma_p t_d = \frac{t_d}{\varepsilon_{ds}}\int_o^{\varepsilon_{ds}}\sigma(\varepsilon)d\varepsilon \tag{7.31}$$

and

$$k_1 = \frac{1}{\sigma_p \varepsilon_{ds}}\int_o^{\varepsilon_{ds}}\sigma(\varepsilon)d\varepsilon \tag{7.32}$$

The coefficient k_2 can be determined by taking the equilibrium of moments about the neutral axis:

$$C(1 - k_2)t_d = \int_o^{t_d}\sigma(x)x\,dx \tag{7.33}$$

Substituting x from Equations (7.29) and dx from Equation (7.30) into (7.33) while changing the integration limit, we obtain

$$C(1 - k_2)t_d = \frac{t_d^2}{\varepsilon_{ds}^2}\int_o^{\varepsilon_{ds}}\sigma(\varepsilon)\varepsilon\,d\varepsilon \tag{7.34}$$

Substituting C from Equation (7.31) into (7.34) gives

$$k_2 = 1 - \frac{1}{\varepsilon_{ds}}\frac{\int_o^{\varepsilon_{ds}}\sigma(\varepsilon)\varepsilon\,d\varepsilon}{\int_o^{\varepsilon_{ds}}\sigma(\varepsilon)d\varepsilon} \tag{7.35}$$

The coefficients k_1 and k_2 can be calculated from Equations (7.32) and (7.35), respectively, if the stress–strain curve $\sigma - \varepsilon$ is given mathematically.

7.1.3.2 Coefficient k_1 for Average Compression Stress

Coefficient k_1 can be obtained from Equation (7.32) using the softened stress–strain curve given in Figure 5.12 and expressed analytically by Equation (5.100) in the ascending branch and by Equation (5.101) in the descending branch. The softened coefficient ζ in these two equations is given in Figure 5.13 and is expressed by Equation (5.102). The two curves, one ending in the ascending branch and one ending in the descending branch, are sketched

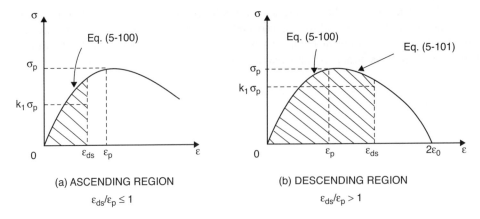

Figure 7.5 Integration of softened stress–strain curve of concrete to determine k_1

in Figure 7.5(a) and (b), respectively. The two expressions for coefficient k_1, one for the ascending branch and another for the descending branch, will be derived separately below:

Ascending branch $\varepsilon_{ds}/\varepsilon_p \leq 1$ (Figure 7.5a)
Inserting Equation (5.100) into (7.32) gives

$$k_1 = \frac{1}{\sigma_p \varepsilon_{ds}} \int_0^{\varepsilon_{ds}} \sigma_p \left[2\left(\frac{\varepsilon}{\varepsilon_p}\right) - \left(\frac{\varepsilon}{\varepsilon_p}\right)^2 \right] d\varepsilon \qquad (7.36)$$

For convenience of integration, the symbols of peak stress σ_p and peak strain ε_p are used in Equation (7.36), rather than $\zeta f_c'$ and $\zeta \varepsilon_o$. The dummy variable is written as ε, rather than ε_d, to avoid confusing with the symbol for the average strain. Integrating Equation (7.36) results in

$$k_1 = \frac{\varepsilon_{ds}}{\varepsilon_p} \left(1 - \frac{1}{3}\frac{\varepsilon_{ds}}{\varepsilon_p} \right) \qquad (7.37)$$

Letting $\varepsilon_p = \zeta \varepsilon_o$ and expressing k_1 in terms of ζ, Equation (7.37) becomes, when $\varepsilon_{ds}/\zeta \varepsilon_o \leq 1$

$$k_1 = \frac{\varepsilon_{ds}}{\zeta \varepsilon_o} \left(1 - \frac{1}{3}\frac{\varepsilon_{ds}}{\zeta \varepsilon_o} \right) \qquad (7.38) \text{ or } [13a]$$

Descending branch $\varepsilon_{ds}/\varepsilon_p > 1$ (Figure 7.5b)
Inserting Equations (5.100) and (5.101) into (7.32) gives

$$k_1 = \frac{1}{\sigma_p \varepsilon_{ds}} \int_0^{\varepsilon_p} \sigma_p \left[2\left(\frac{\varepsilon}{\varepsilon_p}\right) - \left(\frac{\varepsilon}{\varepsilon_p}\right)^2 \right] d\varepsilon + \frac{1}{\sigma_p \varepsilon_{ds}} \int_{\varepsilon_p}^{\varepsilon_{ds}} \sigma_p \left[1 - \left(\frac{\varepsilon - \varepsilon_p}{2\varepsilon_o - \varepsilon_p}\right)^2 \right] d\varepsilon$$

$$(7.39)$$

Table 7.1 k_1 as a function of ζ and ε_{ds} for softened concrete ($\varepsilon_0 = 0.002$)

ζ $\diagdown$ ε_{ds}	0.0005	0.001	0.0015	0.002	0.0025	0.003	0.0035
0.10	0.8654	0.9215	0.9218	0.8994	0.8610	0.8089	0.7439
0.20	0.7333	0.8611	0.8883	0.8806	0.8513	0.8048	0.7429
0.30	0.6018	0.7980	0.8526	0.8604	0.8409	0.8005	0.7419
0.40	0.4948	0.7333	0.8147	0.8385	0.8294	0.7956	0.7407
0.50	0.4167	0.6667	0.7747	0.8148	0.8167	0.7901	0.7394
0.60	0.3588	0.6019	0.7325	0.7891	0.8026	0.7840	0.7379
0.70	0.3146	0.5442	0.6889	0.7613	0.7870	0.7771	0.7362
0.80	0.2799	0.4948	0.6445	0.7314	0.7698	0.7693	0.7342
0.90	0.2521	0.4527	0.6018	0.6997	0.7506	0.7603	0.7319
1.00	0.2292	0.4167	0.5625	0.6667	0.7292	0.7500	0.7292

Integration and simplification of Equation (7.39) result in

$$k_1 = \left[1 - \frac{\varepsilon_p^2}{\left(2\varepsilon_o - \varepsilon_p\right)^2}\right]\left(1 - \frac{1}{3}\frac{\varepsilon_p}{\varepsilon_{ds}}\right) + \frac{\varepsilon_p^2}{\left(2\varepsilon_o - \varepsilon_p\right)^2}\frac{\varepsilon_{ds}}{\varepsilon_p}\left(1 - \frac{1}{3}\frac{\varepsilon_{ds}}{\varepsilon_p}\right) \tag{7.40}$$

Expressing k_1 in terms of ζ by letting $\varepsilon_p = \zeta\varepsilon_o$ gives, when $\varepsilon_{ds}/\zeta\varepsilon_o > 1$:

$$k_1 = \left[1 - \frac{\zeta^2}{\left(2 - \zeta\right)^2}\right]\left(1 - \frac{1}{3}\frac{\zeta\varepsilon_o}{\varepsilon_{ds}}\right) + \frac{\zeta^2}{\left(2 - \zeta\right)^2}\frac{\varepsilon_{ds}}{\zeta\varepsilon_o}\left(1 - \frac{1}{3}\frac{\varepsilon_{ds}}{\zeta\varepsilon_o}\right) \tag{7.41} \text{ or } [13b]$$

Values of the coefficient k_1 as expressed by Equations [13a] and [13b] are tabulated in Table 7.1 as a function of ε_{ds} and ζ, while ε_o is taken as 0.002.

7.1.3.3 Location of Centerline of Shear Flow

Similar to the calculation of coefficient k_1, coefficient k_2 for the location of resultant C can be obtained from Equation (7.35) using the softened stress–strain curve of Figure 5.12 and expressed analytically by Equations (5.100), (5.101) and (5.102). These calculations show that k_2 varies generally in the range 0.40–0.45. The resultant C, therefore, should lie approximately in the range $0.40t_d - 0.45t_d$ from the extreme compression fiber.

The location of the centerline of the shear flow is a more tricky problem to determine. Considering the concrete struts, it should be theoretically located at the position of the resultant C, which is a distance $0.40t_d - 0.45t_d$ from the extreme compression fiber. However, because the shear flow is constituted from the truss action of both the concrete and the steel, the location of the steel bars (or the thickness of the concrete cover) should also have an effect on the location of the shear flow. Fortunately, tests (Hsu and Mo, 1985b) have shown that this effect of the steel bar location is small.

It could also be argued that the average strain ε_d be defined at the location of the resultant C, rather than at the mid-depth of the thickness t_d. Such treatment, of course, will considerably complicate the calculation without any convincing theoretical and experimental justifications.

A simple solution to this tricky problem is to assume that the centerline of shear flow lies at the mid-depth of the thickness t_d, the same location where the average strain has been defined.

In other words, the centerline of the shear flow is located at a distance $0.5t_d$ from the extreme compression fiber. This assumption has three advantages:

1. By using $k_2 = 0.5$, a constant, the tedious process of integration to calculate the coefficient k_2 is avoided. The centerline of shear flow also coincides with the location of the average strain ε_d, Figure 7.4(b).
2. The calculations of A_o and p_o are considerably simplified (see Section 7.1.3.4). This simplification is slightly on the conservative side, because the constant 0.5 is somewhat greater than the actual value of the coefficient k_2 (about 0.4–0.45).
3. In calculating the torque $T = q(2A_o)$, the slight conservatism in A_o is actually desirable to counteract the unconservatism involved in the use of the expression $q = (A_t/s)f_t \tan \alpha_r$. In the expression for q, the ratio A_t/s has been nonconservatively used to describe the transverse steel areas per unit length. (The nonconservative nature of A_t/s is caused by the discrete spacing of the transverse steel bars (explained in detail in Section 4.4.3.1 of Hsu, 1984). Tests have shown (Hsu and Mo, 1985a) that this simplifying assumption of $k_2 = 0.5$ results in good agreement between theory and tests.

7.1.3.4 Formulas for A_o and p_o

Now that the centerline of the shear flow is assumed to coincide with the centerline of the shear flow zone, the formulas for calculating A_o and p_o can be derived. The formulas must be sufficiently accurate for a thick tube, because the thickness of the shear flow zone t_d is usually quite large with respect to the overall dimensions of the cross-section when the softening of concrete is taken into account. For a rectangular cross-section of height a and width b as shown in Figure 7.6(a), A_o and p_o are:

$$A_o = (a - t_d)(b - t_d) = ab - (a + b)t_d + t_d^2 = A_c - \frac{1}{2}p_c t_d + t_d^2 \qquad (7.42)$$

$$p_o = p_c - 4t_d \qquad (7.43)$$

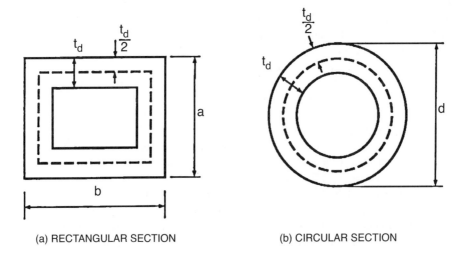

(a) RECTANGULAR SECTION   (b) CIRCULAR SECTION

Figure 7.6   Calculation of A_0 and P_0 (dotted curves represent center lines of shear flow)

where

A_c = cross-sectional area bounded by the outer perimeter of the concrete;

p_c = perimeter of the outer concrete cross-section.

For a circular cross-section of diameter d as shown in Figure 7.6(b), A_o and p_o are:

$$A_o = \frac{\pi}{4}(d - t_d)^2 = \frac{\pi}{4}d^2 - \frac{\pi}{2}dt_d + \frac{\pi}{4}t_d^2 = A_c - \frac{1}{2}p_c t_d + \frac{\pi}{4}t_d^2 \tag{7.44}$$

$$p_o = p_c - \pi t_d \tag{7.45}$$

Comparing Equations (7.42) and (7.44) for A_o, it can be seen that the two formulas are the same, except the last term t_d^2, which has a constant 1 for a rectangular section and a constant $\pi/4$ for a circular section. Similarly, a comparison of Equations (7.43) and (7.45) for p_o shows that the two formulas are the same, except the last term t_d, which has a constant 4 for rectangular section and a constant π for circular section. Therefore, Equations (7.42)–(7.45) can be summarized as follows:

$$A_o = A_c - \frac{1}{2}p_c t_d + \xi t_d^2 \tag{7.46}$$

$$p_o = p_c - 4\xi t_d \tag{7.47}$$

where $\xi = 1$ for a rectangular section and $\xi = \pi/4$ for a circular section. In other words, ξ varies from about 0.8 to 1.

Notice that the last terms in both Equations (7.46) and (7-47) are considerably smaller than the other terms in the formulas. A slight adjustment of these last terms will not produce a significant loss of accuracy for the formulas. Hence, a pair of formulas for A_o and p_o is proposed here for arbitrary bulky cross sections assuming $\xi = 1$:

$$A_o = A_c - \frac{1}{2}p_c t_d + t_d^2 \tag{7.48}$$

$$p_o = p_c - 4t_d \tag{7.49}$$

These two formulas are exact for rectangular sections, and should be quite accurate for any arbitrary bulky cross-sections.

7.1.4 Governing Equations for Torsion

The governing equations for equilibrium condition and compatibility condition are introduced in Sections 7.1.1 and 7.1.2. The softened constitutive equations for concrete struts in bending and compression have been derived in Section 7.1.3. In addition, the stress–strain curve for mild steel has been given in Figure 5.14 and expressed analytically by Equations (5.103)–(5.106), while the stress–strain curve for prestressing steel has been given in Figure 5.15 and expressed analytically by Equations (5.107) and (5.108). All these equations will now be listed in this section so that a strategy for the solution of these equations can be developed.

Equilibrium equations

$$\sigma_\ell - \rho_\ell f_\ell - \rho_{\ell p} f_{\ell p} = \sigma_d \sin^2 \alpha_r \tag{1}$$

$$\sigma_t - \rho_t f_t - \rho_{tp} f_{tp} = \sigma_d \cos^2 \alpha_r \tag{2}$$

$$\tau_{\ell t} = (-\sigma_d) \sin \alpha_r \cos \alpha_r \qquad [3]$$

$$T = \tau_{\ell t}(2A_o t_d) \qquad [4]$$

Compatibility equations

$$\varepsilon_\ell = \varepsilon_r \cos^2 \alpha_r + \varepsilon_d \sin^2 \alpha_r \qquad [5]$$

$$\varepsilon_t = \varepsilon_r \sin^2 \alpha_r + \varepsilon_d \cos^2 \alpha_r \qquad [6]$$

$$\frac{\gamma_{\ell t}}{2} = (\varepsilon_r - \varepsilon_d) \sin \alpha_r \cos \alpha_r \qquad [7]$$

$$\theta = \frac{p_o}{2A_o} \gamma_{\ell t} \qquad [8]$$

$$\psi = \theta \sin 2\alpha_r \qquad [9]$$

$$t_d = \frac{\varepsilon_{ds}}{\psi} \qquad [10]$$

$$\varepsilon_d = \frac{\varepsilon_{ds}}{2} \qquad [11]$$

Constitutive law of concrete in compression

$$\sigma_d = k_1 \zeta f_c' \qquad [12]$$

$$k_1 = \frac{\varepsilon_{ds}}{\zeta \varepsilon_o} \left(1 - \frac{1}{3} \frac{\varepsilon_{ds}}{\zeta \varepsilon_o}\right) \quad \varepsilon_{ds}/\zeta \varepsilon_o \leq 1 \qquad [13a]$$

$$k_1 = \left[1 - \frac{\zeta^2}{(2-\zeta)^2}\right]\left(1 - \frac{1}{3}\frac{\zeta \varepsilon_o}{\varepsilon_{ds}}\right) + \frac{\zeta^2}{(2-\zeta)^2}\frac{\varepsilon_{ds}}{\zeta \varepsilon_o}\left(1 - \frac{1}{3}\frac{\varepsilon_{ds}}{\zeta \varepsilon_o}\right) \quad \varepsilon_{ds}/\zeta \varepsilon_o > 1 \quad [13b]$$

$$\zeta = \frac{0.9}{\sqrt{1 + 600\varepsilon_r}} \qquad [14]$$

Constitutive law of mild steel

$$f_\ell = E_s \varepsilon_\ell \qquad \varepsilon_\ell < \varepsilon_{\ell y} \qquad [15a]$$

$$f_\ell = f_{\ell y} \qquad \varepsilon_\ell \geq \varepsilon_{\ell y} \qquad [15b]$$

$$f_t = E_s \varepsilon_t \qquad \varepsilon_t < \varepsilon_{ty} \qquad [16a]$$

$$f_t = f_{ty} \qquad \varepsilon_t \geq \varepsilon_{ty} \qquad [16b]$$

Constitutive law of prestressing steel

$$f_p \leq 0.7 \, f_{pu} \qquad f_p = E_{ps}(\varepsilon_{dec} + \varepsilon_s) \qquad [17a]\,[18a]$$

$$f_p > 0.7 \, f_{pu} \qquad f_p = \frac{E_{ps}'(\varepsilon_{dec} + \varepsilon_s)}{\left[1 + \left\{\frac{E_{ps}'(\varepsilon_{dec}+\varepsilon_s)}{f_{pu}}\right\}^m\right]^{\frac{1}{m}}} \qquad [17b]\,[18b]$$

where

f_p = stress in prestressing steel, f_p becomes $f_{\ell p}$ or f_{tp} when applied to the longitudinal and transverse steel, respectively;

ε_s = strain in the mild steel, ε_s becomes ε_ℓ or ε_t when applied to the longitudinal and transverse steel, respectively;

ε_{dec} = strain in prestressing steel at decompression of concrete, $\varepsilon_{dec} = \varepsilon_{pi} + \varepsilon_i$; taken as 0.005 for grade 1862 MPa (270 ksi) strands;

ε_{pi} = initial strain in prestressed steel after loss;

ε_i = initial strain in mild steel after loss;

E_{ps} = elastic modulus of prestressed steel, taken as 200 000 MPa (29 000 ksi);

E'_{ps} = tangential modulus of Ramberg–Osgood curve at zero load, taken as 214 000 MPa (31 040 ksi);

f_{pu} = ultimate strength of prestressing steel, taken as 1862 MPa (270 ksi);

m = shape parameter describing the curvature at knee portion, taken as 4.

Notice that the tensile stress of concrete is assumed to be zero, i.e. $\sigma_r = 0$. Thus we have a total of 18 equations, rather than 19.

7.1.5 Method of Solution

The 18 governing equations for a torsional member (Equations [1]–[18]) are listed in Table 7.2 in three categories: the 4 equilibrium equations, the 7 compatibility equations and the 7 constitutive equations. These 18 equations contain 21 unknown variables, which are also divided into three categories in Table 7.2. The 9 stress or force variables include σ_ℓ, σ_t, $\tau_{\ell t}$, σ_d, f_ℓ, f_t, $f_{\ell p}$, f_{tp} and T. The 10 strain or geometry variables include ε_ℓ, ε_t, $\gamma_{\ell t}$, ε_r, ε_d, α_r, θ, ψ, t_d and ε_{ds}; and the 2 material coefficients are ζ and k_1. If 3 unknown variables are given, then the remaining 18 unknown variables can be solved using the 18 equations.

Table 7.2 also lists the 13 governing equation, $\boxed{1}$ to $\boxed{13}$, used in the RA-STM analysis of 2-D elements. The similarity between this set of 13 equations for 2-D elements and the set of

Table 7.2 Summary of Variables and Equations for 2-D Elements and Torsional Members

Category	Variables			Equations		
	Stresses or forces	Strains or geometry	Material	Equilibrium.	Compatibility	Material
2-D element in shear	σ_ℓ	ε_ℓ	ζ	$\boxed{1}$ [1]	$\boxed{4}$ [5]	$\boxed{8}$ [14]
	σ_t	ε_t		$\boxed{2}$ [2]	$\boxed{5}$ [6]	
	$\tau_{\ell t}$	$\gamma_{\ell t}$		$\boxed{3}$ [3]	$\boxed{6}$ [7]	
	σ_d	ε_d				$\boxed{7}$ [12]
		ε_r, α_r				
	f_ℓ					$\boxed{9}$ [15]
	f_t					$\boxed{10}$ [16]
	$f_{\ell p}$					$\boxed{11}$ [17]
	f_{tp}					$\boxed{12}$ [18]
Additional for torsion	T	θ	k_1	[4]	[8]	[13]
		ψ			[9]	
		t_d			[10]	
		ε_{ds}			[11]	
Total for torsion	9	10	2	4	7	7
		Total 21			Total 18	

18 equations for torsion members stems from the fact that a 2-D element in the shear flow zone is subjected to the in-plane truss action of a membrane element. The additional 5 equations for a torsional member and the modifications of the required constitutive relationships are caused by the warping of the 2-D element in the shear flow zone.

For a member subjected to pure torsion the normal stresses σ_ℓ and σ_t acting on a 2-D element in the shear flow zone are equal to zero, i.e. $\sigma_\ell = \sigma_t = 0$. If ε_d is selected as the third variable because it varies monotonically from zero to maximum, then the remaining 18 unknown variables can be solved (Hsu, 1991b, 1993). The series of solutions for various ε_d values allows us to trace the loading history of the torque–twist curve.

7.1.5.1 Characteristics of Equations

An efficient algorithm for solving the 18 equations was derived based on a careful observation of the six characteristics of the equations:

1. The three equilibrium equations [1]–[3], and the three compatibility equations [5] − [7], are transformation-type equations. In other words, the stresses in the $\ell - t$ coordinate ($\sigma_\ell, \sigma_t, \tau_{\ell t}$) are expressed in terms of stresses in the $r - d$ direction ($\sigma_d, \sigma_r = 0$), and the strains in the $\ell - t$ coordinate ($\varepsilon_\ell, \varepsilon_t, \gamma_{\ell t}$) are expressed in terms of strains in the $r - d$ direction ($\varepsilon_r, \varepsilon_d$).
2. Equations [12], [13] and [14] for the constitutive laws of concrete in compression involve only four unknown variables in the $r - d$ coordinate ($\sigma_d, \varepsilon_r, \varepsilon_d$ and ε_{ds}). If the strains ε_r, ε_d and ε_{ds}, are given, then the stresses σ_d can be calculated from these three equations.
3. Equations [3] and [4] are independent from all other equations, because they contain two variables, $\tau_{\ell t}$ and T, which are not involved in any other equations. In other words, these two equations need not be involved in the iteration process of the solution algorithm.
4. The longitudinal steel stresses f_ℓ and $f_{\ell p}$ in equilibrium equation [1] are coupled to the longitudinal steel strain ε_ℓ in compatibility equation [5] through the longitudinal steel stress–strain relationship of Equations [15] and [17]. Similarly, the transverse steel stresses f_t and f_{tp} in equilibrium equation [2] are coupled to the transverse steel strain ε_t in compatibility equation [6] through the transverse steel stress–strain relationships of Equations [16] and [18].
5. Equations [7]–[10] sequentially relate the four variables, $\gamma_{\ell t}$, θ, ψ and ε_{ds}. Hence, these four equations can easily be combined into one equation.
6. The variable t_d is involved in Equations [1], [2], [8] and [10] through the terms A_o, p_o, ρ_ℓ and ρ_t. Since these four equations are coupled, it is necessary to first assume the variable t_d and then check it later. The flow chart for solution algorithm will require a nested DO-loop to determine the variable t_d.

7.1.5.2 Thickness t_d as a Function of strains

The thickness of shear flow zone, t_d, can be expressed in terms of strains using the compatibility equations [5]–[10]. To do this we will first combine the four compatibility equations [7]–[10] into one equation according to characteristic 5 of Section 7.1.5.1. Inserting $\gamma_{\ell t}$ from Equation [7] into [8] gives

$$\theta = \frac{p_o}{A_o}(\varepsilon_r - \varepsilon_d) \sin \alpha_r \cos \alpha_r \qquad (7.50)$$

Inserting θ from Equation (7.50) into [9]:

$$\psi = \frac{p_o}{A_o}(\varepsilon_r - \varepsilon_d)2\sin^2\alpha_r\cos^2\alpha_r \qquad (7.51)$$

Inserting ψ from Equation (7.51) into [10]:

$$\sin^2\alpha_r\cos^2\alpha_r = \frac{A_o}{2p_o t_d}\frac{(-\varepsilon_{ds})}{(\varepsilon_r - \varepsilon_d)} \qquad (7.52)$$

Equation (7.52) is the basic compatibility equation describing the warping of the shear flow zone in a member subjected to torsion. To eliminate α_r in Equation (7.52) we utilize the first type of compatibility equations, (5.46) and (5.47), in Section 5.2.2, to express α_r in terms of strains:

$$\cos^2\alpha_r = \frac{\varepsilon_\ell - \varepsilon_d}{\varepsilon_r - \varepsilon_d} \qquad (7.53)$$

$$\sin^2\alpha_r = \frac{\varepsilon_t - \varepsilon_d}{\varepsilon_r - \varepsilon_d} \qquad (7.54)$$

Substituting $\cos^2\alpha_r$ and $\sin^2\alpha_r$ from Equations (7.53) and (7.54) into Equation (7.52) and taking $\varepsilon_{ds}/2 = \varepsilon_d$ from Equation [11] result in:

$$t_d = \frac{A_o}{p_o}\left[\frac{(-\varepsilon_d)(\varepsilon_r - \varepsilon_d)}{(\varepsilon_\ell - \varepsilon_d)(\varepsilon_t - \varepsilon_d)}\right] \qquad (7.55) \text{ or } [19]$$

It should be noted that the variable t_d is expressed in terms of strains in all the ℓ-, t-, r- and $d-$ directions. The variable t_d is also involved in Equations [1], [2], [8] and [10] through the terms A_o, p_o, ρ_ℓ and ρ_t. Hence, in the solution procedure the variable t_d must first be assumed and then checked by Equation [19].

7.1.5.3 ε_ℓ as a Function of f_ℓ, $f_{\ell p}$

The strain ε_ℓ can be related to the stresses, f_ℓ and $f_{\ell p}$, by eliminating the angle α_r from the equilibrium equation [1] and the compatibility equation (7.52). To do this, we first substitute $\cos^2\alpha_r$ from Equation (7.53) into (7.52) to obtain the compatibility equation:

$$\sin^2\alpha_r = \frac{A_o}{2p_o t_d}\frac{(-\varepsilon_{ds})}{(\varepsilon_\ell - \varepsilon_d)} \qquad (7.56)$$

Then substituting $\sin^2\alpha_r$ from compatibility equation (7.56) into the equilibrium equation [1], and utilizing the definitions of $\rho_\ell = A_\ell/p_o t_d$ and $\rho_{\ell p} = A_{\ell p}/p_o t_d$ result in:

$$\varepsilon_\ell = \varepsilon_d + \frac{A_o(-\varepsilon_{ds})(-\sigma_d)}{2(-p_o t_d\sigma_\ell + A_\ell f_\ell + A_{\ell p}f_{\ell p})} \qquad (7.57)$$

For pure torsion, $\sigma_\ell = 0$. Also $\varepsilon_{ds}/2 = \varepsilon_d$ from Equation [11]. Equation (7.57) becomes:

$$\varepsilon_\ell = \varepsilon_d + \frac{A_o(-\varepsilon_d)(-\sigma_d)}{(A_\ell f_\ell + A_{\ell p}f_{\ell p})} \qquad (7.58) \text{ or } [20]$$

The variables ε_ℓ, f_ℓ, and $f_{\ell p}$ in Equation [20] can be solved simultaneously with the stress–strain relationships of Equations [15] and [17] for longitudinal steel.

7.1.5.4 ε_t as a Function of f_t, f_{tp}

Similarly, the strain ε_t can be related to the stresses f_t and f_{tp} by eliminating the angle α_r from equilibrium equation [2] and the compatibility equation (7.52). To do this, we first substitute $\sin^2 \alpha_r$ from Equation (7.54) into (7.52) to obtain the compatibility equation:

$$\cos^2 \alpha_r = \frac{A_o}{2p_o t_d} \frac{(-\varepsilon_{ds})}{(\varepsilon_t - \varepsilon_d)} \tag{7.59}$$

Then substituting $\cos^2 \alpha_r$ from compatibility equation (7.59) into equilibrium equation [2] and utilizing the definitions of $\rho_t = A_t/s t_d$ and $\rho_{tp} = A_{tp}/s t_d$, result in:

$$\varepsilon_t = \varepsilon_d + \frac{A_o s(-\varepsilon_{ds})(-\sigma_d)}{2p_o(-s t_d \sigma_t + A_t f_t + A_{tp} f_{tp})} \tag{7.60}$$

For pure torsion, $\sigma_t = 0$. Also $\varepsilon_{ds}/2 = \varepsilon_d$. Equation (7.60) becomes:

$$\varepsilon_t = \varepsilon_d + \frac{A_o s(-\varepsilon_d)(-\sigma_d)}{p_o(A_t f_t + A_{tp} f_{tp})} \tag{7.61) or [21]}$$

The variables ε_t, f_t and f_{tp} in Equation [21] can be solved simultaneously with the stress–strain relationships of Equations [16] and [18] for transverse steel.

7.1.5.5 Additional Equations

Four additional equations derived previously are useful in the solution procedure. In Section 7.1.3.4 the cross-sectional properties A_o and p_o, have been expressed as functions of t_d by Equations (7.48) and (7.49), respectively:

$$A_o = A_c - \frac{1}{2} p_c t_d + t_d^2 \tag{7.62) or [22]}$$

$$p_o = p_c - 4 t_d \tag{7.63) or [23]}$$

In Section 5.4.3.4 the tensile strain ε_r and the angle α_r are expressed in terms of strains ε_ℓ, ε_t and ε_d, by Equations $\boxed{15}$ and $\boxed{16}$, respectively:

$$\varepsilon_r = \varepsilon_\ell + \varepsilon_t - \varepsilon_d \tag{7.64) or [24]}$$

$$\tan^2 \alpha_r = \frac{\varepsilon_t - \varepsilon_d}{\varepsilon_\ell - \varepsilon_d} \tag{7.65) or [25]}$$

Equations [22], [23], [24] and [25] are used in Section 7.1.5.6 describing the solution procedure.

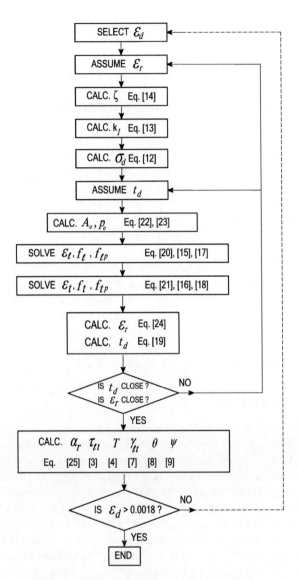

Figure 7.7   Flow chart for torsional analysis

7.1.5.6 Solution Procedure

A set of solution procedures is proposed as shown in the flow chart of Figure 7.7. The procedures are described as follows:

Step 1: select a value of strain in the d-direction, ε_d;

Step 2: assume a value of strain in the r-direction, ε_r;

Step 3: calculate the softened coefficient ζ, the averaging coefficient k_1, and the concrete stresses σ_d from Equations [14], [13], and [12], respectively;

Step 4: assume a value of the thickness of shear flow zone t_d and calculate the cross-sectional properties A_o and p_o by Equations [22] and [23];

Step 5: solve the strains and stresses in the longitudinal steel ε_ℓ, f_ℓ and $f_{\ell p}$, from Equations [20], [15] and [17], and those in the transverse steel ε_t, f_t and f_{tp}, from Equations [21], [16] and [18];

Step 6: calculate the strain $\varepsilon_r = \varepsilon_\ell + \varepsilon_t - \varepsilon_d$ from Equation [24] and t_d from Equation [19]. If ε_r and t_d are the same as assumed, the values obtained for all the strains are correct. If t_d is not the same as assumed, then another value of t_d is assumed and steps 4 to 6 are repeated. If ε_r is not the same as assumed, then another value of ε_r is assumed and steps 3 to 6 are repeated;

Step 7: calculate the angle α_r, the shear stress $\tau_{\ell t}$, the torque T, the shear strain $\gamma_{\ell t}$, the angle of twist θ, the curvature of the concrete struts ψ from Equations [25], [3], [4], [7], [8] and [9], respectively. This will provide one point on the $T - \theta$ curve or on the $\tau_{\ell t} - \gamma_{\ell t}$ curve;

Step 8: select another value of ε_d and repeat steps 2–7. Calculation for a series of ε_d values will provide the whole $T - \theta$ curve and $\tau_{\ell t} - \gamma_{\ell t}$ curve.

These solution procedures have two distinct advantages. First, the variable angle α_r does not appear in the iteration process from step 2 to step 6. Second, the calculation of ε_ℓ and ε_t in step 5 can easily accommodate the nonlinear stress–strain relationships of reinforcing steel, including those for prestressing strands. These advantages were derived from an understanding of the six characteristics of the 18 governing equations. Steps 1–3 are proposed because of characteristic 2. Step 4 is required due to characteristis 6. Steps 5 and 6 are the results of characteristics 1 and 4; and step 7 is possible based on characteristics 1, 3 and 5.

7.1.6 Example Problem 7.1

Analyze the torque–twist behavior of a hollow box girder using the RA-STM. The girder has a trapezoidal cross-section as shown in Figure 7.8. The girder is reinforced with 13 No. 7 longitudinal bars ($A_\ell = 13 \times 387 \text{ mm}^2 = 5031 \text{ mm}^2$) and No. 6 transverse hoop bars ($A_t = 284 \text{ mm}^2$) at 203 mm (8 in.) spacing. Both sizes of mild steel bars have a yield strength of 413 MPa (60 000 psi). $E_s = 200\,000$ MPa (29 000 000 psi). It is also prestressed longitudinally by eight 1862 MPa (270 ksi) strands of 12.7 mm (1/2 in.) diameter ($A_{\ell p} = 8 \times 98.7 \text{ mm}^2 = 790 \text{ mm}^2$). The strain due to decompression, ε_{dec}, is taken as 0.005 and the stress–strain curve can be represented by the Ramberg–Osgood formula (Equations [17a] and [17b]). The cylinder compressive strength of concrete is 41.3 MPa (6000 psi). The softening coefficient of concrete ζ is specified by Equation [14], and the averaging coefficient k_1 by Equations [13a] or [13b]. The net concrete cover is 38.1 mm (1.5 in.).

To illustrate the method of calculation, two values of strain ε_d are selected as examples. The first value of $\varepsilon_d = -0.0005$ is chosen to demonstrate the case at first yield of the transverse steel while the longitudinal steel and the prestressed steel are still in their elastic ranges. In the second case of $\varepsilon_d = -0.0015$, the longitudinal mild steel and the prestressed steel have yielded and the torsional moment obtained is close to the ultimate strength.

Solution

1. Select $\varepsilon_d = -0.0005$

$$\varepsilon_{ds} = 2(-0.0005) = -0.001$$

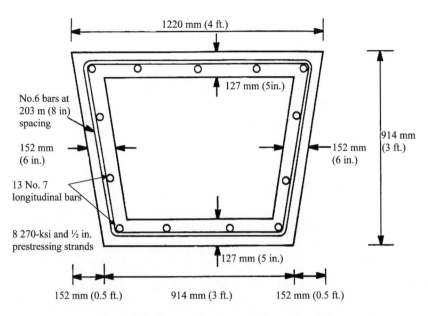

Figure 7.8   Box section for example problem 7.1

Assume $\varepsilon_r = 0.00414$ after several cycles of trial-and-error process:

Equation [14] $\zeta = \dfrac{0.9}{\sqrt{1 + 600\varepsilon_r}} = \dfrac{0.9}{\sqrt{1 + 600(0.00414)}} = 0.482$

$\dfrac{\varepsilon_{ds}}{\zeta\varepsilon_o} = \dfrac{-0.001}{0.482(-0.002)} = 1.037 > 1$ descending branch

$\dfrac{\zeta^2}{(2-\zeta)^2} = \dfrac{0.482^2}{(2-0.482)^2} = 0.1008$

Equation [13b] $k_1 = \left[1 - \dfrac{\zeta^2}{(2-\zeta)^2}\right]\left(1 - \dfrac{1}{3}\dfrac{\zeta\varepsilon_o}{\varepsilon_{ds}}\right) + \dfrac{\zeta^2}{(2-\zeta)^2}\dfrac{\varepsilon_{ds}}{\zeta\varepsilon_o}\left(1 - \dfrac{1}{3}\dfrac{\varepsilon_{ds}}{\zeta\varepsilon_o}\right)$

$= [1 - 0.1008]\left(1 - \dfrac{1}{3}\dfrac{1}{(1.037)}\right) + 0.1008(1.037)\left(1 - \dfrac{1}{3}1.037\right)$

$= 0.6102 - 0.0684 = 0.6786$

k_1, of course, can also be obtained from Table 7.1.

Equation [12] $\sigma_d = k_1\zeta f_c' = 0.6786(0.482)(-41.3) = -13.51 \text{ MPa } (-1963\,\text{psi}).$

Assume $t_d = 96$ mm after two or three trials

$$A_c = \dfrac{914(914 + 1220)}{2} = 975\,000 \text{ mm}^2$$

$$p_c = (914 + 1220 + 2\sqrt{914^2 + 152^2} = 3987 \text{ mm}$$

Equation [22] $A_o = A_c - \dfrac{1}{2}p_c t_d + t_d^2 = 975\,000 - \dfrac{1}{2}(3987)(96) + (96)^2$

$\qquad\qquad\qquad = 793\,000 \text{ mm}^2$

Equation [23] $p_o = p_c - 4t_d = 3987 - 4(96) = 3603 \text{ mm}$

$\qquad\qquad\qquad A_\ell = 13(387) = 5031 \text{ mm}^2$

$\qquad\qquad\qquad A_{\ell p} = 8(98.7) = 790 \text{ mm}^2$

Solve ε_ℓ

Equation [20] $\varepsilon_\ell = \varepsilon_d + \dfrac{A_o(-\varepsilon_d)(-\sigma_d)}{(A_\ell f_\ell + A_{\ell p} f_{\ell p})}$

$\qquad\qquad\qquad = -0.0005 + \dfrac{793\,000(0.0005)(13.51)}{(5031 f_\ell + 790 f_{\ell p})}$

Assume elastic range

Equation [15a] $f_\ell = E_s \varepsilon_\ell = 200 \times 10^3 \varepsilon_\ell$ before yielding

Equation [17a] $f_{\ell p} = E_{ps}(\varepsilon_{dec} + \varepsilon_\ell) = 200 \times 10^3 (0.005 + \varepsilon_\ell)$ before elastic limit

Then $(\varepsilon_\ell + 0.0005)(\varepsilon_\ell + 0.000678) = 4.60 \times 10^{-6}$

$\qquad\qquad\qquad \varepsilon_\ell = 0.00156$

$\qquad\qquad\qquad \varepsilon_y = 0.00207 > 0.00156$ O.K. for mild steel

$\qquad\qquad\qquad \varepsilon_{ps}$ at $0.7 f_{pu} - \varepsilon_{dec} = 0.00652 - 0.005 = 0.00152 \approx 0.00156$

$\qquad\qquad\qquad$ OK for prestressed steel

Solve ε_t

Equation [21] $\varepsilon_t = \varepsilon_d + \dfrac{A_o s(-\varepsilon_d)(-\sigma_d)}{p_o(A_t f_t + A_{tp} f_{tp})}$

$\qquad\qquad\qquad = -0.0005 + \dfrac{793\,000(203)(0.0005)(13.51)}{(3603)(284 f_t + 0)}$

$\qquad\quad \varepsilon_t + 0.0005 = \dfrac{1.063}{f_t}$

Assume elastic range

Equation [16a] $f_t = E_s \varepsilon_t = 200 \times 10^3 \varepsilon_t$ before yielding

Then $\varepsilon_t^2 + 0.0005\varepsilon_t - 5.32 \times 10^{-6} = 0$

$\qquad\qquad\qquad \varepsilon_t = 0.00207 = \varepsilon_y$ (transverse steel just yielded)

Check ε_r

Equation [24] $\varepsilon_r = \varepsilon_\ell + \varepsilon_t - \varepsilon_d = 0.00156 + 0.00207 + 0.0005$

$\qquad\qquad\qquad = 0.00413 \approx 0.00414$ assumed OK

Check t_d

Equation [19] $t_d = \dfrac{A_o}{p_o} \left[\dfrac{(-\varepsilon_d)(\varepsilon_r - \varepsilon_d)}{(\varepsilon_\ell - \varepsilon_d)(\varepsilon_t - \varepsilon_d)} \right]$

$= \dfrac{793\,000}{3603} \left[\dfrac{(0.0005)(0.00413 + 0.0005)}{(0.00156 + 0.0005)(0.00207 + 0.0005)} \right]$

$= 96 \text{ mm} \approx 96 \text{ mm assumed. OK}$

Calculate α_r, $\tau_{\ell t}$, T, $\gamma_{\ell t}$ and θ

Equation [25] $\tan^2 \alpha_r = \dfrac{\varepsilon_t - \varepsilon_d}{\varepsilon_\ell - \varepsilon_d} = \dfrac{0.00207 + 0.0005}{0.00156 + 0.0005} = 1.247$

$\alpha_r = 48.16°$ $\sin \alpha_r = 0.745$ $\cos \alpha_r = 0.667$

Equation [3] $\tau_{\ell t} = (-\sigma_d) \sin \alpha_r \cos \alpha_r = (13.51)(0.745)(0.667) = 6.71 \text{MPa (973 psi)}$

Equation [4] $T = \tau_{\ell t}(2A_o t_d) = 6.71(2)(793\,000)(96) = 1022 \text{ kN m (9040 in. k)}$

Equation[7] $\gamma_{\ell t} = 2(\varepsilon_r - \varepsilon_d) \sin \alpha_r \cos \alpha_r = 2(0.00413 + 0.0005)(0.745)(0.667)$

$= 0.00460$

Equation [8] $\theta = \dfrac{p_o}{2A_o} \gamma_{\ell t} = \dfrac{3603}{2(793\,000)} 0.00460$

$= 0.01045 \text{ rad/m} (0.265 \times 10^{-3} \text{rad/in.})$

2. Select $\varepsilon_d = -0.0015$

$$\varepsilon_{ds} = 2(-0.0015) = -0.0030$$

Assume $\varepsilon_r = 0.00900$ after several cycles of trial-and-error:

Equation [14] $\zeta = \dfrac{0.9}{\sqrt{1 + 600\varepsilon_r}} = \dfrac{0.9}{\sqrt{1 + 600(0.00900)}} = 0.356$

$\dfrac{\varepsilon_{ds}}{\zeta \varepsilon_o} = \dfrac{-0.0030}{0.356(-0.002)} = 4.21 > 1 \text{ descending branch}$

Equation [13b] $k_1 = 0.798 \text{ (from Table 7.1)}$

Equation [12] $\sigma_d = k_1 \zeta f_c' = 0.798(0.356)(-41.3) = -11.73 \text{ MPa } (-1702 \text{ psi})$

Assume $t_d = 127 \text{ mm} (5.00 \text{ in.})$ after two or three trials

Equation [22] $A_o = 975\,000 - \dfrac{1}{2}(3987)(127) + (127)^2 = 738\,000 \text{ mm}^2 (1145 \text{ in.}^2)$

Eq. [23] $p_o = p_c - 4t_d = 3987 - 4(127) = 3479 \text{ mm} (137.0 \text{ in.})$

Solve ε_ℓ

Equation [20] $\varepsilon_\ell = \varepsilon_d + \dfrac{738\,000(0.0015)(11.73)}{(5031 f_\ell + 790 f_{\ell p})}$

Assume $\varepsilon_\ell = 0.00253 > \varepsilon_y$ after a few trials

Equation [15b] $f_\ell = 413$ MPa (60 000 psi)

$E'_{ps}(\varepsilon_{dec} + \varepsilon_\ell) = 214\,000(0.005 + 0.00253) = 1611$ MPa (233.6 ksi)

Equation [17b] $f_{\ell p} = \dfrac{E'_{ps}(\varepsilon_{dec} + \varepsilon_s)}{\left[1 + \left\{\frac{E'_{ps}(\varepsilon_{dec}+\varepsilon_s)}{f_{pu}}\right\}^m\right]^{\frac{1}{m}}} = \dfrac{1611}{\left[1 + \left\{\frac{1611}{1862}\right\}^4\right]^{\frac{1}{4}}}$

$= 1441$ MPa (209 ksi)

Then $\varepsilon_\ell = -0.0015 + \dfrac{738\,000(0.0015)(11.73)}{5031(413) + 790(1441)} = -0.0015 + 0.00404$

$= 0.00254 \approx 0.00253$ assumed. OK

Solve ε_t

Equation [21] $\varepsilon_t = -0.0015 + \dfrac{738\,000(203)(0.0015)(11.73)}{(3479)(284 f_t + 0)}$

Assume yielding

Equation [16a] $f_t = 413$ MPa(60 000 psi)

Then $\varepsilon_t = -0.0015 + 0.00646 = 0.00496 > \varepsilon_y$ OK

Check ε_r

Equation [24] $\varepsilon_r = \varepsilon_\ell + \varepsilon_t - \varepsilon_d = 0.00254 + 0.00496 + 0.0015$

$= 0.00900 = 0.00900$ assumed. OK

Check t_d

Equation [19] $t_d = \dfrac{738\,000}{3479}\left[\dfrac{(0.0015)(0.00900 + 0.0015)}{(0.00254 + 0.0015)(0.00496 + 0.0015)}\right]$

$= 127.8$ mm (5.03 in.) ≈ 127 mm assumed. OK

Calculate α_r, $\tau_{\ell t}$, T, $\gamma_{\ell t}$ and θ

Equation [25] $\tan^2 \alpha_r = \dfrac{0.00496 + 0.0015}{0.00254 + 0.0015} = 1.599$

$\alpha_r = 51.66° \sin\alpha_r = 0.784 \cos\alpha_r = 0.620$

Equation [3] $\tau_{\ell t} = (11.73)(0.784)(0.620) = 5.70$ MPa (827 psi)

Equation [4] $T = 5.70(2)(738\,000)(127) = 1068$ kN m (9453 in. k)

Equation [7] $\gamma_{\ell t} = 2(0.00900 + 0.0015)(0.784)(0.620) = 0.01021$

Equation [8] $\theta = \dfrac{3479}{2(738\,000)}0.01021 = 0.02406$ rad/m (0.611×10^{-3} rad/in.)

Table 7.3 Results of calculations for Example Problem 7.1

Variables		Eqs	Calculated values				
ε_d	10^{-3} selected		−0.250	−0.500	−1.000	−1.500	−1.750
ε_{ds}	10^{-3}	[11]	−0.500	−1.000	−2.000	−3.000	−3.500
ε_r	10^{-3} last assumed		2.25	4.14	7.34	9.00	9.29
ζ		[14]	0.587	0.482	0.387	0.356	0.351
$\varepsilon_d/\zeta\varepsilon_o$			0.426	1.037	2.59	4.21	4.99
k_1		[13]	0.365	0.679	0.842	0.798	0.741
σ_d	MPa	[12]	−8.87	−13.51	−13.48	−11.73	−10.76
t_d	mm last assumed		91	96	109	127	140
A_o	10^3 mm^2	[22]	802	793	770	738	717
p_o	mm	[23]	3624	3602	3553	3479	3429
ε_ℓ	10^{-3}	[20]	0.791	1.56	2.28	2.54	2.48
ε_t	10^{-3}	[21]	1.207	2.07	4.06	4.96	5.06
ε_r	10^{-3} checked	[21]	2.25	4.13	7.34	9.00	9.29
t_d	mm checked	[19]	91.2	96.0	109.0	127.8	140.2
α_r	degrees	[25]	49.79	48.16	51.16	51.66	51.76
$\tau_{\ell t}$	MPa	[3]	4.37	6.73	6.58	5.70	5.23
T	KN m	[4]	638	1027	1103	1068	1048
$\gamma_{\ell t}$	10^{-3}	[7]	2.47	4.60	8.15	10.21	10.73
θ	10^{-3} rad/m	[8]	5.59	10.43	18.78	24.06	25.67
f_ℓ	MPa	[15]	158	312	413	413	413
f_t	MPa	[16]	241	413	413	413	413
$f_{\ell p}$	MPa	[17]	1158	1311	1379	1441	1405
f_{tp}	MPa	[18]	0	0	0	0	0

The results of calculations shown above for $\varepsilon_d = -0.0005$ and -0.0015 are summarized in Table 7.3.

Table 7.3 also gives three additional cases of $\varepsilon_d = -0.00025$, -0.001 and -0.00175. Comparison of all five cases ($\varepsilon_d = -0.00025$, -0.0005, -0.001, -0.0015 and -0.00175) illustrates clearly the trends of all the variables. The relationship between the torsional moment T and the angle of twist θ is plotted in Figure 7.9.

Table 7.3 clearly shows that $\varepsilon_d = -0.0005$ closely represents the point of first yield of the longitudinal mild steel. At $\varepsilon_d = -0.001$ both the longitudinal and transverse mild steel are yielding and the torque resisted reaches a maximum shortly thereafter. When ε_d is increased further, the torsional resistance decreases. At $\varepsilon_d = -0.0015$ the calculated thickness t_d is equal to the actual wall thickness of 127 mm. When ε_d is taken as -0.00175, the calculated thickness $t_d = 140$ mm, i.e. greater than the actual thickness. This situation indicates that the calculation for $\varepsilon_d = -0.00175$ is actually invalid and that the torque–twist curve would drop off more quickly for $\varepsilon_d > -0.0015$ than is shown in Figure 7.9.

The angles of twist at first yield, θ_y, is 0.01043 rad/m (when $\varepsilon_d = -0.0005$) as given in Table 7.3. The maximum angle of twist θ_u for the given wall thickness of 127 mm is 0.02406 rad/m, when $\varepsilon_d = -0.0015$. Let us define the torsional ductility μ_t as the ratio θ_u/θ_y, then the

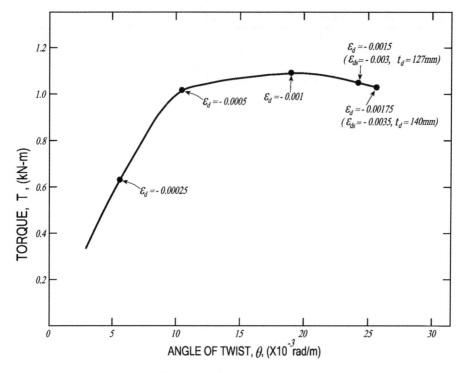

Figure 7.9   Torque–twist curve for example problem 7.1

torsional ductility of this box girder is:

$$\mu_t = \frac{\theta_u}{\theta_y} = \frac{0.02406}{0.01043} = 2.31.$$

It can be seen that the torsional ductility of reinforced concrete box sections is quite small. This is because the warping of the wall elements and the bending of the concrete struts produce large surface compressive strain ($\varepsilon_{ds} = -0.003$) at the maximum angle of twist θ_u.

7.2 Design for Torsion

7.2.1 Analogy between Torsion and Bending

A prismatic tube member of arbitrary cross-section is subjected to a torsional moment T, as shown in Figure 7.1. In Section 2.1.4.1 and Figure 2.6, Bredt (1896) showed that this external torsional moment T is resisted by an internal moment which is the product of the shear flow q in the shear flow zone and twice the lever arm area A_o:

$$T = q(2A_o) \tag{7.66}$$

A reinforced concrete 2-D element can be modeled by a truss, as shown in Figure 2.2 after cracking of concrete and yielding of steel. In Section 2.1.2.1, the shear flow q is expressed by the properties of the transverse steel in Equation (2.5):

$$q = \frac{A_t f_{ty}}{s} \tan \alpha_r \tag{7.67}$$

Inserting q from Equation (7.67) into Equation (7.66) we have

$$T_n = \frac{A_t f_{ty} \tan \alpha_r}{s}(2A_o) \tag{7.68}$$

Equation (7.68) is the fundamental equation for torsion in the truss model. Since A_o is defined by the centerline of shear flow, the crucial problem of finding A_o is to determine the thickness of the shear flow zone, t_d.

The analysis of torsion shown above is analogous to the analysis of bending in a prismatic member discussed in Section 3.2; in Figure 3.9, a rectangular cross-section is subjected to a nominal bending moment M_n. This external moment M_n is resisted by an internal bending moment which is the product of the resultant of the compressive stresses C in the compression zone of depth c and the lever arm jd:

$$M_n = C(jd) \tag{7.69}$$

The equilibrium of the forces requires that $C = T$, where T is the tensile force of longitudinal steel $A_s f_y$ after the cracking of the flexural member and the yielding of steel. Inserting $C = A_s f_y$ into Equation (7.69) gives:

$$M_n = A_s f_y(jd) \tag{7.70}$$

Equation (7.70) is the fundamental equation for bending. Since jd is defined by the position of the resultant C, the crucial problem of finding jd is to determine the depth of the compression zone c.

Equation (7.70) shows that the bending moment capacity M_n is equal to the longitudinal steel force $A_s f_y$ times the resultant lever arm jd. Similarly, in Equation (7.68) the torsional moment capacity T_n is equal to a certain stirrup force per unit length $(A_t f_{ty}/s) \tan \alpha_r$ times twice the lever arm area $2A_o$. In other words, the term of twice the lever arm area $2A_o$ in torsion is equivalent to the resultant lever arm jd in bending, and the shear flow q is similar to the resultant of compressive stresses C.

In bending, an increase of the nominal bending strength M_n due to increasing reinforcement results in an increase of the depth of the compression zone c, and a reduction of the resultant lever arm jd. The relationships among M_n, c and jd can be derived from equilibrium, compatibility and constitutive relationships of materials. Similarly in torsion, an increase of the nominal torsional strength T_n due to increasing reinforcement results in an increase of the thickness of shear flow zone t_d and a reduction of the lever arm area A_o. The relationships among T_n, t_d and A_o can also be derived from the equilibrium, compatibility and constitutive relationships of materials, as shown in Section 7.1. In this section we will derive A_o for torsion design through the determination of t_d.

7.2.2 Various Definitions of Lever Arm Area, A_o

When Rausch (1929) derived the basic torsion equation (7.68) assuming $\alpha_r = 45°$, a reinforced concrete member after cracking was idealized as a space truss with linear, one-dimensional members. Each diagonal concrete strut is idealized as a straight line lying in the center surface of the hoop bars. Hence, the lever arm area A_o is defined by the area within the center surface of the hoop bars. This definition of A_o, commonly denoted as A_1, has been adopted by the German Code (German Standard DIN 4334, 1958) and others. Using the bending analogy, this definition is equivalent to assuming that the resultant lever arm jd is defined as the distance between the centroid of the tension bars and the centerline of the stirrups in the compression zone. In terms of torsional strength this assumption is acceptable near the lower limit of the total steel percentage of about 1%, but becomes increasingly nonconservative with an increasing amount of steel, as shown in Figure 7.10. For a large steel percentage of 2.5–3% near the upper limit of under-reinforcement (both the longitudinal steel and stirrups reach

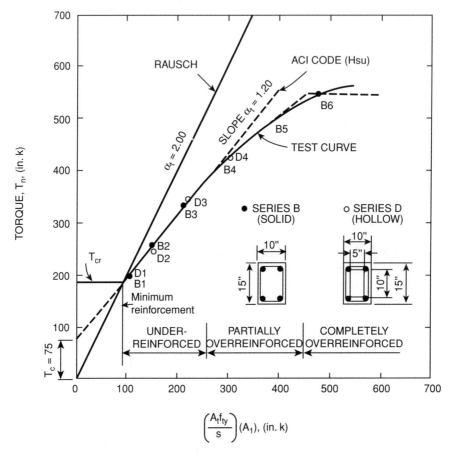

Figure 7.10 Comparison of Rausch's formula and ACI code formula with tests (1 in. = 25.4 mm; 1 in. k = 113 N m)

yielding), the over-prediction of torsional strength by Rausch's equation using A_1 exceeds 30%. This large error has two causes. First, the thickness of the shear flow zone t_d may be very large, of the order of one-quarter of the outer cross-sectional dimension, due to the softening of concrete. Second, in contrast to the bending strength M_n which is linearly proportional to the lever arm jd, the torsional strength T_n is proportional to the lever arm area A_o, which, in turn, is proportional to the square of the lever arm r. The lever arm r is shown in Figure 2.6(a) or Figure 7.2(a).

In order to reduce the nonconservative effect of using A_1 in Rausch's equation, Lampert and Thurlimann (1968) have proposed that A_o be defined as the area within the polygon connecting the centers of the corner longitudinal bars. This lever arm area is commonly denoted as A_2 and has appeared first in the CEB-FIP Model Code (1978). In terms of the bending analogy, this definition is equivalent to assuming that the lever arm jd is defined as the distance between the centroid of the tension bars and the centroid of the longitudinal compression bars. The introduction of A_2 has reduced the nonnconservative nature of Rausch's equation for high steel percentages. However, the assumption of a constant lever arm area (not a function of the thickness of shear flow zone) continues to be unsatisfactory.

Another way of modifying Rausch's equation has been suggested by Hsu (1968a, 1968b) and adopted by the early version of 1971 ACI Building Code (ACI 318-71).

$$T_n = T_c + \frac{A_t f_{ty}}{s}(\alpha_t A_1) \tag{7.71}$$

where:

$\alpha_t = 0.66 + 0.33 y_1/x_1 \le 1.5$;
$x_1 = $ shorter center-to-center dimension of a closed stirrup;
$y_1 = $ longer center-to-center dimension of a closed stirrup;
$T_c = $ nominal torsional strength contributed by concrete
$= 0.8x^2 y \sqrt{f_c'(psi)}$ where x and y are the shorter and longer sides, respectively, of a rectangular section.

Two modifications of Rausch's equation are made in Equation (7.71) based on tests. First, a smaller lever arm area $(\alpha_t/2)A_1$ is specified, where α_t varies from 1 to 1.5. Second, a new term T_c is added. This term represents the vertical intercept of a straight line in the $T_n - (A_t f_{ty}/s)(A_1)$ diagram (Figure 7.10).

These definitions of the lever arm areas A_1, A_2 or $(\alpha_t/2)A_1$ all have a common weakness. They are not related to the thickness of the shear flow zone or the applied torque. A logical way to define A_o must start with the determination of the thickness t_d of shear flow zone.

7.2.3 Thickness t_d of Shear Flow Zone for Design

The thickness of the shear flow zone t_d, given in Equation (7.55) or [19] is suitable for the analysis of torsional strength. It is, however, not useful for the design of torsional members. In design, the thickness t_d should be expressed in terms of the torsional strength T_n. This approach will now be introduced.

The stress in the diagonal concrete struts σ_d can be related to the thickness t_d and the shear flow q by inserting $\tau_{\ell t} = q/t_d$ into the equilibrium Equation (7.3):

$$\sigma_d = \frac{q}{t_d \sin \alpha_r \cos \alpha_r} \tag{7.72}$$

At failure σ_d in Equation (7.72) reaches the maximum $\sigma_{d,\max}$, while the torsional moment reaches the nominal capacity T_n. Substituting $q = T_n/2A_o$ at failure into Equation (7.72) gives:

$$t_d = \frac{T_n}{2A_o \sigma_{d,\max} \sin \alpha_r \cos \alpha_r} \tag{7.73}$$

If t_d is assumed to be small, the last term ξt_d^2 in Equation (7.46) is neglected, and A_o can be expressed by the thin-tube approximation:

$$A_o = A_c - \frac{1}{2} p_c t_d \tag{7.74}$$

Substituting A_o from Equation (7.74) into (7.73) and multiplying all the terms by $2p_c/A_c^2$ result in:

$$\left(\frac{p_c}{A_c} t_d\right)^2 - 2\left(\frac{p_c}{A_c} t_d\right) + \frac{T_n p_c}{A_c^2}\left(\frac{1}{\sigma_{d,\max} \sin \alpha_r \cos \alpha_r}\right) = 0 \tag{7.75}$$

Define:
$$t_{do} = A_c/p_c$$
$$\tau_n = T_n p_c/A_c^2$$
$$\tau_{n,\max} = \sigma_{d,\max} \sin \alpha_r \cos \alpha_r$$

Equation (7.75) becomes

$$\left(\frac{t_d}{t_{do}}\right)^2 - 2\left(\frac{t_d}{t_{do}}\right) + \frac{\tau_n}{\tau_{n,\max}} = 0 \tag{7.76}$$

When t_d/t_{do} is plotted against $\tau_n/\tau_{n,\max}$ in Figure 7.11, Equation (7.76) represents a parabolic curve. Solving t_d from Equation (7.76) gives:

$$t_d = t_{do}\left(1 - \sqrt{1 - \frac{\tau_n}{\tau_{n,\max}}}\right) \tag{7.77}$$

This approach of determining the thickness of the shear flow zone was first proposed by Collins and Mitchell (1980) and was later adopted by the 1984 Canadian Standard (CAN3-A23.3-M84); which gives:

$$t_d = \frac{A_1}{p_1}\left(1 - \sqrt{1 - \frac{T_n p_1}{0.7\varphi_c f_c' A_1^2}\left(\tan \alpha_r + \frac{1}{\tan \alpha_r}\right)}\right) \tag{7.78}$$

In Equation (7.78) A_c and p_c are replaced by A_1 and p_1, respectively, since the concrete cover is considered to be ineffective. $\sigma_{d,\max}$ is assumed to be $0.7\varphi_c f_c'$, in which the material reduction factor φ_c can be taken as 0.6.

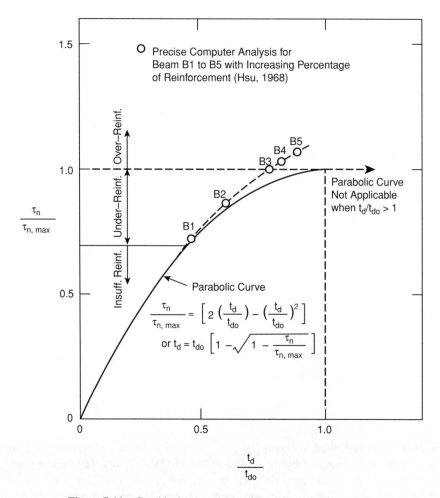

Figure 7.11   Graphical presentation of Equations (7.76) or (7.77)

Equations (7.78) and (7.77) clearly show that the thickness ratio t_d/t_{do} is primarily a function of the shear stress ratio τ_n/f_c'. The thickness ratio t_d/t_{do} is also a function of the cracking angle α_r, but is not sensitive when α_r varies in the vicinity of 45°.

In Equation (7.77), $\tau_n \leq \tau_{n,\max}$ represents the case of under-reinforcement, while $\tau_n > \tau_{n,\max}$ means over-reinforcement. The case of over-reinforcement can not be expressed by Equation (7.77), because it gives a complex number $\sqrt{-1}$. Figure 7.11 shows that Equation (7.77) is applicable when τ_n is less than about $0.9\tau_{n,\max}$. However, when τ_n exceeds $0.9\tau_{n,\max}$, t_d is increasing unreasonably fast. This problem reflects the difficulty in using the thin-tube approximation for A_o (Equation 7.74) to find t_d. When t_d exceeds about $0.7t_{do}$, the tube becomes so thick that the third term ξt_d^2 in Equation (7.46) cannot be neglected.

To avoid this weakness, a different approach is adopted. Using the RA-STM presented in Section 7.1, a computer program was written to analyze the torsional behavior of reinforced concrete members and to calculate the thickness t_d (Hsu and Mo, 1985b). This computer program was used to analyze the 61 eligible torsional members available in the literature.

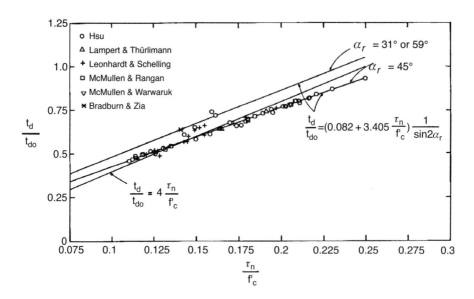

Figure 7.12   Thickness ratio t_d/t_{do} as straight line functions of shear stress ratio, $\tau_n/f'c$

The thicknesses t_d of these test beams are calculated and a linear regression analysis of the thickness ratios t_d/t_{do} is made as a function of τ_n/f'_c. This analysis provides the following expression (Hsu and Mo, 1985b):

$$t_d = \frac{A_c}{p_c}\left(0.082 + 3.405\frac{\tau_n}{f'_c}\right)\frac{1}{\sin 2\alpha_r} \qquad (7.79)$$

Equation (7.79) is plotted in Figure 7.12 for the cases of $\alpha_r = 45°$ and $\alpha_r = 31°$ or $59°$ which were the limits adopted by the CEB-FIP Model Code (1978). The 61 test points are also included and the correlation is shown to be excellent.

Although Equation (7.79) is found to be of excellent accuracy, it is considered to be somewhat unwieldy for practical design. In the next section a simplified expression for t_d is proposed. The simplicity is obtained with a small sacrifice in accuracy.

7.2.4 Simplified Design Formula for t_d

A simple expression for the thickness of shear flow zone t_d can be obtained directly from Bredt's equation, (7.8) or [4], noting that $T = T_n$ at the maximum load:

$$t_d = \frac{T_n}{2A_o\tau_{\ell t}} \qquad (7.80)$$

Assuming that $A_o = m_1 A_c$ and $\tau_{\ell t} = m_2 f'_c$, where m_1 and m_2 are nondimensional coefficients. Substituting them into Equation (7.80) gives

$$t_d = C_m\frac{T_n}{A_c f'_c} \qquad (7.81)$$

where $C_m = 1/(2m_1m_2)$. For under-reinforced members, m_1 varies from 0.55 to 0.85, while m_2 varies from 0.13 to 0.22. These values are obtained from the Appendix of Hsu and Mo's report (1983b). The low values of m_2 are due to the softening of concrete. For an increasing amount of reinforcement, m_2 increases while m_1 decreases. Therefore, the product m_1m_2 can be taken approximately as a constant 0.125, making C_m a constant of 4. Then

$$t_d = \frac{4T_n}{A_c f_c'} \tag{7.82}$$

Equation (7.82) is also plotted in Figure 7.12. Comparison of Equations (7.82) and (7.79) shows the difference to be small. Actually, Equation (7.82) can be considered as a simplification of Equation (7.79) by neglecting the small first term with constant 0.082, and increasing the constant in the second term from 3.405 to 4. The small effect of α_r is also neglected by taking $\sin 2\alpha_r = 1$, which is the exact value when $\alpha_r = 45°$.

Inserting Equation (7.82) into the thin-tube expression of A_o in Equation (7.74) and p_o in Equation (7.49) gives:

$$A_o = A_c - \frac{1}{2}p_c t_d = A_c - \frac{2T_n p_c}{A_c f_c'} \tag{7.83}$$

$$p_o = p_c - 4t_d = p_c - \frac{16T_n}{A_c f_c'} \tag{7.84}$$

The lever arm area A_o in Equation (7.83) is used in conjunction with Equation (7.68) to calculate the torsional strength T_n for the 61 beams available in the literature. The calculated values are plotted in Figures 7.13(a) and (b) and compared with the test values. The average $T_{n,test}/T_{n,calc}$ value is 1.013 and the standard deviation is 0.055.

The maximum thickness of the shear flow zone $t_{d,max}$ to ensure that the beam remains under-reinforced can be obtained by examining the results of six beams in series B of PCA tests

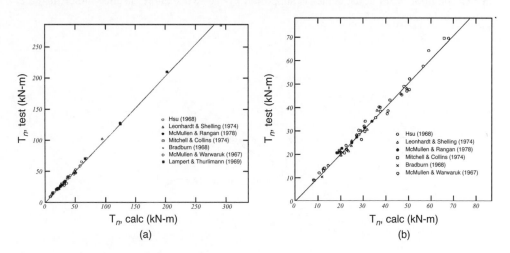

Figure 7.13 Comparison of test strengths with calculated strengths using proposed t_d (Equation 7.82); (b) is an expanded scale for the lower portion of (a)

(Hsu, 1968a). Figure 7.10 shows beams B1–B6 with increasing percentage of reinforcement, and the balanced percentage lies in between beams B3 and B4. The thickness ratios t_d/t_{do} for B3 and B4 are 0.77 and 0.82, respectively, calculated from the computer program. Taking a close average of 0.8 for these two beams gives the maximum thickness $t_{d,\max}$ as:

$$t_{d,\max} = 0.8t_{do} = 0.8\frac{A_c}{p_c} \tag{7.85}$$

The validity of Equation (7.85) can be observed in Figure 7.11 by the dotted curve connecting the test points B1 and B5. The dotted curve crosses from the under-reinforced region into the over-reinforced region at about $0.8t_{do}$. Series B is quite representative of the other series of test points.

7.2.5 Compatibility Torsion in Spandrel Beams

The ACI Code provides a so-called compatibility torsion for the design of spandrel beams in a statically indeterminate structure, when the torsional moment in the spandrel beam can be redistributed to other adjoining members after the formation of plastic hinges. Utilizing the limit design concept, this torsional plastic hinge can maintain a much smaller moment than that calculated by elastic analysis, thus resulting in a cost-effective, and yet very simple, design. This compatibility torsion, which is first introduced in Section 2.3.4.1, will now be studied in a more detailed and systematic manner.

7.2.5.1 Moment Redistribution

The redistribution of moment from a spandrel beam after torsional cracking to the adjoining floor systems will be illustrated by a three-dimensional structural frame, shown in Figure 7.14. The portion of the frame shown includes four columns, two spandrel beams and three floor beams. The floor beam at the center is supported by the two spandrel beams, rather than the

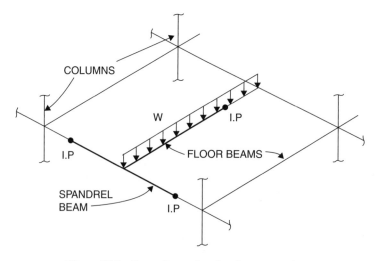

Figure 7.14   Space frame showing the test specimen

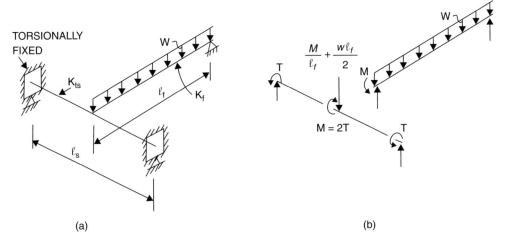

Figure 7.15   Test specimen

columns. When a uniform load w is applied on this floor beam, it will produce a rotation at the ends that in turn induces a torsional moment in the spandrel beams. The interaction of the floor beam and the spandrel beam can be studied using the T-shaped specimen indicated by heavy lines. This test specimen includes a spandrel beam between two inflection points and a floor beam from the joint to the inflection point. The three inflection points, indicated by solid dots, can be simulated by hinges in the tests.

Figure 7.15(a) shows the T-shaped test specimen resting on three spherical hinges and the floor beam loaded by a uniform load w. The ends of the spandrel beams are maintained torsionally fixed. This condition is more severe, and therefore more conservative, than that existing in the frame. It is adopted to simplify the analysis and the testing procedures.

The load w acting on the test specimen will create a negative bending moment at the continuous end of the floor beam due to the torsional restraint of the spandrel beam. If we separate the floor beam from the spandrel beam, as shown in Figure 7.15(b), this negative moment is designated as M acting on the end of the floor beam. The reaction of this moment M becomes a twisting moment acting on the midspan of the spandrel beam, thus creating a uniform torque T in the spandrel beam. The torque T is equal to $M/2$. The bending moment, shear and torque diagrams are drawn in Figure 7.16 from equilibrium conditions. It can be seen that all three diagrams can be plotted if the moment M is known. This moment will be called the joint moment.

The joint moment M can be determined from the compatibility of rotation at the joint while neglecting the vertical deflection of the spandrel beam:

$$M = \frac{\frac{1}{8}w\ell_f^2}{1 + \frac{3}{16}\frac{K_f}{K_{ts}}} \tag{7.86}$$

where:

K_f = flexural stiffness of the floor beam = $4EI/\ell_f$;
K_{ts} = torsional stiffness of spandrel beam = GC/ℓ_s, where C = St. Venant's torsional constant.

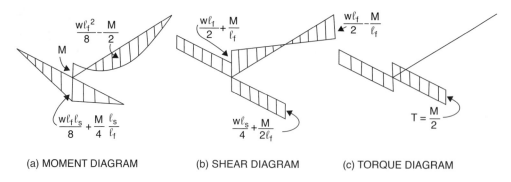

$$\frac{w\ell_f^2}{8} - \frac{M}{2}$$

$$M$$

$$\frac{w\ell_f\ell_s}{8} + \frac{M}{4}\frac{\ell_s}{\ell_f}$$

$$\frac{w\ell_f}{2} + \frac{M}{\ell_f}$$

$$\frac{w\ell_s}{4} + \frac{M}{2\ell_f}$$

$$\frac{w\ell_f}{2} - \frac{M}{\ell_f}$$

$$T = \frac{M}{2}$$

(a) MOMENT DIAGRAM   (b) SHEAR DIAGRAM   (c) TORQUE DIAGRAM

Figure 7.16   Moment, shear and torque diagrams of test specimen

Notice that the joint moment is a function of the stiffness ratio K_f/K_{ts}. If $K_f/K_{ts} = 0$, $M = (1/8)w\ell_f^2$. This is the special case in which the spandrel beam is infinitely rigid in torsion, and the joint moment becomes the fixed end moment. If $K_f/K_{ts} = \infty$, $M = 0$. This gives the other extreme case, in which the spandrel beam has no torsional stiffness. The floor beam is then simply supported, and the joint moment must be zero.

The joint moment M, calculated by Equation (7.86) using the uncracked elastic stiffness properties, should be applicable to the specimen before cracking. After cracking, however, K_{ts} drops drastically to a small fraction of the pre-cracking value (say 5%) while K_f only drops to say, 50% of the pre-cracking value. Therefore, the K_f/K_{ts} ratio increases by an order of magnitude after cracking. According to Equation (7.86), this change will cause a large decrease of the joint moment M compared with the uncracked elastic value. A decrease of M means that the torsional moment in the spandrel beam is reduced, accompanied by a corresponding increase of flexural moment in the floor beam near the midspan. This phenomenon, which can be viewed as a redistribution of moments from the spandrel beam to the floor beam, is known as 'moment redistribution after cracking of a spandrel beam.'

Moment redistribution after torsional cracking of a spandrel beam can be utilized to achieve economy and is the basis of the ACI torsional limit design method. This phenomenon will be illustrated by the tests of T-shaped test specimens, B1, B2 and B5, taken from series B in Hsu and Burton (1974). Figure 7.17 shows a photograph of beam B1 subjected to four concentrated loads evenly spaced to simulate a uniformly distributed load.

A typical T-shaped specimen in Figure 7.17 consists of a spandrel beam connected at midspan to a floor beam. The spandrel beam has a cross-section of 152 × 305 mm (6 × 12 in.) and a span of 2.74 m (9 ft). The floor beam has a cross section of 152 × 229 mm (6 × 9 in.) and a span of 2.74 m (9 ft). A steel torsion arm with a load cell at its tip was attached to one end of a spandrel beam in order to measure the torsional moment. Torsional moment is the load cell force that maintains zero torsional rotation times the distance from the load cell to the center line of the spandrel beam. The angle of twist is the rotation measured by a rotational LVDT at the midspan of spandrel beam divided by one-half of the span.

The three specimens, B1, B2 and B3, were all subjected to a design load $w\ell_f$ of 114 kN (25.6 kips), uniformly distributed along the floor beam. The design torques, however, are different for the three specimens. Specimen B1 was designed by elastic analysis based on uncracked sections. This method requires a nominal torsional stress, $\tau_n = 0.74\sqrt{f'_c(\text{MPa})}$ $(8.9\sqrt{f'_c(\text{psi})})$,

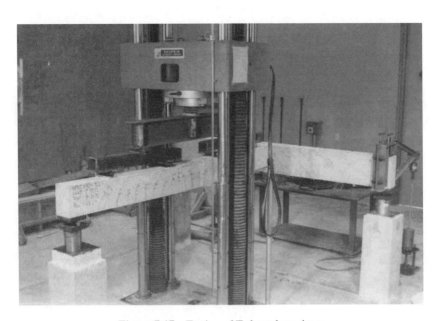

Figure 7.17   Testing of T-shaped specimen

resulting in a high web reinforcement index rf_y of 3.27 MPa (475 psi). In contrast, specimens B2 and B5 were designed by the limit design method using nominal stresses of $0.36\sqrt{f'_c}$(MPa) $(4.4\sqrt{f'_c}$(psi)) and $0.22\sqrt{f'_c}$(MPa) $(2.7\sqrt{f'_c}$(psi)), respectively. These stresses, in turn, give corresponding stirrups indices of 1.14 MPa (165 psi) and 0.41 MPa (60 psi), respectively.

Figure 7.18 shows the variation of torque with increase of load $w\ell_f$ for the three specimens. It can be seen that all three specimens behave in a reasonably elastic manner up to cracking. Upon cracking, the torques remain essentially constant while the load is increased. This confirms the existence of the moment redistribution after cracking. This redistribution of moments can also be observed in Figure 7.19, showing the torque–twist relationships of the three test specimens. Before cracking the torque–twist curve is approximately linear. After cracking, however, the torque–twist curves reach a horizontal plateau, where the angle of twist increases under a constant torque. In other words, plastic hinges are formed in the spandrel beam after cracking. When sufficient stirrups were provided, as in specimens B1 and B2, the plastic hinge in the spandrel beam permitted the floor beam to bend until the longitudinal steel yielded near midspan.

In specimen B1, where an excessive amount of stirrups was provided, the plastic hinges in the spandrel beam not only allowed the floor beam to yield, but also allowed the torque to increase rapidly after the yielding of the floor beam (see Figure 7.18). This phenomenon can be considered as a redistribution of moments from the floor beam back to the spandrel beam and is known as the 'second resdistribution of moment after yielding of a floor beam.' Although this second redistribution of moments may allow us to utilize the ultimate torsional strength of the web reinforcement in the spandrel beam, Figure 7.18 shows that the increase of load is small after the yielding of the floor beam. This small gain in load also occurs at very large angles of twist (as indicated in Figure 7.19) which is undesirable from a serviceability point of view. In short, from an economic and practical point of view, this second redistribution

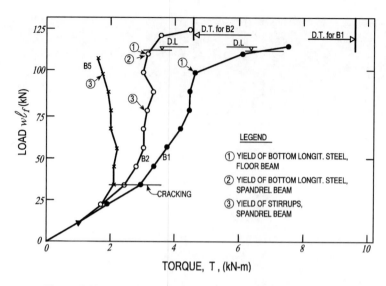

Figure 7.18   Load–torque curves for specimens B1, B2 and B5

of moment after yielding of the floor beam should not be utilized. The excessive amount of web reinforcement in the spandrel beams required by elastic analysis is therefore wasteful and unnecessary.

In specimen B5, an insufficient amount of web reinforcement was provided. Figure 7.19 shows that torques decrease after cracking. Figure 7.20 also shows the brittle torsional failure that occurred in the spandrel beam before the yielding of longitudinal bars in the floor beam.

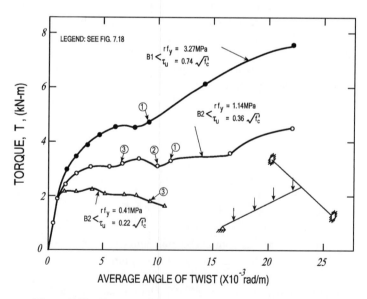

Figure 7.19   Torque–twist curves for specimens B1, B2 and B5

Figure 7.20   Torsion failure of specimen B5

Therefore, the ultimate design load cannot be reached. Such a design would be unsafe from the viewpoint of both strength and ductility.

Comparison of the three specimens shows clearly that specimen B2 has the most desirable design, assuming a torsional design stress τ_n of about $0.33\sqrt{f_c'(\text{MPa})}$ $(4\sqrt{f_c'(\text{psi})})$. This specimen behaves in a ductile manner, can reach its strength, and can satisfy the serviceability condition on crack width control. This limit design method is not only very economical, but also very simple. As a result, it has been adopted in the ACI building Code (ACI 318-77) since 1977.

7.2.5.2 Interaction of Torsion and Shear

Figure 7.21 shows a reinforced concrete space frame, where three floor beams are framing into each span of a continuous spandrel beam. The T-shaped test specimen described in Section 7.2.5.1 and Figure 7.15(a) is indicated by heavy dotted lines. The spandrel beam in this T-shaped specimen represents the portion of the spandrel beam near the midspan where the flexural shear stresses are small. The flexural shear stresses in all three test specimens B1, B2 and B5, are less than $0.166\sqrt{f_c'(\text{MPa})}$ $(2\sqrt{f_c'(\text{psi})})$, the contribution of concrete. The critical sections, however, are near the column faces where large shear stresses v_n as well as large torsional stresses τ_n occur simultaneously. How should we design the web reinforcement to resist the combined actions of large v_n and large τ_n? In order to answer this question, we will now study the portion of the space frame adjacent to a column, i.e. the U-shaped specimen consisting of the negative moment region of the spandrel beam and the two floor beams, as shown by the heavy solid lines in Figure 7.21. Such a test specimen with two floor beams, however, is difficult to manufacture and to test.

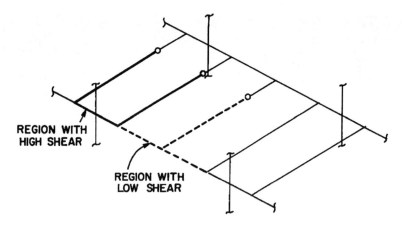

Figure 7.21 High shear region of spandrel beam in frames

Figure 7.22(a) shows the shear diagram, the bending moment diagram and the torque diagram of the spandrel beam in the portion of frame isolated in Figure 7.21. These three diagrams can be created in the spandrel beam of a much simpler T-shaped test specimen with only one floor beam, as shown in Figure 7.22(b). This creation of a new test specimen can be done by moving the two sets of floor beam loading ($P_s/2$ and $w/2$) in Figure 7.22(a) to

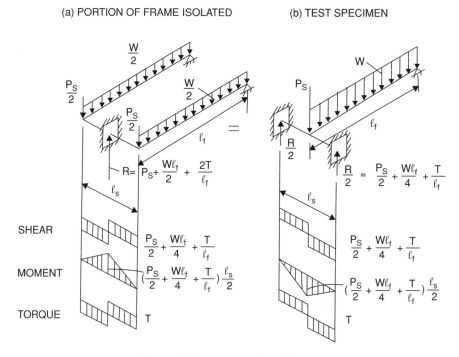

Figure 7.22 Test specimen selected

the midspan of the spandrel beam, while splitting and moving the reaction force R to the two ends of the spandrel beam. When the new set of three diagrams for the spandrel beam in Figure 7.22(b) is compared with the old set in Figure 7.22(a), it can be seen that they are identical, except that all the signs are reversed. These reversals of sign are obviously caused by the interchange of loads and reactions. This interchange of loads and reactions, however, has not changed the magnitudes of the bending moment, shear force and torsional moment in the spandrel beam, nor the relationships among them.

The T-shaped test specimen in Figure 7.22(b) is similar to that given in Figure 7.15(a), except in two aspects. First, an additional concentrated load P_s is applied at the midspan of the spandrel beam, and second, the spandrel beam is much shorter. Both these characteristics are needed to create large shear stresses in the spandrel beam.

Four T-shaped specimens with concentrated load P_s (Figure 7.22b), are taken from series C of Hsu and Hwang (1977) to demonstrate the interaction of high shear and high torque in spandrel beams. Each of the four T-shaped specimens C1, C2, C3 and C4 consists of a spandrel beam and a floor beam. The spandrel beam has a cross section of 152 × 305 mm (6 × 12 in.) and a span of 1.37 m (4.5 ft). The floor beam has a cross-section of 152 × 229 mm (6 × 9 in.) and a span of 2.74 m (9 ft). The design floor beam load $w\ell_f$ was 114 kN (25.6 kips) and the spandrel beam load P_s was 0 kN (0 kips), 62.3 kN (14.0 kips), 173 kN (38.9 kips) and 110 kN (24.7 kips) for specimens C1, C2, C3 and C4, respectively. As a result, the longitudinal reinforcement ratio of the spandrel beams varies from 0.5 to 1.88%, a range normally used in practice.

Specimens C1 and C2 were designed by the torsional limit design method. A design torsional stress τ_n of $0.33\sqrt{f_c'(\text{MPa})}$ $(4\sqrt{f_c'(\text{psi})})$ was assumed for specimen C2, as specified by the ACI Code. The calculated torsional stirrup was added to the shear stirrup, and the calculated torsional longitudinal steel was added to the flexural longitudinal steel. An identical method was used for specimen C1, except that a larger design torsional stress τ_n of $0.50\sqrt{f_c'(\text{MPa})}$ $(6\sqrt{f_c'(\text{psi})})$ was used. In order to maintain a design load $w\ell_f$ of 114 kN (25.6 kips) and the same stirrup index as that of C2, the spandrel beam load P_s for C1 was reduced to zero.

Specimens C3 and C4 were designed by an alternative method with two characteristics: (1) a minimum stirrup index rf_y of 0.98 MPa (142 psi) is specified for torsion; and (2) no interaction between torsion requirement and shear requirement. The stirrup index of the spandrel beam is obtained from the larger of the torsion requirement or the shear requirement.

Specimen C4 was designed by the minimum stirrup index $rf_y = 0.98$ MPa (142 psi) for torsion. The stirrup index required by shear is neglected because it is less than that required by torsion. Specimens C3 was designed by a stirrup index of 186 MPa (270 psi) required by a large shear force. The minimum stirrup index required by torsion is neglected.

The torque–twist curves for these four specimens are given in Figure 7.23. Specimen C2 behaved as a specimen with optimum web reinforcement in the spandrel beam. After diagonal cracking in the spandrel beam, a torsional plastic hinge was formed. The large increase of the angle of twist allowed the moment to be redistributed from the spandrel beam to the floor beam. Failure was preceded by the yielding of stirrups and longitudinal steel in the spandrel beam (represented by the circled numbers 2 and 3 in Figure 7.23). This was followed by the yielding of the bottom longitudinal steel near the midspan of the floor beam (represented by the circled number 1), and the ductile collapse of the whole specimen. Two observations are evident: (1) the development of torsional hinges in the spandrel beams was accompanied by a slight decrease of torque; and (2) the bottom longitudinal reinforcement in the floor beam just reached its yield strain at collapse. In view of these observations, it can be concluded that

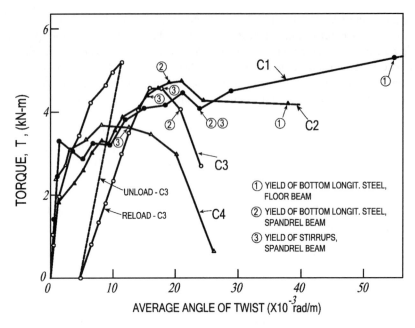

Figure 7.23 Torque–twist curve for series C

the web reinforcement of the spandrel beam designed by the ACI limit design method using a torsional stress τ_n of $0.33\sqrt{f_c'}$(MPa) ($4\sqrt{f_c'}$(psi)) was just enough to provide the required ductility, and that the most economical design had been achieved.

Specimen C1 behaved as a specimen with excessive reinforcement web reinforcement in the spandrel beam. The process of moment redistribution and the mode of failure were similar to those of specimen C2. However, the development of torsional plastic hinges in the spandrel beam was accompanied by a continuous increase of torque, and some of the web reinforcement in the spandrel beam did not yield at failure. Specimen C1 behaved satisfactorily from the viewpoint of the limit design concept, but the design torsional stress τ_n of $0.50\sqrt{f_c'}$(MPa) ($6\sqrt{f_c'}$(psi)) is obviously higher than necessary.

In contrast, specimens C3 and C4, which were designed by neglecting the interaction of torsion and shear, behaved as specimens with inadequate web reinforcement in the spandrel beam. Instead of a flexural failure in the floor beam, collapse of these specimens was caused by a premature shear–torsion failure of the spandrel beam. As shown in Figure 7.23, the torques reached their maximum at the yielding of stirrups (see circled number 3) in the spandrel beam and then decreased rapidly. Since the angle of twist of the torsional plastic hinges was insufficient to allow a redistribution of moment from the spandrel beam to the floor beam, no yielding was observed in the bottom longitudinal steel of the floor beams, and the design floor beam load $w\ell_f$ of 114 kN (25.6 kips) could not be reached.

In summary, we can conclude that the ACI limit design method of using a design torsional stress τ_n of $0.33\sqrt{f_c'}$(MPa) ($4\sqrt{f_c'}$(psi)) is applicable to high shear regions of a continuous spandrel beam near the columns. In the regions of high torsion and high shear, the torsional web reinforcement calculated from the specified torsional stress should be added to the web reinforcement required by flexural shear. The calculated torsional longitudinal steel should be added to the longitudinal steel required by bending moments.

7.2.6 Minimum Longitudinal Torsional Steel

In sections 2.3.1.2 and 2.3.1.3, a pair of equations (2.104) and (2.105), are given for the design of longitudinal torsional steel. Equation (2.104) was derived to ensure the torsional strength, while Equation (2.105) was given to calculate the minimum longitudinal torsional steel to ensure ductility. In this section, we will derive Equation (2.105).

In order to avoid a brittle torsional failure, a minimum amount of torsional reinforcement (including both transverse and longitudinal steel) is required in a member subjected to torsion. The basic criterion for determining this minimum torsional reinforcement is to equate the post-cracking strength T_n to the cracking strength T_{cr}:

$$T_n = T_{cr} \qquad (7.87)$$

Taking the angle $\alpha_r = 45°$, the post-cracking torsional strength T_n of both solid and hollow sections can be predicted from Equation (2.99):

$$T_n = \frac{2 A_o A_t f_{yt}}{s} \qquad (7.88)$$

The cracking torque of solid sections subjected to combined torsion, shear and bending T_{cr} can be predicted by the compatibility torsion given in Section 7.2.5:

$$T_{cr} = 0.33\sqrt{f_c'(\mathrm{MPa})}\frac{A_{cp}^2}{p_{cp}} \text{ or } 4\sqrt{f_c'(\mathrm{psi})}\frac{A_{cp}^2}{p_{cp}} \qquad (7.89)$$

where the ACI notations A_{cp} and p_{cp} are the same as A_c and p_c, respectively, in Section 7.1.3.4. The concrete strength f_c' and $\sqrt{f_c'}$ are in MPa or psi. In the case of hollow sections, Mattock (1995) suggested a simple relationship between the cracking torque of a hollow section $T_{cr,hollow}$ and that of a solid section with the same outer dimensions $T_{cr,solid}$:

$$\frac{T_{cr,hollow}}{T_{cr,solid}} = \frac{A_g}{A_{cp}} \qquad (7.90)$$

where A_g is the cross-sectional area of the concrete only and not including the hole(s), while A_{cp} is the area of the same hollow section including the hole(s). For solid sections, $A_g = A_{cp}$. The cracking torque of solid and hollow sections can then be expressed by one equation:

$$T_{cr} = 0.33\sqrt{f_c'(\mathrm{MPa})}A_g\frac{A_{cp}}{p_{cp}} \text{ or } 4\sqrt{f_c'(psi)}A_g\frac{A_{cp}}{p_{cp}} \qquad (7.91)$$

Inserting Equations (7.91) and (7.88) into Equation (7.87) gives:

$$A_t f_{yt} = \frac{0.167\sqrt{f_c'(\mathrm{MPa})}A_g A_{cp} s}{A_o p_{cp}} \qquad (7.92)$$

Defining p_h as the centerline of the perimeter of the outermost stirrups, and multiplying both sides of Equation (7.92) by p_h/s give

$$A_t f_{yt} \frac{p_h}{s} = 0.167\sqrt{f'_c(\text{MPa})}A_g \left(\frac{A_{cp}}{A_o}\right)\left(\frac{p_h}{p_{cp}}\right) \tag{7.93}$$

The ratios A_{cp}/A_o and p_h/p_{cp} on the right-hand side of Equation (7.93) vary somewhat with the size of cross-sections, because the concrete cover is usually a constant specified by the code for fire and corrosion protections. For cross-sections normally used in buildings, the ratios, A_{cp}/A_o and p_h/p_{cp}, can be taken as 1.5 and 0.83, respectively. Equation (7.93) can then be simplified to:

$$A_t f_{yt} \frac{p_h}{s} = 0.21\sqrt{f'_c(\text{MPa})}A_g \tag{7.94}$$

Minimum torsional reinforcement requires not only transverse steel, but also longitudinal steel. The total yield force of longitudinal torsional steel is obtained from Equation (2.104) assuming $\theta = 45°$:

$$A_\ell f_{y\ell} = A_t f_{yt} \frac{p_h}{s} \tag{7.95}$$

Substituting $A_t f_{yt} p_h/s$ from Equation (7.94) into (7.95) gives

$$A_\ell f_{y\ell} = 0.21\sqrt{f'_c(\text{MPa})}A_g \tag{7.96}$$

Adding Equations (7.94) and (7.96) results in

$$A_\ell f_{y\ell} + A_t f_{yt} \frac{p_h}{s} = 0.42\sqrt{f'_c(\text{MPa})}A_g \tag{7.97}$$

The symbol A_ℓ in Equation (7.97) is the total area of minimum longitudinal steel $A_{\ell,\min}$. Rearranging Equation (7.97) gives the ACI equation of Equation (2.105):

$$A_{\ell,\min} = \frac{0.42\sqrt{f'_c(\text{MPa})}A_g}{f_{y\ell}} - \left(\frac{A_t}{s}\right)p_h \frac{f_{yt}}{f_{y\ell}} \tag{7.98}$$

To limit the value of $A_{\ell,\min}$, the transverse steel area per unit length A_t/s in the second term on the right-hand side of Equation (7.98) needs to be taken not less than $0.172(\text{MPa}) b_w/f_{yt}$.

7.2.7 Design Examples 7.2

A reinforced concrete beam with a rectangular cross-section 0.508 m (20 in.) wide by 0.762 m (30 in.) high is subjected to a nominal torsional moment T_n. The net concrete cover is 38 mm (1.5 in.) and the material strengths are $f'_c = 27.6$ MPa (4000 psi) and $f_y = 413$ MPa (60 000 psi). Design the beam subjected to: (1) $T_n = 136$ kN m (1200 in. kips); and (2) $T_n = 56.5$ kN m (500 in. kips).

Solution (1): $T_n = 136 \, kN \, m \, (1200 \, in. \, kips)$

Sectional properties

$$b = 0.508 \, \text{m}(20 \, \text{in.})$$
$$h = 0.762 \, \text{m}(30 \, \text{in.})$$
$$A_c = A_{cp} = bh = (508)(762) = 387\,100 \, \text{mm}^2 \, (600 \, \text{in.}^2)$$
$$p_c = p_{cp} = 2(b + h) = 2(508 + 762) = 2540 \, \text{mm} \, (100 \, \text{in.})$$
$$\frac{A_c^2}{p_c} = \frac{(387\,100)^2}{2540} = 59.0 \times 10^6 \, \text{mm}^3$$

Check threshold torque

$$\tau_{n,threshold} = 0.083\sqrt{f_c'(\text{MPa})} = 0.083\sqrt{27.6} = 0.436 \, \text{MPa}$$
$$T_{n,threshold} = 0.083\sqrt{f_c'(\text{MPa})}\frac{A_c^2}{p_c} = (0.436)(59 \times 10^6) = 25.7 \, \text{kN} \, \text{m}$$
$$T_n = 136 \, \text{kN} \, \text{m} > 25.7 \, \text{kN} \, \text{m}.$$

Torsion needs to be considered.

$$\tau_n = (136/25.7)(0.083\sqrt{f_c'(\text{MPa})}) = 0.439\sqrt{f_c'(\text{MPa})} > 0.33\sqrt{f_c'(\text{MPa})}$$

The nominal torsional stress $0.439\sqrt{f_c'(\text{MPa})}$ is greater than the cracking torsional stress $0.33\sqrt{f_c'(\text{MPa})}$.

Calculate t_d, A_o and p_o

$$t_d = \frac{4T_n}{A_c f_c'} = \frac{4(136 \times 10^6)}{(387.1 \times 10^3)(27.6)} = 51 \, \text{mm} \, (2.00 \, \text{in.})$$
$$A_o = A_c - \frac{1}{2}p_c t_d = 387\,100 - 2\,540(51)/2 = 322\,000 \, \text{mm}^2 \, (499 \, \text{in.}^2)$$
$$p_o = p_c = 4t_d = 2540 - 4(51) = 2336 \, \text{mm} \, (92.0 \, \text{in.})$$

Design of stirrups

$$\frac{A_t}{s} = \frac{T_n}{2A_o f_{ty}}\cot\alpha_r = \frac{136 \times 10^6}{2(322\,000)(413)}(\cot\alpha_r)$$
$$= 0.511(\cot\alpha_r)\text{mm}^2/\text{mm} \, (0.0201(\cot\alpha_r) \, \text{in.}^2/\text{in.})$$

The angle of cracking direction α_r can be chosen to suit the designer's purpose. The ACI Code requires that $30° \leq \alpha_r \leq 60°$ to prevent excessive cracking. For best crack control, α_r should be taken as 45°.

Select $\alpha_r = 45°$, $\cot\alpha_r = 1$:

$$\frac{A_t}{s} = 0.511 \, \text{mm}^2/\text{mm} \, (0.0201 \, \text{in.}^2/\text{in.})$$

Select No. 4 bars $s = \dfrac{129}{0.511} = 252 \, \text{mm} \, (10.0 \, \text{in.})$

Check ACI maximum stirrup spacing: $s < p_h/8$

$$p_h = 2540 - 8(38 + 6) = 2188 \text{ mm},$$
$$p_h/8 = 2188/8 = 273 \text{ mm} > 252 \text{ mm OK}$$

Use No. 4 stirrups at 250 mm (10.0 in.) spacing.

Design of longitudinal steel

$$A_\ell = \left(\frac{A_t}{s}\right) p_h \frac{f_{yt}}{f_{y\ell}} \tan^2 \alpha_r = \left(\frac{A_t}{s}\right) p_h (1)(1) = (0.511)(2188)$$
$$= 1118 \text{ mm}^2 (1.73 \text{ in.}^2) \text{ governs}$$

$$A_{\ell,min} = \frac{0.42\sqrt{f_c'(\text{MPa})} A_{cp}}{f_{y\ell}} - \left(\frac{A_t}{s}\right) p_h \frac{f_{yt}}{f_{y\ell}}$$
$$= \frac{0.42\sqrt{27.6}(387.1 \times 10^3)}{413} - (0.511)(2188)(1)$$
$$= 2068 - 1118 = 950 \text{ mm}^2 < 1118 \text{ mm}^2 \text{ does not govern}$$

Use 10 No. 4 longitudinal bars so that spacing will be less than 305 mm (12 in.) specified by the ACI Code.

$$A_\ell = 10(129) = 1290 \text{ mm}^2 (2.0 \text{ in.}^2) > 1118 \text{ mm}^2 (1.73 \text{ in.}^2) \text{ OK}$$

This example shows that the minimum torsional longitudinal steel $A_{\ell,min}$ does not govern, when the applied torsional stress (τ_n) is *greater* than the cracking torsional stress $0.33\sqrt{f_c'(\text{MPa})}$ $(4\sqrt{f_c'(\text{psi})})$. When τ_n is much greater than $0.33\sqrt{f_c'(\text{MPa})}$, $A_{\ell,min}$ could become negative, but still have the same meaning: 'does not govern'.

Solution (2): $T_n = 56.5 \text{ kN m}$ (500 in. kips)
Sectional properties: same as (1)
Check threshold torque

$$T_{n,threshold} = 0.083\sqrt{f_c'(\text{MPa})} \frac{A_c^2}{p_c} = (0.436)(59.0 \times 10^6) = 25.7 \text{ kN m}$$
$$T_n = 56.5 \text{ kN m} > 25.7 \text{ kN m}$$

Torsion needs to be considered.

$$\tau_n = (56.5/25.7)(0.083\sqrt{f_c'(\text{MPa})} = 0.183\sqrt{f_c'(\text{MPa})} < 0.33\sqrt{f_c'(\text{MPa})}$$

The nominal torsional stress, $0.183\sqrt{f_c'(\text{MPa})}$, is much less than the compatibility torsional stress of $0.33\sqrt{f_c'(\text{MPa})}$.

Calculate t_d, A_o and p_o

$$t_d = \frac{4T_n}{A_c f_c'} = \frac{4(56.5 \times 10^6)}{(387.1 \times 10^3)(27.6)} = 21 \text{ mm} (0.83 \text{ in.})$$
$$A_o = A_c - \frac{1}{2} p_c t_d = 387\,100 - 2540(21)/2 = 360\,000 \text{ mm}^2 (558 \text{ in.}^2)$$
$$p_o = p_c - 4t_d = 2540 - 4(21) = 2456 \text{ mm} (96.7 \text{ in.})$$

Design of stirrups

Select $\alpha_r = 45°$, $\cot \alpha_r = 1$:

$$\frac{A_t}{s} = \frac{T_n}{2A_o f_{ty}} = \frac{56.5 \times 10^6}{2(360\,000)(413)}$$

$$= 0.190 \text{ mm}^2/\text{mm } (0.0075 \text{ in.}^2/\text{in.})$$

Select No. 3 bars

$$s = \frac{71}{0.190} = 373 \text{ mm } (14.7 \text{ in.})$$

Check ACI maximum stirrup spacing: $s < p_h/8$

$$p_h = 2540 - 8(38 + 6) = 2188 \text{ mm},$$
$$p_h/8 = 2188/8 = 273 \text{ mm governs}$$
$$s_{\max} = 305 \text{ mm } (12 \text{ in.})$$

Use No. 3 stirrups at 270 mm (10.6 in.) spacing.

Design of longitudinal steel

$$A_\ell = \left(\frac{A_t}{s}\right) p_h \frac{f_{yt}}{f_{y\ell}} \tan^2 \alpha_r = \left(\frac{A_t}{s}\right) p_h (1)(1) = (0.190)(2188)$$

$$= 416 \text{ mm}^2 (0.64 \text{ in.}^2)$$

$$A_{\ell,\min} = \frac{0.42\sqrt{f_c'(\text{MPa})} A_{cp}}{f_{y\ell}} - \left(\frac{A_t}{s}\right) p_h \frac{f_{yt}}{f_{y\ell}}$$

$$= \frac{0.42\sqrt{27.6}(387.1 \times 10^3)}{413} - (0.190)(2188)(1)$$

$$= 2068 - 416 = 1652 \text{ mm}^2 (2.56 \text{ in.}^2) \text{ governs}$$

Use 10 No. 5 longitudinal bars so that spacing will be less than 305 mm (12 in.) specified by the ACI Code.

$$A_{\ell,\min} = 10(200) = 2000 \text{ mm}^2 (3.1 \text{ in.}^2) > 1652 \text{ mm}^2 (2.56 \text{ in.}^2) \text{ OK}$$

This example shows that the minimum torsional longitudinal steel $A_{\ell,\min}$ will govern, when the applied torsional stress (τ_n) is *less* than the cracking torsional stress of $0.33\sqrt{f_c'(\text{MPa})}$ $(4\sqrt{f_c'(\text{psi})})$. In this low range of torsional stress τ_n, the ACI Code allows the volume of stirrups to be reduced with decreasing τ_n to $A_t/s = 0.190 \text{ mm}^2/\text{mm}$. However, a compensating volume of longitudinal steel must be increased to $A_{\ell,\min} = 1652 \text{ mm}^2$ in order to ensure torsional ductility. This way of treating limit design is economical, because the cost of stirrups is high while the cost of longitudinal steel is low.

8

Beams in Shear

8.1 Plasticity Truss Model for Beam Analysis

8.1.1 Beams Subjected to Midspan Concentrated Load

A plasticity truss model of a beam subjected to a concentrated load at midspan is shown in Figure 8.1(a). The truss model is made up of two parallel top and bottom stringers, a series of transverse steel bars spaced uniformly at a spacing of s, and a series of diagonal concrete struts. The beam can be divided into two types of regions (main and local) depending on the α_r angle of the diagonal concrete struts. The main region is the one where the α_r angle is a constant, so that the series of diagonal struts are parallel to each other. Regularity of the truss in this region makes simple analysis possible based on the sectional actions of M, V, T and P. Analysis of elements from such regions has been made in Sections 2.1 and 2.2. The local region is the one where the α_r angle varies. Such regions lie in the vicinity of the concentrated loads, including the end reaction region and the midspan region under the load.

In the end reaction region, the local effect is shown in Figure 8.1(a) as the 'fanning' of the compression struts from the application point of the concentrated load. In this case the angle varies from 90° to α_r and each concrete strut has a narrow triangular shape. Since the angle of the principal compressive stress is assumed to coincide with the angle of a concrete strut, the compressive forces will radiate upward from the load application point to form truss action with the forces in the top stringer and in the transverse steel bars. Similarly, Figure 8.1(a) also shows the 'fanning' of the compression struts under the concentrated load at midspan. Because the load application point is on the top surface, the compressive forces radiate downward and form truss action with the forces in the bottom stringer and in the transverse steel bars.

Now let us analyze how the forces vary in the transverse and longitudinal reinforcement in the main regions of the beam. The depth of the beam is represented by d_v, the distance between the centerlines of top and bottom stringers. To simplify the analysis we select the following three geometric relationships: (1) angle of concrete struts $\tan \alpha_r = 5/3$; (2) spacing of stirrups $s = d_v/3$; and (3) half-span length $\ell/2 = 4(d_v \tan \alpha_r) = (20/3)d_v = 20s$. The α_r angle is close to 60°, the maximum allowed by the ACI Code and should be the most economical. The spacing s and the half-span $\ell/2$ are typical in practice.

Unified Theory of Concrete Structures   Thomas Hsu and Yi-Lung Mo
© 2010 John Wiley & Sons, Ltd

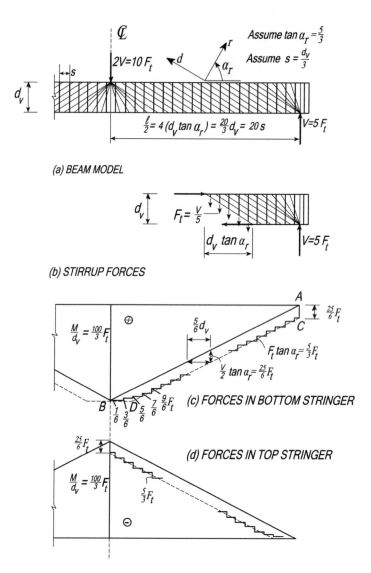

Figure 8.1   Plasticity truss model for a beam with midspan concentrated load

8.1.1.1 Transverse Stirrup Force

In the plasticity truss model theory for beams, we have assumed that the shear force in a cross-section is uniformly distributed along the depth of the beam, i.e. shear flow is constant in the transverse direction. In a beam subjected to concentrated load at midspan, the shear force should also be constant along the length of the beam within one half-span, i.e. shear flow should also be constant in the longitudinal direction. Since any element in the web in the main regions of the beam is subjected to an identical pure shear condition, the stirrup force should be constant along its own length (transverse direction), as well as throughout the length of the beam (longitudinal direction).

By definition, the force in one stirrup F_t should be $n_t s$. Recalling $n_t = (V/d_v)\cot\alpha_r$ from beam shear equation in Table 2.1 we have

$$F_t = n_t s = V\left(\frac{s}{d_v}\right)\cot\alpha_r \qquad (8.1)$$

Adopting the assumed geometric relationship of $\cot\alpha_r = 3/5$ and $s = d_v/3$, a very simple equation for F_t results:

$$F_t = \frac{V}{5} \qquad (8.2)$$

This simple value of F_t can also be obtained by taking vertical equilibrium of the free body shown in Figure 8.1(b). Since F_t is a constant throughout the beam, we will use it as a reference force to measure other forces in the beam.

8.1.1.2 Forces in Bottom and Top Stringers

Forces in the bottom stringers are contributed by the bending moment and the shear force according to Equation (2.52): $N_{b\ell} = M/d_v + (V/2)\tan\alpha_r$. Since the bending moment varies linearly along the length of a beam subjected to a midspan load, the force caused by bending should have a triangular shape in one half-span as shown by the solid line AB in Figure 8.1(c). Adopting the length of one half-span as $(20/3)d_v$, the maximum stringer force at midspan is $V(\ell/2)/d_v = 5F_t(20/3)d_v/d_v = (100/3)F_t$.

Since the shear force V is a constant along the beam, the stringer force due to shear should also be a constant and equal to $(V/2)\tan\alpha_r = (5F_t/2)(5/3) = (25/6)F_t$. The sum of the two stringer forces due to bending and shear is then represented by the dotted line CD, which is displaced vertically from the solid line by a distance of $(25/6)F_t$. In actuality, of course, the stirrups are not uniformly smeared, but are concentrated at discrete points with spacing s. Therefore, the stringer force contributed by shear should change at each stirrup and should have a stepped shape as indicated. Each step of change should introduce a stringer force of $F_t\tan\alpha_r$. For the main region of the beam, $\tan\alpha_r = 5/3$ and each step is $(5/3)F_t$.

When the midspan is approached, however, $\tan\alpha_r$ gradually decreases, and the last five steps in the local region decrease in the following sequence: $(9/6)F_t$, $(7/6)F_t$, $(5/6)F_t$, $(3/6)F_t$ and $(1/6)F_t$. This stepped curve near the midspan can be approximated conservatively by a horizontal dotted line DB, which is commonly used in design.

Using the same logic, the forces in the top stringer are plotted in Figure 8.1(d). It can be seen that the compressive force in the top stringer due to bending is reduced by the tensile forces due to shear.

8.1.1.3 Shift Rule

As stated above, the dotted line CD in Figure 8.1(c) for bottom stringer force is displaced downward by a distance $(25/6)F_t$ from the solid line AB. It is also possible to view the dotted line CD as having displaced horizontally toward the support by a distance $(5/6)d_v$ from the solid line AB. This horizontal distance can be determined from the geometry of similar triangles. Since $(25/6)F_t$ is 1/8 of $(100/3)F_t$ at midspan, then 1/8 of one half- span $(20/3)d_v$ is $(5/6)d_v$.

This shifting of the bending moment diagram horizontally toward the support has been recognized in many codes and is known as the 'shift rule'. In determining the cut-off points for longitudinal bars, the ACI code requires that the design moment diagram be shifted a distance d (effective depth) toward the support. This detailing measure is an indirect recognition of the additional longitudinal tensile force demanded by the applied shear force, which has not been taken into account in the flexural design of bottom longitudinal steel. The code required shift of a distance d is obviously conservative, because d is greater than the theoretically required shift of $(5/6)d_v$, which has been calculated from the largest α_r angle (60°) generally accepted.

8.1.2 Beams Subjected to Uniformly Distributed Load

The same model beam analyzed in Section 8.1.1 will now be subjected to a uniformly distributed load (Figure 8.2a). According to the plasticity truss model the following two assumptions are made:

1. The forces in all the truss members are transferred only at the levels of the top and bottom stringers. There will be no transfer of forces between the vertical stirrups and the diagonal concrete struts in the main body (or web) of the beam between the top and bottom stringers. Hence, the tensile force in a stirrup and the compressive force in a concrete strut are uniform throughout their lengths.
2. The vertical stirrups and the diagonal concrete struts should have infinite plasticity to deform freely. This second assumption is necessary, because the first assumption implies that no bond or compatibility exists between the concrete struts and the stirrups in the main body. The first assumption is applicable only if the second assumption is valid.

Based on these two assumptions, the stresses in the beam would flow from the midspan to the support in a manner shown by the discontinuous banded stress field in Figure 8.2(b) and the struts-and-ties model in Figure 8.2(c). Each strut or tie in the latter figure is located in the centerline of each band of stress field in the former figure. The force in each strut or tie represents the resultant of the stresses in the band. Analysis of the stress flow in the stirrups (to be discussed below) will result in a staggered shear diagram, as shown in Figure 8.2 (d).

8.1.2.1 Stirrup Forces and Staggered Shear Diagram

Following the stress flow from midspan to the support, we start with the applied load w between points a and b (Figure 8.2(a) and (b)). The total load within this distance is $w\ell/8$. It is resisted by the truss actions of the concrete struts and the top stringer. The total force in the concrete struts from a to b should be $w\ell/8\cos\alpha_r$. This strut force is transmitted diagonally from the top stringer to the bottom stringer within line b'c', while crossing the section b–b' en route. At the bottom stringer level between b' and c', the diagonal force in the concrete struts is resisted by a stringer force, as well as a stirrup force of $w\ell/8$. This stirrup force is equal to the shear force at section b-b', which is designated as $V_{b-b'} = w\ell/8$. Because the stirrup force is distributed

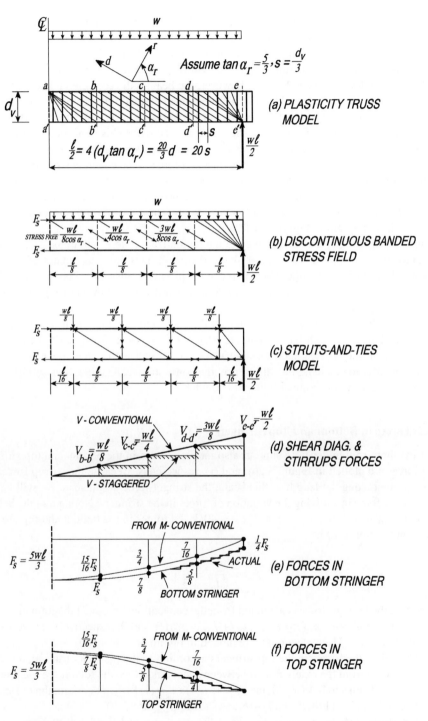

Figure 8.2   Plasticity truss model for a beam with uniformly distributed load

uniformly between points b′ and c′, the required shear capacity should also be constant from b′and c′. This constitutes the first step of the staggered shear diagram plotted in Figure 8.2(d).

In the second step, the uniform stirrup force at the bottom stringer from b′ to c′ is transmitted vertically to the top stringer within b and c. This stirrup force at the top stringer level remains at $w\ell/8$. At this level between b and c, however, the beam will receive an additional load of $w\ell/8$ from the externally applied load w. The combined vertical force of $w\ell/4$ between b and c should then be resisted by the truss actions of the concrete struts and the top stringer. The force in the concrete struts from b to c should be $w\ell/4\cos\alpha_r$. This strut force is transmitted diagonally to the bottom stringer within c′ and d′, while crossing section c–c′ en route. At the bottom stringer level between c′ and d′, the diagonal force in the concrete struts is resisted by a stringer force as well as a stirrup force of $w\ell/4$. This stirrup force is equal to the shear force at section c–c′ and is designated as $V_{c-c'} = w\ell/4$. Since the stirrup force is uniform between c′ and d′ the shear diagram is a constant in this region. This constitutes the second step of the staggered shear diagram in Figure 8.2(d).

Using the same logic, the stirrup force in the region from d′ to e′ should be $3w\ell/8$, which is equal to the shear force at section d-d′, $V_{d-d'} = 3w\ell/8$. This third step completes the entire staggered shear diagram. It should be noted that the plasticity truss model predicts zero forces in the stirrup near the midspan from a to b. In practice, however, a minimum amount of stirrup should always be provided.

In Figure 8.2(d), it is interesting to compare the staggered shear diagram with the conventional shear diagram, which has a triangular shape. Obviously, the staggered shear diagram based on the plasticity truss model is much less conservative than the conventional shear diagram. In order to elucidate this difference, the conventional shear diagram will be studied in Section 8.2.2 using the compatibility truss model.

8.1.2.2 Forces in Bottom and Top Stringers

The stress flow from midspan to support also induces forces in the bottom and top stringers. These forces are plotted in Figure 8.2(e) and (f), respectively. Because the stirrup force is no longer constant along the length of this beam, the stringer force at midspan, F_s, will be used as a reference force in studying the variation of forces in the stringers. Noting that the bending moment at midspan is $w\ell/8$ and that $\ell/d_v = 40/3$, the forces in the bottom and top stringers at midspan (section a-a′) is:

$$F_s = \frac{w\ell^2}{8d_v} = \frac{5}{3}w\ell \tag{8.3}$$

Similarly, the stringer forces calculated from the conventional moment diagram at sections b–b′, c–c′, d–d′ and e–e′ are $(15/16)F_s$, $(3/4)F_s$, $(7/16)F_s$ and 0, respectively. These stringer forces, designated as M-conventional, are plotted in Figure 8.2(e) and (f).

Now, let's find the forces in the top stringers according to the plasticity truss model. Since the vertical force from the external load between a and b is $w\ell/8$, the total force increment in the top stringer from a to b is $(w\ell/8)\tan\alpha_r = (5/24)w\ell = (1/8)F_s$. Therefore, the top stringer force at point b is $F_s - (1/8)F_s = (7/8)F_s$, as shown in Figure 8.2(f).

Further down the route of stress flow, the total vertical load on the top stringer from b to c is $w\ell/4$, and the total force increment in the top stringer from b to c is

$(w\ell/4)\tan\alpha_r = (5/12)w\ell = (1/4)F_s$. Then the top stringer force at point c is $(7/8)F_s - (1/4)F_s = (5/8)F_s$. Similar logic would give the top stringer force at point d to be $(5/8)F_s - (3/8)F_s = (1/4)F_s$, and at point e to be 0. These forces in top stringer at points a, b, c, d and e are connected by straight lines in Figure 8.2(f). In actuality, of course, the top stringer forces should change at the discrete points of the stirrups. In the region c to d, for example, each stirrup force increment is $(3w\ell/8)(1/5)\tan\alpha_r = (1/8)w\ell = (3/40)F_s$.

In the local 'fanning' region between d and e, however, each discrete vertical force at the stirrup locations is $(w\ell/2)(1/5) = w\ell/10$ and the increments of the top stringer force are $(w\ell/10)\tan\alpha_r$, depending on the variable angle α_r. The values of $\tan\alpha_r$ changes from d to e in the following sequence: $(9/10)(5/3)$, $(7/10)(5/3)$, $(5/10)(5/3)$, $(3/10)(5/3)$ and $(1/10)(5/3)$. Therefore, the stirrup force increments change from d to e as follows: $(9/100)F_s$, $(7/100)F_s$, $(5/100)F_s$, $(3/100)F_s$ and $(1/100)F_s$.

The force diagram for the top stringer derived from the plasticity truss model is compared with the force diagram calculated from the conventional moment diagram in Figure 8.2(f). It can be seen that the former is smaller than the latter due to the effect of shear. This shear effect is frequently ignored in design practice.

Now, let us find the forces in the bottom stringers according to the plasticity truss model. Looking at the bottom stringer in Figure 8.2(a), the vertical force from a' to b' is zero. Therefore, no increment of force will occur in the bottom stringer between these two points and the stringer force remains a constant of F_s from a' to b' (Figure 8.2e). In the next region from b' to c', however, the vertical force is $w\ell/8$. The force increment in the bottom stringer is $(w\ell/8)\tan\alpha_r = (5/24)w\ell = (1/8)F_s$. Therefore, the bottom stringer force at c' is $F_s - (1/8)F_s = (7/8)F_s$. Similar reasoning gives the bottom stringer forces at points d' and e' to be $(5/8)F_s$ and $(1/4)F_s$, respectively.

Figure 8.2(e) compares the forces in the bottom stringer obtained from plasticity truss model with those calculated from the conventional moment diagram. It can be seen that the shear force increases the bottom stringer force. Therefore, the 'shift rule' given in Section 8.1.1.3 for detailing of bottom reinforcement remains valid.

8.1.2.3 Comments on Plasticity Truss Model

The beam example in Sections 8.1.1 and 8.1.2, analyzed by the plasticity truss model, provides a very clear concept of stress flow in the beam and the resulting forces in the stirrups, stringers and concrete struts. This model is quite valid for the case of a beam subjected to midspan concentrated load as shown in Section 8.1.1. However, the solution may not be conservative in the case of a beam subjected to uniformly distributed load as shown in Section 8.1.2.

The plasticity truss model satisfies only the force equilibrium condition, not the strain compatibility condition. Because the second assumption of infinite plastic deformation can not be guaranteed for concrete, a design based on the staggered shear diagram (Section 8.1.2.1) may not be conservative. In Section 8.2, a compatibility truss model will be presented, based on the elastic behavior of concrete and steel. As far as the material property is concerned, the plasticity truss model provides an upper bound solution, while the compatibility truss model yields the lower bound solution. For a nonlinear material of limited plasticity, such as concrete, the solution should lie in between these two limits.

8.2 Compatibility Truss Model for Beam Analysis

8.2.1 Analysis of Beams Subjected to Uniformly Distributed Load

In Section 8.1.2 we have analyzed a beam subjected to an uniformly distributed load according to the plasticity truss model (Figure 8.2). By assuming an infinite plasticity for the materials we obtained a discontinuous banded stress field as shown in Figure 8.2(b). From this stress field a staggered shear diagram was derived as shown in Figure 8.2(d).

In this section we will apply the compatibility truss model to the same beam, Figure 8.3(a), subjected to the same uniformly distributed load w. In this model a continuous stress field will be introduced (Hsu, 1982). This continuous stress field will satisfy not only the equilibrium

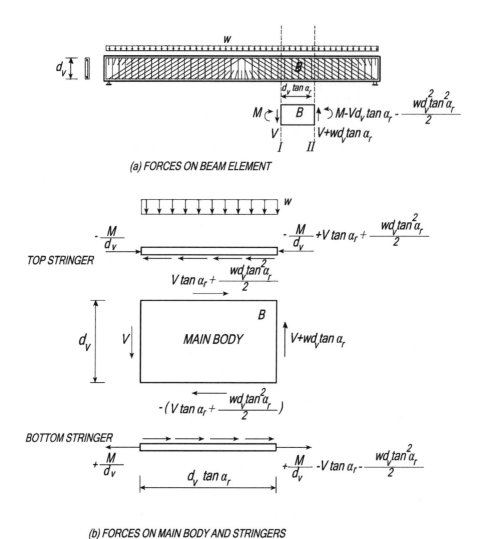

(a) FORCES ON BEAM ELEMENT

(b) FORCES ON MAIN BODY AND STRINGERS

Figure 8.3   Forces in beam under uniform load

condition, but also the compatibility condition and the stress–strain relationship of materials according to Hook's law (Hsu, 1983a).

The continuous stress field can be illustrated by describing the variation of the stresses in the stirrups and longitudinal bars in a typical element B isolated from the beam, as shown in Figure 8.3(a). The length of the element B is taken as $d_v \tan \alpha_r$. Since $\tan \alpha_r = 5/3$ and $d_v = 3s$ in the model beam, the length of the beam element B is equal to $5s$. This means the element contains five uniformly distributed stirrups.

The beam element B shown in Figure 8.3(a) is subjected to a shear force V and a bending moment M on the left face. On the right face the shear force receives an additional increment of $wd_v \tan \alpha_r$ from the uniform load w. Similarly, the bending moment M receives two additional increments: $-Vd_v \tan \alpha_r$ from the shear force V and $-(1/2)wd_v^2 \tan^2 \alpha_r$ from the uniform load w.

According to the truss model concept, the bending resistance of a beam element will be supplied by the stringers (flanges) and the shear resistance by the main body (web). The forces resisted by the stringers and by the main body are shown separately in Figure 8.3(b). It can be seen that the main body is subjected to shear stresses caused by both the shear force V and the uniform load w. The shear force on the right face of the main body is $V + wd_v \tan \alpha_r$ and the shear force on the top and bottom faces is $V \tan \alpha_r + (1/2)wd_v \tan^2 \alpha_r$.

To find a continuous stress field that satisfies both the equilibrium and the compatibility conditions in the main body of element B, we make two assumptions. First, the shear stresses are assumed to be *uniformly* distributed over the depth of the beam element B at each cross section. Second, in view of the uniformly distributed load w, it is logical to assume that the cross-sectional shear force due to w varies *linearly* along the length of beam element B from zero at the left face to $wd_v \tan \alpha_r$ at the right face. As a result, the part of the shear force due to w, i.e. $(1/2)wd_v \tan^2 \alpha_r$, on the top and bottom faces, should also be distributed in a linear fashion from zero at the left face of element B to a maximum at the right face.

With these two assumptions we can now analyze the stresses in the stirrups and in the longitudinal web bars within the main body. The variation of the stirrup forces and the longitudinal web steel forces will be studied in Sections 8.2.2 and 8.2.3, respectively. The two assumptions we have made can be proven by the theory of elasticity to satisfy the compatibility condition (Section 6.2.4 of Hsu, 1993).

8.2.2 Stirrup Forces and Triangular Shear Diagram

The main body of the element as shown in Figure 8.3(b) is subjected to two sets of self-equilibrating forces, one caused by the shear force V and the other by the uniform load w. These two sets of forces due to V and w will produce two distinct stress fields. The stress field in the main body is the sum of these two stress fields.

The stress field due to V and the stress field due to w are shown separately in Figure 8.4(a) and (b). The stress field due to V (Figure 8.4a), is one of pure shear, identical to that given in Figure 2.5(b) (Section 2.1.3) for a simple beam subjected to a midspan concentrated load. In the pure shear stress field of Figure 8.1 (Section 8.1.1) the stirrup forces, F_t, are uniformly distributed and are equal to a constant $V/5$, as given by Equation (8.2).

The stress field due to w Fig. 8.4 (b), is the same as that of a cantilever beam under a uniform load. The stirrup forces are nonuniformly distributed, i.e. F_t is not a constant. To illustrate this stress field, the beam element B is separated into 25 equal subelements, as shown

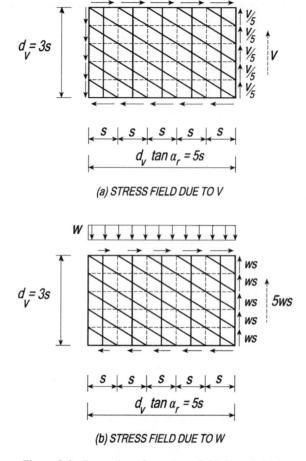

(a) STRESS FIELD DUE TO V

(b) STRESS FIELD DUE TO W

Figure 8.4   Separation of two stress fields in main body

in Figure. 8.5. Each subelement contains a vertical steel bar at its center. Forces acting on each subelement are indicated and vary from subelement to subelement in accordance with the two assumptions.

Take, for example, the subelement at the upper right corner. It is subjected to a shear force of ws on the right face and $0.8ws$ on the left face. This change is in accordance with the second assumption. On the top face we have vertical force ws due to external load acting along the centerline of the subelement. (This force is drawn to the right of the centerline for clarity.) To maintain vertical equilibrium, the vertical force on the bottom face must be $0.8ws$ acting along the centerline (also drawn to the right of the centerline). This compressive force must be carried by the vertical stirrup, because vertical stresses cannot pass through the diagonal cracks, except through connecting steel bars.

To maintain moment equilibrium and horizontal force equilibrium, shear forces on the top and bottom faces must both be $(0.9ws) \tan \alpha_r$ and act in the opposite direction. The equality of the top and bottom shear forces is in accordance with the first assumption. Each of these

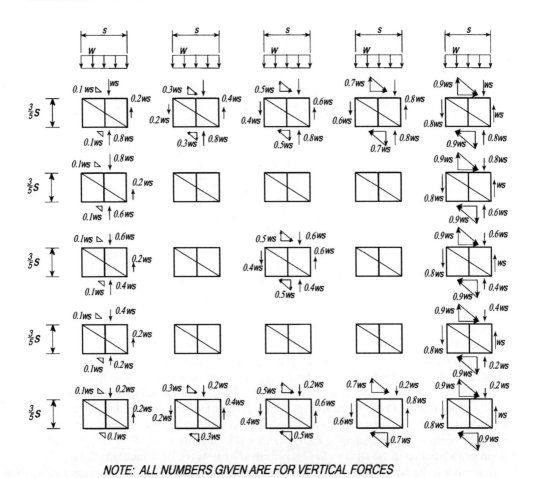

NOTE: ALL NUMBERS GIVEN ARE FOR VERTICAL FORCES

Figure 8.5   Equilibrium within a beam element B (uniformly distributed load on top surface)

two shear forces can be resolved into a diagonal compressive force in the concrete struts and a vertical tensile force of $0.9ws$ in the stirrup. Summing the two vertical forces in the stirrup on the top face gives $0.9ws - ws = -0.1ws$. On the bottom face, however, the sum of the two vertical forces is $0.9ws - 0.8ws = 0.1ws$, which is different from the top face.

If we observe the equilibrium of the five subelements of the extreme right column in Figure 8.5, it can be seen that the stirrup forces due to w on the top faces increase linearly downward in the following sequence: $-0.1ws$, $0.1ws$. $0.3ws$, $0.5ws$ and $0.7ws$. The stirrup forces on the bottom faces increase linearly downward in the sequence of $0.1ws$, $0.3ws$, $0.5ws$, $0.7ws$ and $0.9ws$.

If we also look at the top faces of the top five subelements, the stirrup forces due to w increase from right to left in the following sequence: $-0.9ws$. $-0.7ws$, $-0.5ws$, $-0.3ws$ and $-0.1ws$. Similarly, the sequence of stirrup forces, $-0.7ws$, $-0.5ws$, $-0.3ws$, $-0.1ws$ and $0.1ws$, can be observed on the bottom faces of the five extreme left subelements from

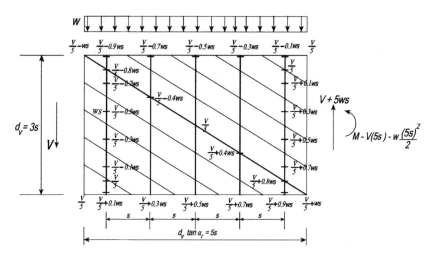

Figure 8.6   Variation of stirrup forces in compatibility truss model

top to bottom. Finally, the sequence of stirrup forces on the bottom faces of the five bottom subelements from left to right are $0.1ws$, $0.3ws$, $0.5ws$, $0.7ws$ and $0.9ws$.

The nonuniform stress field of stirrup forces due to w (Figure 8.4b and Figure 5) can now be added to the uniform tensile stress field of stirrup forces due to V (Figure 8.4a), which is a constant of $V/5$. The summation of the two stress fields is shown in Figure 8.6. This figure clearly shows that the forces in the stirrups increase linearly downward and that the maximum forces are located at the bottom of the bars. Stirrup forces at these lowest locations vary linearly along the beam length according to the *conventional triangular shear diagram*. This pattern of stress distribution in the stirrups has been verified by tests (Belarbi and Hsu, 1990).

In summary, the nonuniform stirrup forces of Figure 8.6 are based on the two assumptions which satisfy not only the equilibrium condition, but also the compatibility conditions and the linear stress–strain relationship of materials. This nonuniform continuous stress field derived from compatibility truss model is quite different from the discontinuous banded stress field assumed in the equilibrium (plasticity) truss model of Figure 8.2(b). The difference between these two models has two significant consequences. First, while the equilibrium (plasticity) truss model gives the upper bound solution as far as material is concerned, the compatibility truss model provides the lower bound solution. Second, while the equilibrium (plasticity) truss model suggests a staggered shear diagram for design of stirrups, the compatibility truss model requires the conventional triangular shear diagram.

8.2.3 Longitudinal Web Steel Forces

Distribution of horizontal forces in the main body of the beam element can also be derived from the two stress fields in Figure 8.4(a) and (b). For the stress field due to V (Figure 8.4a), the shear force on the vertical face of each subelement is $V/5$. This vertical force can be resolved into a diagonal compressive force and a longitudinal tensile force $V/3$. This longitudinal tensile force should be taken by one longitudinal web bar at the center of each subelement.

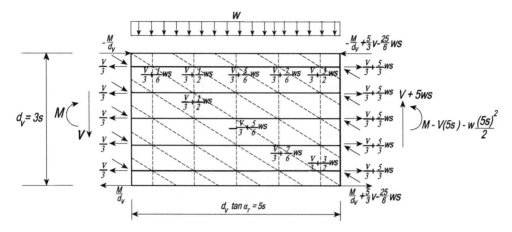

Figure 8.7   Variation of longitudinal steel forces (compability truss model)

This longitudinal tensile force $V/3$ should be uniform throughout the main body of the beam element.

For the stress field due to w (Figures 8.4b and 8.5) an upward shear force ws is shown to act on the right face of each of the five subelements in the far right-hand column. Each of these five forces can be resolved into a diagonal tensile force and a longitudinal tensile force of $ws \tan \alpha_r = (5/3)ws$. This longitudinal tensile force should be taken by each of the longitudinal web bar on the right face of the element, and should decrease linearly from $(5/3)ws$ at the right face to zero at the left face of the element.

The summation of the longitudinal steel forces in the two stress fields due to V and w is given in Figure 8.7. It can be seen that the tensile force in each longitudinal web bar is $V/3$ on the left face of the element. This tensile force increases linearly from left to right until it becomes $V/3 + (5/3)ws$ on the right face of the element. The tensile forces at the centers of the five subelements in each row increases linearly in the following sequence: $V/3 + (1/6)ws$, $V/3 + (1/2)ws$, $V/3 + (5/6)ws$, $V/3 + (7/6)ws$ and $V/3 + (3/2)ws$.

8.2.4 Steel Stresses along a Diagonal Crack

Now that the forces in the stirrups and in the longitudinal bars of a beam element are clarified, we can observe the variation of these forces along a diagonal crack. A triangular free body of the beam element is shown in Figure 8.8 and all the forces acting on the free body are indicated. Along the diagonal crack, the stirrup forces increase from left to right in the following linear sequence: $V/5 - 0.8ws$, $V/5 - 0.4ws$, $V/5$, $V/5 + 0.4ws$, $V/5 + 0.8ws$. The longitudinal web steel forces increase linearly from top to bottom in the sequence $V/3 + (1/6)ws$, $V/3 + (1/2)ws$, $V/3 + (5/6)ws$, $V/3 + (7/6)ws$ and $V/3 + (3/2)ws$.

If the five longitudinal web rebars are not available, a redistribution of the forces will occur. The longitudinal forces in the web rebars can be replaced by the longitudinal forces in the top and bottom stringers to ensure the equilibrium of the total longitudinal forces. This replacement can easily take care of the uniform longitudinal forces ($V/3$ in each web rebar) produced by the shear force V. The equilibrium of the nonuniform longitudinal forces produced by the

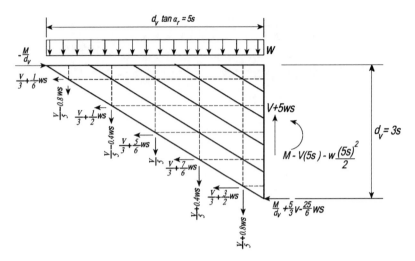

Figure 8.8   Stirrups and longitudinal steel stresses along a diagonal crack (compatibility truss model)

distributed load w is more difficult. Equilibrium of these local areas at the intersections of vertical stirrups and concrete struts will involve the dowel action of the vertical stirrups, the shear resistance of the concrete struts and the angle change of the concrete struts.

8.3 Shear Design of Prestressed Concrete I-beams

8.3.1 Background Information

Prestressed concrete I-beams are used extensively as the primary superstructure components of highway bridges. One of the most troublesome problems in the design of prestressed concrete beams is the shear problem. In fact, there is at present no rational model to predict the shear behavior of prestressed concrete structures and the various modes of shear failures. Because of this deficiency, all the shear design provisions, such as those in the ACI Code and AASHTO Specifications, are empirical, complicated and have severe limitations.

8.3.1.1 ACI (2008) Shear Design Provisions

$$V_n = V_c \text{ (the lesser of } V_{ci} \text{ or } V_{cw}) + V_s \tag{8.4}$$

$$V_{ci} = 0.05\sqrt{f_c'(MPa)}b_w d + V_d + \frac{V_i M_{cre}}{M_{max}} > 0.14\sqrt{f_c'(MPa)}b_w d \tag{8.5}$$

$$M_{cre} = \frac{I}{y_t}\left(0.5\sqrt{f_c'\,(MPa)} + f_{pe} - f_d\right) \tag{8.6}$$

$$V_{cw} = \left(0.29\sqrt{f_c'(MPa)} + 0.3f_{pc}\right)b_w d + V_p \tag{8.7}$$

$$V_s = \frac{A_v f_y d}{s} \tag{8.8}$$

Equations (8.4)–(8.8) show that the ACI shear provisions are quite complicated. The complexity stems from two sources. First, the V_c term is taken as the lesser of V_{ci} or V_{cw}. This is because the ACI provisions recognize two types of shear failure forces, V_{ci} for flexural shear failure and V_{cw} for web shear failure. Second, V_{cw} is assumed to be the cracking strength and V_{ci} is assumed to be the sum of a cracking strength plus an empirical post-cracking strength. Since the cracking strengths for V_{ci} or V_{cw} are strongly affected by the prestress force, the ACI shear strength becomes a complicated function of the prestress force as shown above.

8.3.1.2 AASHTO (2007) Shear Design Provisions

$$V_n = V_c + V_s \tag{8.9}$$
$$V_c = 0.083\beta\sqrt{f'_c(MPa)}b_v d_v \tag{8.10}$$

where β is a function of v_u/f'_c and ε_x

$$V_s = \frac{A_v f_y d_v \cot\theta}{s} \tag{8.11}$$

where θ is a function of v_u/f'_c and ε_x

An engineer must calculate a factor β for the V_c term and a crack angle θ for the V_s term. Both β and θ are functions of the shear stress ratio v_u/f'_c and the longitudinal strain ε_x in the web. The strain ε_x is calculated at the mid-depth under the combined action of bending moment, shear force and axial force acting on the member:

$$\varepsilon_x = \frac{\frac{|M_u|}{d_v} + 0.5N_u + 0.5\left|V_u - V_p\right|\cot\theta - A_{ps}f_{po}}{2\left(E_s A_s + E_{ps} A_{ps}\right)} \tag{8.12}$$

and the shear stress v_u is calculated by

$$v_u = \frac{\left|V_u - \phi V_p\right|}{\phi b_w d_v} \tag{8.13}$$

Then, β and θ are obtained from Table 8.1 using an iterative procedure.

Two points should be noted: First, the AASHTO shear provisions do not distinguish between web-shear and flexural-shear failure modes. As a result, beams in shear are always critical in web-shear failure mode near the supports, where the largest shear stresses are located. Second, even without recognizing the two modes of failure, Equations (8.9)–(8.13) and Table 8.1 show that the AASHTO shear provisions are more complicated than the ACI shear provisions. To make matters worse, the tables (or graphs) provided by the AASHTO specifications give no physical meaning.

8.3.2 Prestressed Concrete I-Beam Tests at University of Houston

8.3.2.1 Test Set-up and Test Variables

Five full-scale prestressed concrete I-beams, B1–B5, were tested at the University of Houston (UH), as shown in Figure 8.9. The cross-sections of these beams, known as TxDOT Type A, are shown in Figure 8.10. Each beam has twelve 7-wire, low-lax prestressing strands of

Table 8.1 Values of β and θ for sections with transverse reinforcement (AASHTO Specifications, 2007)

$\dfrac{v_u}{f'_c}$		Longitudinal strain, $\varepsilon_x \times 1000$								
		≤ -0.20	≤ -0.10	≤ -0.05	≤ 0	≤ 0.125	≤ 0.25	≤ 0.50	≤ 0.75	≤ 1.00
≤ 0.075	θ	22.3	20.4	21.0	21.8	24.3	26.6	30.5	33.7	36.4
	β	6.32	4.75	4.10	3.75	3.24	2.94	2.59	2.38	2.23
≤ 0.100	θ	18.1	20.4	21.4	22.5	24.9	27.1	30.8	34.0	36.7
	β	3.79	3.38	3.24	3.14	2.91	2.75	2.50	2.32	2.18
≤ 0.125	θ	19.9	21.9	22.8	23.7	25.9	27.9	31.4	34.4	37.0
	β	3.18	2.99	2.94	2.87	2.74	2.62	2.42	2.26	2.13
≤ 0.150	θ	21.6	23.3	24.2	25.0	26.9	28.8	32.1	34.9	37.3
	β	2.88	2.79	2.78	2.72	2.60	2.52	2.36	2.21	2.08
≤ 0.175	θ	23.2	24.7	25.5	26.2	28.8	29.7	32.7	35.2	36.8
	β	2.73	2.66	2.65	2.60	2.52	2.44	2.28	2.14	1.96
≤ 0.200	θ	24.7	26.1	26.7	27.4	29.0	30.6	32.8	34.5	36.1
	β	2.63	2.59	2.52	2.51	2.43	2.37	2.14	1.94	1.79
≤ 0.225	θ	26.1	27.3	27.9	28.5	30.0	30.8	32.3	34.0	35.7
	β	2.53	2.45	2.42	2.40	2.34	2.14	1.86	1.73	1.64
≤ 0.250	θ	27.5	28.6	29.1	29.7	30.6	31.3	32.8	34.3	35.7
	β	2.39	2.39	2.33	2.33	2.12	1.93	1.70	1.58	1.50

Figure 8.9 General view of prestressed concrete I-beam tests

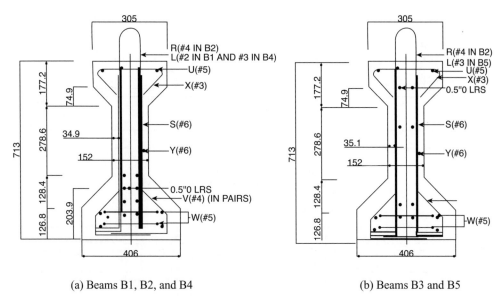

(a) Beams B1, B2, and B4   (b) Beams B3 and B5

Figure 8.10   End cross-section of Beams B1–B5 (all dimensions are in mm)

12.7 mm (0.5 in.) diameter, with an ultimate strength of 1862 MPa (270 ksi). The strands in beams B1, B2 and B4 are straight (Figure 8.10a) and those in beams B3 and B5 are draped (Figure 8.10b). The total length of each of the beams was 7.62 m (25 ft) while the center-to-center span length was 7.32 m (24 ft), as shown in Figure 8.11.

The primary purpose of these experiments was to study the effect of three variables on the ultimate shear strength of prestressed I-Beams, namely, the shear-span-to-depth ratio a/d, the transverse steel ratio ρ_t, and the presence of draped strands. As such, the positions of the vertical loads on the beams together with the support positions are shown in Figure 8.11. In B1, B2 and B3 (Figure 8.11a), the loads from actuators B and C were applied at 0.914 m (3 ft) from the supports (both north and south supports) to create an a/d ratio of 1.61. At this low a/d ratio, the shear capacity is expected to be governed by web-shear failure mode. In B4 and B5 (Figure 8.11b), the loads from actuators B and C were applied at 2.44 m (8 ft) from the supports to create an a/d ratio of 4.29. At this high a/d ratio, the shear capacity is expected to be governed by flexural-shear failure mode.

The second variable in the test program was the percentage of shear reinforcement ρ_t. B1, B4 and B5 have a ρ_t value of 0.17%, close to the minimum stirrup requirement, while B2 and B3 have a normal ρ_t value of 0.95%.

The third variable was the presence of draped strands. B3, with draped strands, can be compared with B2, with straight strands, in the case of web-shear failure with $\rho_t = 0.95\%$; while B5, with draped strands, can be compared with B4, with straight strands. in the case of flexural-shear failure with $\rho_t = 0.17\%$.

During testing, linear voltage displacement transducers (LVDTs) were used to measure the displacements within the failure regions of the beam adjacent to the points of load application, as shown in Figure 8.11. Electrical resistance strain gauges were installed on both legs of the

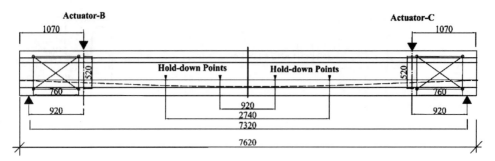

All Dimensions are in mm
(a) Beams B1, B2 and B3 (draped strands for B3 only)

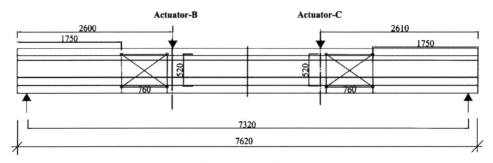

All dimensions are in mm
(b) Beams B4 and B5 (draped strands for B5, same as B3)

Figure 8.11   Loading positions and LVDT rosettes in beams B1–B5 (all dimensions are in mm)

vertical stirrups inside the beams to monitor the steel strains during the load test. The locations of strain gages on the stirrups were selected so that they intersected the predicted shear failure planes of the beams. It was found that stirrup strains reached yielding and beyond at failure, with many stirrups far into the strain hardening range. In short, the stirrups in all five beams, B1–B5, can be assumed to yield at shear failure.

8.3.2.2 Test Results

The load–deflection curves of the five beams B1–B5 are shown in Figure 8.12. Curves 1–6 represent the north and south ends of the group of three beams B1–B3, which have $a/d = 1.61$. All three experimental curves follow the theoretical curve 11 based on flexural analysis until the beams failed in web-shear mode. Similarly, curves 7–10 represent the north and south ends of the group of two beams B4–B5, which have $a/d = 4.29$. All the experimental curves of these two beams followed the theoretical curve 12 based on flexural analysis until the beams failed in flexural-shear mode.

Shear-span-to-depth ratio (a/d)
Figure 8.12 clearly shows that the beams designed for web-shear failure (B1, B2 and B3) had much higher shear capacities compared with the beams designed to fail in flexural-shear

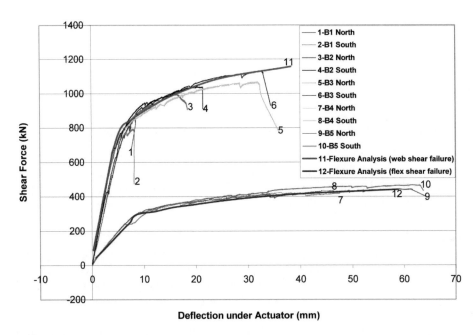

Figure 8.12   Load deflection curves of beams B1–B5 (1kN = 0.225 kips, 1 mm = 0.0394 in.)

(B4 and B5). In other words, the shear-span-to-depth ratio (a/d) has a strong effect on the shear strength of the beams. Unfortunately, this very important variable (a/d) is neglected in the AASHTO (2007) Shear Design Provisions (see Section 8.3.1.2). The situation in the ACI (2008) Shear Design Provisions (see Section 8.3.1.1), however, is more subtle. The variable a/d does not appear directly in the ACI shear equations, but is indirectly taken into account in the determination of V_c by choosing the lesser of V_{ci} and V_{cw}. V_{ci} is based on flexural-shear failure mode with high a/d ratio, while V_{cw} is based on web-shear failure mode with low a/d ratio. In short, the variable a/d must be incorporated into a shear strength equation, in order to create a simple and accurate shear design method.

Percentage of stirrups ρ_t
The effect of this second variable can be observed by comparing curves 3 and 4 for beam B2 with curves 1 and 2 for beam B1 in Figure 8.12. The shear strength of B2 with $\rho_t = 0.95\%$ is 20% higher than that of B1 with $\rho_t = 0.17\%$. Two trends can be observed from this comparison. First, the stirrups are indeed effective in increasing the shear capacity of prestressed I-beams at low a/d ratios. Second, the V_c term could be almost four times larger than the V_s term in the web-shear failure mode with $a/d = 1.61$. This means that the concrete contribution to shear capacity increases rapidly with decreasing a/d ratio in the range of low a/d ratios, resulting in a rapid increase of the overall shear capacity.

Draped strands
The effect of the third variable, can be observed in Figure 8.12 by comparing curves 5 and 6 for beam B3 with curves 3 and 4 for beam B2 in the case of web-shear failure mode,

and by comparing curves 9 and 10 for beam B5 with curves 7 and 8 for beam B4 in the case of flexural failure mode. It can be seen that the shear strength of beams with draped strands is about 10% higher than that of beams with straight strands. In other words, the vertical force component V_p of the draped strands could be taken into account in assessing the shear capacity.

8.3.2.3 Variables Affecting the Shear Strength of Prestressed I-beams

In addition to the shear-span-to-depth ratio a/d, the ρ_t ratio, and the vertical force component V_p of draped strands discussed in Section 8.3.2.2, the shear capacity of prestressed I-beams should also be a function of the web area $b_w d$ and the strength of concrete $\sqrt{f_c'}$. These last two variables can be construed as follows:

Web area $b_w d$
Comparison of the five UH test specimens with the nine beams tested by Lyngberg (1976) provided an important revelation. The two groups of test specimens were similar in size and shape, except that Lyngberg's specimens had wide top flanges (700 mm) and the UH specimens had narrow top flanges (305 mm). Since good agreement in shear strengths was observed between these two groups of tests, it was concluded that the top flange width was not a significant variable with regard to the shear strength, and that the web region, defined as $b_w d$, was the primary shear-resisting component. This observation is consistent with the basic concept of the mechanics of materials for I-beams, i.e. the bending moment is carried by the top and bottom flanges, while the shear force is carried by the web.

Concrete strength $\sqrt{f_c'}$
From the study of RC shear elements with concrete strengths up to 100 MPa (Zhang and Hsu, 1998), it was observed that the softened compressive strength of concrete struts is proportional to the square root of the compressive strength $\sqrt{f_c'}$ (see Sections 2.3.3.1 and 6.1.7.2). Therefore, the shear force that causes the crushing of concrete in the web of a beam must be proportional to $\sqrt{f_c'} b_w d$. This parameter could be used to normalize a shear force V_u into a nondimensional quantity, $V_u/\sqrt{f_c'} b_w d$.

In conclusion, in the new shear equation the shear capacity V_n should be a function of five variables: the a/d ratio, the strength of concrete $\sqrt{f_c'}$, the web area $b_w d$, the ρ_t ratio and the vertical force component V_p of draped strands.

8.3.2.4 Negligible Effects of Prestress Force and Failure Crack Angle

As shown in Section 8.3.1, the ACI shear provisions consider prestress force as a variable, while the AASHTO shear provisions take into account both the prestress force and the failure crack angle as variables. In contrast, the UH study showed that these two variables had no significant effect on the shear capacity of prestressed I-beams.

Prestress force
It is well-known that prestress has a strong effect on the cracking strength of prestressed beams. However, if we bypass the cracking strength and go directly to investigate the ultimate

Figure 8.13   Beam B4 showing 45° failure crack in flexural-shear failure zone

shear strength, then the concept of decompression will show that the ultimate shear strength is only slightly affected by the prestress force. This concept of prestressed beam behavior after decompression was substantiated by Lyngberg (1976). Lyngberg's test program consisted of nine beams, in which the major variable was the intensity of prestress. The cross-section, web reinforcement, ultimate moment, and shear span were held constant. The test results showed convincingly that the influence of prestress force on the shear strength can be neglected. As such, the prestress force need not be taken as a variable in an equation for shear strength.

Failure crack angle
Figure 8.13 shows the failure crack angle of beam B4 that failed in flexural-shear. In this case, the AASHTO shear provisions would assume a crack angle much smaller than 45° according to the principal compressive direction of concrete in the prestressed webs. Such a crack with small angle is shown in Figure 8.13 by the major crack that passed through grid 4 at the bottom edge of the web. However, this crack with small angle did not develop to form a failure surface at ultimate load. In fact, Figure 8.13 shows that another major crack with angle of approximately 45° developed and passed through grid 8 at the bottom edge of the web. This approximately 45° crack was observed to cause failure.

The assumption of 45° failure crack can also be supported by a study of energy dissipations in the failure zone. According to Laskar (2009), a crack with inclination close to 45° would result in minimum shear energy dissipation.

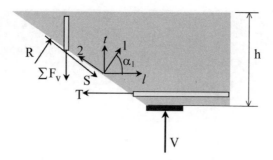

Figure 8.14   Analytical model used for calculating web-shear capacities of beams

In Section 8.3.3, we will derive a simple formula for shear strength V_n that does not involve the prestressing force and the failure crack angle.

8.3.3 UH Shear Strength Equation

8.3.3.1 Shear Model

The concept of shear resistance developed by Loov (2002) (Figure 8.14) was used to derive the ultimate shear capacity. According to this model, the contribution of concrete to the shear capacity of the beams stems from the shear stress in the concrete along a failure crack, represented by the force S. Loov's concept is very different from the concept of existing shear design methods (ACI, 2008; AASHTO, 2007) which assume that the concrete contribution to the shear capacity of beams is derived from the tensile stress across the cracks.

Assuming the failure surface to be an inclined plane, and taking the force equilibrium of the free body along the crack direction (Figure. 8.14), the shear capacity of the beam V_n can be calculated as:

$$V_n = \frac{S - T \sin \alpha_1}{\cos \alpha_1} + \sum F_V \qquad (8.14)$$

where $\sum F_V$ is the summation of vertical forces of the stirrups intersected by the failure crack at the ultimate load. T is the tensile force in the prestressing tendons at the ultimate load of the beams; and α_1 is the angle between the normal to the failure crack and the longitudinal axis. The α_1 angles of beams B1–B3, which failed in web-shear, were observed in tests to be approximately 45°.

The term $(S - T \sin \alpha_1)/ \cos \alpha_1$ in Equation (8.14) is the 'contribution of concrete in shear' (V_c). In order to avoid the excessive complexity involved in the calculation of S, T and α_1, it was decided to derive the V_c term directly from tests. In Equation (8.14) $\sum F_V$ is the 'contribution of steel in shear' denoted as V_s. Thus,

$$V_n = V_c + V_s \qquad (8.15)$$

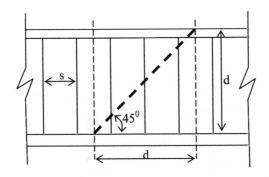

(a) Smeared Stirrup Method

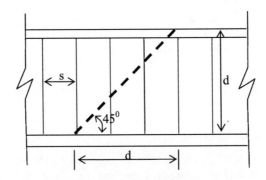

(b) Minimum Shear Resistance Method

Figure 8.15   Determination of number of stirrups for 'contribution of steel' V_s

8.3.3.2 'Contribution of Steel' (V_s)

The V_s term in Equation (8.15) must be based on the observed failure crack of approximately 45° as shown in Figure 8.13. In the ACI Code, the 45° crack is located in the manner shown in Figure 8.15(a). This ACI concept of smearing the stirrups results in an average number of stirrups d/s crossing the crack. A more realistic concept, however, is to seek a crack that produces a 'minimum shear resistance' among a series of individual stirrups, as shown in Figure 8.15(b). This concept by Loov gives the number of stirrups crossing the crack as $(d/s - 1)$, and the V_s term as:

$$V_s = A_v f_y \left(\frac{d}{s} - 1 \right) \tag{8.16}$$

Besides being more rational and conservative, Equation (8.16) has an additional practical advantage: all the tedious 'stirrup spacing limitations' to guard against the unsafe nature of the 'smeared stirrups' assumption can be eliminated.

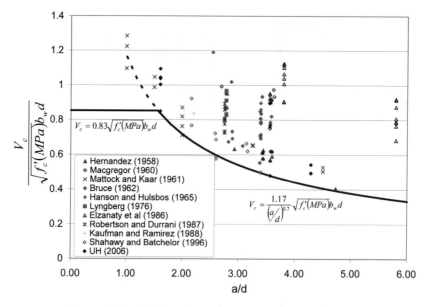

Figure 8.16   Variation of normalized concrete shear with a/d

8.3.3.3 'Contribution of Concrete' (V_c)

As discussed in Section 8.3.2.2, the V_c term in Equation (8.15) must be a function of the shear-span-to-depth ratio a/d, the strength of concrete $\sqrt{f_c'}$ and the web area $b_w d$. We can now implement the variable a/d into a new shear strength equation using the UH test results of beams B1–B5, as well as the beams of Hernandez (1958), MacGregor et al. (1960), Mattock and Kaar (1961), Bruce (1962), Hanson and Hulsbos (1965), Lyngberg (1976), Elzanaty et al. (1986), Robertson and Durrani (1987), Kaufman and Ramirez (1988), and Shahawy and Batchelor (1996).

The concrete shear contribution, V_c, of all the specimens were calculated by subtracting the steel contribution V_s as per Equation (8.16), from the total shear capacities of the beams. The normalized concrete shear stress $V_c/\sqrt{f_c'}b_w d$ of the specimens was obtained thereafter and its variation versus a/d was plotted in Figure 8.16. A conservative $V_c/\sqrt{f_c'}b_w d$ versus a/d curve can be expressed as:

$$V_c = \frac{1.17}{(a/d)^{0.7}}\sqrt{f_c'\,(\text{MPa})}b_w d \leq 0.833\sqrt{f_c'\,(\text{MPa})}b_w d \qquad (8.17)$$

where

b_w = width of the web of the prestressed beam

d = effective depth from the centroid of the tendons to the top compression fiber of the prestressed beam. The value of d is not taken to be less than 80% of the total depth of the beam.

The $V_c = 1.17\,(a/d)^{-0.7}\sqrt{f_c'}b_w d$ expression shown in Equation (8.17) was substantiated by the large-size test specimens of Mattock and Kaar (1961) with beam height of 648 mm

(25.5 in.). The shear strengths of their beams agreed very well with the proposed expression in Figure 8.16 for a/d ratios of 4.5, 3.25, 2.0, 1.5, and all the way down to 1.0.

The upper limit of $0.833\sqrt{f_c'}\,(\text{MPa})b_w d$ in Equation (8.17) is imposed for simply supported beams to ensure that the end anchorage will be sufficient to produce yielding of the stirrups, as observed in the web-shear failure of beams B1–B3. Mattock and Kaar tested continuous beams, which did not have the anchorage problems at beam ends associated with simply supported beams.

8.3.3.4 Shear/bond Failure at Beam Ends

Shahawy and Batchelor (1996) showed that the prestressed strands at the ends of AASHTO Type II pretensioned concrete beams would slip before the web crushing of shear failures. They referred to this type of failure as 'shear/bond failure'. When a/d ratio is greater than about 1.6, the bond slip did not appear to have a noticeable effect on the shear strengths. However, in their B1 series of 10 beams, where a/d ratios varied from 1.37 to 1.52, the bond slip appeared to have reduced the shear strengths. It was not clear, however, whether the transverse stirrups in the shear span had yielded.

Ma et al. (2000) tested two NU1100 pretensioned beams with a/d ratios varying from 1.16 to 1.28. At both ends of their beam A, the strands were extended beyond the end face, bent upward, and securely anchored into a large concrete diaphragm attached to the end face. Since bond slip was not allowed to occur, beam A failed in web crushing, a typical shear failure, at both ends. One failed end of beam A was then sawed off, leaving the ends of strands flush with the end face. Application of shear loads then produced a 'shear/bond' failure, rather than a web crushing failure. The 'shear/bond' failure load of the saw-cut end was 25% less than the web-crushing failure load of the ends with perfectly anchored strands.

Since shear failures at beam ends (a/d ratios less than 1.6) are significantly affected by the bond slip of prestressing strands, the limitation in Equation (8–17) at low a/d ratios must be a function of the anchorage length of the strands. A satisfactory anchorage length for shear should be one that ensures the yielding of the transverse stirrups at shear failure in the web. The upper limit in Equation (8.17) for a/d less than 1.6 appears to be valid for the TxDOT Type A beams tested at UH. The UH beams have twelve 7-wire strands of 12.7 mm (0.5 in.) diameter, with an extension of 152 mm (6 in.) beyond the support resultant and a supporting plate of 152 mm (6 in.) wide. More research is required to determine a satisfactory anchorage length of prestressed strands to prevent shear/bond failure.

8.3.3.5 Total Shear Strengths

Assuming that the transverse steel yields at failure, the final equation for the shear strength of prestressed I-beams is obtained by substituting Equations (8.16) and (8.17) into Equation (8.15) and by generalizing the ratio a/d to become $M_u/V_u d$:

$$V_u = 1.17\left(\frac{V_u d}{M_u}\right)^{0.7}\sqrt{f_c'\,(\text{MPa})}b_w d + A_v f_y\left(\frac{d}{s}-1\right)+V_p \tag{8.18}$$

where $1.17\,(V_u d/M_u)^{0.7}\le 0.833$.

Using the US customary system, Equation (8.18) becomes:

$$V_u = 14 \left(\frac{V_u d}{M_u} \right)^{0.7} \sqrt{f_c' \, (\text{psi})} b_w d + A_v f_y \left(\frac{d}{s} - 1 \right) + V_p \qquad (8.18a)$$

where $14 \, (V_u d / M_u)^{0.7} \leq 10$.

The additional shear strength contributed by the vertical component of prestressing force V_p of draped strands is also taken into account in Equation (8.18) (see last paragraph of Section 8.3.2.2). Following the ACI Code (2008) and the AASHTO Specifications (2007), V_p is calculated from the effective prestressing force after loss, because this force appears to give a conservative estimate of the observed additional shear strength.

It should be noted that the V_p term in Equation (8.18) is applicable to prestressed beams failing in web-shear as well as in flexural-shear. This is not the same as the provisions of the ACI Code, which utilizes the vertical component of the prestressing force in draped strands in the case of web-shear failure only, and not in the case of flexural-shear failure.

8.3.4 Maximum Shear Strength

8.3.4.1 Maximum Shear Strengths in Code Provisions

Since Equation (8.18) is based on the yielding of transverse steel reinforcement, a maximum shear strength $V_{n,max}$ must be defined to ensure that transverse steel will yield before the web crushing of concrete. However, $V_{n,max}$ in the ACI Code and the AASHTO Specifications are quite different.

ACI Code
ACI-ASCE Committee 326, Shear (1962) first proposed the following equation, which was later incorporated into the ACI Code (1963) for nonprestressed beams:

$$V_{n,max} = 0.83\sqrt{f_c' \, (\text{MPa})} b_w d \text{ or } 10\sqrt{f_c'(\text{psi})} \, b_w d \qquad (8.19)$$

Equation (8.19) was slightly liberalized in 1971 for application to both nonprestressed and prestressed concrete beams (ACI, 1971), in the form of limiting the V_s term to a 'maximum contribution of steel':

$$V_{s,max} = 0.66\sqrt{f_c' \, (\text{MPa})} b_w d \text{ or } 8\sqrt{f_c'(\text{psi})} b_w d \qquad (8.20)$$

If V_c is taken conservatively as $0.17\sqrt{f_c' \, (\text{MPa})} b_w d$ (or $2\sqrt{f_c'(\text{psi})} \, b_w d$), then the $V_{n,max}$ in 1971 Code is identical to that in the 1963 Code. Since the V_c term could be greater than $0.17\sqrt{f_c' \, (\text{MPa})} b_w d$ (or $2\sqrt{f_c'(\text{psi})} \, b_w d$), the $V_{n,max}$ for prestressed beams in 1971 ACI Code is allowed to be somewhat greater than $0.83\sqrt{f_c' \, (\text{MPa})} b_w d$ (or $10\sqrt{f_c'(\text{psi})} \, b_w d$). The ACI formula for $V_{s,max}$ has continued to be used up to the present time (ACI, 2008).

AASHTO Specifications
The current AASHTO Specifications (2007) require:

$$V_{n,max} = 0.25 f_c' b_w d_v \text{ or } 0.225 f_c' b_w d, \qquad (8.21)$$

if d_v is assumed to be $0.9d$.

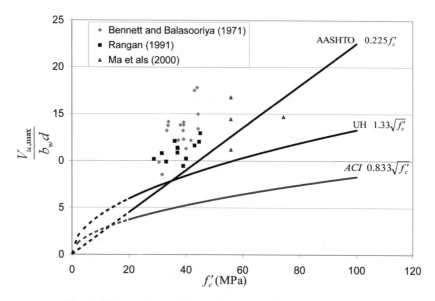

Figure 8.17   Variation of ultimate strength with concrete strength

Equation (8.21) was introduced into the first edition of the AASHTO LRFD Specifications (1994), based on the truss model concept first introduced by the Canadian Standard (1977) and the CEB-FIP Model Code (1978). No change was made in the second edition of AASHTO (1998), up to the current fourth edition of AASHTO (2007).

The fact that $V_{n,max}$ is proportional to $\sqrt{f'_c}$ in the ACI Code, Equation (8.20), and is proportional to f'_c in the ASSHTO Specifications, Equation (8.21), testifies to the confusion surrounding the formulas for $V_{n,max}$. These two equations are plotted in Figure 8.17.

In Figure 8.17, the ACI and AASHTO formulas are checked by the prestressed beams of Bennett and Balasooriya (1971), Rangan (1991), and Ma *et al.* (2000), which are over-reinforced in shear. It can be seen that the ACI formula is way too conservative. The AASHTO formula is more reasonable when compared with the test data. However, the AASHTO formula is expected to be unsafe for beams with concrete strength higher than 60 MPa (8700 psi),

8.3.4.2 UH Maximum Shear Strength

According to Sections 2.3.3.1 and 6.1.7.2, the maximum shear strength $V_{n,max}$ can be expressed as a function of the concrete compression strength $\sqrt{f'_c}$ and the web area $b_w d$:

$$V_{n,max} = C_1 \sqrt{f'_c} b_w d \tag{8.22}$$

where C_1 is a constant to be determined by the shear tests of prestressed I-beams.

Before deciding on the constant C_1 for the maximum shear strength, we must first calibrate the balanced condition defined as:

$$V_{n,b} = C_b \sqrt{f'_c} b_w d \tag{8.23}$$

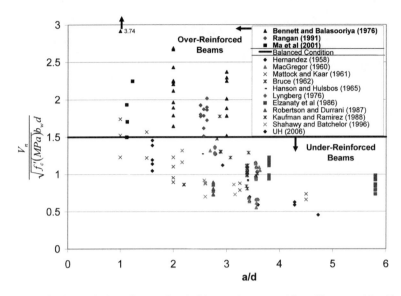

Figure 8.18   Variation of normalized ultimate shear capacities of beams with a/d

where $V_{n,b}$ is the balanced shear force and C_b is the constant corresponding to the balanced condition. The balanced condition occurs when the amount of shear reinforcement in a beam is such that the yielding of the transverse steel occurs simultaneously with the web crushing of the concrete. When $V_n < V_{n,b}$, the beam is defined as under-reinforced in shear, where the transverse steel yields before the crushing of concrete. When $V_n > V_{n,b}$, the beam is defined as over-reinforced in shear, where the concrete crushes without the yielding of steel.

The balanced constant C_b can be calibrated by comparing the over-reinforced beams versus the under-reinforced beams, as shown in Figure 8.18, which plots $V_n/\sqrt{f_c'}b_w d$ against a/d. Over-reinforced beams have been tested by three groups of researchers (Bennett and Balasooriya, 1971; Rangan, 1991; Ma *et al.*, 2000). Under-reinforced beams have been tested by all the other researchers. It can be seen that all the over-reinforced beams had a $V_n/\sqrt{f_c'}b_w d$ value above 1.5 for f_c' in MPa (or 18 for f_c' in psi) and almost all the under-reinforced beams have $V_n/\sqrt{f_c'}b_w d$ value below 1.5 for f_c' in MPa (or 18 for f_c' in psi). Therefore, the balanced constant can be taken as $C_b = 1.5$ for f_c' in MPa.

In order to provide some ductility in shear failure, the constant C_1 for maximum shear strength must be taken as less than C_b. $C_1 < C_b$ is also desirable because Rangan's over-reinforced beams have large web stiffeners under the loads to prevent local compression failures. In view of the fact that prestressed beams used in practice do not contain web stiffeners, it was decided to choose a conservative C_1 value of 1.33 for f_c' in MPa (or 16 for f_c' in psi) and $V_{n,max}$ can thus be expressed as

$$V_{n,\max} = 1.33\sqrt{f_c'\,(\text{MPa})}b_w d \text{ or } 16\sqrt{f_c'(\text{psi})}\,b_w d \tag{8.24}$$

Equation (8.24) is the equation for UH maximum shear strength, and is also plotted in Figure 8.17. Comparison of the UH formula with available test data shows that Equation (8.24) is the most reasonable for concrete strengths up to 100 MPa (14 500 psi). Obviously, more over-reinforced beam tests with $f'_c \geq 60$ MPa (8700 psi) are desirable. However, the ACI formula is evidently too conservative, and the AASHTO formula may be seriously unsafe when f'_c is greater than 60 MPa (8700 psi).

8.3.5 Minimum Stirrup Requirement

8.3.5.1 Minimum Shear Requirements in Code Provisions

Minimum shear reinforcement in prestressed I-girders is required to prevent the brittle failure of the girders due to the fracture of the shear reinforcement shortly after the formation of the inclined shear cracks. The ACI Code (2008) states that the minimum area of shear reinforcement $A_{v,min}$ provided at a spacing s in a member having width b_w and effective depth d, should be greater than the smaller of the two quantities expressed below:

$$A_{v,min} = 0.0625\sqrt{f'_c \text{(MPa)}}\frac{b_w s}{f_y} \text{ or } \left(0.75\sqrt{f'_c \text{(psi)}}\frac{b_w s}{f_y}\right)$$

$$\geq 0.35\frac{b_w s}{f_y} \text{ or } \left(50\frac{b_w s}{f_y}\right) \tag{8.25}$$

$$A_{v,min} = \frac{A_{ps}f_{pu}s}{80f_y d}\sqrt{\frac{d}{b_w}} \tag{8.26}$$

The minimum shear reinforcement in the AASHTO Specifications (2007) is:

$$A_{v,min} = 0.083\sqrt{f'_c \text{(MPa)}}\frac{b_w s}{f_y} \text{ or } \left(\sqrt{f'_c \text{(psi)}}\frac{b_w s}{f_y}\right) \tag{8.27}$$

Equations (8.25) and (8.27) are evaluated by the test beams found in the literature as follows: First, all the I-beams having $\rho_t f_y$ less than 1.38 MPa (200 psi) were selected. Second, the nondimensional reinforcement indices $\rho_t f_y/\sqrt{f'_c}$ of these beams were plotted against their corresponding a/d ratios, as shown in Figure 8.19. Third, the beams which failed in a brittle manner due to the low amount of shear reinforcement were represented by solid points, while those which failed in a ductile manner were represented by hollow points.

8.3.5.2 UH Minimum Shear Requirement

Figure 8.19 shows that some beams tested by Hernandez (1958), MacGregor (1960), Hanson and Hulsbos (1965) and Robertson and Durrani (1987), failed in a brittle manner, even though they had shear stirrups greater than that required by Equations (8.25) and (8.27). The beams that failed in a brittle manner had a/d ratios in the range from 2.0 to 4.0, the region known as 'Kani's Valley'.

Two recommendations can be drawn from examining Figure 8.19. (a) ACI Equagtion (8.25) is applicable to all shear-span ratios a/d, except in the range 2–4. In other words, when $0.25 < V_u d/M_u < 0.5$, the ACI minimum stirrup required by Equation (8.25) should be doubled. (b) ACI Equation (8.26) should be removed as in the AASHTO Specifications, because this formula could be nonnconservative for prestressed I-beams.

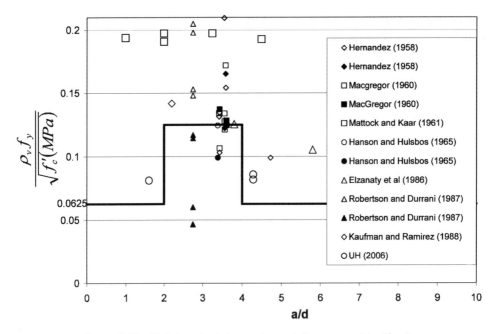

Figure 8.19   Variation of minimum shear reinforcement with *a/d* ratio

The UH minimum stirrup requirement at the chosen sections can be checked in terms of the maximum stirrup spacing s_{max} as follows:

$$s_{max} = \frac{A_v f_y}{0.0625\sqrt{f_c'} \text{(MPa)} b_w} \text{ or } \left(\frac{A_v f_y}{0.75\sqrt{f_c'} \text{(psi)} b_w} \right) \quad \begin{array}{l} V_u d/M_u \leq 0.25 \\ \\ V_u d/M_u \geq 0.5 \end{array} \qquad (8.28a)$$

$$s_{max} = \frac{A_v f_y}{0.125\sqrt{f_c'} \text{(MPa)} b_w} \text{ or } \left(\frac{A_v f_y}{1.5\sqrt{f_c'} \text{(psi)} b_w} \right) \quad 0.25 < V_u d/M_u < 0.5 \qquad (8.28b)$$

8.3.6 Comparisons of Shear Design Methods with Tests

8.3.6.1 UH Shear Design Method

The analytical shear strengths V_{cal} calculated from Equations (8.18), (8.24) and (8.28a or b) are compared with the experimental shear strengths V_{exp} for 148 prestressed I-beams tested by 14 groups of researchers available in the literature. The V_{exp}/V_{cal} ratios for the UH method are plotted against the span-to-depth ratio *a/d* in Figure. 8.20. It can be seen that the V_{exp}/V_{cal} ratios for all 148 beams vary from 1.0 to about 2.0.

Closer examination of the V_{exp}/V_{cal} ratios in Figure 8.20 reveals that this strength ratio is strongly influenced by the sizes of the test specimens. Let's define beams with a height greater

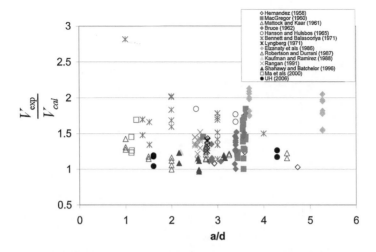

Figure 8.20 V_{exp}/V_{cal} ratios of beams using UH method

than 500 mm (20 in.) as large, and those less than 500 mm (20 in.) as small. The V_{exp}/V_{cal} ratios of the 58 large specimens (Mattock and Kaar, 1961; Lyngberg, 1976; Robertson and Duranni, 1987; Kaufman and Ramirez, 1988; Rangan, 1991; Shahawy and Batchelor, 1996; Ma *et al.* 2000; and Laskar *et al.* 2006) are then plotted in Figure 8.21. It can be seen that the V_{exp}/V_{cal} ratios of large specimens vary from 1.0 to about 1.5, much less than the scatter observed in Figure 8.20. Apparently, the strength ratios of the 90 small specimens are more conservative and have larger scatter. Besides the effect of specimen size on the shear strength, the wide scatter of V_{exp}/V_{cal} ratios is anticipated, because the smaller beams tend to be those tested earlier, while the larger beams were tested more recently.

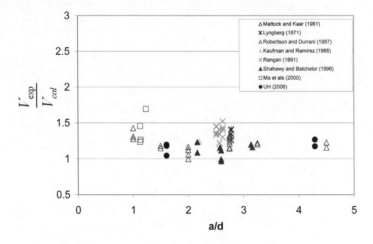

Figure 8.21 V_{exp}/V_{cal} ratios of large beams using UH method

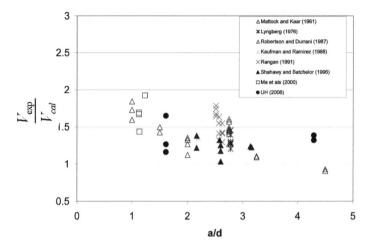

Figure 8.22 V_{exp}/V_{cal} ratios of large beams using ACI method

8.3.6.2 ACI and AASHTO Shear Design Methods

The V_{exp}/V_{cal} ratios of the 58 large specimens according to the ACI method (see Section 8.3.1.1) are plotted in Figure 8.22. The V_{exp}/V_{cal} ratios vary from 1.0 to about 2.0. Obviously, the accuracy of the ACI method in predicting the shear strengths is not as good as the UH method. This is because the ACI method is strongly influenced by the small and older test specimens, while the UH method places more emphasis on the large and more recent specimens.

The V_{exp}/V_{cal} ratios of the 58 large specimens according to the AASHTO method (see Section 8.3.1.2) are plotted in Figure 8.23. The V_{exp}/V_{cal} ratios in this figure have very large

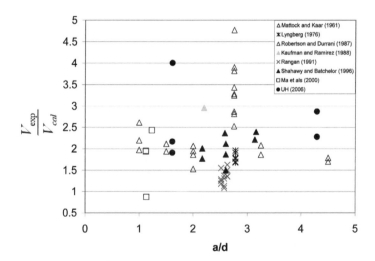

Figure 8.23 V_{exp}/V_{cal} ratios of large beams using AASHTO method

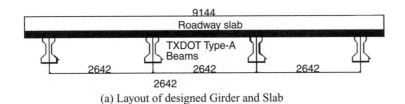

(a) Layout of designed Girder and Slab

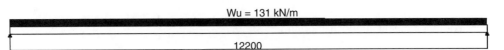

(b) Elevation of Designed Girder

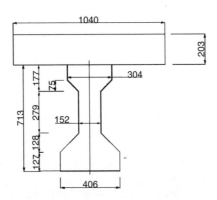

(c) Typical Cross Section of Designed Girder

Figure 8.24 Design example 1 using TxDOT type A beams (units are in mm unless otherwise mentioned) (Note: 1 mm = 0.039 inches, 1 kN/m = 0.069 kips/ft)

scatter, ranging from 0.872 to 4.77. This scatter is much larger than for the UH method (Figure 8.21) and the ACI method (Figure 8.22).

8.3.7 Shear Design Example

8.3.7.1 Problem Statement

TxDOT Type A girders, 12.2 m (40 ft) long, are designed to support a deck slab 9.14 m (30 ft) wide and 203 mm (8 in.) thick, as shown in Figure 8.24(a–c). A uniform load of 49.6 kN per square meter (1.04 kips per square foot) requires four girders spaced at 2.64 m (8.67 ft) c/c. Each girder is reinforced with 18 seven-wire, low-lax strands of 12.7 mm (1/2-in.) diameter and a strength of 1862 MPa (270 ksi). The moment capacity of a composite girder is 2617 kN m (1929 kip ft). The reduction factors ϕ are based on ACI-ASCE Joint Committee 343 Report (1995, R2004).

The values of various quantities required for shear design are as follows:

$h_{girder} = 711.2$ mm (28 in.)	$h_{composite} = 914.4$ mm (36 in.)	$l = 12.2$ m (40 ft)
$b_w = 152.4$ mm (6 in.)	$f_{c'} = 68.95$ MPa (10 ksi)	$f_y = 413.7$ MPa (60 ksi)
$A_v = 142.0$ mm^2 (0.22 in.^{2})	$A_{ps} = 1777$ mm^2 (2.75 in.)	$w_u = 131$ kN/m (9 kips/ft)
$V_p = 0$	ϕ for shear $= 0.9$	ϕ for bending $= 0.95$

Maximum moment at midspan of girder: $M_u = \dfrac{1}{\phi}\dfrac{w_u l^2}{8} = \dfrac{1}{0.95}\dfrac{131 \times 12.2^2}{8}$

$$= 2\,565 \text{ kN m } (1\,891 \text{ kip ft}) < 2\,617 \text{ kN m } (1\,929 \text{ kip ft}) \text{ OK}$$

8.3.7.2 UH Shear Design Solution

*Calculate effective depth **d***

Distance of the centroid of prestressing strands from bottom of beam $= 107$ mm (4.2 in.)
Effective depth of the composite section, $d = h - 107 = 914.4 - 107 = 807$ mm (31.8 in.)
$d_{min} = 0.8h = 0.8 \times 914.4 = 732$ mm (28.8 in.) < 807 mm (31.8 in.)
$d = 807$ mm (31.8 in) governs

*Check cross section at distance **d** from support*

The size of beam cross-section is checked at the critical section a distance d from the support.
  Shear force at critical section

$$V_u = \left(\frac{w_u l}{2} - w_u d\right) = \left(\frac{131 \times 12.2}{2} - 131 \times \frac{807}{1000}\right) = 693 \text{ kN (156 kips)}$$

Maximum shear capacity at critical section

$$\phi V_{n,max} = \phi 1.33\sqrt{f_c'(\text{MPa})}b_w d = 0.9 \times 1.33 \times \sqrt{68.95} \times 152.4 \times 807$$
$$= 1222 \text{ kN (275 kips)} > 693 \text{ kN (156 kips) OK}$$

The size of the chosen cross section is adequate.

Calculate V_c

The amounts of transverse steel at all the chosen sections along one-half of the beam are calculated using a spreadsheet, and the crucial values are shown in Table 8.2 under the column UH. The section that requires the smallest stirrup spacing is found to be at a distance 3.05 m (10 ft) from the support. Physically, failure at the quarter point of a simple span represents a typical flexural-shear failure mode. Calculations of V_c at this critical section are as follows:
  Factored shear force at the section

$$V_u = \left(\frac{w_u l}{2} - w_u x\right) = \left(\frac{131 \times 12.2}{2} - 131 \times 3.05\right) = 400 \text{ kN (90 kips)}$$

Table 8.2 Comparison of three design methods (UH, ACI and AASHTO) for design example 1

				UH				ACI					AASHTO					
x (m)	V_u (kN)	M_u (kN m)	$\dfrac{V_u d}{M_u}$	V_c (kN)	V_s (kN)	s (#3) (mm)	s_{prov} (mm)	V_{ci} (kN)	V_{cw} (kN)	V_s (kN)	s (#3) (mm)	s_{prov} (mm)	θ (°)	β	V_c (kN)	V_s (kN)	s (#3) (mm)	s_{prov} (mm)
0.457[1]	739	352						2481	369	452	105	102	22.5	3.14	263	509	222	203
0.797[2]	695	595																
0.807[3]	693	603	0.928	848			305	1160	400	310	153	102	22.5	3.14	263	448	253	203
1.22	639	877	0.589	825			305	736	400	221	214	203	21.8	3.75	314	308	381	305
1.83	559	1243	0.363	589	33	519	305	519	400	132	358	203	26.6	2.94	246	287	326	305
2.44	479	1560	0.248	451	82	338	305	384	400	60	788	610	30.5	2.59	217	227	350	305
3.05	400	1828	0.177	356	88	323	305	289	400	66	713	610	33.7	2.38	199	156	450	305
3.66	320	2047	0.126	281	74	356	305	216	400	51	934	610	36.4	2.23	187	80	797	305
4.27	240	2218	0.087	217	49	439	305	155	400	23	2103	610	40.8	1.95	163	14	3776	559
4.88	160	2340	0.055	157	20	600	406	145	400			610	40.8	1.95	163			559
5.49	80	2413	0.027	95			406					610	40.8	1.95	163			559
6.1		2437					406											559

[1] Section $h/2$; [2] section d_v; [3] section d

Note: 1 m = 3.28 ft, 1 kN = 0.225 kips, 1 kN m = 0.74 kip ft, 1 mm = 0.039 in.

Factored moment at the section

$$M_u = \left(\frac{w_u l}{2}x - \frac{w_u x^2}{2}\right) = \left(\frac{131 \times 12.2}{2} \times 3.05 - \frac{131 \times 3.05^2}{2}\right)$$

$$= 1828 \text{ kN m } (1347 \text{ kip ft})$$

$$\frac{V_u d}{M_u} = \frac{400 \times 807}{1828 \times 1000} = 0.177$$

$$V_c = 1.17 \left(\frac{V_u d}{M_u}\right)^{0.7} \sqrt{f_c'(\text{MPa})}b_w d = 1.17 \times 0.177^{0.7} \times \sqrt{68.95} \times 152.4 \times 807$$

$$= 356 \text{ kN } (80 \text{ kips})$$

Maximum concrete contribution to shear capacity of the beam

$$V_{c,\max} = 0.83\sqrt{f_c'}b_w d = 0.83 \times \sqrt{68.95} \times 152.4 \times 807 = 848 \text{ kN } (191 \text{ kips})$$

$$> 356 \text{ kN } (80 \text{ kips}).$$

$$V_c = 356 \text{ kN } (80 \text{ kips}) \text{ governs}$$

Calculate amount of transverse steel

$$V_s = V_u/\phi - V_c - V_p = 400/0.9 - 356 - 0 = 88 \text{ kN } (20 \text{ kips})$$

Using two-legged #3 rebar ($A_v = 142.0 \text{ mm}^2$ or 0.22 in.^2) as shear reinforcement, the spacing required to provide the required V_s is:

$$V_s = A_v f_y \left(\frac{d}{s} - 1\right)$$

$$\Rightarrow s = \frac{d}{\left(\dfrac{V_s}{A_v f_y} + 1\right)} = \frac{807}{\left(\dfrac{88 \times 10^3}{142 \times 413.7} + 1\right)} = 323 \text{ mm } (12.7 \text{ in.})$$

Provide two-legged #3 rebars at 305 mm (12 in. c/c).
$s_{\max} = d/2 = 807/2 \approx 406 \text{ mm } (16 \text{ in.}) < 610 \text{ mm } (24 \text{ in.})$
$s_{\max} = 406 \text{ mm } (16 \text{ in.})$ is the maximum governing stirrup spacing throughout the length of the girder.

Check minimum stirrup requirement
At the critical section 3.05 m from the support:

$$V_u d/M_u = 0.177 < 0.25$$

Minimum stirrup requirement gives s_{max} as:

$$s_{max} = \frac{A_v f_y}{0.0625\sqrt{f_c'\,(\text{MPa})}b_w} = \frac{142(413.7)}{0.0625\sqrt{68.95}(152.4)}743 \text{ mm } (29.3 \text{ in.}) > 406 \text{ mm}$$

(16 in.) > 305 mm (12 in.) does not govern

Arrangement of stirrups along the beam

Table 8.2 shows the shear design calculations over one half-span of the beam under the column UH. Since this girder is critical in flexural-shear at a section 3.05 m (10 ft) from the support, we adopt the concept that 'transverse steel should be the same from the flexural-shear critical section to the support'. Then we should provide #3 rebars with a spacing of 305 mm (12 in.) in the whole region from the support to a distance of 4.27 m (14 ft) from the support. Beyond 4.27 m (14 ft), the spacing is increased to 406 mm (16 in.).

8.3.7.3 ACI and AASHTO Shear Design Solutions

The shear design calculations of the above example using the UH method are compared in Table 8.2 with those designed by the ACI method (see Section 8.3.1.1) and the AASHTO method (see Section 8.3.1.2). Table 8.2 records the crucial steps of shear design calculations of these two methods under the column ACI and the column AASHTO. Detailed calculations of shear design using ACI and AASHTO methods can be found in a report by Hsu *et al.* (2008).

In Table 8.2 under the column ACI, the required stirrups spacing is most critical at a distance $h/2$ from the support. The provided stirrup spacing of the #3 rebars is 102 mm (4 in. c/c) up to a distance 1.22 m (4 ft) from the support. The spacing is thereafter increased to 203 mm (8 in.) between 1.22 m (4 ft) and 2.44 m (8 ft). Beyond 2.44 m (8 ft) the spacing is further increased to 610 mm (24 in.) up to midspan.

In Table 8.2 under the column ASHTO, the required stirrup spacing is most critical at a distant d_v from the support. The provided stirrup spacings of the #3 rebars is 203 mm (8 in. c/c) up to a distance 1.22 m (4 ft) from the support. The spacing is thereafter increased to 305 mm (12 in. c/c) up to a distance 3.66 m (12 ft) from the support. Beyond 3.66 m, the spacing is 559 mm (22 in. c/c).

In summary, the ACI and AASHTO methods would predict the beam to fail in web-shear failure near the support. The provided stirrup spacing will be small near the support and will increase rapidly from the support to the one-quarter point of the beam. In other words, these two shear design methods will result in excessive stirrups near the support and insufficient stirrups near the quarter point of the beam. In contrast, the UH shear design method requires a constant and moderate stirrup spacing from the support to about the one-third point of the beam.

8.3.8 Three Shear Design Examples

The report by Hsu *et al.* (2008) also includes two more bridge design examples solved in detail by the three shear design methods (UH, ACI, AASHTO). The bridge example given in Section 8.3.7 is called Example 1, and the other two examples are called Examples 2 and 3.

In contrast to the long and lightly loaded girder in Example 1, the girder in Example 2 is short and is subjected to heavy load. The UH method indicates that this girder in Example 2 fails in web-shear at a critical section at distance d from the support. The stirrup spacing increases slowly from the support toward the quarter-point of the span. Using the ACI and AASHTO methods, the girder also fails in web-shear near the support, but the stirrup spacing increases much faster from the support toward the quarter-point of the span.

Example 3 uses two large AASHTO Type VI prestressed girders, each with a large height of 1.83 m (6 ft) and a large span of 33.5 m (110 ft), made continuous for live loads (Modjeski and Masters Inc., 2003). The shear forces and bending moments in this third example are calculated using influence lines for HS-20 truck load and uniform lane load defined by the AASHTO-LRFD Specifications. Such girders are found to be critical in flexural-shear near the quarter point of a span. In contrast, the ACI and AASHTO methods would indicate that these girders would fail in web-shear near the support.

In short, the shear designs of girders in Examples 1 and 3 using the ACI and AASHTO methods would wrongly identify these girders as critical in web-shear failure near the supports, rather than flexural-shear failure near the quarter-points of their spans. As a result, these two shear design methods would over-reinforce the regions near the support and would under-reinforce the regions near the quarter point of the beam.

9

Finite Element Modeling of Frames and Walls

9.1 Overview

9.1.1 Finite Element Analysis (FEA)

The rapid development of computers has completely revolutionalized research and practice in every scientific and engineering field. The dream that every office and home would have a computer terminal and/or personal computers has become a reality. Since the 1990s personal computers have become as popular as pocket calculators were in the 1970s and slide rules in the 1950s. Following this trend, analysis and design methods that provide computerized solutions to scientific and engineering problems have been developing rapidly for increasingly routine use. In this chapter, we focus on one such significantly developed method, the finite element method. Although this method is applicable to many scientific and engineering fields, we deal only with the field of structural analysis and design.

The finite element method has long been a fertile research field. It has also been increasingly used as a research tool for numerical experiment. Most importantly, the finite element method has become an analysis and design tool used routinely by structural engineers.

The finite element method in structural analysis is a technique that first divides a structure into a set, or different sets, of structural components. The components of a particular set would share some similar geometric pattern and physical assumption. Each pattern of such components is called a specific kind of *finite element*. Each kind of finite element has a specific type of structural shape and is interconnected with the adjacent elements by nodal points.

Acting at each node point are nodal forces, and the node is subjected to displacements (degrees of freedom). In a general sense, these nodal physical quantities are not limited to being forces and displacements, but extend to cases involving thermal, fluid, electrical, and other problems. Thus for each element, a standard set of simultaneous equations can be formulated to relate these physical quantities. Physically assembling these elements to form the whole structure is equivalent to superimposing these element equations mathematically. The result is a large set of simultaneous equations, which are suited for solution by computer. Upon implementing the loading and boundary conditions for structural problems, the assembled set

Unified Theory of Concrete Structures   Thomas Hsu and Yi-Lung Mo
© 2010 John Wiley & Sons, Ltd

of equations can be solved and the unknown parameter found. Substituting these values back to each element formulation provides the distributions of stress and displacement everywhere within each element. This solution procedure is called 'finite element analysis' (FEA). The concept of FEA comes from the idea of discretization and numerical approximation.

In many reinforced concrete (RC) structures, beams and columns can be simulated by 1-D fiber elements. Some other RC structures, such as shear walls, deep girders, nuclear containment vessels, and offshore platforms, can be visualized as assemblies of plane stress (membrane) elements, as shown in Figure 9.1. The behavior of each whole structures can be predicted by performing FEA.

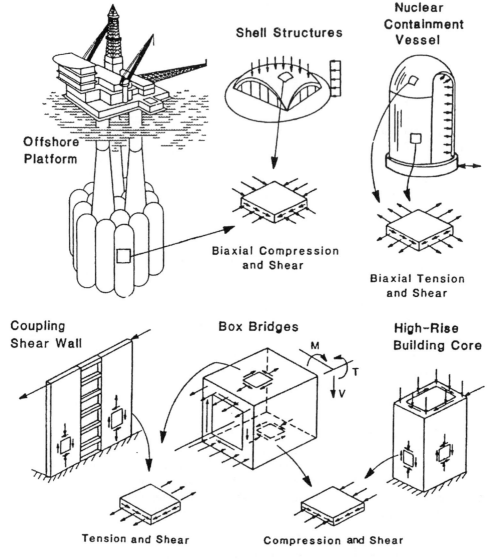

Figure 9.1   Discretization of RC structures into membrane elements

FEA of reinforced concrete structures is a highly nonlinear problem because of the nonlinear constitutive relationships of reinforcement and concrete. The nonlinear problem can be solved using the iterative solution algorithm until convergence is achieved with acceptable accuracy. During the past 20 years many advanced algorithms, such as the line-searches method, the quasi-Newton method with accelerations, the arc-length method, automatic increments and the re-starts method have been developed for nonlinear analysis (Crisfield, 1997). By applying the advanced solution algorithm, fast and stable convergence can be achieved, and acceptable accuracy can be obtained. Today, the establishment of the material constitutive relationships, improvement of the finite element method, and rapidly increasing computer power make it possible to perform nonlinear FEA on reinforced concrete structures subjected to reversed cyclic or dynamic loading.

The finite element computer programming of reinforced concrete structures represents an enormous amount of work. Most nonlinear finite element computer programs involve three similar procedures: (1) building finite element models such as nodes, elements, materials, and loads; (2) solving nonlinear equations using the iterative solution algorithm; and (3) recording the results of analysis. It would be an extremely laborious task for individual researchers to develop their own finite element programs. Moreover, it would not be economical for researchers to invest the time and effort in developing the general finite element code. Thus, an increasing number of researchers tend to use general finite element programs or frameworks such as FEAP, ABAQUS, or OpenSees, which offer users the ability to implement their own elements and constitutive laws. Traditional general finite element programs, such as FEAP (Taylor, 1999), were usually written in FORTRAN, which was characterized as a function-oriented language. In recent years, however, object-oriented languages such as C++ and Java have been proven to significantly improve the extendability and usability of the software over the function-oriented languages. Most notable was the finite element framework OpenSees (Fenves, 2005) in applying objected-oriented languages to FEA.

9.1.2 OpenSees – an Object-oriented FEA Framework

OpenSees stands for Open System for Earthquake Engineering Simulation (Fenves, 2005). It was developed at the Pacific Earthquake Engineering Center (PEER), the University of California, Berkeley. OpenSees is an object-oriented framework for simulation applications in earthquake engineering using finite element methods. Key features of OpenSees include the interchangeability of components and the ability to integrate existing libraries and new components into the framework without the need to change the existing code (Fenves, 2005). This makes it convenient to implement new classes of elements, materials and other components. Many advanced finite element techniques that are suitable for the nonlinear FEA have already been implemented in OpenSees. As it stands, however, OpenSees is not readily applicable to reinforced concrete plane stress structures because no suitable constitutive models for such structures are included. That was the situation until recent studies done at the University of Houston (Mo *et al.*, 2008)

The OpenSees framework is adopted in this book. The constitutive model CSMM is implemented and a nonlinear finite element computer program, Simulation of Concrete Structures (SCS), is developed. This program is capable of predicting the nonlinear behavior of reinforced concrete structures subjected to static, reversed cyclic, and dynamic loadings. The program is then validated with studies on different types of structures, as reported in the literature, including panels, prestressed beams, framed shear walls, post-tensioned bridge columns, and a seven-story wall building, as demonstrated in Chapter 10.

9.1.3 Material Models

There are two key points that need to be addressed in the application of nonlinear FEA to reinforced concrete (RC) structures: the establishment of nonlinear constitutive models of RC elements and the development of nonlinear finite element methods specifically for reinforced concrete structures. The behavior of reinforced concrete elements should be thoroughly understood before being applied to the finite element method. Many researchers have developed different types of analytical models of reinforced concrete such as truss models, orthotropic models, nonlinear elastic models, plastic models, micro models, etc. The orthotropic model stands out both in accuracy and in efficiency as compared with the other models. Since the 1980s, orthotropic models have been developed to predict the shear behavior of reinforced concrete elements by many researchers: Vecchio and Collins (1981, 1982), Balakrishnan and Murray (1988), Crisfield and Wills (1989), Izumo et al. (1992), Shin et al. (1991), Hsu (1993), Belarbi and Hsu (1995), Pang and Hsu (1995), Sittipunt and Wood (1995), Pang and Hsu (1996), Hsu and Zhang (1997), Ayoub and Fillippou (1998), Kaufmann and Marti (1998), Vecchio (2000, 2001), Ile and Reynouard (2000), Belletti et al. (2001), Hsu and Zhu (2002), Kwon and Spacone (2002), Palermo and Vecchio (2003), Foster et al. (2004), and Mansour and Hsu (2005a, 2005b).

Over the past 20 years, extensive experimental and theoretical studies on the behavior of reinforced concrete have been carried out by the research group at the University of Houston (UH). A series of four analytical models were established to predict the nonlinear shear behavior of reinforced concrete membrane elements. The models are: (1) the RA-STM by Hsu (1988, 1993), Belarbi and Hsu (1995), and Pang and Hsu (1995); (2) the FA-STM by Pang and Hsu (1996) and Hsu and Zhang (1997); (3) the SMM by Hsu and Zhu (2002); and (4) the CSMM by Mansour and Hsu (2005a, 2005b) and Laskar et al. (2008). All four models are rational because they satisfy Navier's three principles of mechanics of materials: the stress equilibrium, the strain compatibility, and the constitutive relationships of materials.

Among these constitutive models, the cyclic softened membrane model (CSMM) presented in Section 6.3 is the most versatile and accurate. It is capable of predicting the cyclic shear behavior of reinforced and prestressed concrete membrane elements (Hsu and Mansour, 2005; Laskar et al., 2008) including the stiffness, ultimate strength, descending branch, ductility, and energy dissipation. In this book, the CSMM is employed and implemented in the finite element method to predict the behavior of reinforced concrete structures.

9.1.4 FEA Formulations of 1-D and 2-D Models

We adopt the fiber model for 1-D structures, such as beams, columns and frames in this book. In the fiber model, the force-based finite element formulation is employed by following the classical flexibility approach. The formulation for FEA of 1-D structures has been reported in detail (Taucer et al. 1991). For 2-D structures, such as plane stress problems, CSMM is very appropriate, and its finite element formulation is briefly introduced in this Chapter (Zhong, 2005; Hsu et al., 2006; Mo et al., 2008). Basically, the coordinate systems and the corresponding transformation matrix between any two coordinate systems for finite element formulation need to be set up as the first step. Then the equilibrium and compatibility equations are derived, as shown in Section 6.1.1. Afterwards, the relationship between biaxial strains and uniaxial strains is needed, as introduced in Section 6.1.4.4. Finally, uniaxial constitutive relationships of the reinforced concrete materials developed in Section 6.1.6 to 6.1.10 are employed, including uniaxial constitutive relationships of concrete in compression and tension, uniaxial constitutive relationship of mild steel bars embedded in concrete, and constitutive relationship of concrete in shear. The uniaxial constitutive relationships required for prestressed concrete were derived by Wang (2006) and Laskar (2008).

9.2 Material Models for Concrete Structures

In both the fiber model (FM) and CSMM, all the required material models for concrete, reinforcing bars and prestressing tendons of reinforced/prestressed concrete have been implemented in OpenSees (Fenves, 2005). These material models are summarized in Table 9.1 and are presented in the following section.

Table 9.1   Uniaxial material models employed

Model name	Material type	Source
Steel01	Steel	Existing in OpenSees
Concrete01	Concrete	Existing in OpenSees
SteelZ01	Steel	Developed based on CSMM
TendonL01	Tendons	Developed based on CSMM-PC
ConcreteL01	Concrete	Developed based on CSMM-PC

9.2.1 Material Models in OpenSees

9.2.1.1 Steel01

The steel module Steel01 needed in the FM is based on a uniaxial bilinear material model as shown in Figure 9.2. It has an elastic phase before yielding and a strain-hardening part after yielding. The unloading and reloading paths follow a bilinear pattern in which the slopes of

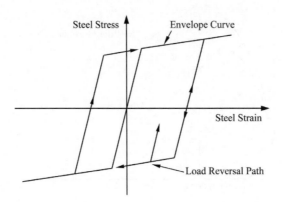

Figure 9.2   Steel01 material model

the paths are the same as the elastic modulus before yielding, and the hardening modulus after yielding. Defining the Steel01 steel model requires the yield stress and Young's modulus of the steel before and after yielding. This model has been implemented in OpenSees by researchers at UC Berkeley (Fenves, 2005).

9.2.1.2 Concrete01

The concrete module Concrete01 needed in the FM incorporates the modified Kent and Park material model (Scott *et al.* 1982) proposed for confined concrete with degraded linear unloading and reloading paths according to the work of Karsan and Jirsa (1969). No tensile strength is considered in Concrete01. The modified Kent and Park model is illustrated in Figure 9.3 and is summarized as follows:

$$\text{For } \varepsilon \leq 0.002K, \ \ f_c = Kf_c' \left[\frac{2\varepsilon}{0.002K} - \left(\frac{\varepsilon}{0.002K} \right)^2 \right], \tag{9.1}$$

$$\text{For } \varepsilon > 0.002K, \ \ f_c = Kf_c' \left[1 - Z \left(\varepsilon - 0.002K \right) \right] \geq 0.2Kf_c, \tag{9.2}$$

where

$$K = 1 + \frac{\rho_s f_{yh}}{f_c'}, \tag{9.3}$$

$$Z = \frac{0.5}{\dfrac{3 + 0.29 f_c'}{145 f_c' - 1000} + \dfrac{3}{4} \rho_s \sqrt{\dfrac{b''}{S}} - 0.002K} \tag{9.4}$$

f_c = longitudinal compressive stress in concrete (MPa)

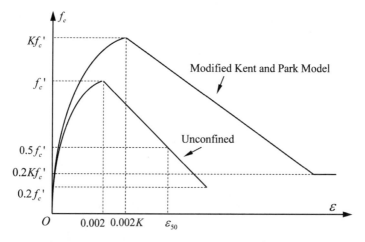

Figure 9.3 Modified Kent and Park model for monotonic stress–strain relationship of confined concrete

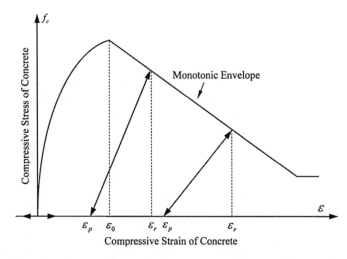

Figure 9.4   Unloading and reloading paths of stress–strain relationship of concrete in concrete01

f_c' = concrete compressive cylinder strength (MPa)
ε = longitudinal compressive strain in concrete
ρ_s = ratio of volume of hoop reinforcement to volume of concrete core
b'' = width of concrete core measured to outside of peripheral hoop (mm)
f_{yh} = yield stress of the confining stirrups (MPa)
S = spacing of the confining stirrups (mm)

When $K = 1$, the confined model will change to the unconfined model, as shown in Figure 9.3. The unloading and reloading paths of concrete01 illustrated in Figure 9.4 are described as follows:

1. Unloading from a point on the envelope curve takes place along a straight line connecting the point where the unloading commences (strain $=\varepsilon_r$) and the point where stress equals zero (strain $=\varepsilon_p$). The straight line defined by Equations (9.5) and (9.6) was proposed by Karsan and Jirsa (1969).

$$\frac{\varepsilon_p}{\varepsilon_0} = 0.145 \left(\frac{\varepsilon_r}{\varepsilon_0} \right)^2 + 0.13 \left(\frac{\varepsilon_r}{\varepsilon_0} \right), \quad \text{for} \ \left(\frac{\varepsilon_r}{\varepsilon_0} \right) < 2 \tag{9.5}$$

$$\frac{\varepsilon_p}{\varepsilon_0} = 0.707 \left(\frac{\varepsilon_r}{\varepsilon_0} - 2 \right)^2 + 0.834, \quad \text{for} \ \left(\frac{\varepsilon_r}{\varepsilon_0} \right) \geq 2 \tag{9.6}$$

2. The concrete stress is equal to zero for strains smaller than the strain at complete unloading (strain $= \varepsilon_p$). On reloading, the stress is zero as long as the strain is smaller than ε_p. For strains larger than ε_p, the reloading path follows the same straight line as the unloading path.

9.2.2 Material Models Developed at UH

9.2.2.1 SteelZ01

The steel module SteelZ01 needed in the CSMM incorporates both the envelope and the unloading/reloading pattern of uniaxial constitutive relationships of embedded mild steel (Figure 9.5). The equations for the envelope are given as Equations (9.7)–(9.11). The nonlinear unloading and reloading paths are described in Equations (9.12)–(9.15).

(Stage 1)
$$f_s = E_s \bar{\varepsilon}_s, \quad \left(\bar{\varepsilon}_s \leq \varepsilon'_y \right), \tag{9.7}$$

(Stage 2T)
$$f_s = f_y \left[(0.91 - 2B) + \left(0.02 + 0.25 B \frac{\bar{\varepsilon}_s}{\varepsilon_y} \right) \right], \quad \left(\bar{\varepsilon}_s > \varepsilon'_y \right) \tag{9.8}$$

$$\text{where } B = \frac{1}{\rho} \left(\frac{f_{cr}}{f_y} \right)^{1.5} \quad f_{cr} = 0.31 \sqrt{f'_c (\text{MPa})} \quad \text{and} \quad \rho \geq 0.15\% \tag{9.9}$$

$$\varepsilon'_y = \varepsilon_y (0.93 - 2B) \tag{9.10}$$

(Stage 2C)
$$f_s = -f_y \tag{9.11}$$

(Stage 3 and Stage 4)
$$\bar{\varepsilon}_s - \bar{\varepsilon}_{si} = \frac{f_s - f_i}{E_s} \left[1 + A^{-R} \left| \frac{f_s - f_i}{f_y} \right|^{R-1} \right] \tag{9.12}$$

where:

$$A = 1.9 k_p^{-0.1} \tag{9.13}$$

$$R = 10 k_p^{-0.2} \tag{9.14}$$

$$k_p = \frac{\bar{\varepsilon}_p}{\varepsilon'_y} \tag{9.15}$$

$\bar{\varepsilon}_p$ = smeared uniaxial plastic strain of rebars

Because the strains are expressed in terms of the stresses in Equation (9.12), iteration would be needed to calculate a stress based on a given strain. To bypass this iteration, a multilinear simplification was proposed by Jeng (2002) to approximate the nonlinear curves using straight-line segments.

In Figure 9.5, the dashed curves are the unloading and reloading paths defined by the CSMM, and the solid curves are the linear simplifications implemented in SteelZ01. Two turning points, $(\varepsilon_{m1}, f_{m1})$ and $(\varepsilon_{m2}, f_{m2})$ in Figure 9.5, are selected as the points at which $f_{m1} = \pm 0.65 f_y$ and $f_{m2} = 0$, and ε_{m1} and ε_{m2} are calculated by substituting f_{m1} and f_{m2} into f_s of Equation (9.12), respectively. Once the turning points are determined, the stress of the point on the line segments is a linear function of the strains and stresses of the turning points

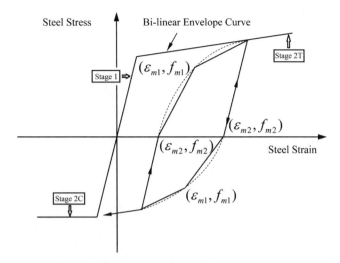

Figure 9.5   SteelZ01 material model

and can be easily calculated from a given strain. In this way, the iteration is bypassed with acceptable deviation from the original curves. To calculate A and R in Equations (9.13) and (9.14), default values of the coefficients are set as 1.9 and 10, respectively. Users can also define the values for the coefficients.

9.2.2.2   TendonL01

The tendon module TendonL01 needed in the CSMM for prestressed concrete incorporates both the stress–strain relationships of embedded tendons in tension, as predicted by SMM for prestressed concrete (Figure 9.6) along with the unloading/reloading pattern of uniaxial

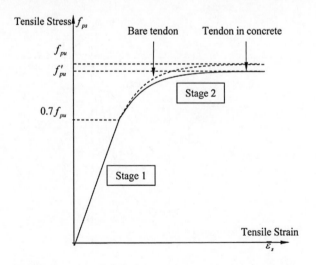

Figure 9.6   Constitutive relationship of prestressing tendons

constitutive relationships of embedded mild steel. The tensile stress–strain relationships of embedded tendons as per SMM for prestressed concrete are given in Equations (9.16) and (9.17). The initial prestressing strain in the tendons is also implemented as an input parameter in this module.

$$f_{ps} = E_{ps}\bar{\varepsilon}_s, \quad \bar{\varepsilon}_s < \frac{0.7 f_{pu}}{E_{ps}} \tag{9.16}$$

$$f_{ps} = \frac{E''_{ps}\bar{\varepsilon}_s}{\left[1 + \left(\dfrac{E''_{ps}\bar{\varepsilon}_s}{f'_{pu}}\right)^5\right]^{\frac{1}{5}}}, \quad \bar{\varepsilon}_s \geq \frac{0.7 f_{pu}}{E_{ps}} \tag{9.17}$$

where:

f_{ps} = stress in prestressing tendons for a given uniaxial strain $\bar{\varepsilon}_s$
E_{ps} = elastic modulus of prestressing tendons taken as 200 GPa (29 000 ksi)
f_{pu} = ultimate strength of prestressing tendons taken as 1862 MPa (270 ksi),
E''_{ps} = modulus of prestressing tendons, used in plastic region, taken as 209 GPa (30 345 ksi)
f'_{pu} = revised strength of prestressing tendons taken as 1793 MPa (260 ksi)

9.2.2.3 ConcreteL01

The concrete module ConcreteL01 was developed according to the cyclic uniaxial concrete model of CSMM for prestressed concrete (Figure 9.7). This cyclic uniaxial concrete model considers the softening effect due to the perpendicular tensile strain on the concrete struts.

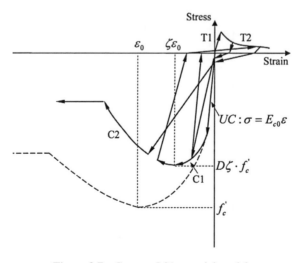

Figure 9.7   ConcreteL01 material model

The envelopes of the ConcreteL01 model in tension and compression are similar to those of the concrete constitutive model in CSMM for reinforced concrete. Only the compression envelope prior to the peak stress is changed in this module to keep the initial slope of the compressive stress–strain envelope lower than $2f_c'/\varepsilon_0$. The equations defining the envelopes are expressed as below.

Stage UC : $\qquad \sigma_1^c = E_c' \bar{\varepsilon}_1 + \sigma_{ci}, \ \bar{\varepsilon}_1 \le (\bar{\varepsilon}_{cx} - \bar{\varepsilon}_{ci})$ $\qquad$ (9.18)

Stage T1 : $\qquad \sigma_1^c = E_c''(\bar{\varepsilon}_1 + \bar{\varepsilon}_{ci} - \bar{\varepsilon}_{cx}), \ (\bar{\varepsilon}_{cx} - \bar{\varepsilon}_{ci}) < \bar{\varepsilon}_1 \le (\varepsilon_{cr} - \bar{\varepsilon}_{ci})$ $\qquad$ (9.19)

Stage T2 : $\qquad \sigma_1^c = f_{cr} \left(\dfrac{\varepsilon_{cr}}{\bar{\varepsilon}_1 + \bar{\varepsilon}_{ci}} \right)^{0.5}, \ \bar{\varepsilon}_1 > (\varepsilon_{cr} - \bar{\varepsilon}_{ci})$ $\qquad$ (9.20)

$$E_c = 3875\sqrt{f_c'(\text{MPa})}, \ f_{cr} = 0.31\sqrt{f_c'(\text{MPa})}, \ \varepsilon_{cr} = 0.00008$$

Stage C1 : $\qquad \sigma_2^c = D\zeta f_c' \left[2\left(\dfrac{\bar{\varepsilon}_2}{\zeta \varepsilon_0} \right) - \left(\dfrac{\bar{\varepsilon}_2}{\zeta \varepsilon_0} \right)^2 \right] \le \dfrac{kDf_c'}{\varepsilon_0} \bar{\varepsilon}_2, \ \dfrac{\bar{\varepsilon}_2}{\zeta \varepsilon_0} \le 1$ $\qquad$ (9.21)

Stage C2 : $\qquad \sigma_2^c = D\zeta f_c' \left[1 - \left(\dfrac{\bar{\varepsilon}_2/\zeta \varepsilon_0 - 1}{4/\zeta - 1} \right)^2 \right], \ \dfrac{\bar{\varepsilon}_2}{\zeta \varepsilon_0} > 1$ $\qquad$ (9.22)

$$\zeta = f(f_c') f(\bar{\varepsilon}_1) f(\beta) W_p \le 0.9$$ $\qquad$ (9.23)

$$f(f_c') = \dfrac{5.8}{\sqrt{f_c'}} \le 0.9 \ (f_c' \text{ in MPa})$$ $\qquad$ (9.24)

$$f(\bar{\varepsilon}_1) = \dfrac{1}{\sqrt{1 + 400\bar{\varepsilon}_1}}$$ $\qquad$ (9.25)

$$f(\beta) = 1 - \dfrac{|\beta|}{24°}$$ $\qquad$ (9.26)

$$W_p = 1.15 + \dfrac{|\beta|(0.09|\beta| - 1)}{6}$$ $\qquad$ (9.27)

$$\beta = \dfrac{1}{2} \tan^{-1} \left[\dfrac{\gamma_{12}}{(\varepsilon_1 - \varepsilon_2)} \right]$$ $\qquad$ (9.28)

where:

σ_1^c = smeared (average) concrete stress in the 1–direction
σ_2^c = smeared (average) concrete stress in the 2–direction
σ_{ci} = initial compressive stress in concrete due to prestress force
$\bar{\varepsilon}_1$ = uniaxial smeared (average) principal strain in the 1–direction

$\bar{\varepsilon}_2 =$ uniaxial smeared (average) principal strain in the 2−direction

$\bar{\varepsilon}_{ci} =$ initial strain in concrete

$\bar{\varepsilon}_{cx} =$ extra strain in concrete after decompression

$\varepsilon'_c =$ maximum compression strain normal to the compression direction under consideration (always negative)

$\varepsilon_{cr} =$ concrete cracking strain

$\varepsilon_0 =$ concrete cylinder strain corresponding to peak compressive stress f'_c

$\varepsilon_1 =$ biaxial smeared (average) principal strain in the 1−direction

$\varepsilon_2 =$ biaxial smeared (average) principal strain in the 2−direction

$\gamma_{12} =$ smeared (average) shear strain in the 1 − 2 coordinate

$E'_c =$ decompression modulus of concrete, given as kf'_c/ε_0

$k = 1.4$–1.5

$E''_c =$ modulus of concrete in tension before cracking

$f_{cr} =$ cracking tensile strength of concrete

$D =$ damage coefficient $= 1 - 0.4(\varepsilon'_c/\varepsilon_o) \leq 1.0$

$f'_c =$ cylinder compressive strength of concrete

$\zeta =$ softening coefficient

The constant, 400 in Equation (9.25), is changed to 250 when sequential loading is applied. The unloading and reloading paths defined in the concrete constitutive model in CSMM for reinforced concrete were simplified in ConcreteL01 and are illustrated in Figure 9.7. The unloading and reloading paths from the ascending branch of the compressive envelope are simplified as one straight line and the slope is taken as the initial modulus of concrete E_{c0}, where

$$E_{c0} = kf'_c/\varepsilon_0 \qquad (9.29)$$

The slope of the unloading and reloading paths from the descending branch of the compressive envelope are simplified as one straight line with slope of $0.8E_{c0}$. The reloading paths from tension to compression reflect the full closing of the concrete cracks before reaching the peak point in the compressive envelope and the partial closing of the concrete cracks after exceeding the peak point.

9.3 1-D Fiber Model for Frames

The fiber element approach, originally developed by Professor Filippou and his co-workers (Taucer *et al.*, 1991) has been recognized as one of the most promising methods for static and dynamic analysis of RC frame structures since the late 1980s. By using the fiber element approach, the hysteretic behavior of RC members can be captured directly from cyclic stress and strain relationships of the materials; the axial force–biaxial bending interaction can be rationally accounted for. Also, it has been found that the state-of-the-art force-based approach generally has superior robustness and requires fewer degrees of freedom for comparable accuracy when dealing with strength softening elements such as RC members. Hence the force-based fiber element approach is utilized in the model.

A schematic figure from Taucer *et al.* (1991) to conceptually illustrate the configuration of a fiber beam–column element is replicated in Figure 9.8. Each of the control cross-sections is subdivided into many fibers or filaments; some are concrete fibers while the others may be steel fibers. There are a number of control cross-sections along the element. These sections are located at the control points of the numerical integration scheme. The characteristics of the

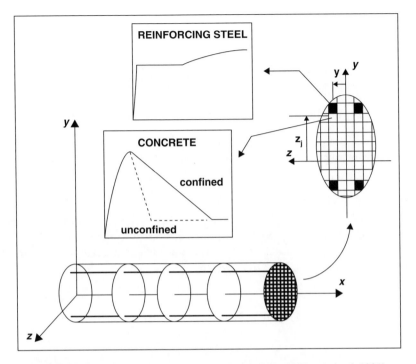

Figure 9.8   Illustration of fiber beam–column element (Taucer *et al.* 1991)

element, such as element flexibility and element displacements, are obtained from integrating their sectional components along the element. The corresponding sectional components, such as section flexibility and section forces, are calculated from summing the contributions of all the fibers on the specific cross-section. The section force and deformation relation of RC fiber section, the fiber element formulation, and the analysis procedures of the fiber model can be found in the project report by Taucer *et al.* (1991).

9.4 2-D CSMM Model for Walls

As introduced previously, the CSMM is a rational model that can readily be applied to a plane stress problem, such as reinforced concrete walls subject to reversed cyclic loading. Using this model the coordinate systems, implementation and analysis procedures are presented below.

9.4.1 Coordinate Systems for Concrete Structures

Three Cartesian – coordinates, x–y, 1–2, and si–ti, are defined in the reinforced concrete elements, as demonstrated in Figure 9.9. Coordinate x–y represents the local coordinate of the elements. The coordinate 1–2 defines the principal stress directions of the applied stresses, which have an angle α_{1x} with respect to the x–y coordinate. Steel bars can be distributed in different directions in the concrete. Coordinate si–ti shows the ith direction of the reinforcing steel bars, where the ith steel bars are located in the direction of axis si at an angle α_{ix} to the x–y coordinate.

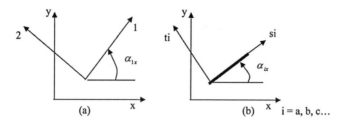

Figure 9.9 Coordinate systems for reinforced concrete elements: (a) applied principal stresses in local coordinate (b) reinforcement component in local coordinate

The stress and strain vectors in x–y coordinates and 1–2 coordinates are denoted as $\left\{ \sigma_x \ \sigma_y \ \tau_{xy} \right\}^{\mathrm{T}}$, $\left\{ \varepsilon_x \ \varepsilon_y \ \tfrac{1}{2}\gamma_{xy} \right\}^{\mathrm{T}}$, $\left\{ \sigma_1 \ \sigma_2 \ \tau_{12} \right\}^{\mathrm{T}}$ and $\left\{ \varepsilon_1 \ \varepsilon_2 \ \tfrac{1}{2}\gamma_{12} \right\}^{\mathrm{T}}$, respectively. Here $\tau_{12} = 0$ because the 1–2 coordinate represents the principal stress directions.

By using the transformation matrix $[T(\alpha)]$, the stresses and strains can be transformed between different coordinates. $[T(\alpha)]$ is given by

$$[T(\alpha)] = \begin{bmatrix} c^2 & s^2 & 2sc \\ s^2 & c^2 & -2sc \\ -sc & sc & c^2 - s^2 \end{bmatrix}. \qquad (9.30)$$

where $c = \cos(\alpha)$ and $s = \sin(\alpha)$, and the angle α is the angle between the two coordinates. For example, if the angle α is from the x–y coordinate to the 1–2 coordinate the angle α will be written as α_{1x}. If the angle α is from the 1–2 coordinate to the x–y coordinate the angle α will be written as α_{x1}. Of course, $\alpha_{x1} = -\alpha_{1x}$.

The stresses and strains transformed from the x–y coordinate to the 1–2 coordinate using the transformation matrix are expressed as follows:

$$\begin{Bmatrix} \sigma_1 \\ \sigma_2 \\ 0 \end{Bmatrix} = [T(\alpha_{1x})] \begin{Bmatrix} \sigma_x \\ \sigma_y \\ \tau_{xy} \end{Bmatrix} \qquad (9.31)$$

$$\begin{Bmatrix} \varepsilon_1 \\ \varepsilon_2 \\ \tfrac{1}{2}\gamma_{12} \end{Bmatrix} = [T(\alpha_{1x})] \begin{Bmatrix} \varepsilon_x \\ \varepsilon_y \\ \tfrac{1}{2}\gamma_{xy} \end{Bmatrix} \qquad (9.32)$$

9.4.2 Implementation

For a 2-D membrane element, a 'material constitutive matrix' is needed to relate the state of stresses and strains. The material constitutive matrix (which may also be known as the 'material stiffness matrix') can be expressed in terms of secant or tangent formulations. The secant material constitutive matrix relates the absolute values of strains and stresses of the element, while the tangent material constitutive matrix defines the relationship between the increment of the stresses and strains of the element. In this section, the material constitutive matrix for the reinforced concrete membrane material using CSMM is derived in terms of tangent formulations. The details of the derivation of the secant and tangent material constitutive matrices are presented in (Zhong, 2005). The tangent material constitutive matrix is implemented in the FEA in this book since OpenSees adopts the tangent stiffness formulation.

A tangent material constitutive matrix $[D]$ for a reinforced concrete plane stress element is formulated as:

$$[D] = \cfrac{d\left\{ \begin{array}{c} \sigma_x \\ \sigma_y \\ \tau_{xy} \end{array} \right\}}{d\left\{ \begin{array}{c} \varepsilon_x \\ \varepsilon_y \\ \frac{1}{2}\gamma_{xy} \end{array} \right\}} \tag{9.33}$$

$[D]$ is evaluated by

$$[D] = [T\,(\alpha_{x1})]\,[D_c]\,[V]\,[T\,(\alpha_{1x})] + \sum_i [T\,(\alpha_{xi})]\,[D_{si}]\,[T\,(\alpha_{i1})]\,[V]\,[T\,(\alpha_{1x})] \tag{9.34}$$

In Equation (9.34), $[D_c]$ is the uniaxial tangent constitutive matrix of concrete, $[D_{si}]$ is the uniaxial tangent constitutive matrix of steel, and $[V]$ is the matrix defined previously in Equation (6.37), which converts the biaxial strains into uniaxial strains using the Hsu/Zhu ratios. $[T\,(\alpha_{1x})]$ is the transformation matrix from the x–y coordinate to the 1–2 coordinate; $[T\,(\alpha_{x1})]$ is the transformation matrix from the 1–2 coordinate to the x–y coordinate; $[T\,(\alpha_{i1})]$ is the transformation matrix from the 1–2 coordinate to the si–ti coordinate; $[T\,(\alpha_{xi})]$ is the transformation matrix from the si–ti coordinate to the x–y coordinate;

The uniaxial constitutive matrix of concrete $[D_c]$ is given by

$$[D_c] = \begin{bmatrix} \overline{\overline{E}}_1^c & \dfrac{\partial \sigma_1^c}{\partial \overline{\varepsilon}_2} & 0 \\[2ex] \dfrac{\partial \sigma_2^c}{\partial \overline{\varepsilon}_1} & \overline{\overline{E}}_2^c & 0 \\[2ex] 0 & 0 & G_{12}^c \end{bmatrix} \tag{9.35}$$

In Equation (9.35), $\overline{\overline{E}}_1^c$ and $\overline{\overline{E}}_2^c$ are the uniaxial tangent moduli of concrete in the 1 and 2 directions, respectively, evaluated at a certain stress/strain state. The off-diagonal terms

$$\frac{\partial \sigma_1^c}{\partial \overline{\varepsilon}_2} \quad \text{and} \quad \frac{\partial \sigma_2^c}{\partial \overline{\varepsilon}_1}$$

are obtained by using the uniaxial constitutive relationships and taking into account the states of the concrete stresses and uniaxial strains in the 1–2 directions, which are not zero because the stress and strain of the concrete in compression is softened by the orthogonal tensile strains.

$$G_{12}^c = \frac{\sigma_1^c - \sigma_2^c}{\varepsilon_1 - \varepsilon_2}$$

is the shear modulus of concrete as described previously.

The uniaxial stiffness matrix of rebars $[D_{si}]$ is evaluated as follows:

$$[D_{si}] = \begin{bmatrix} \rho_{si} \cdot \overline{\overline{E}}_{si} & 0 & 0 \\ 0 & 0 & 0 \\ 0 & 0 & 0 \end{bmatrix} \tag{9.36}$$

where $\overline{\overline{E}}_{si}$ is the uniaxial tangent modulus for the rebars, as determined for a particular stress/strain state.

9.4.3 Analysis Procedures

After the tangent material constitutive matrix $[D]$ is determined, the tangent element stiffness matrix can be evaluated using the basic finite element procedure and can be expressed as:

$$[K]_e = \int_V [B]^T [D] [B] \, dV \tag{9.37}$$

where $[B]$ is a matrix that depends on the assumed element displacement functions of the elements.

An iterative tangent stiffness procedure was developed to perform nonlinear analyses of reinforced concrete structures. A flow chart for an iterative analysis solution under load increment using the Newton–Raphson method is described in Figure 9.10. Throughout the procedure, the tangent material constitutive matrix $[D]$ is determined first, and then the tangent element stiffness matrix $[k]$ and the element resisting force increment vector $\{\Delta f\}$ are calculated. After that, the global stiffness matrix $[K]$ and global resisting force increment vector $\{\Delta F\}$ are assembled. In each iteration, the material constitutive matrix $[D]$, the element tangent stiffness matrix $[k]$, and the global stiffness matrix $[K]$ are iteratively refined until convergence criterion is achieved.

The procedure for establishing the material constitutive matrix using CSMM is shown in the grey block in Figure 9.10. It is noted that an additional iterative loop is defined to obtain the material constitutive matrix $[D]$ for RC plane stress elements because the principal stress direction θ_1 is an unknown value before $[D]$ is established. The procedure for the stiffness calculation of RC plane stress elements is outlined by the outer white block in Figure 9.10.

This flow chart shows the simple analysis procedure of RC plane stress structures using load increment. It can be incorporated with other static integrators such as displacement control, and dynamic integrators such as the Newmark and the Wilson methods for different kinds of nonlinear finite element analysis. The integrator schemes for static and dynamic analyses used in this chapter are presented in Section 9.6.

9.5 Equation of Motion for Earthquake Loading

9.5.1 Single Degree of Freedom versus Multiple Degrees of Freedom

Any concrete structure can be represented as a single-degree-of-freedom (SDOF) system in dynamic analysis; and the dynamic response of the structure can be evaluated by the solution

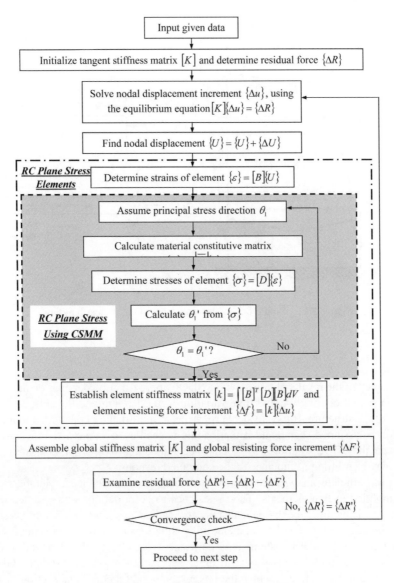

Figure 9.10   Nonlinear analysis algorithm

of a single differential equation of motion. If the physical properties of the structure are such that its motion can be described by a single coordinate and no other motion is possible, it is actually a SDOF system, and the solution of the equation of motion provides the exact dynamic response. On the other hand, if the structure actually has more than one possible mode of displacement and it is reduced mathematically to a SDOF approximation by assuming its deformed shape, the solution of the equation of motion is only an approximation of the true dynamic behavior.

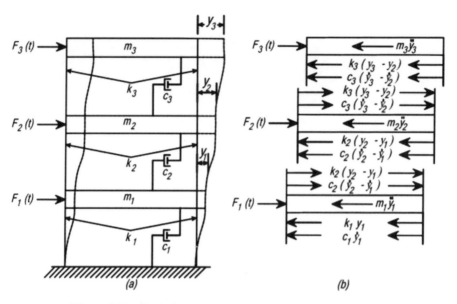

Figure 9.11   Single-bay model representation of a shear building

The accuracy of the results obtained with a SDOF approximation depends on many factors, such as the spatial distribution and time variation of the loading and the stiffness and mass properties of the structure. If the physical properties of the structure constrain it to move most easily with the assumed shape, and if the loading is such as to excite a significant response in this shape, the SDOF solution will probably be a good approximation. Otherwise, the true behavior may bear little resemblance to the computed response.

To perform efficient analysis, the actual structure must be idealized. Members or walls can be simplified to line elements. Their resistance to deformation is represented by material properties. How these idealizing decisions are made is extremely important and involves considerable judgment. In the development of the equations of motion of a general multiple-degree-of-freedom (MDOF) system, it is convenient to refer to the three-story structure modeled as a shear building, as shown in Figure 9.11(a), as a typical example. Given that a shear building is a structure in which there is no rotation of a horizontal section at the level of the floors. The deflected building would have many of the features of a cantilever beam that is deflected by shear forces only (Paz 1997). To accomplish such deflection in a building the following need to be assumed: (1) the mass is concentrated at the levels of the floors; (2) the girders on the floors are infinitely rigid; and (3) the deformation of the structure is independent of the axial forces present in the columns. Hence, a three-story structure modeled as a shear building (Figure 9.11a) would have three degrees of freedom in the horizontal direction. The first assumption is based on the concept of lumped mass. The second assumption means that the joints between girders and columns are fixed against rotation. The third assumption leads to the condition that the rigid girders will remain horizontal during motion. Note that the building may have any number of bays.

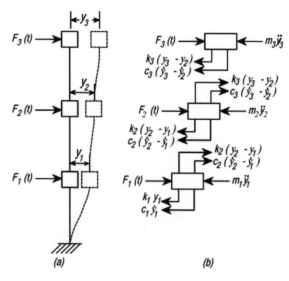

Figure 9.12   Single-column model representation of a shear building

9.5.2 A Three-degrees-of-freedom Building

Actually, we can further idealize the shear building as a single column (Figure 9.12a). In each of the two representations (Figures 9.11 and 9.12) for a viscously damped shear building, the stiffness and damping coefficients shown between any two consecutive masses are the forces required to produce relative unit displacements and velocities, respectively, of the two adjacent floor levels.

From the free body diagrams in Figure 9.11(b) or Figure 9.12(b), we obtain

$$m_1\ddot{y}_1 + c_1\dot{y}_1 + k_1 y_1 - c_2(\dot{y}_2 - \dot{y}_1) - k_2(y_2 - y_1) = F_1(t)$$

$$m_2\ddot{y}_2 + c_2(\dot{y}_2 - \dot{y}_1) + k_2(y_2 - y_1) - c_3(\dot{y}_3 - \dot{y}_2) - k_3(y_3 - y_2) = F_2(t) \qquad (9.38)$$

$$m_3\ddot{y}_3 + c_3(\dot{y}_3 - \dot{y}_2) + k_3(y_3 - y_2) = F_3(t)$$

In matrix notation

$$[M]\{\ddot{y}\} + [C]\{\dot{y}\} + [K]\{y\} = \{F(t)\} \qquad (9.39)$$

where

$$[M] = \begin{bmatrix} m_1 & 0 & 0 \\ 0 & m_2 & 0 \\ 0 & 0 & m_3 \end{bmatrix} \qquad (9.40)$$

$$[K] = \begin{bmatrix} k_1 + k_2 & -k_2 & 0 \\ -k_2 & k_2 + k_3 & -k_3 \\ 0 & -k_3 & k_3 \end{bmatrix} = \begin{bmatrix} k_{ij} \end{bmatrix} \tag{9.41}$$

$$[C] = \begin{bmatrix} c_1 + c_2 & -c_2 & 0 \\ -c_2 & c_2 + c_3 & -c_3 \\ 0 & -c_3 & c_3 \end{bmatrix} \tag{9.42}$$

$$\{y\} = \begin{Bmatrix} y_1 \\ y_2 \\ y_3 \end{Bmatrix} \tag{9.43}$$

$$\{\dot{y}\} = \begin{Bmatrix} \dot{y}_1 \\ \dot{y}_2 \\ \dot{y}_3 \end{Bmatrix} \tag{9.44}$$

$$\{\ddot{y}\} = \begin{Bmatrix} \ddot{y}_1 \\ \ddot{y}_2 \\ \ddot{y}_3 \end{Bmatrix} \tag{9.45}$$

$$\{F\} = \begin{Bmatrix} F_1(t) \\ F_2(t) \\ F_3(t) \end{Bmatrix} \tag{9.46}$$

Equation (9.40) indicates the mass matrix from lumped masses.

Equation (9.41) is the stiffness matrix for the three-degrees-of-freedom building. k_i in this matrix represents the stiffness of story i. k_{ij} is the force at degree of freedom i when a unit displacement is given at degree of freedom j. For example, $k_{22} = k_2 + k_3 =$ the force required at the second floor when a unit displacement is given to this floor. For framed shear walls, k_i can be a certain number of degree of freedom in either beam–column elements for beams/columns or plane stress elements for walls.

9.5.3 Damping

For the damping matrix in Equation (9.42), the format is the same as the stiffness matrix. However, in the damping term, Rayleigh damping (Chopra, 2005) can be used which includes both mass-proportional damping and stiffness-proportional damping, $c = a_0 m + a_1 k$. In the analytical model, the mass-proportional damping is not included ($a_0 = 0$) because the mass-proportional damping, if applied to the structure in the analysis, would act as external dampers which do not physically exist in base-supported structures such as structures under earthquake action. The damping proportional to the converged stiffness at each time step is applied to model the energy dissipation arising from story deformations. The coefficient a_1 is given by Equation (9.47).

$$a_1 = \frac{2\zeta}{(\omega_1 + \omega_2)} \tag{9.47}$$

where ζ is the damping ratio and ω_1 and ω_2 are the natural frequencies associated with the first and second modes at the initial condition.

The damping ratios were determined based on the different damage levels of the structures. The recommended damping values for reinforced concrete structures at different stress levels, given by Chopra (2005), are 3–5% for reinforced concrete with considerable cracking; and 7–10% at the stress level at or just below the yield point. However, the values are roughly estimated for reinforced concrete structures and are not specified for different types of structures, such as frames or shear walls. Since the damping in structures is determined by both material damping and system damping, a specific damping value for reinforced concrete shear walls should be used for wall-related structures. Farrar and Baker (1995) conducted experiments to measure the damping ratio of low-rise reinforced concrete shear walls. They found that the damping ratio was 1–2% for undamaged low-rise shear walls. With an increasing damage level of a wall, from moderate to severe, the measured damping ratio correspondingly increased from 2 to 8 %. The measured damping ratio of the wall at final failure was as high as 22%. In the seismic tests of reinforced concrete shear walls performed by Ile and Reynouard (2000), the damping ratio was found to increase from slightly more than 1% at the first test run to approximately 4% before the last test run at failure.

In this book, the damping ratios for reinforced concrete shear walls found by Farrar and Baker (1995) and Ile and Reynouard (2000) have been adopted. In general, a damping ratio of 2% is used when the structure is in the elastic stage and only minor cracking is observed. A damping ratio of 4% is used for the case that the steel in the wall yields and considerable cracking is observed. If minor crushing of concrete is observed, a damping ratio of 8% is used. When severe crushing of the concrete is observed and the structure has failed, high damping ratios of 15 and 25% are used, respectively, in order to consider the damaged condition of the specimen.

In the case under earthquake excitations, Equation (9.46) is replaced by Equation (9.48).

$$\{F\} = -\begin{Bmatrix} m_1 \ddot{y}_g \\ m_2 \ddot{y}_g \\ m_3 \ddot{y}_g \end{Bmatrix} \tag{9.48}$$

where $\ddot{y}_g$ is the ground acceleration.

9.6 Nonlinear Analysis Algorithm

An incremental procedure is usually used to study nonlinear finite element problems. For nonlinear problems, however, the incremental procedure would lead to a build-up of error. An iterative procedure using a certain solution algorithm should be employed to correct the build-up errors. Therefore, the combination of the incremental approach and iterative procedure are used as the basis for most of the nonlinear finite element analyses in our work. In the analysis procedure, the integrator determines the next predictive step during the analysis procedure, and specifies the tangent matrix and residual vector at any iteration. In this book, the commonly used integrators for static and dynamic analyses such as load control, displacement control, Newmark's method, and Wilson's method are introduced and the analysis procedure of each integrator is described. The algorithm determines the sequence of steps taken to solve the nonlinear equations during the iterative procedures. Hence, different algorithms are also described.

9.6.1 Load Control Iteration Scheme

To study the problem of a structure under monotonic loading, an incremental procedure can be used. The structural tangent stiffness matrix is related to the increments of loads to increments of displacements, which incorporates a tangential material constitutive matrix relating the increments of stresses to the increments of strains. Under load control, the total load is divided into small load increments. Each load step is applied in turn and iterations are performed until convergence is achieved at the structural level. Then the next load step is processed.

In solving nonlinear equations, the commonly used solution algorithm is the full Newton–Raphson method. In each iteration, the stiffness matrix is iteratively refined until the convergence criterion is achieved. The stiffness matrix is computed from the last iterative solution during the iterative procedure until convergence is achieved. A modified Newton–Raphson procedure is also used in the solution algorithms. Different from the full Newton–Raphson method, the stiffness matrix from the last converged equilibrium was used during the iterative procedure until convergence is achieved. The Newton–Raphson method uses the initial stiffness matrix throughout the iterative procedure.

When compared with theNewton–Raphson method and the modified Newton–Raphson method, the full Newton–Raphson method converges more rapidly, and the process will converge in fewer iterations and give smaller residual force at each iteration. However, it requires that the tangent stiffness matrix be evaluated at each iteration, which can be significant for large structures. In contrast to the full Newton–Raphson method, the initial stiffness matrix in the Newton–Raphson method is calculated at the beginning of the load step and the stiffness matrix remains the same throughout the procedure. A large number of iterations are required to achieve convergence. The modified Newton–Raphson method shows the balance between the computation and iteration numbers. Many algorithms, such as the KrylovNewton method (Carlson and Miller, 1998), have been developed by improving the Newton-type methods with the acceleration technology. The KrylovNewton method is a modified Newton–Raphson method with 'Krylov subspace' acceleration, which greatly decreases the number of iterations in the solution.

9.6.2 Displacement Control Iteration Scheme

Loads can be applied to a structure using either of two different methods: under load control or displacement control. To simulate the seismic behavior of a reinforced concrete structure subjected to reversed cyclic loading, the entire load–displacement curve, including the ascending branch, descending branch, and the hysteresis loops, can be obtained using the displacement control scheme. On the other hand, the displacement control method also has advantages over load control in the analysis procedure as described below.

1. Under load control it is impossible to indicate the behavior of the structure at a local limit such as the temporary drop of force due to the initial concrete cracking. More importantly, load control is incapable of producing the ultimate strength of the structure, or to trace the behavior of the structure in the post-peak region. Under load control the tangent stiffness matrix becomes nearly singular at the peak point of the load–displacement curve. It was pointed out by Ayoub (1995) and Ayoub and Filippou (1998) that the failure of the solution to converge is not an indication that the structure has reached its collapse point, but rather a failure of the solution convergence. Under the displacement control, especially the displacement control with arc length scheme (Batoz and Dhatt, 1979), it is possible to obtain the behavior of the structure beyond the crack point and the maximum point and to determine the entire response including ascending, descending and cyclic branches.
2. When there is no preference of load control or displacement control, the displacement control method shows faster convergence and is more stable than the load control method. This is observed in the nonlinear finite element analyses of reinforced concrete plane stress structures in this book.

Many researchers (e.g. Zienkiewicz, 1971; Haisler *et al.,* 1977; Batoz and Dhatt, 1979) have proposed the displacement control scheme to overcome the limits of the load control method. Meanwhile, the arc length method has been developed to overcome the local and global limit points in the nonlinear analysis, which treated the load factor as a variable. The arc length method was originally proposed by Riks (1972) and was improved by Crisfield (1981). A displacement control with an arc length scheme originally proposed by Batoz and Dhatt (1979) is available.

9.6.3 Dynamic Analysis Iteration Scheme

9.6.3.1 Newmark's Method

Newmark's method (Newmark, 1959) and Wilson's method (Wilson *et al.,* 1973) are two well-known methods for dynamic analysis. These two methods have been implemented as a class of objects in the OpenSees. The theoretical backgrounds of Newmark's and Wilson's methods are briefly summarized in this and the next section, respectively. The iterative procedures using these two methods for nonlinear systems are described as well.

The explicit relationships of displacements, velocities, and accelerations from step i to step $i + 1$ are:

$$\dot{u}_{i+1} = \dot{u}_i + (\Delta t)\ddot{u}_i, \tag{9.49}$$

$$u_{i+1} = u_i + (\Delta t)\dot{u}_i + 0.5(\Delta t)^2 \ddot{u}_i, \tag{9.50}$$

where u_i, $\dot{u}_i$, and $\ddot{u}_i$ are approximations to the displacement, velocity, and acceleration at step i; Δt is the time interval.

Newmark (1959) proposed one of the most popular algorithms for the solution of structural dynamics. The method is based on the following interpolations of displacements, velocities, and accelerations from step i to step $i + 1$:

$$\dot{u}_{i+1} = \dot{u}_i + [(1 - \gamma)\,\Delta t]\,\ddot{u}_i + (\gamma\,\Delta t)\,\ddot{u}_{i+1}, \tag{9.51}$$

$$u_{i+1} = u_i + (\Delta t)\,\dot{u}_i + \left[(0.5 - \beta)(\Delta t)^2\right]\ddot{u}_i + \left[\beta\,(\Delta t)^2\right]\ddot{u}_{i+1} \tag{9.52}$$

where u_{i+1}, $\dot{u}_{i+1}$, and $\ddot{u}_{i+1}$ are approximations to the displacement, velocity, and acceleration at step $i + 1$; β and γ are the parameters that define the variation of the Newmark method.

Typical selection for γ and β is $\gamma = \frac{1}{2}$ and $\frac{1}{6} \le \beta \le \frac{1}{4}$, respectively. When $\gamma = \frac{1}{2}$ and $\beta = \frac{1}{4}$, the Newmark method becomes a special case called the 'average acceleration' method. The average acceleration method is unconditionally stable. When $\gamma = \frac{1}{2}$ and $\beta = \frac{1}{6}$ the Newmark method becomes a special case called the 'linear acceleration' method. The linear acceleration method is stable if

$$\frac{\Delta t}{T_n} \le 0.551$$

where T_n is the shortest natural period of the structure.

For the incremental formulation required by the nonlinear system, Equations (9.51) and (9.52) can be rewritten as

$$\Delta \dot{u}_i = (\Delta t)\,\ddot{u}_i + (\gamma\,\Delta t)\,\Delta \ddot{u}_i \tag{9.53}$$

$$\Delta u_i = (\Delta t)\,\dot{u}_i + \frac{(\Delta t)^2}{2}\ddot{u}_i + \beta\,(\Delta t)^2\,\Delta \ddot{u}_i \tag{9.54}$$

Equation (9.54) can be solved as

$$\Delta \ddot{u}_i = \frac{1}{\beta\,(\Delta t)^2}\Delta u_i - \frac{1}{\beta\,(\Delta t)}\dot{u}_i - \frac{1}{2\beta}\ddot{u}_i \tag{9.55}$$

Substituting Equation (9.55 into (9.53) gives

$$\Delta \dot{u}_i = \frac{\gamma}{\beta\,\Delta t}\Delta u_i - \frac{\gamma}{\beta}\dot{u}_i + \Delta t\left(1 - \frac{\gamma}{2\beta}\right)\ddot{u}_i \tag{9.56}$$

The incremental equation of motion is given by

$$m\,\Delta \ddot{u}_i + c\,\Delta \dot{u}_i + k\,\Delta u_i = \Delta p_i \tag{9.57}$$

Substituting Equations (9.55) and (9.56) into Equation (9.57) gives

$$\hat{k}_i\,\Delta u_i = \Delta \hat{p}_i \tag{9.58}$$

where

$$\hat{k}_i = k_i + \frac{\gamma}{\beta\,\Delta t}c + \frac{1}{\beta\,(\Delta t)^2}m \tag{9.59}$$

$$\Delta \hat{p}_i = \Delta p_i + \left(\frac{1}{\beta \Delta t} m + \frac{\gamma}{\beta} c \right) \dot{u}_i + \left[\frac{1}{2\beta} m + \Delta t \left(\frac{\gamma}{2\beta} - 1 \right) c \right] \ddot{u}_i, \quad (9.60)$$

where the k_i is the stiffness matrix in static analysis and Δp_i is the load increment applied to the structure. $\hat{k}_i$ is not a constant value for a nonlinear system.

It can be seen that, although Equation (9.58) shows a process similar to static analysis, the system properties m, c, algorithm parameters β and γ, and the velocity and the acceleration need to be taken into account.

For a nonlinear system, the incremental procedure described in Equations (9.57) and (9.58) needs to be accomplished using an iterative procedure. The analysis procedure for the time step i of Newmark's method is presented below, where the modified Newton–Raphson method is incorporated.

1. Calculate $\Delta \hat{p}_i$ based on Equation. (9.60)
2. Determine the tangent stiffness k_i
3. Calculate $\hat{k}_i$ based on Equation (9.59)
4. Solve $\hat{k}_i \Delta u_i = \Delta \hat{p}_i$ and obtain trial Δu_i, trial $u_i = u_i + \Delta u_i$
5. Determine the resisting force Δf and unbalanced force $\Delta R' = \Delta R - \Delta f$
6. Check convergence criterion; if yes, go to step 7; if no, go to step 4
7. Determine $\Delta \dot{u}_i$ and $\Delta \ddot{u}_i$ based on Equations (9.56) and (9.55)
8. Calculate u_{i+1}, $\dot{u}_{i+1}$, and $\ddot{u}_{i+1}$ by using $u_{i+1} = u_i + \Delta u_i$, $\dot{u}_{i+1} = \dot{u}_i + \Delta \dot{u}_i$, and $\ddot{u}_{i+1} = \ddot{u}_i + \Delta \ddot{u}_i$, respectively
9. Proceed to the next time step

9.6.3.2 Wilson's Method

Wilson's method (Wilson *et al.*, 1973) is a modification of a special case of Newmark's method called the linear acceleration method

$$\left(\gamma = \frac{1}{2} \text{ and } \beta = \frac{1}{6} \right)$$

Wilson's method assumes the acceleration varies linearly over an extended time step $\delta t = \theta \Delta t$, which makes it unconditionally stable. Substituting

$$\gamma = \frac{1}{2} \text{ and } \beta = \frac{1}{6}$$

into Equations (9.53)–(9.56) as well as Equations (9.59) and (9.60), and replacing the time interval Δt by δt gives

$$\delta \dot{u}_i = (\delta t) \ddot{u}_i + \left(\frac{\delta t}{2} \right) \Delta \ddot{u}_i, \quad (9.61)$$

$$\delta u_i = (\delta t) \dot{u}_i + \frac{(\delta t)^2}{2} \ddot{u}_i + \frac{(\delta t)^2}{6} \delta \ddot{u}_i, \quad (9.62)$$

$$\delta \ddot{u}_i = \frac{6}{(\delta t)^2} \delta u_i - \frac{6}{(\delta t)} \dot{u}_i - 3 \ddot{u}_i. \tag{9.63}$$

$$\delta \dot{u}_i = \frac{3}{\delta t} \delta u_i - 3 \dot{u}_i - \frac{1}{2} \delta t \ddot{u}_i \tag{9.64}$$

$$\hat{k}_i = k_i + \frac{3}{\theta \Delta t} c + \frac{6}{(\theta \Delta t)^2} m \tag{9.65}$$

$$\delta \hat{p}_i = \theta \Delta p_i + \left(\frac{6}{\theta \Delta t} m + 3c \right) \dot{u}_i + \left[3m + \frac{\theta \Delta t c}{2} \right] \ddot{u}_i \tag{9.66}$$

A summary of iterative procedure of Wilson's method by the use of the modified Newton–Raphson method is described as follows:

1. Calculate $\delta \hat{p}_i$ based on Equation (9.66)
2. Determine the tangent stiffness k_i
3. Calculate $\hat{k}_i$ based on Equation (9.65)
4. Solve $\hat{k}_i \delta u_i = \delta \hat{p}_i$ and obtain trial δu_i, trial $u_i = u_i + \delta u_i$
5. Determine the resisting force δf and unbalanced force $\delta R' = \delta R - \delta f$
6. Check convergence criterion; if yes, go to step 7; if no, go to step 4
7. Determine $\delta \dot{u}_i$ based on Equations (9.63) and $\Delta \ddot{u}_i = \frac{1}{\theta} \delta \ddot{u}_i$ for $\delta t = \theta \Delta t$
8. $\Delta \dot{u}_i = (\Delta t) \ddot{u}_i + \left(\frac{\Delta t}{2} \right) \Delta \ddot{u}_i; \Delta u_i = (\Delta t) \dot{u}_i + \frac{(\Delta t)^2}{2} \ddot{u}_i + \frac{(\Delta t)^2}{6} \Delta \ddot{u}_i$
9. $u_{i+1} = u_i + \Delta u_i; \dot{u}_{i+1} = \dot{u}_i + \Delta \dot{u}_i; \ddot{u}_{i+1} = \ddot{u}_i + \Delta \ddot{u}_i$
10. Proceed to the next time step

If $\theta = 1$, Wilson's method will be the same as the linear acceleration method, which is stable if $\Delta t < 0.551 T_n$. If $\theta \geq 1.37$, Wilson's method becomes unconditionally stable.

9.7 Nonlinear Finite Element Program SCS

This chapter has presented the theoretical background of integrators for static and dynamic analyses including the load control method, the displacement control method, Newmark's and Wilson's methods. The integrators are incorporated with algorithms for nonlinear analysis. These integrators and algorithms have already been implemented as classes of objects in the OpenSees (Fenves, 2005), which makes it capable of performing nonlinear finite element analysis. To perform analysis on reinforced concrete plane stress structures such as shear walls, OpenSees is modified to extend its capability.

1. Appropriate uniaxial materials of steel and concrete are added to OpenSees for reinforced/prestressed concrete plane stress structures because of the following reasons. In OpenSees the available uniaxial modules for steel and concrete are Steel01 and Concrete01. The features of Steel01 and Concrete01 have been introduced in Section 9.1. Steel01 does not consider the smeared yield strain and stress of embedded steel, and the unloading and reloading paths do not take into account of the Bauschinger effect. Concrete01 does not

consider the softening effect on the compressive strain and stress due to the tensile strain in the perpendicular direction. Also the tensile stress of concrete in Concrete01 is ignored.

2. Reinforced/prestressed concrete plane stress material is added to OpenSees for analysis of reinforced/prestressed concrete plane stress structures such as panels, deep beams, and walls.

3. In addition, the displacement control scheme implemented in OpenSees is only for monotonic displacement control with uniform displacement increment. The displacement control needs to be modified for reversed cyclic loading.

In order to perform analysis on reinforced/prestressed concrete plane stress structures several new material classes are implemented into the OpenSees framework, and some analysis classes in the OpenSees are modified. Details of the implementation and modification are presented as follows:

Three uniaxial material classes, SteelZ01, TendonL01, and ConcreteL01 are created and implemented into OpenSees. The features of these three material modules have been described in Section 9.1. An object made of SteelZ01 needs four input parameters: yield stress, Young's modulus, concrete compressive strength, and steel ratio. The latter two parameters are used to calculate the smeared yield stress and strain of embedded rebars. In addition, the coefficients A and R that determine the shapes of unloading and reloading paths (Equations 9.13 and 9.14) are set as two additional input parameters. The default values are defined as 1.9 and 10. An object made of TendonL01 needs similar inputs and has similar attributes as that of SteelZ01. In addition it also needs the initial prestressing strains in the tendons as input. An object of ConcreteL01 needs two input parameters: ultimate compressive strength f_c' and the compressive strain ε_0 corresponding to f_c'.

A 2D material class, PCPlaneStress, is created and implemented into OpenSees. The PCPlaneStress is a class for plane stress concrete material using CSMM-PC. In PCPlaneStress, the material constitutive matrix derived in this chapter is evaluated and the stress vector is calculated. An object made of PCPlaneStress material needs the tags of the created uniaxial steel and concrete objects of SteelZ01, TendonL01, and ConcreteL01, the directions of the steel grids, and the steel ratio for the steel in each direction. Two uniaxial concrete objects are needed in defining one PCPlaneStress object, which represents the concrete in the two principal stresses directions. The steel orientations are not necessary in the horizontal and vertical directions, and the user can define the arbitrary angles of the steel. The steel directions are also not necessarily orthogonal. Users can define the steel in the reinforced/prestressed concrete element in up to four directions.

The implementation of the PCPlaneStress into OpenSees is shown in Figure 9.13. The Analysis and Recorder objects in OpenSees are omitted in this figure. The PCPlaneStress is implemented with the Quadrilateral element to represent the PC plane stress four node elements. The PCPlaneStress is related with SteelZ01, TendonL01 and ConcreteL01 to determine the tangent material constitutive matrix and to calculate the stress of the elements. For each trial displacement increment in the analysis procedure, PCPlaneStress will receive the strains of the elements, determine the uniaxial strains of the concrete, tendons and steel – then send the uniaxial strain of concrete and the tensile perpendicular strain to the two uniaxial concrete objects. After receiving the uniaxial strain and corresponding tensile strain, the concrete object will produce the tangent stiffness and stress and then send the values back to the PCPlaneStress object. Similarly, PCPlaneStress will send the uniaxial strain of the steel and tendon to the

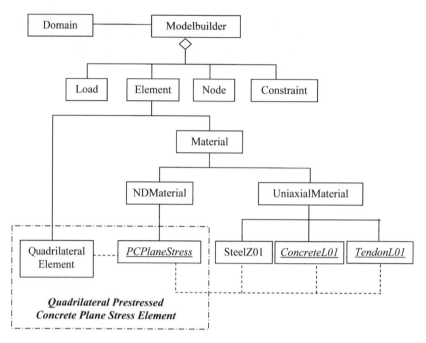

Figure 9.13   Implementation of PCPlaneStress module into OpenSees

uniaxial steel and tendon objects, respectively, and receive the tangent stiffness and stress from the related uniaxial steel and tendon objects. After receiving the uniaxial stiffness and stress of each of concrete, tendon and steel, the tangent stiffness matrix can be evaluated and the stress vector calculated.

Meanwhile, the displacement control scheme in the OpenSees was modified for an arbitrary displacement path. The user can define the displacement paths of the degree-of-freedom of the node whose response controls the solution. Or, the displacement increment for each path of the displacement scheme may be varied. Changes in the size of the displacement increment make it possible to overcome some numerical problems in the nonlinear analysis.

In addition, some classes are modified to overcome or bypass the numerical problems occurring in the analysis of nonlinear problems. For example, the analytical models about the convergence check are modified. Instead of giving one maximum iterative number for all increments, the user can decide different maximum iterative numbers for different paths in the displacement control history. Users can increase the maximum iterative number in the specified path that may have numerical problems. Lower iterative numbers in the rest of the displacement paths are defined to save computing time.

After implementing the new material models and modifying relevant existing classes in OpenSees, a finite element program named SCS (Simulation of Concrete Structures) was developed by adopting the OpenSees as the finite element framework. This program is able to perform nonlinear finite element analysis of concrete structures under static, reversed cyclic, or dynamic loading. Chapter 10 supports the application of SCS to reinforced/prestressed

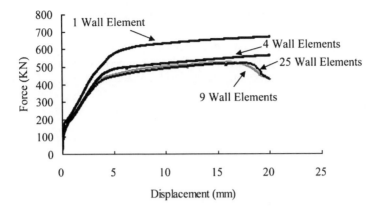

Figure 9.14   Analytical results of a shear wall with various sizes of finite element meshes

concrete structures by first validating the program by comparing the predictions to the tests of panels, beams, framed shear walls, bridge columns, and wall buildings.

Four example input files of SCS are shown in Appendix A. The first example is the static analysis of the prestressed concrete beam B1, as described in Section 10.2. The second is the reversed cyclic analysis of post-tensioned column C5C, as described in Section 10.4. The third one is the reversed cyclic analysis of the framed shear wall FSW13, as described in Section 10.3.1. The last is the dynamic analysis of the shear wall STN under seismic loading, as presented in Section 10.5.

For nonlinear finite element analysis of RC/PC plane stress structures, the following discussion is worthy of study.

1. Mesh size. Because cracks are fully smeared in the CSMM, the finite element that addresses cracked concrete with embedded steel bars and tendons can be considered as a continuum. To examine the proper size of the elements used in finite element analysis, a reinforced concrete shear wall FSW6 described in Section 10.3.1was analyzed using different mesh sizes. The analytical results are shown in Figure 9.14.

   Figure 9.14 shows that the analytical result obtained using only one wall element gives much higher prediction than the remaining three curves with 4, 9 and 25 elements. The prediction using 4 wall elements is slightly greater than those of 9 elements and 25 elements. The two analytical curves using 9 elements and 25 elements are barely distinguishable.

   The rapid convergence of the analytical curves with increasing number of elements was also observed by Okamura and Maekawa (1991) when the smeared crack model was applied to the finite element analysis of shear walls. Thus, a relatively small number of RC plane stress elements is adequate to obtain accurate outcomes using the developed SCS program and also saves precious computation time.

2. Nonlinear solutions. The modified Newton–Raphson method with Krylov subspace acceleration (Carlson and Miller, 1998) was found to converge faster and was more stable

than other solution algorithms such as the full Newton–Raphson method and the modified Newton–Raphson method.

3. Numerical problems. Numerical problems may be encountered in nonlinear finite element analyses, especially in reversed cyclic analysis. These numerical problems can be overcome in most cases by changing the solution algorithm, the increment size, or by increasing the iteration number.

10

Application of Program SCS to Wall-type Structures

This chapter reports the successful application of the newly developed finite element program Simulation of Concrete Structures (SCS) to real-life concrete structures. The validity of this application is demonstrated by correlations between analytical outcomes from SCS and physical data from testing full-scale concrete specimens subjected to monotonic or reversed cyclic loading, or to shake table excitations.

10.1 RC Panels Under Static Load

A series of four panels (1.4 × 1.4 × 0.18 m) previously tested by Pang and Hsu (1995) at the University of Houston were analyzed using SCS. The test panels were subjected to monotonically increasing shear stresses until failure. The steel grids in all panels were set parallel to the plane of pure shear, as shown in Figure 10.1. The four panels were reinforced with the same amount of steel in the two orthogonal steel directions, but the reinforcing ratios varied from one panel to the other. Reinforcing ratios of panels A1, A2, A3, and A4 were 0.77, 1.19, 1.79, and 2.98%, respectively. The concrete compressive strength was approximately 42 MPa and the yield stress of the reinforcing bars was approximately 460 MPa.

Since the stress conditions and the material properties were uniform throughout all panels, it was adequate to model the entire panel using one 2-D RCPlaneStress element, as shown in Figure 10.2. The applied load pattern and boundary conditions are defined to simulate the pure shear on the panel. The predicted monotonic responses of panels are compared with the experimental responses in Figure 10.3.

It is clear the SCS analysis can, indeed, accurately predict the ascending branch of panel A1. The maximum shear stress of panel A1 appeared slightly overestimated by the analytical results. This is because the CSMM is based on a fully smeared crack concept, but the failure of panel A1 occurred by steel yielding in the vicinities of cracks widely spaced. The predicted results of panels A2 and A3 agree very well with the experimental results in the whole loading history in both the ascending and descending branches.

Unified Theory of Concrete Structures   Thomas Hsu and Yi-Lung Mo
© 2010 John Wiley & Sons, Ltd

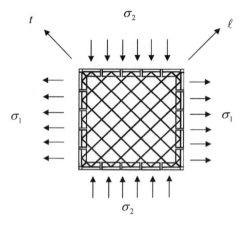

Figure 10.1   Specimen

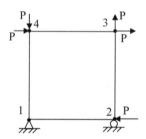

Figure 10.2   Finite element mesh

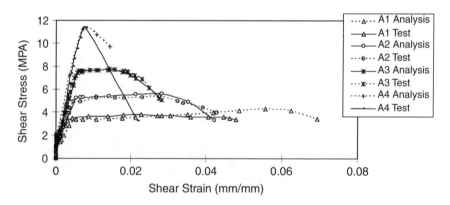

Figure 10.3   Predicted and experimental stress–strain curves of panels

In panel A4, the SCS analysis can accurately predict the ascending branch and the peak point of the test curve. In the physical test, the concrete of panel A4 crushed, and the specimen failed right after reaching the peak point. Only one point of experimental data in the post-peak region was measured. Therefore, the experimental descending branch might not represent the real behavior of the panel after peak point. This may explain the discrepancy between the predicted and experimental descending branch of the panel.

10.2 Prestresed Concrete Beams Under Static Load

The five beams tested at the University of Houston (UH) have been described in Section 8.3.2 in Chapter 8. Figure 8.9 shows the test setup of a 7.62 m long, simply supported, prestressed concrete I-beam subjected to two monotonically applied concentrated loads (Laskar *et al.*, 2006). The cross-section of the beam is shown in Figure 8.10. The loads were applied at distances 0.92 m from each end of three of the beams B1, B2, and B3, as shown in Figure 8.11(a), and at distances of 2.41 m from each end of the other two beams B4 and B5, as shown in Figure 8.11(b). Beams B1–B3 having a shear span of 0.92 m were designed to fail in web-shear mode, while beams B4 and B5 having a shear span of 2.41 m were designed to fail in flexural-shear mode. Details of the individual specimens are shown in Table 10.1.

In Figure 8.9, two of the four actuators (B and C), each attached to a vertical steel frame, were used to apply the vertical loads on the beams. Each of these two actuators had a capacity of 1423 kN. The beam specimens were each placed on top of two load cells placed at each end. The load cells of 2224 kN capacity were placed on top of the steel pedestals fixed to a strong floor. On top of the load cells, bearing plates to support the beams were placed with a roller at one end and a hinge at the other, thus allowing the beam to rotate freely at the supports and to expand freely along its length.

The loads and displacements of the actuators were precisely controlled by the MTS MultiFlex System. Actuators B and C were first programmed with a load control mode of 22.2 kN/min. When the slope of the load–displacement curve started decreasing, the control mode was switched to a displacement control of 5.08 mm/hour (0.2 inch/hour). This step continued until shear failure occurred at either end of the beam. This displacement control feature was essential in capturing the ductility/brittleness behavior of beams failing in shear. During testing, linear voltage differential transformers (LVDTs) were used to measure the displacements at the failure regions of the beam adjacent to the points of load application. A set of six LVDTs forming a rosette was installed on both faces of the beams to get the average deformations at the failure zone, as shown in Figure 8.11. Several LVDTs were also

Table 10.1 Test variables of beams

Beam	a/d	Tendon profile (straight/draped)	Transverse steel (%)	Concrete strength (MPa)	Failure mode
B1-North	1.61	Straight	0.17	72.4	Web-shear
B1-South	1.61	Straight	0.17	72.4	Web-shear
B2-North	1.61	Straight	0.95	74.5	Web-shear
B2-South	1.61	Straight	0.95	74.5	Web-shear
B3-North	1.61	Draped	0.95	64.6	Flexure/web-shear
B4-South	4.29	Straight	0.17	71.0	Flexural-shear
B5-North	4.29	Draped	0.17	64.5	Flexure/Flexural-shear

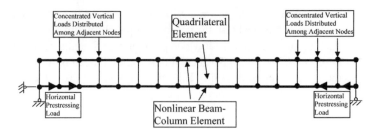

Figure 10.4   Finite element modeling of the beam

placed vertically under the beam, both at the supports and at the point of loading, to measure the total and net deflections of the beam.

   Finite element analyses of the tested beams were conducted using Program SCS. Each of the specimens was modeled by the finite element mesh, as shown in Figure 10.4. The SCS analytical results of the shear force-displacement relationships of Beams B1 and B5 are illustrated by the thin black curve in Figure 10.5.

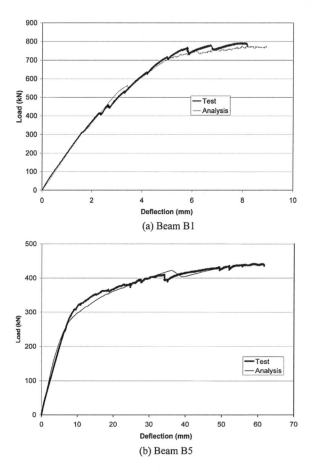

Figure 10.5   Comparison of analytical and experimental results of beams B1 and B5

In Figure 10.4, the web region of the beam was modeled using PCPlaneStress quadrilateral elements. The top and bottom flanges were modeled using Nonlinear Beam-Column elements. The prestressing load acting on the beam was applied as horizontal nodal forces, which remain constant in the analysis. The initial strain in the tendons were applied while defining the tendons using the TendonL01 module. The analysis was performed in two steps. In the first step, prestressing loads were applied to the beam using load control. After that, prestressing loads were kept constant and monotonic vertical loads were applied by a predetermined displacement control scheme. The nodal displacement and corresponding vertical forces were recorded at each converged displacement step, and the stress and strain of the elements were also monitored.

In Figure 10.5, the thin black curve is compared to the thick black curve, representing the experimental results. It can be seen that good agreements were obtained for the load-deflection curves, the initial stiffness, the yield point, and the ultimate strength. Figure 10.5 shows that the theoretical simulation based on the CSMM for prestressed concrete can accurately predict the true behavior of prestressed concrete beams with a failure mode of shear.

10.3 Framed Shear Walls under Reversed Cyclic Load

10.3.1 Framed Shear Wall Units at UH

Nine 1/3-scale framed shear walls, as listed in Table 10.2, were tested at the University of Houston (Gao, 1999). Each framed shear wall unit, as shown in Figure 10.6, represents a typical unit taken from a multi-story, multi-bay building and in-filled with shear walls. Each unit consists of a 914.4 mm (36 in) by 914.4 mm (36 in) frame made up of two boundary columns and two boundary beams and infilled with a shear wall. The cross-section of both the columns and the beams was 152.4 mm (6 in) square, and the thickness of the wall was 76.2 mm (3 in).

A framed shear wall unit was subjected to a constant vertical axial load at the top of each column and a horizontal, reversed, cyclic load at the level of the top beam, as shown in Figure 10.7. The bottom left and right corners of the specimen were idealized by a hinge and a roller, respectively. The framed shear wall units with low axial load simulates those units taken from the upper stories and units with high axial load simulates those taken from the lower stories. Two variables are planned for the nine-specimen test series: The first variable is the magnitude of vertical load on each column P/P_o, which varies from 0.07 to 0.46. The second variable is the steel ratio ρ_w in the shear wall, which varies from 0.23 to 1.10.

Finite element analyses were conducted on the nine specimens. The specimens were modeled by the finite element mesh, as shown in Figure 10.7. The wall panel was defined by nine RCPlaneStress quadrilateral elements. Each of the boundary columns and beams were modeled using three Nonlinear Beam-Column elements. The axial loads acting on the columns were applied using load control, as vertical nodal forces, which remain constant in the analysis. Reversed cyclic horizontal loads were then applied by a predetermined displacement control scheme. The nodal displacements and corresponding horizontal forces were recorded at each converged displacement step, and the stress and strain of each of the elements were also monitored.

The analytical results of the shear force-drift relationships of two shear walls are illustrated by the dashed hysteretic loops in Figure 10.8. These dashed loops are compared to the solid loops, representing the experimental results. It can be seen that good agreements were obtained

Table 10.2 Dimensions and properties of specimens

Specimen name	f'_c (MPa[a])	Column and beam			Wall panel		Vertical load	
		Hoop steel (mm[b])	Longitudinal steel	Longitudinal steel (%)	Panel pteel (mm[b])	Panel steel (%) ρ_w	P (KN[c])	$\frac{P}{P_o}$ ratio
FSW-13	56.91	#2@63.5	6#4	3.33	W2@152.4	0.23	89	0.07
FSW-6	49.75	#2@63.5	6#4	3.33	#2@152.4	0.55	89	0.08
FSW-11*	56.99	#2@63.5 #2@31.75	6#4	3.33	#2@76.2	1.10	89	0.07
FSW-8	48.29	#2@63.5	6#4	3.33	W2@152.4	0.23	267	0.24
FSW-5	56.34	#2@63.5	6#4	3.33	#2@152.4	0.55	267	0.20
FSW-9	50.24	#2@63.5 #2@31.75	6#4	3.33	#2@76.2	1.10	267	0.23
FSW-12	57.07	#2@63.5	6#4	3.33	W2@152.4	0.23	534	0.40
FSW-4	49.51	D3@63.5	6#4	3.33	#2@152.4	0.55	534	0.46
FSW-10*	55.85	#2@63.5 #2@31.75	6#4	3.33	#2@76.2	1.10	534	0.41

[a] 1 MPa = 0.145 ksi.
[b] 1 mm = 0.0394 in.
[c] 1 KN = 0.22 kip.

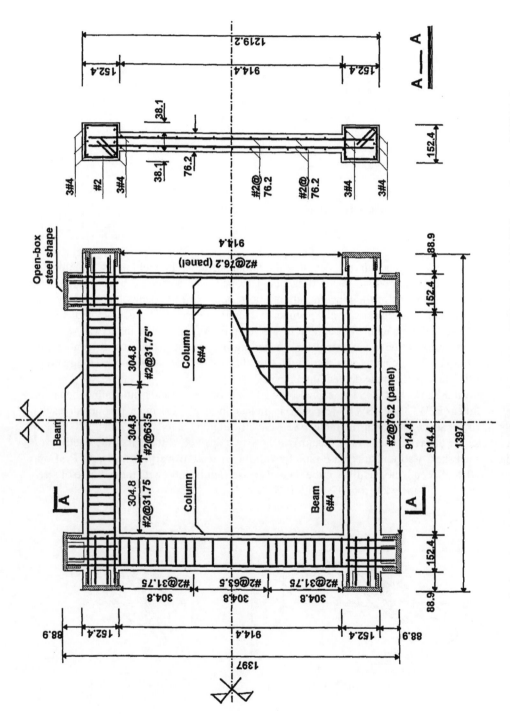

Figure 10.6 Dimensions and steel arrangement of specimens FSW-9, 10 and 11 (Unit: mm, 1 mm = 0.0394 in.)

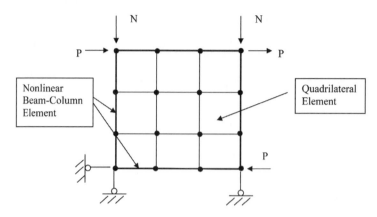

Figure 10.7   Finite element modeling of the wall

for the primary backbone curves, including the initial stiffness, the yield point, the ultimate strength, and the failure state in the descending branch. The hysteretic behavior provides accurate measurements of the pinching effect, the residual displacements, the ductility and the energy dissipation capacity in all specimens. Even the failure modes can be predicted by the CSMM-based finite element program. In specimen FSW13 steel bars in the walls yielded significantly prior to the concrete crushing, resulting in long yield plateaus. In contrast, in specimen FSW12 the concrete crushed right after the steel yielded, which caused an abrupt drop of the shear force in the descending branch.

Figure 10.8 shows that SCS was capable of capturing the ductile and brittle failure behavior of specimens FSW13 and FSW12, respectively. In fact, the analytical and experimental results for the other seven specimens in Table 10.2 were also in good agreement (Zhong, 2005). As a whole, the behavior of the nine specimens show two distinct trends. First, the ductility of the framed shear wall units decrease rapidly with the increase of vertical loads, P/P_o, from 0.07 to 0.46. Second, the ductility of the units increase significantly with the increase of steel percentage, ρ_w, in the wall from 0.23 to 1.10%.

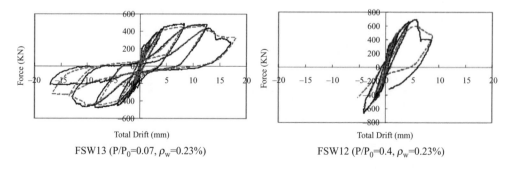

Figure 10.8   Shear force – drift displacement of specimens FSW13 and FSW12

10.3.2 Low-rise Framed Shear Walls at NCREE

A low-rise shear wall (RLB) was tested under reversed cyclic horizontal loading at NCREE (Zhong, 2005). Figure 10.9 shows the height, length, and thickness of the wall to be 1.4, 2.8 and 0.12 m, respectively. Other dimensions and reinforcements of the specimen are also given. The compressive strength of the concrete was 36.0 MPa, and the yield stress of rebars in the wall was 329 MPa. The steel ratio for the specimen is 0.48%. The end regions of the shear wall RLB were provided with a 240 × 240 mm boundary element having longitudinal bars and stirrups. Reversed cyclic horizontal loads were applied on the top of the shear wall. The test procedure is controlled by the horizontal displacement at the top of the wall. Analysis of the reversed cycle tests from this specimen was performed by program SCS.

The specimen was modeled using the finite element meshes, as shown in Figure 10.10. The mesh was divided into three regions consisting of the web panel, the top beam and the boundary elements. For simplification of the analyses, the foundation was omitted and the wall was modeled fixed to the ground. The wall panel was modeled by 8 RCPlaneStress quadrilateral elements. The top beam was modeled using four nonlinear beam–column elements and each boundary element of the specimen was modeled using two nonlinear beam–column elements. Each nonlinear beam–column element used to model the boundary elements was defined with three control sections; and each section was discretized into 42 fibers. The configuration of the section discretization is shown in Figure 10.11. The white cells represent the unconfined concrete fibers (24 fibers), the gray cells represent the confined concrete fibers (10 fibers), and the black cells represent the reinforcing steel fibers (8 fibers). The stress and strain of the confined concrete was determined based on the modified Kent and Park model (Scott *et al.*, 1982). The horizontal loads were uniformly imposed as the nodal forces along the nodes of top

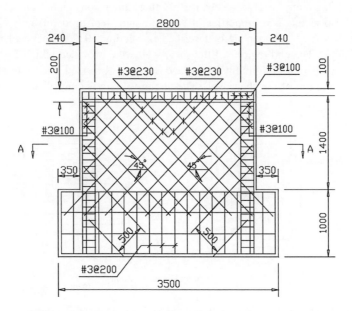

Figure 10.9   Dimensions and reinforcement of specimen RLB

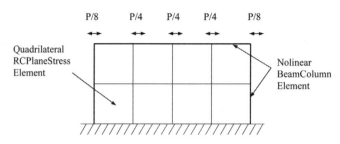

Figure 10.10   Finite element mesh of specimens RLN and RLB

beam. The horizontal loads were applied according to a predetermined lateral displacement scheme. The displacement increment used in the analysis was 0.01 mm.

The observed and calculated load–displacement relationships for specimen RLB are compared and shown in Figure 10.12. Compared with the experimental results, the analyses accurately predicted the load–versus displacement characteristics including pre-cracking stiffness, post-cracking stiffness, ultimate strength, residual displacement, and energy dissipation. The envelopes including ascending and descending branches of the specimen, which show the typical type of shear failure, were accurately predicted by the analytical results.

10.3.3 Mid-Rise Framed Shear Walls at NCREE

Now we address a mid-rise shear wall (RMB) which was tested under reversed cyclic horizontal loading at NCREE. (Zhong, 2005). Figure 10.13 shows the height, length, and thickness of the wall to be 4.2, 2.8 and 0.12 m, respectively. Other dimensions and reinforcements of the specimen are also given. The properties of concrete and steel used for the specimen were similar to the low-rise specimen RLB described in Section 10.3.2. The steel ratio for the specimen is 0.48%. The end regions of the shear wall were provided with a 240 × 240 mm boundary element having longitudinal bars and stirrups. Reversed cyclic horizontal loads were applied on the top of the shear wall. The test procedure is controlled by the horizontal

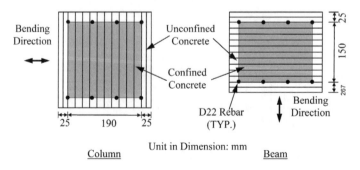

Figure 10.11   Section discretization of the beam and columns of specimens RLB

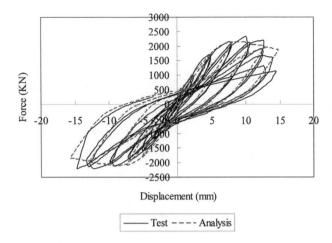

Figure 10.12   Predicted and experimental load versus displacement curves of specimen RLB

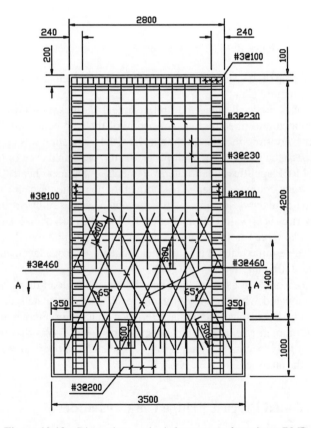

Figure 10.13   Dimensions and reinforcement of specimen RMB

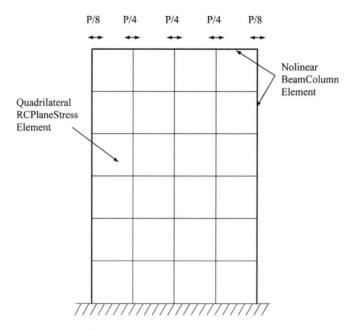

Figure 10.14   Finite element mesh of specimen RMB

displacement at the top of the wall. Reversed cyclic analyses were conducted on this mid-rise specimen using program SCS.

This specimen was modeled using the finite element meshes, as shown in Figure 10.14. The mesh was divided into three regions consisting of the web panel, the top beam and the boundary elements. For simplification of the analyses, the foundation was omitted and the wall was modeled fixed to the ground. The wall panel, the top beam and each of the columns were modeled using 24 RCPlaneStress quadrilateral elements, 4 nonlinear beam–column elements, and 6 nonlinear beam–column elements, respectively. The nonlinear beam–column elements used to model the boundary elements of the specimen were similar to the ones used for modeling the low-rise shear wall RLB in Section 10.3.2. The horizontal loads were uniformly imposed as the nodal forces along the nodes of top beam. The horizontal loads were applied according to a predetermined lateral displacement scheme. The displacement increment used in the analysis was 0.01 mm.

Figure 10.15 compares the experimental and the calculated load-displacement curves for the specimen. Compared with the experimental results, the calculated analyses accurately predicted the load versus displacement characteristics including pre-cracking stiffness, post-cracking stiffness, ultimate strength, residual displacement, and energy dissipation. The nearly flat-top envelopes of the specimen, which is a typical behavior of the flexure mechanism, were also predicted by the analyses.

10.4 Post-tensioned Precast Bridge Columns under Reversed Cyclic Load

Two post-tensioned precast columns tested at SUNY Buffalo (Ou, 2007) were analyzed using program SCS. Each Specimen was 5.7 m in total height and consisted of a foundation, four

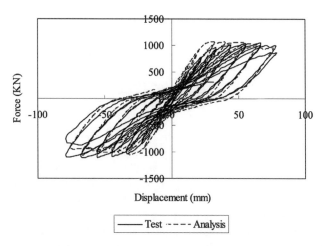

Figure 10.15   Predicted and experimental load versus displacement curves of specimen RMB

precast column segments with hollow cross sections of 200 mm thick walls and a precast cap beam. A schematic diagram of the test specimen is shown in Figure 10.16. The major design parameters of the specimens are shown in Table 10.3, while Table 10.4 lists their material properties. The designed gravity load of each specimen equaled 1456 kN. The concrete compressive strengths of the specimens ranged from 30 to 55 MPa as shown in Table 10.4. Each specimen had four prestressing tendons. Each tendon comprised of two seven-wire strands made of steel equivalent to ASTM A416 Grade 270. Each strand had a nominal diameter of 15.24 mm. The total design prestressing force in the two specimens (C5C and C8C) were 1042 kN. Energy dissipation (ED) bars, having diameters of 16 and 25 mm, were provided across the joints of the precast segments. The ED bars were made of steel equilavalent to A760 Grade 60 steel. The average yield strength and the ultimate strength of the ED bars were 434 kN and 653 kN, respectively.

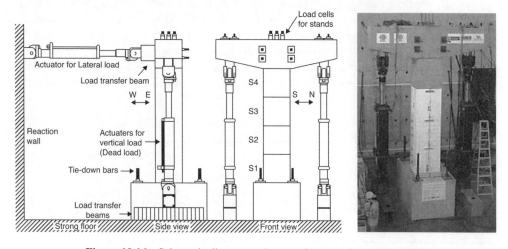

Figure 10.16   Schemetic diagram and setup of test specimens (Ou 2007)

Table 10.3 Major design parameters of column specimens

Specimen	Prestressing force (kN, f_{pu}, f_cA_g)	ED bar ratio (%)	ED bar detail Bar diameter (mm)
C5C	1042, 0.5, 0.072	0.5	16
C8C	1042, 0.5, 0.072	1.0	25

The specimens were modeled using the finite element mesh illustrated in Figure 10.17. The two flange sides of the bridge piers which carry the bending moment, are subjected mainly to compression and tension. They are modeled as nonlinear beam–column elements with fiber sections. The two web sides of the bridge piers which are parallel to the bending direction and thus resisting the shear force, are modeled by PCPlaneStress Quadrilateral elements. The cap beam on the top of the column is defined as a rigid body in the finite element model. The prestressing tendons in the center of the specimens were modeled separately using nonlinear beam–column elements consisting only of fibers of TendonL01 material.

The boundary condition and load pattern in the finite element model were defined according to the test condition, as shown in Figure 10.17. The axial loads acting on the columns were applied as vertical nodal forces on the cap beam. The prestressing force was applied as a vertical nodal load acting at the top and bottom of the column. The direction and magnitude of the axial loads and the prestressing remain constant in the analysis. The horizontal forces were changed according to the displacement control scheme.

The analysis procedure was separated into two steps. In the first step, the axial loads were applied to the columns using load control by 10 load increments. In the second step, the axial loads were kept constant and the reversed cyclic horizontal loads were applied by the predetermined displacement control on the drift displacement. The common displacement increment used in the analysis was 1.0 mm. The nodal displacement and corresponding horizontal forces were recorded at each converged displacement step, and the stress and strain of each of the elements were also monitored.

The SCS calculated load–drift relationships for the two specimens are compared with the experimental results in Figures 10.18 and 10.19. It can be seen that the finite element analyses successfully predicted the load–drift characteristics of the specimen, including post-cracking stiffness, yield drift, ultimate strength, and energy dissipation. As observed in the experiment, the finite element analysis could predict the specimens reaching its peak load in the 3% drift loading cycle. The strength degradation in the post-peak region was also well predicted in the analyses, both in the positive and the negative directions. The nearly flat-top envelopes of the specimens (a typical behavior of the flexure mechanism) was also predicted by the analyses.

Table 10.4 Material properties of column specimens

Specimen	Concrete compressive strength (MPa)	Prestressing steel		ED bar	
		Yield strength (MPa)	Ultimate strength (MPa)	Yield strength (MPa)	Ultimate strength (MPa)
C5C	55	1670	1860	434	653
C8C	30	1670	1860	434	653

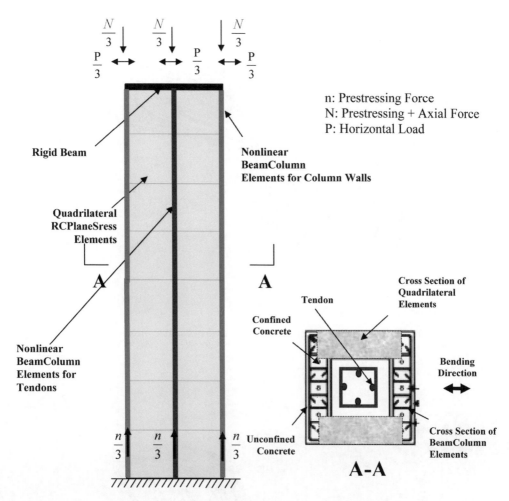

n: Prestressing Force
N: Prestressing + Axial Force
P: Horizontal Load

Figure 10.17   Finite element model of post-tensioned columns

10.5 Framed Shear Walls under Shake Table Excitations

Two low-rise shear walls with a height-to-width ratio of 0.5 were tested on a shake table simulating seismic excitations at NCREE. Both specimens were designed with a scale factor of 1:2 according to the capacity limit of the shake table, and both had the dimensions of 625 mm in height, 1400 mm in length, and 60 mm in thickness. The two specimens were designed identically, with the exception of the steel grid orientation in the walls. While one specimen (STC) was designed as a conventional shear wall with the steel bars in the horizontal and vertical directions (Figure 10.20a), the wall reinforcement in the other specimen (STN) was oriented at 45° to the horizontal (Figure 10.20b). The wall panels in both specimens were designed with a 120 × 120 mm boundary column with four D16 bars (diameter 16 mm) as the longitudinal steel. The walls in the specimens had a steel ratio of 0.5%. The concrete compressive strength for the two specimens was 34.0 MPa. The yield stress and ultimate

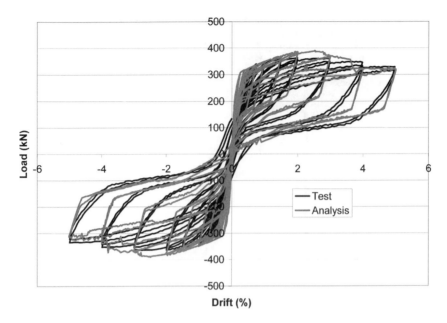

Figure 10.18   Experimental and analytical load drift diagram of specimen C5C

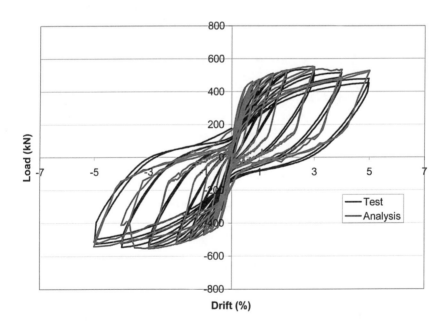

Figure 10.19   Experimental and analytical load drift diagram of specimen C8C

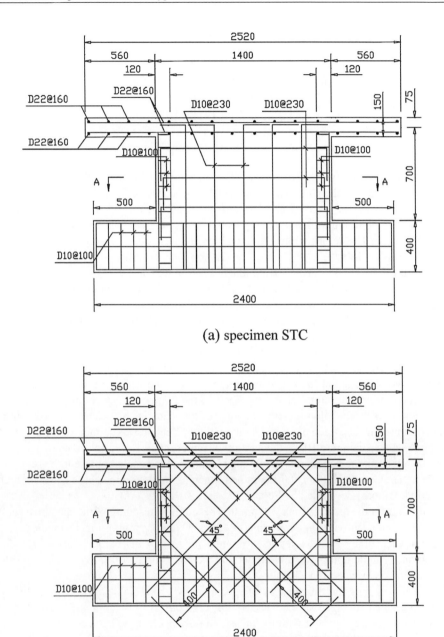

(a) specimen STC

(b) specimen STN

Figure 10.20   Dimensions and steel arrangement

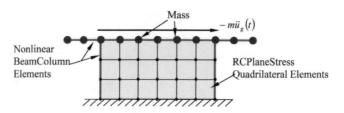

Figure 10.21   Wall elevation in finite element mesh

strength of the steel bars were in the range 370–510 MPa and 550–650 MPa, respectively. Accelerometers were placed on the shake table and the top slab to measure the actual acceleration of the shake table and the response acceleration of the specimen. LVDTs were connected to the foundation, the top of the wall, and the top slab to measure the displacement of the specimen at these locations. The deformations of the wall panels were also measured by LVDTs.

A total mass of 18000 kg was placed on the top slab. In order to avoid the undesirable rocking of the mass blocks during testing, the test set-up was designed by providing four steel columns under the four corners of the top slab. A roller was placed on the top of the column, which would transfer the gravity force of the mass block only in the vertical direction and avoid any horizontal force. The test set-up proved very effective in preventing the top slab from rocking during the shake table excitations. The tcu078Eji seismogram of the 1999 Taiwan earthquake was used as the uniaxial horizontal ground motion acceleration for the shake table.

The finite element mesh of the structure was divided into three zones: the web panel, the boundary columns, and the top slab (Figure 10.21). The wall panel was modeled using 18 RCPlaneStress quadrilateral elements. Each of the two boundary columns was modeled using 3 nonlinear beam–column elements, and the top slab was modeled using 10 nonlinear beam–column elements. Each nonlinear beam–column element was defined with three control sections. As shown in Figure 10.22, the white cells represent the unconfined concrete fibers, the gray cells represent the confined concrete fibers, and the black cells represent the reinforcing steel fibers. The stress and strain of the confined concrete was determined based on the modified Kent and Park model developed by Scott *et al.*, (1982).

The analysis for the first two runs was omitted because the response of the specimen was too small when compared with the remaining test runs. Damping proportional to the converged stiffness at each time step was applied to model the energy dissipation arising from story deformations. The damping ratios were determined based on the different damage levels of the specimens, as discussed in Chapter 9.

The calculated drift and time history of specimens STC and STN for the third to the sixth runs are presented in Figures 10.23 and 10.24, respectively. The computed drift and time histories show good agreement with the measured responses for both specimens. In the sixth run, the analyses slightly overestimated the drifts for both specimens. The results also show that the damping ratios used in the analyses were appropriate to take into account the different damage levels of the structures.

10.6 A Seven-story Wall Building under Shake Table Excitations

A full-scale seven-story reinforced concrete wall building (Figure 10.25) was tested on the shake table located at UCSD's Engelkirk Structural Engineering Center (Zhong *et al.*, 2006). The building was composed of a web wall, a flange wall, a post-tensioned precast pier, gravity

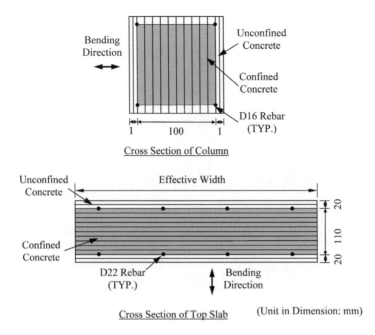

Cross Section of Column

Cross Section of Top Slab   (Unit in Dimension: mm)

Figure 10.22   Cross-section discretization of wall

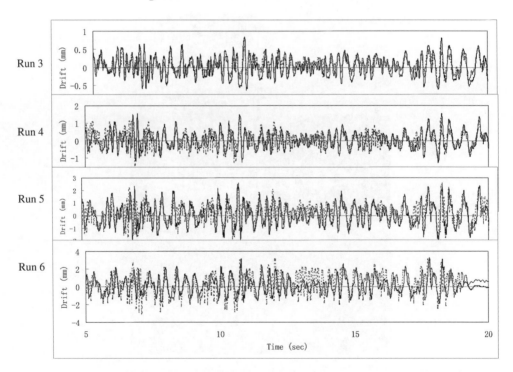

Figure 10.23   Measured and computed drift time history of specimen STC

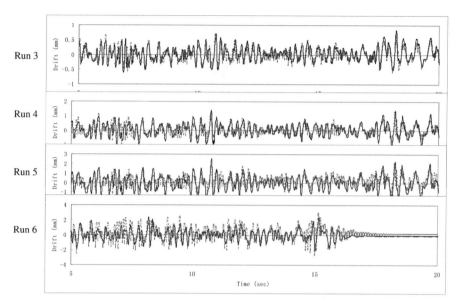

Figure 10.24   Measured and computed drift time history of specimen STN

Figure 10.25   A seven-story wall building tested at UCSD (http://nees.ucsd.edu/7Story.html)

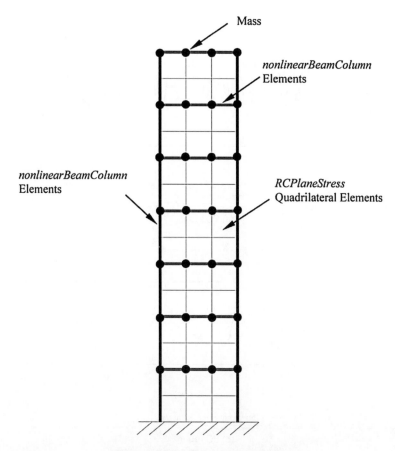

Figure 10.26   Finite element model

columns, and slabs at each floor. The building was subjected to four uniaxial earthquake ground motions of increasing intensity(EQ1, EQ2, EQ3 and EQ4). The peak accelerations of the four ground motions were $0.152g$, $0.224g$, $0.334g$ and $0.486g$, respectively.

The finite element model was composed of total 42 RCPlaneStress quadrilateral elements and 49 nonlinear beam–column elements with fiber sections. Since the web wall was the main lateral force resisting system in the building, the building was simplified to a 2-D finite element model, as shown in Figure 10.26. The web wall at each level was modeled as six RCPlaneStress quadrilateral elements. The steel grid orientations in the wall elements were defined in the horizontal and vertical directions and the steel ratios in the horizontal and vertical directions were calculated according to the construction drawings from the website provided in Zhong *et al.* (2006).

The contribution of the flange wall, post-tensioned precast pier, gravity columns, and foundation slab to the stiffness of the whole building was considered in the following way. The mass of the flange walls and precast pier was applied at the nodes at each floor at the two ends of the web wall. They were modeled as nonlinear beam–column elements with fiber sections. The section of the column elements at Level 1 was discretized into the confined

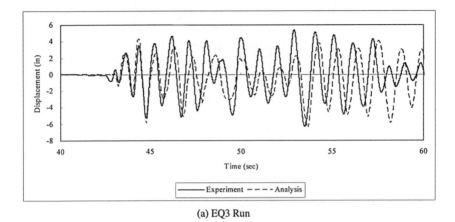

(a) EQ3 Run

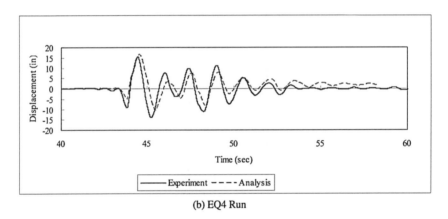

(b) EQ4 Run

Figure 10.27   Time histories of measured and calculated displacement at building top

concrete, unconfined concrete and steel fibers, because the longitudinal steel in the boundary elements at this level were confined with the stirrups. For the boundary elements at levels 2–6, the section was composed of the unconfined concrete and steel fibers. The slabs at each floor were also modeled as nonlinear beam–column elements with fiber sections.

The analysis procedure was separated into two steps, the static analysis and dynamic analysis. In the first step, the gravity loads were applied to the building. It was assumed that half the gravity load was supported by the web wall and the other half was taken by the gravity columns. In the second step, the earthquake excitation was applied. Different damping ratios were applied for the four test runs. The assumed damping ratios used for the runs EQ1–EQ4 were 2, 3, 6, and 8% respectively.

For simplicity, time histories of the predicted displacement at the top of the building for EQ3 and EQ4 are presented in Figure 10.27. The measured test data is also shown in the figures for comparison. It can be seen from Figure 10.27 that the prediction of the displacement is close to the measured results both in the vibration frequency and displacement magnitude.

Appendix A

Example 1

Given

A series of five simply supported PC I-girders with a span of 24 feet were tested at the University of Houston. Out of these five girders, beam B1 was selected for the purpose of demonstration and has been analyzed using SCS. Input.tcl is the input file to create the FE Model. OpenSees.exe is the executable used to run the input file. An Output file called B1_disp_Output.out is created during the execution of OpenSees.exe with the input file. B1_disp_Output.out contains the concentrated vertical loads acting on each of both ends of the beam in kN along with the vertical displacements (in mm) at the three nodes numbered 3, 9, and 15 in Figure A.1.1.

FE Model

The finite element model of the beam is shown in Figure A.1.1. The two flanges of the beams, which resist the bending moment acting on the beams, are modeled as NonlinearBeamColumn elements with fiber sections previously developed in OpenSees. The web of the beams, which resists the shear force acting on the beams, is defined by PCPlaneStress Quadrilateral elements. In the model of the beam, one set of nodes is defined for the beam column elements and another set of nodes is defined for the quadrilateral elements. Nodes 1–34 are for defining beam column elements. Nodes 35 onwards are for defining Quadrilateral elements. So nodes 1 and 35 are at the same location. Similarly nodes 2 and 36 are at the same location and so on. Equal degrees of freedom have been assigned to these nodes lying at the same locations.

In the beginning of the analysis, the prestressing loads acting on the beam have been applied as nodal forces adjacent to the ends of the beams, using load control. Thereafter the prestressing loads have been kept constant and monotonic vertical loads have been applied by a predetermined displacement control scheme.

Unified Theory of Concrete Structures   Thomas Hsu and Yi-Lung Mo
© 2010 John Wiley & Sons, Ltd

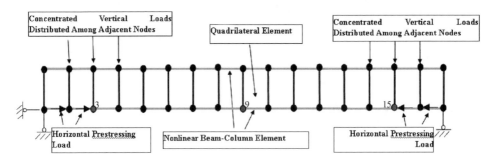

Figure A.1.1   FE model of I-girder B1

Results

The load and displacement curve of the beam as obtained from the above output is shown in Figure A.1.2. The analytical curve has also been compared with the experimental load–displacement curve in Figure A.1.2. The maximum load-carrying capacity of the specimen as observed from the test was 792 kN. The load-carrying capacity of the beam as predicted from analysis is 773 kN. The analysis could well predict the load–deformation characteristics of all five specimens, including the initial stiffness, post-cracking stiffness, yield displacement, and ultimate strength.

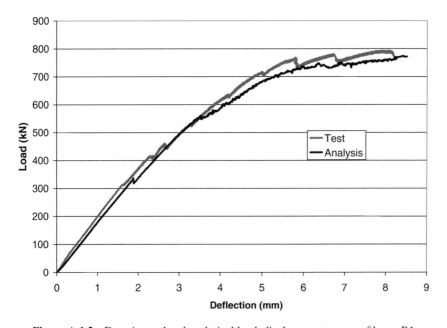

Figure A.1.2   Experimental and analytical load–displacement curves of beam B1

Solution

The input and output data files for the above described beam are shown below:

```
# Input file for beam B1
# ------------------------------
# Units: N, mm, sec, MPa
# ------------------------------

# ------------------------------
# Start of model generation
# ------------------------------


# --------------------------------------------------------------
# Create ModelBuilder for top and bottom flanges (with two-
dimensions and 3 DOF/node)
# --------------------------------------------------------------
   model basic -ndm 2 -ndf 3


# --------------------------------------------------------------
# Create nodes needed for top and bottom flanges (3dof)
# --------------------------------------------------------------

# set dimension of the beam and mesh

   set L 7312; #24 feet
   set H 546;  #21.5 inches
   set deltaL 457;
   set deltaH 546;
   set nL [expr $L/$deltaL];  #16 elements along length
   set nH [expr $H/$deltaH];  # 1 element along depth of web
   set topH 546;
   set botH 0;

# Create nodes for top and bottom flanges
# tag X Y
   set nodeStartID [expr ($nH+1)*($nL+1)+1];   # 35
   set eleStartID  [expr $nH*$nL+1];           # 17

   set j 0;

   while {$j < [expr $nL+1]} {
      node [expr $nodeStartID+$j] [expr $j*$deltaL] [expr $botH]
      node [expr $nodeStartID+$nL+1+$j] [expr $j*$deltaL] [expr
$topH]
      set j [expr $j+1]
   }
```

```
# Provide roller and hinge supports at two ends of beams
# tag DX DY RZ

   fix [expr $nodeStartID]      1 1 0
   fix [expr $nodeStartID+$nL] 0 1 0


# ----------------------------------------------------
# Define nonlinear materials for flanges
# ----------------------------------------------------

# CONCRETE tag f'c ec0 f'cu ecu

   uniaxialMaterial Concrete01 1 -72.4 -0.0024 -14.0 -0.008

# STEEL
# Reinforcing steel

   set E 200000.0; # Young's modulus

# tag fy E0 b

   uniaxialMaterial Steel01 3 413.7 $E 0.001

# for strands fu=270ksi=1861.65MPa
# for strands fy=0.7fu = 1303.0 MPa
   uniaxialMaterial Steel01 4 1303.0 $E 0.001

# ------------------------------------------------
# Define cross-section for top and bottom flanges
# ------------------------------------------------

# set some paramaters

# #5 bar
   set As1   200
# 1/2 inch seven-wire strands
   set As2   99
# for top flange

   section Fiber 1 {
   patch rect 1 10 4 -152 -70 152 70

# Creating the reinforcing fibers
   layer straight 3 2 $As1   -127.0   40.0 127.0 40.0
}

# for bottom flange
   section Fiber 2 {

   patch rect 1 10 4  -203 -95 203 95
```

```
# Creating the prestressing strands
  layer straight 4 2 $As2   -153 -49 153 -49

}


# ------------------------------------------------------
# Define flange elements
# ------------------------------------------------------

  geomTransf Linear 2

  set np 3;
  set iterNum 10;
  set iterTol 1e-2;

  set i 0;

  while {$i < [expr $nL]} {
    # bottom flange elements
    element nonlinearBeamColumn [expr $eleStartID+$i] [expr
$nodeStartID+$i] [expr $nodeStartID+1+$i] $np 2 2 -iter
$iterNum $iterTol
    # top flange elements
    element nonlinearBeamColumn [expr $eleStartID+$nL+$i] [expr
$nodeStartID+$nL+1+$i] [expr $nodeStartID+$nL+2+$i] $np 1 2
-iter $iterNum $iterTol
    set i [expr $i+1]
}

# ------------------------------------------------------
# Create ModelBuilder for 2D web elements (with two-
dimensions and 2 DOF/node)
# ------------------------------------------------------

 model basic -ndm 2 -ndf 2

# Create nodes & add to Domain - command: node nodeId xCrd
yCrd

  set j 0;
  while {$j < [expr $nH+1]} {
    set i 0;
    while {$i < [expr $nL+1]} {
      node [expr $j*($nL+1)+$i+1] [expr $i*$deltaL] [expr
$j*$deltaH]
      set i [expr $i+1]
    }
    set j [expr $j+1]
    }
```

```
# Provide roller and hinge supports at two ends of beams
  fix 1     1 1
  fix 17    0 1

# tie nodes between flange elements and web elements

# tying nodes along bottom flange
  equalDOF    2 36    1 2
  equalDOF    3 37    1 2
  equalDOF    4 38    1 2
  equalDOF    5 39    1 2
  equalDOF    6 40    1 2
  equalDOF    7 41    1 2
  equalDOF    8 42    1 2
  equalDOF    9 43    1 2
  equalDOF    10 44    1 2
  equalDOF    11 45    1 2
  equalDOF    12 46    1 2
  equalDOF    13 47    1 2
  equalDOF    14 48    1 2
  equalDOF    15 49    1 2
  equalDOF    16 50    1 2

# tying nodes along top flange
  equalDOF    18 52    1 2
  equalDOF    19 53    1 2
  equalDOF    20 54    1 2
  equalDOF    21 55    1 2
  equalDOF    22 56    1 2
  equalDOF    23 57    1 2
  equalDOF    24 58    1 2
  equalDOF    25 59    1 2
  equalDOF    26 60    1 2
  equalDOF    27 61    1 2
  equalDOF    28 62    1 2
  equalDOF    29 63    1 2
  equalDOF    30 64    1 2
  equalDOF    31 65    1 2
  equalDOF    32 66    1 2
  equalDOF    33 67    1 2
  equalDOF    34 68    1 2


# ---------------------------------------------------------
#  Define materials for 2D PrestressConcretePlaneStress
element
# ---------------------------------------------------------
```

```
# set fc fy E
  set wfc 72.4;
  set wfpu 1862;
  set wfy 413.7;
  set wE  200000.0;
  set rou1 0.0055;
  set rou2 0.00164;
  set ec 0.002;
  set t 152.4;

# UniaxialMaterial: steelZ01
#                        tag     fy       E0     fpu      rou      epsi
  uniaxialMaterial    TendonL01  11   [expr 0.7*$wfpu]       $wE
                        $wfpu   $rou1  0.006
#                        tag    fy       E0    fpu      rou
  uniaxialMaterial    SteelZ01   12    $wfy     $wE  $wfc  $rou2

# UniaxialMaterial: concreteL01
# ConcreteL01                    tag    f'c       ec0
  uniaxialMaterial ConcreteL01   13  [expr -$wfc] [expr -$ec]
  uniaxialMaterial ConcreteL01   14  [expr -$wfc] [expr -$ec]

set pi 3.141592654
# NDMaterial: FAPrestressConcretePlaneStress
#                          tag  rho s1 s2 c1 c2      angle1
angle2        rou1  rou2   fpc  fpy                fy   E0  ec
  nDMaterial FAPrestressConcretePlaneStress 15  0.0 11 12 13
14 [expr 1.0*$pi] [expr 0.5*$pi] $rou1 $rou2 -0.0 $wfc [expr
0.7*$wfpu] $wfy $wE $ec
  nDMaterial FAPrestressConcretePlaneStress 16  0.0 11 12 13
14 [expr 0.0*$pi] [expr 0.5*$pi] $rou1 $rou2 -0.0 $wfc [expr
0.7*$wfpu] $wfy $wE $ec


# --------------------------------------------------------
#  Define 2D ReinforceConcretePlaneStress element
# --------------------------------------------------------
  set j 0;
  while {$j < $nH} {
  set i 0;
  while {$i < [expr $nL]} {
  # Create quad elements - command:
  # element  quad  eleID  node1  node2  node3  node4  thick
type matID

  }     element quad [expr $j*$nL+$i+1] [expr $j*($nL+1)+$i+1]
[expr $j*($nL+1)+$i+2] [expr ($j+1)*($nL+1)+$i+2] [expr
($j+1)*  ($nL+1)+$i+1] $t PlaneStress 15
```

```
    #element quad [expr $j*$nL+($i+$nL/2)+1] [expr $j*($nL+1)
+($i+$nL/2)+1] [expr $j*($nL+1)+($i+$nL/2)+2] [expr ($j+1)*
($nL+1)+($i+$nL/2)+2] [expr ($j+1)*($nL+1)+($i+$nL/2)+1]
$t PlaneStress 16

    set i [expr $i+1]
    }
  set j [expr $j+1]
  }

  model basic -ndm 2 -ndf 3

# --------------------------
# Define prestress loads
# --------------------------

# set pForce 1.654e6;   # 372kips=1.654e6N
  set pForce 1.654e6;

# Creating a Plain load pattern with a linear TimeSeries
 pattern Plain 1 "Linear" {
  # Create the nodal load - command: load nodeID xForce yForce
  load 36   [expr $pForce/2] 0 0;
  load 37   [expr $pForce/2] 0 0;
  load 49   [expr -$pForce/2] 0 0;
  load 50   [expr -$pForce/2] 0 0;
 }      element quad [expr $j*$nL+$i+1] [expr $j*($nL+1)+$i+1]
[expr $j*($nL+1)+$i+2] [expr ($j+1)*($nL+1)+$i+2] [expr
($j+1)*($nL+1)+$i+1] $t PlaneStress 15

# ------------------------------
# Start of analysis generation
# ------------------------------

# Creating the system of equation, a sparse solver with
partial  pivoting
   system BandGeneral

# Creating the constraint handler
 constraints Plain

# Creating the DOF numberer
 numberer Plain

# Creating the convergence test
 test NormDispIncr 1.0e-3 100 5

# Creating the solution algorithm
#    algorithm Newton
```

```
#    algorithm NewtonLineSearch 0.8
  algorithm KrylovNewton

# Creating the integration scheme, the DisplacementControl
scheme
  integrator LoadControl 0.1 '

# Creating the analysis object
  analysis Static

# initialize in case we need to do an initial stiffness
iteration
  initialize

# -----------------------------
# End of analysis generation
# -----------------------------
# perform the analysis
 analyze 10

# Print out the state of nodes
  puts "After Prestress Force"

  print node 3 9 15
  print node 37 43 49

# Set the prestress loads to be constant & reset the time in
the domain

  loadConst -time 0.0

# ------------------------------------------------------
# Start of modeling for vertical loads to 200kips
# ------------------------------------------------------
  set P1    1000;

# Create a Plain load pattern with a linear TimeSeries
  pattern Plain 2 "Linear" {
  # Create the nodal load - command: load nodeID xForce yForce
  load 53  0 [expr -$P1/3] 0;
  load 54  0 [expr -$P1/3] 0;
  load 55  0 [expr -$P1/3] 0;
  load 65  0 [expr -$P1/3] 0;
  load 66  0 [expr -$P1/3] 0;
  load 67  0 [expr -$P1/3] 0;

 }
# -----------------------------
# End of model generation
# -----------------------------
```

```
# ------------------------------
# Start of analysis generation
# ------------------------------

# Create the system of equation, a sparse solver with p
artial pivoting
   system BandGeneral

# Create the constraint handler
   constraints Plain

# Create the DOF numberer
   numberer Plain

# Creating the convergence test
   test NormDispIncrVaryIter 0.1 1 5 numStep 2000 numIter 100

# Creating the solution algorithm
   algorithm KrylovNewton

# Creating the integration scheme, the DisplacementControl
   scheme
   integrator DisplacementPath 66 2 1 numStep 2000 increment
-0.01

# Creating the analysis object
   analysis Static

# puts "analysis performed"

# initialize in case we need to do an initial stiffness
iteration
#   initialize

# ------------------------------
# End of analysis generation
# ------------------------------

# Creating a recorder to monitor nodal displacements
   recorder Node -file B1_disp_Output.out -time -node 3 9 15 -
dof 2 disp

# perform the analysis
   analyze 2000

# Print out the state of nodes
   print node 3 9 15 54
```

Output File

The output file obtained after running the input file using the developed program is shown below. It contains the load acting on the beam in kN along with the vertical displacements (in mm) at the nodes 3, 9, and 15 indicated in the finite element model of the beam in Figure A.1.1.

```
1.32289 1.62192 3.81952 1.62755
2.81607 1.61287 3.79747 1.61714
4.26667 1.60291 3.77638 1.60697
5.78098 1.59281 3.75297 1.59685
7.22307 1.58008 3.72696 1.58678
8.62239 1.56956 3.70658 1.57655
10.2415 1.55975 3.68262 1.5664
11.6689 1.54916 3.65882 1.55617
13.1751 1.5402 3.63586 1.54619
14.67 1.53114 3.6147 1.53601
16.4757 1.51872 3.58938 1.52594
. . . . .
. . . . .
... (continued..)
```

Example 2. Post-tensioned Bridge Column under Reversed Cyclic Loads

Given

A precast post-tensioned bridge column (tested at SUNY Buffalo) as shown in the Figure A.2.1 has been analyzed using SCS. The finite element model of the column is also shown in Figure A.2.2. C8C.tcl is the input file to create the FE Model shown below. OpenSees.exe is the executable used to run the input file. An Output file called C8C_disp_Output.out is created during the execution of OpenSees.exe with the input file. C8C_disp_Output.out contains the load acting on the bridge column in kN along with the horizontal displacements (in mm) at the top of the column.

FE Model

The specimen was modeled using the finite element mesh illustrated in Figure A.2.2. The two flange sides of the bridge column perpendicular to the bending direction, which are mainly under compression and tension due to bending, are modeled as NonlinearBeamColumn elements with fiber sections. The two web sides of the bridge column parallel to the bending direction, which resist the shear force, are defined by PCPlaneStress Quadrilateral elements. The cap beam on the top of the column is defined as a rigid body in the finite element model. The prestressing tendons in the center of the specimens were modeled separately using NonlinearBeamColumn elements consisting only of fibers of TendonL01 material.

The analysis procedure was separated into two steps. In the first step, axial loads were applied to the columns using load control by 10 load increments. In the second step, axial loads were kept constant and reversed cyclic horizontal loads were applied by the predetermined displacement control on the drift displacement.

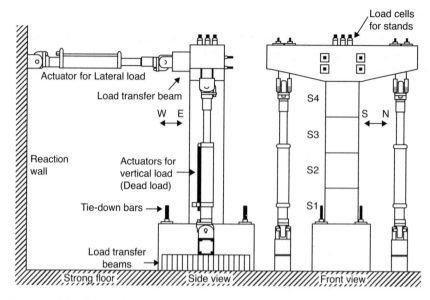

Figure A.2.1   Schematic diagram of post-tensioned bridge column specimen (Ou, 2007)

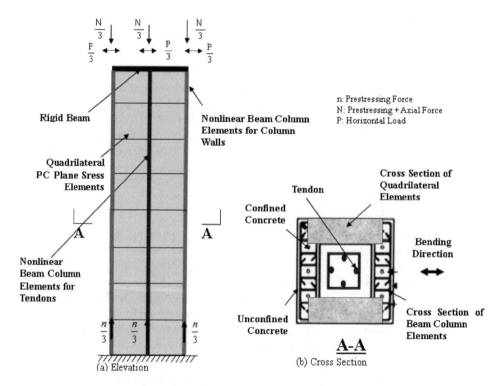

Figure A.2.2   FE model of the bridge column

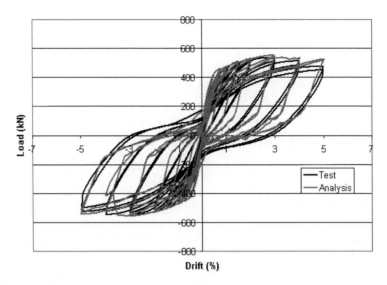

Figure A.2.3   Experimental and analytical load–displacement curves of column specimen

Results

The measured and calculated load–drift relationship for the bridge column specimen is shown in Figure A.2.3. Compared with the experimental results, the analyses predicted the load–drift characteristics of the specimen including post-cracking stiffness, yield drift, ultimate strength, and energy dissipation.

   As observed in the experiment, the analysis could predict the specimen reaching its peak load in the 3% drift loading cycle. The peak loads carried by the specimen during testing and as predicted from the analysis are 535 and 544 kN, respectively. The strength degradation of the specimen in the post-peak region is well predicted in the analysis both in the positive as well as the negative directions. The nearly flat-top envelopes of specimen, which is a typical behavior of the flexure mechanism, is also captured by the analysis.

Input File for Post-tensioned Bridge Column

The input file for the above described column is shown below:

```
# Input File for Specimen C8C

# ------------------------------
# Units: N, mm, sec, MPa
# ------------------------------

# ------------------------------
# Start of model generation
# ------------------------------
```

```
# 4.1
# -------------------------------------------------------------------
# Create ModelBuilder for moment carrying walls of hollow bridge
columns (with two-dimensions and 3 DOF/node)
# -------------------------------------------------------------------
model basic -ndm 2 -ndf 3

# -------------------------------------------------------------------
# Create nodes needed for moment carrying walls of hollow bridge
columns (3dof)
# -------------------------------------------------------------------

# set dimension of the wall and mesh

  set L 660;
  set H 4000;
  set deltaL 330;
  set deltaH 500;
  set nL [expr $L/$deltaL];
  set nH [expr $H/$deltaH];
  set t 400;

# Creating nodes for moment carrying walls of hollow bridge columns
# tag X Y

  set nodeStartID [expr ($nH+1)*($nL+1)+1];
  set eleStartID  [expr $nH*$nL+1];
  set j 0;

  while {$j < [expr $nH+1]} {
    if { $j < [expr $nH] } {
      node [expr $nodeStartID+$j*2]     0.0        [expr $deltaH*$j]
      node [expr $nodeStartID+$j*2+1] [expr $L] [expr $deltaH*$j]
    } else {
      set i 0;
      while {$i < [expr $nL+1]} {
        node [expr $nodeStartID+$j*2+$i] [expr $i*$deltaL] [expr
$H]
        set i [expr $i+1]
      }
    }
    set j [expr $j+1]
  }


# ---------------------------------------
# Create nodes for tendons
# ---------------------------------------
```

```
  set tendonnodeStartID [expr ($nH+1)*($nL+1)+$nH*2+($nL+1)
+1];
  set tendoneleStartID [expr $nH*$nL+$nH*2+$nL+1];

  set j 0;

  while {$j < [expr $nH+1]} {
     node [expr $tendonnodeStartID+$j]  [expr $deltaL] [expr
$deltaH*$j]
   set j [expr $j+1]
   }

# Fix supports at base of columns
# tag DX DY RZ
  fix [expr $nodeStartID]    1 1 1
  fix [expr $nodeStartID+1] 1 1 1
  fix [expr $tendonnodeStartID] 1 1 1


# ---------------------------------------------------------------
# Define nonlinear materials for moment carrying walls
and tendons
# ---------------------------------------------------------------

# CONCRETE tag f'c ec0 f'cu ecu
# Cover concrete (unconfined)
  uniaxialMaterial Concrete01 1 -30 -0.003 -6 -0.01

# Core concrete (confined)
  uniaxialMaterial Concrete01 2 -45 -0.003 -9 -0.03

# STEEL
# Longitudinal Reinforcing steel

  set fy 434; # Yield stress for bare bar
  set E 200000.0; # Young's modulus

#                                tag  fy   E0  fpc  rou
  uniaxialMaterial  Steel01   4   $fy $E    0.01

#                                tag  fy    E0  fpc  rou  epsi
  uniaxialMaterial  TendonL01  5   1670  $E 1860 0.001 0.0036

# -------------------------------------------------------
# Define cross-section for nonlinear columns
# -------------------------------------------------------

# set some parameters

  set colWidth 860.0
```

```
set colDepth 200.0
set cover 50.0
set As 430;
```

some variables derived from the parameters

```
set cy1 [expr $colDepth/2.0]
set cz1 [expr $colWidth/2.0]
```

the section 1 is for stirrup confinement is #4@80
```
section Fiber 1 {
```

Create the concrete core fibers
mat num num
```
  patch rect 2  10  1 [expr $cover-$cy1] [expr $cover-$cz1]
[expr $cy1-$cover] [expr $cz1-$cover]
```

Create the concrete cover fibers (top, bottom, left, right)
```
  patch rect 1 10 1 [expr -$cy1] [expr $cz1-$cover] $cy1 $cz1
  patch rect 1 10 1 [expr -$cy1] [expr -$cz1] $cy1 [expr
$cover-$cz1]
  patch rect 1 2 1 [expr -$cy1] [expr $cover-$cz1] [expr
$cover-$cy1] [expr $cz1-$cover]
  patch rect 1 2 1 [expr $cy1-$cover] [expr $cover-$cz1]
$cy1 [expr $cz1-$cover]
```

Create the reinforcing fibers (2 layers)
```
  layer straight 4 2 $As [expr $cy1-$cover] [expr $cz1
-$cover] [expr $cy1-$cover] [expr $cover-$cz1]
  layer straight 4 2 $As [expr $cover-$cy1] [expr $cz1
-$cover] [expr $cover-$cy1] [expr $cover-$cz1]

 }
```

```
# -----------------------------------------------
# Define cross-section for top beam
# -----------------------------------------------
```

```
#                        E    A    I
  section Elastic 3 2e5 1e6 1e14
```

```
# -----------------------------------------------
# Define cross-section for tendons
# -----------------------------------------------
```

```
  section Fiber 4 {
```

Create the concrete core fibers

Create the reinforcing fibers (2 layers)

```
   layer straight 5 2 280 -185 0 185 0
   layer straight 5 2 280 0 -185 0 185

   }


# ------------------------------------------------------------
# Define column elements
# ------------------------------------------------------------

   geomTransf Linear 1

   set np 2
   set iterNum 10
   set iterTol 1e-3

   set j 0;
   while {$j < [expr $nH]} {

      if {$j < [expr $nH-1]} {

         element nonlinearBeamColumn [expr $eleStartID+$j*2] [expr
$nodeStartID+$j*2]    [expr $nodeStartID+$j*2+2] $np 1 1 -iter
$iterNum $iterTol
         element nonlinearBeamColumn [expr $eleStartID+$j*2+1]
[expr $nodeStartID+$j*2+1] [expr $nodeStartID+$j*2+3] $np 1 1
-iter $iterNum $iterTol

      } else {
         element nonlinearBeamColumn [expr $eleStartID+$j*2] [expr
$nodeStartID+$j*2]    [expr $nodeStartID+$j*2+2] $np 1 1 -iter
$iterNum $iterTol
         element nonlinearBeamColumn [expr $eleStartID+$j*2+1]
[expr $nodeStartID+$j*2+1] [expr $nodeStartID+$nH*2+$nL]
$np 1 1 -iter $iterNum $iterTol
      }
   set j [expr $j+1]
   }


# ------------------------------------------------------------
# Define tendon elements
# ------------------------------------------------------------

#  geomTransf Linear 1

   set i 0;

   while {$i < [expr $nH]} {

      element nonlinearBeamColumn [expr $tendoneleStartID+$i]
[expr $tendonnodeStartID+$i] [expr $tendonnodeStartID+1+$i]
$np 4 1 -iter $iterNum $iterTol
```

```
   set i [expr $i+1]
}

# --------------------------------------------------------
# Define beam elements
# --------------------------------------------------------

   geomTransf Linear 2

   set j [expr $nH];
   set i 0;

   while {$i < [expr $nL]} {

      element nonlinearBeamColumn [expr $eleStartID+$j*2+$i]
[expr $nodeStartID+$j*2+$i] [expr $nodeStartID+$j*2+1+$i] $np
3 2 -iter $iterNum $iterTol

   set i [expr $i+1]
}

# 4.2
# --------------------------------------------------------
# Create ModelBuilder for 2D elements representing shear car-
rying walls of the column (with two-dimensions and 2 DOF/node)
# --------------------------------------------------------

 model basic -ndm 2 -ndf 2

# Create nodes & add to Domain - command: node nodeId xCrd
yCrd

   set j 0;
   while {$j < [expr $nH+1]} {
   set i 0;
   while {$i < [expr $nL+1]} {
     node [expr $j*($nL+1)+$i+1] [expr $i*$deltaL] [expr
$j*$deltaH]
     set i [expr $i+1]
   }
   set j [expr $j+1]
 }

# Set the boundary conditions - command: fix nodeID xResrnt?
yRestrnt?

   set i 0;
   while {$i < [expr $nL+1]} {
    fix [expr $i+1] 1 1
    set i [expr $i+1]
}
```

```
# tying nodes between moment carrying walls and 2D elements
representing shear carrying walls of the column

equalDOF   4 30 1 2
equalDOF   6 31 1 2
equalDOF   7 32 1 2
equalDOF   9 33 1 2
equalDOF  10 34 1 2
equalDOF  12 35 1 2
equalDOF  13 36 1 2
equalDOF  15 37 1 2
equalDOF  16 38 1 2
equalDOF  18 39 1 2
equalDOF  19 40 1 2
equalDOF  21 41 1 2
equalDOF  22 42 1 2
equalDOF  24 43 1 2

equalDOF 25 44 1 2
equalDOF 26 45 1 2
equalDOF 27 46 1 2

equalDOF 55 45 1 2


# ------------------------------------------------------------
#  Define materials for 2D ReinforceConcretePlaneStress
element
# ------------------------------------------------------------

# set fc fy E
  set wfc 30.0;
  set wfyv 434;
  set wfyh1 413;
  set wE  200000.0;
  set rou1 0;
  set rou2 0;
  set rouv  0.01;
  set rouh1 0.0246;  # #10@70

# UniaxialMaterial: steelZ01
#                    tag   fy        E0   fpc     rou
uniaxialMaterial     SteelZ01  11   $wfyv      $wE  $wfc  $rouv
uniaxialMaterial     SteelZ01  12   $wfyh1     $wE  $wfc  $rouh1

# UniaxialMaterial: concreteZ01
# ConcreteZ01                  tag   f'c       ec0
uniaxialMaterial ConcreteL01  14  [expr -$wfc]  -0.003
uniaxialMaterial ConcreteL01  15  [expr -$wfc]  -0.003
```

```
  set pi 3.141592654
# NDMaterial: FAFourSteelPCPlaneStress
#             tag  rho p1   p2 s1 s2 c1 c2   angle1    angle2
angle3    angle4  roup1 roup2  rous1  rous2    fpc  fpy  fy  E0
   nDMaterial FAFourSteelPCPlaneStress 21   0.0 11   12 11 12 14
15 [expr 1.0*$pi] [expr 0.966*$pi] [expr 0.5*$pi] [expr 0.0*$pi]
$rou1 $rou2   $rouv $rouh1 0.0 0.0 $wfc $wfyv $wfyv $wE 0.003


# ------------------------------------------------
#  Define 2D ReinforceConcretePlaneStress element
# ------------------------------------------------

  set j 0;
  while {$j < $nH} {
    set i 0;
    while {$i < $nL} {
    # Create quad elements - command:
    # element quad eleID node1 node2 node3 node4 thick   type
matID
    element quad [expr $j*$nL+$i+1] [expr $j*($nL+1)+$i+1]
[expr $j*($nL+1)+$i+2] [expr ($j+1)*($nL+1)+$i+2] [expr
($j+1)*($nL+1)+$i+1] $t PlaneStress 21
    set i [expr $i+1]
    }
    set j [expr $j+1]
  }


# 4.3
# ----------------------------------------
# Define prestress and gravity loads
# ----------------------------------------

# Create a Plain load pattern with a linear TimeSeries
  pattern Plain 1 "Linear" {

  # Create the nodal load - command: load nodeID xForce yForce
  load 4 0 [expr   346e3]
  load 5 0 [expr   346e3]
  load 6 0 [expr   346e3]
  load 22 0 [expr  -346e3]
  load 23 0 [expr  -346e3]
  load 24 0 [expr  -346e3]
  load 25 0 [expr  -486e3]
  load 26 0 [expr  -486e3]
  load 27 0 [expr  -486e3]
}
```

```
#  ------------------------------
#  End of model generation
#  ------------------------------

#  4.4
#  ------------------------------
#  Start of analysis generation
#  ------------------------------

#  Creating the system of equation, a sparse solver with
partial pivoting
   system BandGeneral

#  Creating the constraint handler
   constraints Plain

#  Creating the DOF numberer
   numberer Plain

#  Creating the convergence test
   test NormDispIncr 1.0e-3 100 5

#  Creating the solution algorithm
   algorithm KrylovNewton

#  Creating the integration scheme, the DisplacementControl
scheme
   integrator LoadControl 0.1

#  Creating the analysis object
   analysis Static

#  ------------------------------
#  End of analysis generation
#  ------------------------------

#  perform the analysis
   analyze 10

#  Print out the state of nodes
   print node 25 26 27 44 45 46

#  4.5
#  Set the gravity and prestress loads to be constant & reset
the time in the domain

   loadConst -time 0.0
```

```
# ----------------------------------------------------------
# End of Model Generation & Initial Gravity and Prestress
Load Analysis
# ----------------------------------------------------------

# -----------------------------------------------------------
# Start of additional modeling for lateral loads
# -----------------------------------------------------------

# ---------------------------------
# Define horizontal loads
# ---------------------------------

set P 1000.0;

# Create a Plain load pattern with a linear TimeSeries
  pattern Plain 2 "Linear" {

  # Create the nodal load - command: load nodeID xForce yForce
  load 25 [expr    $P/3] 0
  load 26 [expr    $P/3] 0
  load 27 [expr    $P/3] 0
 }


# ------------------------------
# End of model generation
# ------------------------------

# 4.6
# ------------------------------
# Start of analysis generation
# ------------------------------

# Creating the system of equation, a sparse solver with
partial pivoting
  system BandGeneral

# Creating the constraint handler
  constraints Plain

# Creating the DOF numberer
  numberer Plain

# Creating the convergence test
  test NormDispIncrVaryIter 0.001 45 5 numStep 10 20 20 20 25
30 30 30 35 40 40 40 50 60 60 30 70 80 80 80 100 120 120
120 140 160 160 160 31 15 240 240 240 140 80 40 80 45 50
50 50 53 4 48 30 numIter 100 0 0 0 0 0 0 0 0 0 0 0 0 0 0
0 0 0 0 0 0 0 0 0 0 0 0 0 0 0 0 0 0 0 0 0 0 0 0 0 0 0 0
```

```
# Creating the solution algorithm
 algorithm KrylovNewton

# Creating the integration scheme, the DisplacementControl
scheme
   integrator DisplacementPath 26 1 45 numStep 10 20 20 20 25
30 30 30 35 40 40 40 50 60 60 30 70 80 80 80 100 120 120
120 140 160 160 160 31 15 240 240 240 140 80 40 80 45 50
50 50 53 4 48 30 increment 1.0 -1.0 1.0 -1.0 1.0 -1.0 1.0
-1.0 1.0 -1.0 1.0 -1.0 1.0 -1.0 1.0 -2.0 1.0 -1.0 1.0
-1.0 1.0 -1.0 1.0 -1.0 1.0 -1.0 1.0 -1.0 5.0 3.0 -1.0 1.0
-1.0 2.0 -4.0 8.0 -4.0 8.0 -8.0 8.0 -8.0 8.0 4.0 -10.0 8.0

# Creating the analysis object
   analysis Static

# initialize in case we need to do an initial stiffness
iteration
   initialize

# -----------------------------
# End of analysis generation
# -----------------------------

# Creating a recorder to monitor nodal displacements
   recorder Node -file C8C_disp_Output.out -time -node 26 -dof
1 disp

# perform the analysis
   analyze 3366

# Print out the state of nodes
   print node 25 26 27
```

Output File

The output file obtained after running the input file using the developed program is shown below. It contains the load acting on the bridge column in kN along with the horizontal displacements (in mm) at the top of the column.

```
46.8304 0.999735
93.8233 1.99973
140.121 2.99973
179.901 3.99973
209.193 4.99973
234.1 5.99973
255.115 6.99973
275.302 7.99973
```

```
294.726 8.99973
313.518 9.99973
291.245 8.99973
268.562 7.99973
245.114 6.99973
220.444 5.99973
196.097 4.99973
168.155 3.99973
135.281 2.99973
91.9684 1.99973
46.1333 0.999735
-0.0932907 -0.000265051
.....
.....
... (continued..)
```

Example 3. Framed RC Shear Walls under Reversed Cyclic Loading

Given

A   one-third-scale framed shear walls model, subjected to a constant axial load at the top of each column and a reversed cyclic load at the top beam, (tested at the University of Houston) has been analyzed in this example. Figure A.3.1 demonstrates the details of

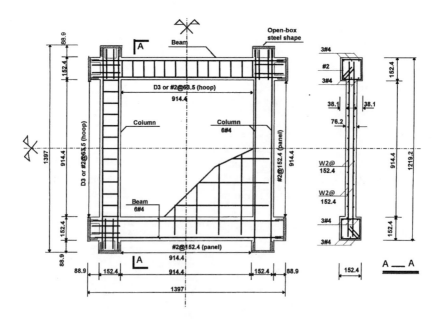

Figure A.3.1   Dimensions and steel arrangement of specimens FSW13

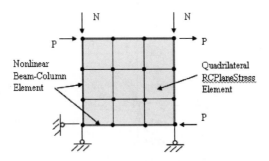

Figure A.3.2 FE model of the framed shear wall

dimensions and reinforcement of specimens 8, 12, and 13. The finite element model of the framed shear walls are also shown in Figure A.3.2. N_FSW13.tcl is the input file to create the FE model shown below. OpenSees.exe is the executable used to run the input file. An Output file called P_FSW13_n13.out is created during the execution of OpenSees.exe with the input file. P_FSW13_n13.out contains the lateral load acting on the wall specimen in kN along with the displacements (in mm) at the top of the wall.

FE Model

The specimen was modeled using the finite element mesh illustrated in Figure A.3.2. The wall panel was modeled as nine Quadrilateral elements with RCPlaneStress material. The steel ratio and the steel grid orientations of the wall panel were defined in the RCPlaneStress material. SteelZ01 and ConcreteZ01 were used to create the uniaxial constitutive laws of steel and concrete, respectively. Each of the boundary beams and columns were modeled as three NonlinearBeamColumn elements.

The analysis procedure was separated into two steps. In the first step, axial loads were applied to the columns using load control by 10 load increments. In the second step, axial loads were kept constant and reversed cyclic horizontal loads were applied by the predetermined displacement control on the drift displacement. The common displacement increment used in the analysis was 0.01 mm.

Results

The analytical results of the shear force–drift relationships of the shear walls are illustrated by the dashed curves in Figure A.3.3. For ease of comparison, the corresponding experimental results, indicated by the solid curves, are also plotted in the figure. It can be seen from the comparison that for the primary curves (backbone curves) the predicted outcomes agree very well with the experimental results in the initial stiffness, yield point, and ultimate state for the specimen. The predictions for the hysteretic behavior simulate the energy dissipation, residual displacement and pinching effect very closely in the specimen.

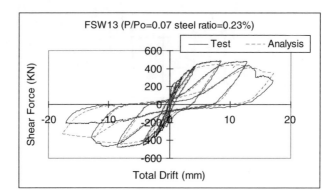

Figure A.3.3   Experimental and analytical load–displacement curves of specimen FSW13

Input File for Framed RC Shear Wall

The input file for the above described specimen is shown below:

```
# N_FSW13.tcl
#
# Apply axial load first

# -----------------------------
# Units: N, mm, sec, MPa
# -----------------------------

# -----------------------------
# Start of model generation
# -----------------------------

#4.1
# ---------------------------------------------------------------
# Create ModelBuilder for beams and columns (with two-dimensions
and 3 DOF/node)
# ---------------------------------------------------------------
  model basic -ndm 2 -ndf 3

# ---------------------------------------------------------------
# Create nodes needed for beams and columns (3dof)
# ---------------------------------------------------------------

# set dimension of the wall and mesh

  set L 1068;
  set H 1068;
  set deltaL 356;
  set deltaH 356;
  set nL [expr $L/$deltaL];
```

```
  set nH [expr $H/$deltaH];
  set t 76.2;


# Create nodes for beams and columns
# tag X Y

  set nodeStartID [expr ($nH+1)*($nL+1)+1];
  set eleStartID  [expr $nH*$nL+1]

  set j 0;
  while {$j < [expr $nH+1]} {
    if{$j < [expr $nH]} {
    node [expr $nodeStartID+$j*2]    0.0   [expr $deltaH*$j]
    node [expr $nodeStartID+$j*2+1] [expr $L] [expr $deltaH*$j]
  } else{
    set i 0;
    while {$i < [expr $nL+1]} {
      node [expr $nodeStartID+$j*2+$i] [expr $i*$deltaL] [expr
$H]
    set i [expr $i+1]
    }
  }
  set j [expr $j+1]
  }

  node 27 356 0.0
  node 28 712 0.0

# ----------------------------------------------------
# Define materials for nonlinear columns and beams
# ----------------------------------------------------

# CONCRETE tag f'c ec0 f'cu ecu
# Core concrete (confined)

  uniaxialMaterial Concrete01 1 -64.7 -0.0024 -13.0 -0.006

# Cover concrete (unconfined)

  uniaxialMaterial Concrete01 2 -57.0 -0.002 -0.0 -0.005

# STEEL
# Reinforcing steel

  set fy 370.0; # Yield stress for #7 bar
  set E 216082.0; # Young's modulus

# tag fy E0 b
```

```
   uniaxialMaterial Steel01 3 $fy $E 0.023
#  uniaxialMaterial SteelZ01  3  $fy $E  49.75 0.033

# -----------------------------------------------
# Define cross-section for nonlinear columns
# -----------------------------------------------

# set some paramaters
   set colWidth 152.4
   set colDepth 152.4
   set cover 20.0
   set As 126.7; # area of no. 4 bars

# some variables derived from the parameters

   set cy1 [expr $colDepth/2.0]
   set cz1 [expr $colWidth/2.0]

   section Fiber 1 {

   # Create the concrete core fibers
   patch rect 1 10 1 [expr $cover-$cy1] [expr $cover-$cz1]
[expr $cy1-$cover] [expr $cz1-$cover]

   # Create the concrete cover fibers (top, bot-
tom, left, right)
   patch rect 2 10 1 [expr -$cy1] [expr $cz1-$cover] $cy1 $cz1
   patch rect 2 10 1 [expr -$cy1] [expr -$cz1] $cy1 [expr
$cover-$cz1]
   patch rect 2 2 1 [expr -$cy1] [expr $cover-$cz1]
[expr $cover-$cy1] [expr $cz1-$cover]
   patch rect 2 2 1 [expr $cy1-$cover] [expr $cover-$cz1] $cy1
[expr $cz1-$cover]

   # Create the reinforcing fibers (4 layers)
   layer straight 3 3 $As [expr $cy1-$cover] [expr $cz1-$cover]
[expr $cy1-$cover] [expr $cover-$cz1]
   layer straight 3 3 $As [expr $cover-$cy1] [expr $cz1-$cover]
[expr $cover-$cy1] [expr $cover-$cz1]

   }

# -----------------------------------------------
# Define cross-section for nonlinear beams
# -----------------------------------------------

#  set some paramaters

   set beamWidth 152.4
   set beamDepth 152.4

# some variables derived from the parameters
```

```
  set by1 [expr $beamDepth/2.0]
  set bz1 [expr $beamWidth/2.0]

  section Fiber 2 {

  # Create the concrete core fibers
  patch rect 1 10 1 [expr $cover-$by1] [expr $cover-$bz1]
[expr $by1-$cover] [expr $bz1-$cover]

  # Create the concrete cover fibers (top, bottom, left,
right)
   patch rect 2 10 1 [expr -$by1] [expr $bz1-$cover] $by1 $bz1
   patch rect 2 10 1 [expr -$by1] [expr -$bz1] $by1 [expr
$cover-$bz1]
   patch rect 2 2 1 [expr -$by1] [expr $cover-$bz1] [expr
$cover-$by1] [expr $bz1-$cover]
   patch rect 2 2 1 [expr $by1-$cover]
[expr $cover-$bz1] $by1 [expr $bz1-$cover]

  # Create the reinforcing fibers (2 layers)
   layer straight 3 3 $As [expr $by1-$cover] [expr $bz1-$cover]
[expr $by1-$cover] [expr $cover-$bz1]
   layer straight 3 3 $As [expr $cover-$by1] [expr $bz1-$cover]
[expr $cover-$by1] [expr $cover-$bz1]
}

# ----------------------------------------------------------
# Define column elements
# ----------------------------------------------------------

 geomTransf Linear 1

  set np 3
  set iterNum 10
  set iterTol 1e-3

  set j 0;
  while {$j < [expr $nH]} {
   if {$j < [expr $nH-1]} {
      # define the columns elements
      element nonlinearBeamColumn [expr $eleStartID+$j*2]
[expr $nodeStartID+$j*2] [expr $nodeStartID+$j*2+2] $np 1 1
-iter $iterNum $iterTol
      element nonlinearBeamColumn [expr $eleStartID+$j*2+1]
[expr $nodeStartID+$j*2+1] [expr $nodeStartID+$j*2+3] $np 1 1
-iter $iterNum $iterTol
   } else {
      element nonlinearBeamColumn [expr $eleStartID+$j*2]
[expr $nodeStartID+$j*2] [expr $nodeStartID+$j*2+2] $np 1 1
-iter $iterNum $iterTol
```

```
      element nonlinearBeamColumn [expr $eleStartID+$j*2+1]
[expr $nodeStartID+$j*2+1] [expr $nodeStartID+$nH*2+$nL]
$np 1 1 -iter $iterNum $iterTol
    }
     set j [expr $j+1]
   }

# ----------------------------------------------------
# Define beam elements
# ----------------------------------------------------

  geomTransf Linear 2

  set j [expr $nH];

  set bA 48000;
  set bE 2e15;
  set bI 1e10;

  set i 0;

  while {$i < [expr $nL]} {

    element nonlinearBeamColumn [expr $eleStartID+$j*2+$i]
[expr $nodeStartID+$j*2+$i] [expr $nodeStartID+$j*2+1+$i]
$np 2 2 -iter $iterNum $iterTol
    #   element   elasticBeamColumn [expr $eleStartID+$j*2+$i]
[expr $nodeStartID+$j*2+$i] [expr $nodeStartID+$j*2+1+$i]
$bA $bE $bI 2

  set i [expr $i+1]
 }

    element nonlinearBeamColumn 19 17 27 $np 2 2 -iter $iterNum
  $iterTol
    element nonlinearBeamColumn 20 27 28 $np 2 2 -iter $iterNum
$iterTol
    element nonlinearBeamColumn 21 28 18 $np 2 2 -iter $iterNum
$iterTol

# 4.2
# ----------------------------------------------------
# Create ModelBuilder for 2D element (with two-dimensions
and 2 DOF/node)
# ----------------------------------------------------

  model basic -ndm 2 -ndf 2

# Create nodes & add to Domain - command: node nodeId
xCrd yCrd
  set j 0;
```

```
  while {$j < [expr $nH+1]} {
  set i 0;
  while {$i < [expr $nL+1]} {
    node [expr $j*($nL+1)+$i+1] [expr $i*$deltaL] [expr
$j*$deltaH]
    set i [expr $i+1]
  }
  set j [expr $j+1]
 }
# fix one end as a pin, the other end as a roller
fix 1 1 1
fix 4 0 1

# tie nodes between beam, column and 2D elements

equalDOF 1 17 1 2
equalDOF 4 18 1 2
equalDOF 5 19 1 2
equalDOF 8 20 1 2
equalDOF 9 21 1 2
equalDOF 12 22 1 2
equalDOF 13 23 1 2
equalDOF 14 24 1 2
equalDOF 15 25 1 2
equalDOF 16 26 1 2

equalDOF 2 27 1 2
equalDOF 3 28 1 2


# -----------------------------------------------------------
#  Define materials for 2D ReinforceConcretePlaneStress
element
# -----------------------------------------------------------

# set fc fy E
  set wfc 57.0;
  set wfy 419.2;
  set wE  187544.0;
  set rou1 0.0023;
  set rou2 0.0023;

# UniaxialMaterial: steelZ01
#                     tag   fy   E0    fpc      rou
  uniaxialMaterial    SteelZ01 11  $wfy   $wE  $wfc  $rou1
  uniaxialMaterial    SteelZ01 12  $wfy   $wE  $wfc  $rou2

# UniaxialMaterial: concreteZ01
```

```
# ConcreteZ01                           tag       f'c     ec0
  uniaxialMaterial ConcreteZ01  13 [expr -$wfc] -0.0025
  uniaxialMaterial ConcreteZ01  14 [expr -$wfc] -0.0025

  set pi 3.141592654
# NDMaterial: ReinforceConcretePlaneStress
#                         tag   rho s1 s2 c1 c2    angle1      angle2
rou1    rou2   fpc   fy   E0
 nDMaterial FAReinforceConcretePlaneStress 15  0.0 11 12 13 14
[expr 0.0*$pi]   [expr 0.5*$pi] $rou1 $rou2 $wfc $wfy $wE 0.002


# ------------------------------------------------------------
#  Define 2D ReinforceConcretePlaneStress element
# ------------------------------------------------------------

  set j 0;
  while {$j < $nH} {
  set i 0;
  while {$i < $nL} {
  # Create quad elements - command:
  # element quad eleID node1 node2 node3 node4 thick   type
matID

    element quad [expr $j*$nL+$i+1]
[expr $j*($nL+1)+$i+1] [expr $j*($nL+1)+$i+2]
[expr ($j+1)*($nL+1)+$i+2] [expr ($j+1)*($nL+1)+$i+1]
     $t PlaneStress 15

   set i [expr $i+1]
  }
  set j [expr $j+1]
 }
#4.3
# --------------------------
# Define horizontal loads
# --------------------------

  set N 89000.0;

# Create a Plain load pattern with a linear TimeSeries
  pattern Plain 1 "Linear" {

  # Create the nodal load - command: load nodeID xForce yForce
  load 13 0 [expr  -$N]
  load 16 0 [expr  -$N]
  }

#4.4
```

```
# -----------------------------
# End of model generation
# -----------------------------


# -----------------------------
# Start of analysis generation
# -----------------------------

# Create the system of equation, a sparse solver with partial
pivoting
   system BandGeneral

# Create the constraint handler
   constraints Plain

# Create the DOF numberer
   numberer Plain

# Create the convergence test
   test NormDispIncr 1.0e-3 100 5

# Create the solution algorithm
   algorithm KrylovNewton

# Create the integration scheme, the DisplacementCon-
trol scheme
   integrator LoadControl 0.1

# Create the analysis object
   analysis Static

# initialize in case we need to do an initial stiffness
iteration
#   initialize

# -----------------------------
# End of analysis generation
# -----------------------------

# Create a recorder to monitor nodal displacements
#   recorder Node -file N_FSW13.out -time -node 15 -dof
1 2 3 disp

# perform the analysis
   analyze 10

# Print out the state of nodes, if wanted
   print node 13 14 15 16 26 27 28 18
```

```
# Print out the state of elements, if wanted
# print ele 4

# 4.5
# Set the gravity loads to be constant & reset the time
in the domain

   loadConst -time 0.0

# -------------------------------------------------------
# End of Model Generation & Initial Gravity Analysis
# -------------------------------------------------------

# -------------------------------------------------------
# Start of additional modeling for lateral loads
# -------------------------------------------------------


# ------------------------------
# Define horizontal reference load
# ------------------------------

   set P 1000.0;

# Create a Plain load pattern with a linear TimeSeries
   pattern Plain 2 "Linear" {

   # Create the nodal load - command: load nodeID xForce yForce
   load 13 [expr    $P/2] 0
   load 16 [expr    $P/2] 0
   load 4  [expr   -$P/2] 0
   }


# ------------------------------
# End of model generation
# ------------------------------


# ------------------------------
# Start of analysis generation
# ------------------------------

# Create the system of equation, a sparse solver with
partial pivoting
   system BandGeneral

# Create the constraint handler
   constraints Plain

# Create the DOF numberer
   numberer Plain
```

```
# Create the convergence test
  test NormDispIncrVaryIter 0.01 14 5 numStep 100 400 300 400
600 800 1200 1600 2050 2500 2960 3470 1000 10 numIter 100
0 0 0 0 100 0 0 0 0 0 0 0 0

# Create the solution algorithm
  algorithm KrylovNewton

# Create the integration scheme, the DisplacementControl
scheme
  integrator DisplacementPath 13 1 14 numStep 100 400 300 400
600 800 1200 1600 2050 2500 2960 3470 1000 10 increment 0.01
-0.005 0.01 -0.01 0.01 -0.01 0.01 -0.01 0.01 -0.01 0.01
-0.01 0.01 -0.01

# Create the analysis object
  analysis Static

# initialize in case we need to do an initial stiffness
iteration
  initialize

# -----------------------------
# End of analysis generation
# -----------------------------

# Create a recorder to monitor nodal displacements
  recorder Node -file P_FSW13_n13.out -time -node 13 -dof
1 disp

# perform the analysis
  analyze 17380

# Print out the state of elements, if wanted
# print ele 1 2 3 4 5 6 7 8 9

# Print out the state of nodes, if wanted
  print node 27 28 18 25 13 14 15 16
```

Output File

The output file obtained after running the input file using the developed program is shown below. It contains the lateral load acting on the wall in kN along with the displacements (in mm) at the top of the wall.

```
8.2451   9.93E-03
15.9972  1.99E-02
23.9965  2.99E-02
31.7533  3.99E-02
39.5082  4.99E-02
```

```
47.2173    5.99E-02
54.9102    6.99E-02
62.5003    7.99E-02
70.0604    8.99E-02
77.6282    9.99E-02
84.9838    1.10E-01
92.3075    1.20E-01
99.1569    1.30E-01
104.988    1.40E-01
109.29     1.50E-01
.....
.....
... (continued..)
```

Example 4. Shear Wall under Shake Table Excitations

Given

A low-rise shear wall (STN) subjected to seismic excitations on a shake table, which is located at the National Center for Research on Earthquake Engineering (NCREE), Taipei, Taiwan, has been analyzed in this example.

Figure A.4.1 demonstrates the details of dimensions and reinforcement of the specimen. The test set-up is shown in Figure A.4.2. The tcu078Eji seismogram of the 1999 Taiwan earthquake was used as the ground motion acceleration for the shake table. The normalized seismogram is shown in Figure A.4.3. In each test run, a scale factor was applied to the input ground motion acceleration such that the peak ground acceleration (PGA) would reach the predetermined value. A sequence of dynamic tests (a total of six runs) was applied to the specimen. The predetermined PGA for the six test runs is $0.05g$, $0.4g$, $0.8g$, $1.2g$, $1.6g$, and $2.0g$, respectively.

The finite-element model of the specimen is shown in Figure A.4.4. There are separate input files for each test run. The input file, called STN_Run3.tcl, is the input file for the third test run. The ground acceleration file for the third text run, called STN_run3_input_accel.txt, is also required. OpenSees.exe is the executable file used to run the input files. An output file called STN_run3_disp.out is created during the execution of OpenSees.exe with the input files. To run Example 4, a seismogram with the file name of "STN_run3_input_accel.txt" needs to be downloaded from the UH website: <http://www.egr.uh.edu/structurallab/SCS.htm?e=SCS>. The output contains the computed time history of the drift of the wall panel. Analytical drift and time histories for the third to the sixth test runs are presented.

FE Model

The specimen was modeled using the finite element mesh illustrated in Figure A.4.4. The wall panel was modeled as eighteen Quadrilateral elements with RCPlaneStress material. The steel ratio and the steel grid orientations of the wall panel were defined in the RCPlaneStress material. SteelZ01 and ConcreteZ01 were used to create the uniaxial constitutive laws of steel and concrete for the wall panel, respectively. The boundary column at each side of the

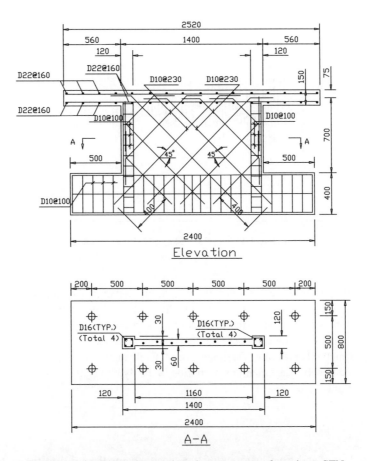

Figure A.4.1   Dimensions and steel arrangement of specimen STN

Figure A.4.2   Test set-up

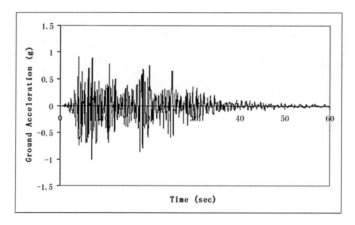

Figure A.4.3   Normalized tcu078Eji seismogram

wall was modeled using three NonlinearBeamColumn elements. The top slab was modeled using ten NonlinearBeamColumn elements. Steel01 and Concrete01 were used to create the uniaxial constitutive laws of steel and concrete for the boundary elements, respectively. The mass blocks on the top of the wall were modeled as nodal masses.

The analysis methods built into OpenSees for dynamic analysis were adopted. The Wilson method was used to solve the equations of motion in the dynamic analysis. Rayleigh damping was considered in the analysis.

Results

The calculated drift and time histories of the specimen for the third to the sixth runs are presented in Figure A.4.5. Experimental and analytical results in the figure are shown as solid and dashed curves, respectively. The analytical results for the first and second runs were not reported here because the PGAs were small.

For the third run to the fifth test run, the computed drift and time histories show good agreements with the measured responses. In the sixth run, the analyses slightly overestimated the drift of the wall panel.

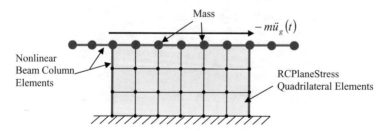

Figure A.4.4   FE model of the specimen

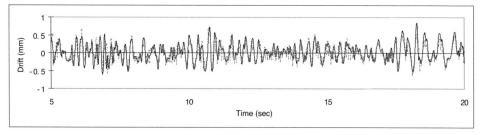

(a) Run 3

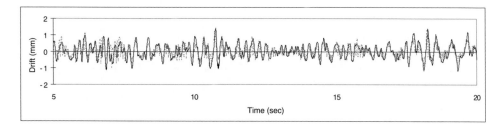

(b) Run 4

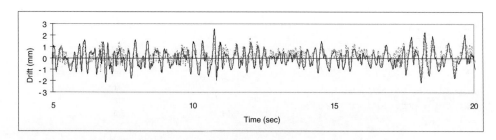

(c) Run 5

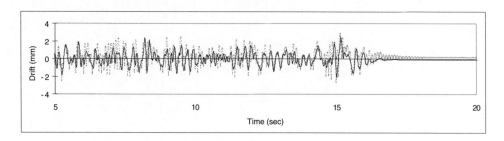

(d) Run 6

Figure A.4.5 Experimental and analytical drift time history of specimen (experimental and analytical results are shown as solid and dashed curves, respectively)

Input File for Specimen

The input file for the third test run is shown below as an example:

```
# STN_Run3.tcl

# ------------------------------
# Units: N, mm, sec, MPa
# ------------------------------

# ------------------------------
# Start of model generation
# ------------------------------

# -----------------------------------------------------------
# Create ModelBuilder for beams and columns
# -----------------------------------------------------------
  model basic -ndm 2 -ndf 3

# -----------------------------------------------------------
# Create nodes for beams and columns
# -----------------------------------------------------------

# set dimension of the wall and mesh

  set L 1158;
  set H 699;
  set deltaL 193;
  set deltaH 233;
  set nL [expr $L/$deltaL];
  set nH [expr $H/$deltaH];
  set t 60;

# Create nodes for beams and columns
# tag X Y

  set LB 1278;
  set deltaLB 213;

  set nodeStartID [expr ($nH+1)*($nL+1)+1];
  set eleStartID  [expr $nH*$nL+1]

  set j 0;

  while {$j < [expr $nH+1]} {
     if {$j < [expr $nH] } {
         node [expr $nodeStartID+$j*2]    0.0 [expr $deltaH*$j]
         node [expr $nodeStartID+$j*2+1] [expr $LB] [expr
$deltaH*$j]
     } else {
         set i 0;
         while {$i < [expr $nL+1]} {
```

```
            node [expr $nodeStartID+$j*2+$i] [expr $i*$deltaLB]
[expr $H]
            set i [expr $i+1]
        }
    }
    set j [expr $j+1]
    }


# ----------------------------------------------------
# Define materials for nonlinear columns and beams
# ----------------------------------------------------

    set ec 0.003;

# CONCRETE tag f'c ec0 f'cu ecu
# Core concrete (confined)

    uniaxialMaterial Concrete01 1  -34.0 [expr -$ec] -6.8 -0.008

# Cover concrete (unconfined)

    uniaxialMaterial Concrete01 2 -34.0 [expr -$ec] -0.0 -0.006

# STEEL
# Reinforcing steel

    set fy 360.0; # Smeared yield stress for steel bars
    set E 200000.0; # Young's modulus

# tag fy E0 b

    uniaxialMaterial Steel01 3 $fy $E 0.02

# ----------------------------------------------------
# Define cross-section for columns
# ----------------------------------------------------

# set some parameters
    set colWidth 117.0
    set colDepth 117.0
    set cover 20.0
    set As1 197.9;

# some variables derived from the parameters

    set cy1 [expr $colDepth/2.0]
    set cz1 [expr $colWidth/2.0]

    section Fiber 1 {

# Create the concrete core fibers
    patch rect 1 10 1 [expr $cover-$cy1] [expr $cover-$cz1]
[expr $cy1-$cover] [expr $cz1-$cover]
```

```
# Create the concrete cover fibers (top, bottom, left, right)
  patch rect 2 10 1 [expr -$cy1] [expr $cz1-$cover] $cy1 $cz1
  patch rect 2 10 1 [expr -$cy1] [expr -$cz1] $cy1
[expr $cover-$cz1]
  patch rect 2 2 1 [expr -$cy1] [expr $cover-$cz1]
[expr $cover-$cy1] [expr $cz1-$cover]
  patch rect 2 2 1 [expr $cy1-$cover] [expr $cover-$cz1] $cy1
[expr $cz1-$cover]

# Create the reinforcing fibers (4 layers)
  layer straight 3 2 $As1 [expr $cy1-$cover]
[expr $cz1-$cover] [expr $cy1-$cover] [expr $cover-$cz1]
  layer straight 3 2 $As1 [expr $cover-$cy1]
[expr $cz1-$cover] [expr $cover-$cy1] [expr $cover-$cz1]

}

# -------------------------------------------------
# Define cross-section for beams
# -------------------------------------------------

# set some parameters

  set beamWidth 480
  set beamDepth 150
  set As2 387.9;

# some variables derived from the parameters

  set by1 [expr $beamDepth/2.0]
  set bz1 [expr $beamWidth/2.0]

  section Fiber 2 {

 # Create the concrete core fibers
   patch rect 1 10 1 [expr $cover-$by1] [expr $cover-$bz1]
[expr $by1-$cover] [expr $bz1-$cover]

 # Create the concrete cover fibers (top, bottom, left, right)
   patch rect 2 10 1 [expr -$by1] [expr $bz1-$cover] $by1 $bz1
   patch rect 2 10 1 [expr -$by1] [expr -$bz1] $by1
[expr $cover-$bz1]
   patch rect 2 2 1 [expr -$by1] [expr $cover-$bz1]
[expr $cover-$by1] [expr $bz1-$cover]
   patch rect 2 2 1 [expr $by1-$cover] [expr $cover-$bz1] $by1
[expr $bz1-$cover]

 # Create the reinforcing fibers (2 layers)
   layer straight 3 3 $As2 [expr $by1-$cover] [expr $bz1-
$cover]
```

```
[expr $by1-$cover] [expr $cover-$bz1]
  layer straight 3 3 $As2 [expr $cover-$by1] [expr $bz1-
$cover]
[expr $cover-$by1] [expr $cover-$bz1]

}

# ------------------------------------------------------------
# Define column elements
# ------------------------------------------------------------

 geomTransf Linear 1

  set np 2
  set iterNum 10
  set iterTol 1e-3

  set j 0;
  while {$j < [expr $nH]} {
      if {$j < [expr $nH-1]} {
        # define the columns elements
        element nonlinearBeamColumn [expr $eleStartID+$j*2]
[expr $nodeStartID+$j*2] [expr $nodeStartID+$j*2+2]
$np 1 1 -iter $iterNum $iterTol
        element nonlinearBeamColumn [expr $eleStartID+$j*2+1]
[expr $nodeStartID+$j*2+1] [expr $nodeStartID+$j*2+3]
$np 1 1 -iter $iterNum $iterTol
      } else {
        element nonlinearBeamColumn [expr $eleStartID+$j*2]
[expr $nodeStartID+$j*2] [expr $nodeStartID+$j*2+2]
$np 1 1 -iter $iterNum $iterTol
        element nonlinearBeamColumn [expr $eleStartID+$j*2+1]
[expr $nodeStartID+$j*2+1] [expr $nodeStartID+$nH*2
+$nL] $np 1 1 -iter $iterNum $iterTol
      }
    set j [expr $j+1]
 }

# ------------------------------------------------------------
# Define beam elements
# ------------------------------------------------------------

  geomTransf Linear 2

  set j [expr $nH];

  set bA 48000;
  set bE 2e15;
  set bI 1e10;
```

```
  set i 0;
  while {$i < [expr $nL]} {

       element nonlinearBeamColumn [expr $eleStartID+$j*2+$i]
[expr $nodeStartID+$j*2+$i] [expr $nodeStartID+$j*2
+1+$i] $np 2 2 -iter $iterNum $iterTol
       set i [expr $i+1]
 }


# -------------------------------------------
# Define cantilever beams on both sides
# -------------------------------------------
  node 42   -500.0 699.0
  node 43   -250.0 699.0
  node 44   1530.0 699.0
  node 45   1780.0 699.0

  element nonlinearBeamColumn 31 42 43 $np 2 2 -iter
$iterNum $iterTol
  element nonlinearBeamColumn 32 43 35 $np 2 2 -iter
$iterNum $iterTol
  element nonlinearBeamColumn 33 41 44 $np 2 2 -iter
$iterNum $iterTol
  element nonlinearBeamColumn 34 44 45 $np 2 2 -iter
$iterNum $iterTol


# ----------------------------------------------------------
# Create ModelBuilder for wall elements
# ----------------------------------------------------------

  model basic -ndm 2 -ndf 2

# Create nodes & add to Domain - command: node nodeId xCrd
yCrd
  set j 0;
  while {$j < [expr $nH+1]} {
  set i 0;
  while {$i < [expr $nL+1]} {
       node [expr $j*($nL+1)+$i+1] [expr $i*$deltaL] [expr
$j*$deltaH]
       set i [expr $i+1]
       }
   set j [expr $j+1]
 }

# Set the boundary conditions - command: fix nodeID xResrnt?
```

```
yRestrnt?
  set i 0;
  while {$i < [expr $nL+1]} {
      fix [expr $i+1] 1 1
      set i [expr $i+1]
  }

# tie nodes between beams, columns and wall elements
equalDOF  1 29   1 2
equalDOF  7 30   1 2
equalDOF  8 31   1 2
equalDOF 14 32   1 2
equalDOF 15 33   1 2
equalDOF 21 34   1 2
equalDOF 22 35   1 2
equalDOF 23 36   1 2
equalDOF 24 37   1 2
equalDOF 25 38   1 2
equalDOF 26 39   1 2
equalDOF 27 40   1 2
equalDOF 28 41   1 2

# ---------------------------------------------------------
#  Define materials for wall elements
# ---------------------------------------------------------

# set fc fy E
  set wfc 34.0;
  set wfy 371.2;
  set wE  200000.0;
  set rou1 0.005;
  set rou2 0.005;

# UniaxialMaterial: steelZ01
#                              tag    fy     E0   fpc     rou
  uniaxialMaterial    SteelZ01   11    $wfy   $wE  $wfc  $rou1
  uniaxialMaterial    SteelZ01   12    $wfy   $wE  $wfc  $rou2

# UniaxialMaterial: concreteZ01
# ConcreteZ01                 tag    f'c      ec0

  uniaxialMaterial ConcreteZ01  13 [expr -$wfc] [expr -$ec]
  uniaxialMaterial ConcreteZ01  14 [expr -$wfc] [expr -$ec]

  set pi 3.141592654
# nDMaterial FAReinforceConcretePlaneStress
# tag rho s1 s2 c1 c2 angle1 angle2 rou1 rou2  fpc  fy  E0 ec
  nDMaterial FAReinforceConcretePlaneStress 15 0.0 11 12 13 14
```

```
[expr 0.25*$pi] [expr 0.75*$pi]   $rou1 $rou2   $wfc $wfy $wE
$ec

#  ----------------------------------------------------------
#  Define wall elements
#  ----------------------------------------------------------

  set j 0;
  while {$j < $nH} {
    set i 0;
    while {$i < $nL} {
      # Create quad elements - command:
      # element quad eleID node1 node2 node3 node4 thick  type
matID

        element quad [expr $j*$nL+$i+1] [expr $j*($nL+1)+$i+1]
[expr $j*($nL+1)+$i+2] [expr ($j+1)*($nL+1)+$i+2]
[expr ($j+1)*($nL+1)+$i+1] $t PlaneStress 15

    set i [expr $i+1]
    }
    set j [expr $j+1]
 }


#  ----------------------------------------------------------
#  Define dynamic analysis parameters
#  ----------------------------------------------------------

  model basic -ndm 2 -ndf 3

# define masses
  set m 1.86;

  mass 35 $m $m 0
  mass 36 $m $m 0
  mass 37 $m $m 0
  mass 38 $m $m 0
  mass 39 $m $m 0
  mass 40 $m $m 0
  mass 41 $m $m 0

  mass 42 $m $m 0
  mass 43 $m $m 0
  mass 44 $m $m 0
  mass 45 $m $m 0

# define earthquake excitation

  set inputAccel "Path -filePath STN_run3_input_accel.TXT -dt
0.005 -factor [expr -10000.0]"
```

```
  pattern UniformExcitation 2 1 -accel $inputAccel
# define damping

  set dampingratio 0.04;
  set a1 [expr $dampingratio*0.0073];
  rayleigh 0.0 0.0 0.0 $a1

# -----------------------------
# End of model generation
# -----------------------------

# -----------------------------
# Start of analysis generation
# -----------------------------

# Create the system of equation
  system BandGeneral

# Create the constraint handler
  constraints Plain

# Create the DOF numberer
  numberer Plain

# Create the convergence test
  test NormDispIncr 0.01 100 5

# Create the solution algorithm
  algorithm KrylovNewton

# Create the integration scheme
  integrator WilsonTheta 1.42

# Create the analysis object
  analysis Transient

# -----------------------------
# End of analysis generation
# -----------------------------
# -----------------------------------
# Create recorder and perform analysis
# -----------------------------------

# Create a recorder to record lateral displacement on the
top of the central wall
  recorder Node -file STN_run3_disp.out -time -node 38 -dof
1 disp

# Perform analysis
  analyze 5000 0.005
```

Output File

The output file obtained after running the input file using the developed program is shown below. It contains the time history of the lateral displacement on the top of the wall panel. The first column is the time step (in seconds) and the second column is the displacement (in mm).

```
0.005    -3.0194968907484093e-005
0.01     -1.8768681774086542e-004
0.015    -3.8277650269579979e-004
0.02     -4.8107632832309450e-004
0.025    -4.2891415484089907e-004
0.03     -2.8493710734939443e-004
0.035    -1.6676445398986021e-004
0.04     -1.5571472982844791e-004
0.045    -2.4190859369522210e-004
0.05     -3.1697629469170562e-004
.....
.....
.....
```

(continued)

References and Further Reading

AASHTO (1994) 'AASHTO LRFD Bridge Design Specifications,' 1st edn, American Association of State Highway and Transportation Officials (AASHTO), Washington, DC.

AASHTO (1998) 'AASHTO LRFD Bridge Design Specifications,' 2nd edn, American Association of State Highway and Transportation Officials (AASHTO), Washington, DC.

AASHTO (2007) 'AASHTO LRFD Bridge Design Specifications,' 4th edn, American Association of State Highway and Transportation Officials (AASHTO), Washington, DC.

ACI Committee 318 (1910) 'Building Code Requirements for Reinforced Concrete (ACI 318–10),' American Concrete Institute, Detroit, MI, 10 pp.

ACI Committee 318 (1963) 'Building Code Requirements for Reinforced Concrete (ACI 318–63),' American Concrete Institute, Detroit, MI, 144 pp.

ACI Committee 318 (1971) 'Building Code Requirements for Reinforced Concrete (ACI 318–83),' American Concrete Institute, Detroit, MI, 78 pp.

ACI Committee 318 (1983) 'Commentary on Building Code Requirements for Reinforced Concrete (ACI 318-83),' American Concrete Institute, Detroit, MI, 155 pp.

ACI Committee 318 (1995) 'Building Code Requirements for Reinforced Concrete (ACI 318-95),' American Concrete Institute, Detroit, MI, 369 pp.

ACI Committee 318 (2008) 'Building Code Requirements for Reinforced Concrete (ACI 318-08),' American Concrete Institute, Detroit, MI.

ACI Committee 340 (1973) 'Design Handbook in Accordance with Strength Design Method of ACI 318-71,' Volume 1, Special Publication SP-17(73), American Concrete Institute, Detroit, MI, 403 pp.

ACI Committee 435 (1974) 'Deflection of Concrete Structures,' Special Publication SP-43, American Concrete Institute, Detroit, MI, 637 pp.

ACI-ASCE Committee 326 Report (1962) 'Shear and Diagonal Tension,' *Journal of the American Concrete Institute, Proceedings*, **59**, 1–30; 277–334; 352–396.

ACI-ASCE Joint Committee 343 Report (1995) 'Analysis and Design of Reinforced Concrete Bridge Structures (343R-95) (Reapproved 2004),' American Concrete Institute, Farmington Hills, MI, 158 pp.

Ayoub, Amir (1995) 'Nonlinear Finite Element Analysis of Reinforced Concrete Sub Assemblages,' PhD Dissertation, Department of Civil Engineering, University of California, Berkeley, 1995.

Ayoub, Amir and Filippou, F.C. (1998) 'Nonlinear Finite-Element Analysis of RC Shear Panels and Walls,' *Journal of Structural Engineering*, ASCE, **124** (3) 298–308.

Bai, S.L. *et al.* (1991) 'Experimental Investigation of the Static and Seismic Behavior of Knee Joints in In-Situ Reinforced Concrete Frames,' Research Report, Chongching Architectural and Civil Engineering Institute, Chongching, Szechuang, and Central Design Institute of Metallurgy, Beijing, China, 203 pp. (in Chinese).

Balakrishnan, S. and Murray, D.W. (1988) 'Prediction of R/C Panels and Deep Beam Behavior by NLFEA,' *Journal of Structural Engineering*, ASCE, **114** (10) 2323–2342.

Batoz, J.L. and Dhatt, G. (1979) 'Incremental Displacement Algorithms for Nonlinear Problems,' *International Journal for Numerical Methods in Engineering*, **14**, 1262–1266.

Baumann, T. (1972) 'Zur Frage der Netzbewehrung von Flachentragwerken,' *Der Bauingenieur*, **46** (6) 367–377.

Belarbi, A. and Hsu, T.T.C. (1990) 'Stirrup Stresses in Reinforced Concrete Beams,' *Structural Journal of the American Concrete Institute*, **87** (5) 350–358.

Belarbi, A. and Hsu, T.T.C. (1994) 'Constitutive Laws of Concrete in Tension and Reinforcing Bars Stiffened by Concrete,' *ACI Structural Journal*, **91** (4) 465–474.

Belarbi, A. and Hsu, T.T.C. (1995) 'Constitutive Laws of Softened Concrete in Biaxial Tension-Compression,' *ACI Structural Journal*, **92** (5) 562–573.

Belletti, B., Cerioni, R. and Iori, I. (2001) 'Physical Approach for Reinforced-Concrete (PARC) Membrane Elements,' *Journal of Structural Engineering*, ASCE, **127** (12) 1412–1426.

Bennett, E.W. and Balasooriya, B.M.A. (1971) 'Shear Strength of Prestressed Beams with Thin Webs Failing in Inclined Compression, *ACI Journal, Proceedings*, **69** (3) 204–212.

Boresi, A.P., Schmidt, R.J. and Sidebottom, O.M. (1993) Advanced Mechanics of Materials, John Wiley & Sons, Inc., New York, 110 pp, Problem 3.3.

Branson, D.E. (1965) 'Instantaneous and Time-Dependent Deflections on Simple and Continuous Reinforced Concrete Beams,' HPR Report No. 7, Part 1, Alabama Highway Department, Bureau of Public Roads, Birmingham, Alabama, August, pp. 1–78.

Branson, D.E. (1977) Deformation of Concrete Structures, McGraw-Hill, New York, 546 pp.

Bredt, R. (1896) 'Kritische Bemerkungen zur Drehungselastizitat,' *Zeitschrift des Vereines Deutscher Ingenieure*, **40** (28) 785–790; **40** (29) 813–817 (in German).

Bruce, R.N. (1962) 'An Experimental Study of the Action of Web Reinforcement in Prestressed Concrete Beams,' PhD Dissertation, University of Illinois Urbana.

Canadian Standard Association (1977) *Design of Concrete Structures and Buildings*, CAN3-A23.3-M77, Canadian Standard Association, Rexdale (Toronto), Ontario, 281 pp.

Canadian Standard Association (1984) *Design of Concrete Structures and Buildings*, CAN3-A23.3-M84, Canadian Standard Association, Rexdale (Toronto), Ontario, 281 pp.

Carlson, N.N. and Miller, K. (1998) 'Design and Application of A Gradient-weighted Moving Finite Element Code I: in One Dimension,' *SIAM Journal of Science Computing*, **19** (3) 728–765.

CEB-FIP (1978) Model Code for Concrete Structures, CEB-FIP International Recommendation, 3rd edn, Comite Euro-International du Beton (CEB), 348 pp.

Chintrakarn, R. (2001) 'Minimum Shear Steel and Failure Modes Diagram of Reinforced Concrete Membrane Elements,' M.S. Thesis, Department of Civil and Environmental Engineering, University of Houston, Houston, Texas, USA.

Chopra, A.K. (2000) Dynamics of Structures: Theory and Application to Earthquake Engineering, Prentice Hall, New Jersey.

Collins, M.P. (1973) 'Torque–Twist Characteristics of Reinforced Concrete Beams,' Inelasticity and Non-Linearity in Structural Concrete, Study No. 8, University of Waterloo Press, Waterloo, Ontario, Canada, pp. 211–232.

Collins, M.P. and Mitchell, D. (1980) 'Shear and Torsion Design of Prestressed and Non-Prestressed Concrete Beams,' *Journal of the Prestressed Concrete Institute*, **25** (5) 32–100.

Crisfield, M.A. (1981) 'A Fast Incremental/iterative Solution Procedure That Handles "Snap Through",' *Computers and Structures*, **13**, 55–62.

Crisfield, M.A. and Wills, J. (1989) 'Analysis of R/C Panels Using Different Concrete Models,' *Journal of Engineering Mechanics*, ASCE, **115** (3) 578–597.

Crisfield, M. (1997) Nonlinear Finite Element Analysis of Solids and Structures, Advanced Topics, John Wiley & Sons, Inc., New York, NY.

Delhumeau, G. (1999) L'Invention du Beton arme: Hennebique, 1890–1914, Editions Horma, Institut Francais d'Architecture, Paris, 344 pp.

Derecho, A.T., Ghosh, S.K. and Iqbal, M. (1979) 'Strength, Stiffness, and Ductility Required in Reinforced Concrete Structural Walls for Earthquake Resistance,' *ACI Journal*, **76** (8) 875–895.

Elfgren, L. (1972) 'Reinforced Concrete Beam Loaded in Torsion, Bending and Shear', *Publications 71:3*, Division of Concrete Structures, Chalmers University of Technology, Goteborg, Sweden, 249 pp.

Elzanaty, A.H., Nilson, A.H. and Slate, F.O. (1986) 'Shear Capacity of Prestressed Concrete Beams Using High-Strength Concrete,' *ACI Journal, Proceedings*, **83** (3) 359–368.

Farrar, C.R. and Baker, W.E. (1992) 'Measuring the Stiffness of Concrete Shear Walls during Dynamic Tests,' *Experimental Mechanics*, **32** (2) 179–183.

Fenves, G.L. (2005) Annual Workshop on Open System for Earthquake Engineering Simulation, Pacific Earthquake Engineering Research Center, UC Berkeley, http://opensees.berkeley.edu/.

Fialkow, M.N. (1985) 'Design and Capacity Evaluation of Reinforced Concrete Shell Membranes,' *Journal of the American Concrete Institute, Proceedings*, **82** (6) 844–852.

Foster, S.J. and Marti, P. (2003) 'Cracked Membrane Model: Finite Element Implementation,' *Journal of Structural Engineering*, ASCE, **129** (9) 1155–1163.

Gao, X.D. (1999) 'Framed Shear Walls under Cyclic Loading,' PhD Dissertation, Department of Civil and Environmental Engineering, University of Houston, Houston, TX.

German Standard DIN 4334 (1958) Bemessung im Stahlbetonbau (Design of Reinforced Concrete), Ernst & Sohn, Berlin, Germany, 57 pp.

Gupta, A.K. (1981) 'Membrane Reinforcement in Shells,' *Journal of the Structural Division*, ASCE, **107** (ST1) 41–56.

Gupta, A.K. (1984) 'Membrane Reinforcement in Concrete Shells: A Review,' *Nuclear Engineering and Design*, **82**, 63–75.

Haisler, W.E., Stricklin, J.A. and Key, J.E. (1977) 'Displacement Incrementation in Non-linear Structural Analysis by the Self-Correcting Method,' *International Journal for Numerical Methods in Engineering*, **11**, 3–10.

Han, K.J. and Mau, S.T. (1988) 'Membrane Behavior of R/C Shell Elements and Limits on the Reinforcement,' *Journal of Structural Engineering*, ASCE, **114** (2) 425–444.

Hanson, J.M. and Hulsbos, C.L. (1965) 'Overload Behavior of Pretensioned Prestressed Concrete I-Beams with Web Reinforcement,' *Highway Research Record 76*, Highway Research Board, pp. 1–31.

Hernandez, G. (1958) 'Strength of Prestressed Concrete Beams with Web Reinforcement,' *Report*, The Engineering Experiment Station, University of Illinois, Urbana, IL.

Hognestad, E., Hanson, N.W. and McHenry, D. (1955) 'Concrete Stress Distribution in Ultimate Strength Design,' *ACI Journal*, **52** (4) 455–479.

Hsu, T.T.C. (1968a) 'Torsion of Structural Concrete – Behavior of Reinforced Concrete Rectangular Members,' Torsion of Structural Concrete, Special Publication SP-18, American Concrete Institute, Detroit, MI, pp. 261–306.

Hsu, T.T.C. (1968b) 'Ultimate Torque of Reinforced Rectangular Beams,' *Journal of the Structural Division*, ASCE, **94** (ST2) 485–510.

Hsu, T.T.C. and Burton, K. (1974) 'Design of Reinforced Concrete Spandrel Beams,' *Journal of the Structural Division*, ASCE, **100** (ST1) 209–229.

Hsu, T.T.C. and Hwang, C.S. (1977) 'Torsional Limit Design of Spandrel Beams,' *ACI Journal, Proceedings*, **74** (2) 71–79.

Hsu, T.T.C. (1982) 'Is the Staggering Concept of Shear Design Safe?' *Journal of the American Concrete Institute, Proceedings*, **79** (6) 435–443.

Hsu, T.T. C (1983a) Author's closure to the paper, 'Is the "Staggering Concept" of Shear Design Safe?,' *Journal of the American Concrete Institute, Proceedings*, **80** (5) 450–454.

Hsu, T.T.C. and Mo, Y.L. (1983b) 'Softening of Concrete in Torsional Members,' *Research Report No. ST-TH-001-83*, Department of Civil Engineering, University of Houston, Houston, Texas, 107 pp.

Hsu, T.T.C. (1984) Torsion of Reinforced Concrete, Van Nostrand Reinhold, Inc., New York, 544 pp.

Hsu, T.T.C. and Mo, Y.L. (1985a) 'Softening of Concrete in Torsional Members – Theory and Tests,' *Journal of the American Concrete Institute, Proceedings*, **82** (3) 290–303.

Hsu, T.T.C. and Mo, Y.L. (1985b) 'Softening of Concrete in Torsional Members – Design Recommendations,' *Journal of the American Concrete Institute, Proceedings*, **82** (4) 443–452.

Hsu, T.T.C. (1988) 'Softening Truss Model Theory for Shear and Torsion,' *Structural Journal of the American Concrete Institute*, **85** (6) 624–635.

Hsu, T.T.C. (1990) 'Shear Flow Zone in Torsion of Reinforced Concrete,' *Journal of the Structural Division*, ASCE, **116** (11) 3205–3225.

Hsu, T.T.C. (1991a) 'Nonlinear Analysis of Concrete Membrane Elements,' *Structural Journal of the American Concrete Institute*, **88** (5) 552–561.

Hsu, T.T.C. (1991b) 'Nonlinear Analysis of Concrete Torsional Members,' *Structural Journal of the American Concrete Institute*, **88** (6) 674–682.

Hsu, T.T.C. (1993) Unified Theory of Reinforced Concrete, CRC Press Inc., Boca Raton, FL, 329 pp.

Hsu, T.T.C., Belarbi, A. and Pang, X.B. (1995) 'A Universal Panel Tester,' *Journal of Testing and Evaluations*, ASTM, **23** (1) 41–49.

Hsu, T.T.C., Zhang, L.X. and Gomez, T. (1995) 'A Servo-Control System for Universal Panel Tester,' *Journal of Testing and Evaluations*, ASTM, **23** (6) 424–430.

Hsu, T.T.C. (1996) 'Toward a Unified Nomenclature for Reinforced Concrete Theory,' *Journal of Structural Engineering*, ASCE, **122** (3) 275–283.

Hsu, T.T.C. and Zhang, L.X. (1996) 'Tension Stiffening in Reinforced Concrete Membrane Elements,' *Structural Journal of the American Concrete Institute*, **93** (1) 108–115.

Hsu, T.T.C. and Zhang, L.X. (1997) 'Nonlinear Analysis of Membrane Elements by Fixed-Angle Softened-Truss Model,' *Structural Journal of the American Concrete Institute*, **94** (5) 483–492.

Hsu, T.T.C. (1997) 'ACI Shear and Torsion Provisions for Prestressed Hollow Girders,' *Structural Journal of the American Concrete Institute*, **94** (6) 787–799.

Hsu, T.T.C. (1998) 'Stresses and Crack Angles in Concrete Membrane Elements,' *Journal of Structural Engineering*, ASCE, **124** (12) 1476–1484.

Hsu, T.T.C. and Mansour, M.Y. (2002) 'Failure Mechanism of Reinforced Concrete Elements under Cyclic Loading,' ACI Special Publication 207: *Concrete: Material Science to Applications – A Tribute of Surendra Shah*, American Concrete Institute, Farmington Hill, MI, pp. 1–24.

Hsu, T.T.C. and Zhu, R.R.H. (2002) 'Softened Membrane Model for Reinforced Concrete Elements in Shear,' *Structural Journal of the American Concrete Institute*, **99** (4) 460–469.

Hsu, T.T.C. and Mansour, M.Y. (2005) 'Stiffness, Ductility, and Energy Dissipation of RC Elements under Cyclic Shear,' *Earthquake Spectra*, EERI, **21** (4) 1093–1112.

Hsu, T.T.C., Mansour, M.Y., Mo, Y.L. and Zhong, J. (2006) 'Cyclic Softened Membrane Model for Nonlinear Finite Element Analysis of Concrete Structures,' ACI SP-237, *Finite Element Analysis of Reinforced Concrete Structures*, American Concrete Institute, Farmington, MI, pp. 71–98.

Hsu, T.T.C., Laskar, A. and Mo, Y.L. (2008) 'Shear Design of Prestressed Concrete Beams,' *Research Report UHCEE 08-1*, Department of Civil and Environmental Engineering, University of Houston, TX.

Ile, Nicolae and Reynouard, J.M. (2000) 'Nonlinear Analysis of Reinforced Concrete Shear Wall under Earthquake Loading,' *Journal of Earthquake Engineering*, **4** (2) 183–213.

Izumo, J., Shin, H., Maekawa, K. and Okamura, H. (1992) 'An Analytical Model for RC Panels subjected to In-plane Stresses,' *Concrete Shear in Earthquake*, T.T.C. Hsu and S.T. Mo (eds), (Proceedings of the International Workshop on Concrete Shear in Earthquake, Jan. 14–16, 1991, Houston), Elsevier Science Publishers, Inc., London-New York, 206–215.

Jeng, C.H. (2002) 'Development of Nonlinear Finite Element Software for Static and Dynamic Simulation of RC/PC Frames: Modeling, Implementation, and Application,' PhD Dissertation, Department of Civil and Environmental Engineering, University of Houston, July 2002.

Jeng, C.H. and Hsu, T.T.C. (2009) 'A Softened Membrane Model for Torsion in Reinforced Concrete Members,' *Engineering Structures*, **31** (9) 1944–1954.

Karsan, I.D. and Jirsa, J.O. (1969) 'Behavior of Concrete under Compressive Loadings,' *Journal of Structural Engineering*, ASCE, **95** (ST12) 2543–2563.

Kaufman, M.K. and Ramirez, J.A. (1988) 'Re-evaluation of Ultimate Shear Behavior of High-Strength Concrete Prestressed I-Beams,' *ACI Structural Journal*, **85** (3) 295–303.

Kaufmann, W. and Marti, P. (1998) 'Structural Concrete: Cracked Membrane Model,' *Journal of Structural Engineering*, **124** (12) 1467–1475.

Kwon, M.H. and Spacone, E. (2002) 'Three-dimensional Finite Element Analyses of Reinforced Concrete Columns,' *Computers and Structures*, **80** (2) 199–212.

Laskar, A., Wang, J., Hsu, T.T.C. and Mo, Y.L. (2006) 'Rational Shear Provisions for AASHTO LRFD Specifications,' *Technical Report 0-4759-1*, Department of Civil and Environmental Engineering, University of Houston, Houston, TX, 216 pp.

Laskar, A., Wang, J., Hsu, T.T.C. and Mo, Y.L. (2008) 'Cyclic Softened Membrane Model for Prestressed Concrete,' *18th Analysis and Computation Specialty Conference, ASCE Structure Congress*, Vancouver, Canada, April 24–26.

Laskar, A. (2009) 'Shear Behavior and Design of Prestressed Concrete Members,' PhD Dissertation, Department of Civil and Environmental Engineering, University of Houston, Houston Texas, USA.

Lampert, P. and Thurlimann, B. (1968, 1969) 'Torsionsversuch an Stahlbetonbalken,' (Torsion Tests of Reinforced Concrete Beams), *Bericht 6506-2*, 101 pp; 'Torsion-Biege-Versuche an Stahlbetonbalken,' (Torsion-Bending Tests on Reinforced Concrete Beams), *Bericht Nr. 6506-3*, Institute of Baustatik, ETH, Zurich, Switzerland (in German).

Loov, R.E. (2002) 'Shear Design of Uniformly Loaded Beams,' Presented at the 6th International Conference on Short and Medium Span Bridges, Vancouver, Canada, 31 July–2 August 2002.

Lyngberg, B.S. (1976) 'Ultimate Shear Resistance of Partially Prestressed Reinforced Concrete I-Beams,' *ACI Journal, Proceedings*, **73** (4) 214–222.

Ma, Z.J., Tadros, M.K. and Baishya, M. (2000) 'Shear Behavior of Pretensioned High-Strength Concrete Bridge I-Girders,' *ACI Structural Journal*, **97** (1) 185–193.

MacGregor, J.G., Sozen, M.A. and Siess, C.P. (1960) 'Strength and Behavior of Prestressed Concrete Beams with Web Reinforcement,' *Report*, The Engineering Experiment Station, University of Illinois, Urbana, IL.

MacGregor, J.G. and Ghoneim, M.G. (1995) 'Design for Torsion,' *ACI Structural Journal*, **92**, 211–218.

Mansour, M. (2001) 'Behavior of Reinforced Concrete Membrane Elements Under Cyclic Shear: Experiments to Theory,' PhD Dissertation, Department of Civil and Environmental Engineering, University of Houston, Houston Texas, USA.

Mansour, M., Hsu, T.T.C. and Lee J.Y. (2001a) 'Pinching Effect in Hysteretic Loops of R/C Shear Elements,' ACI Special Publication 205: Finite Element Analysis of Reinforced Concrete Structures, K. Willam and A. Tanabe (eds), American Concrete Institute, Farmington Hill, MI, pp. 293–321.

Mansour, M., Lee, J.Y. and Hsu, T.T.C. (2001b) 'Constitutive Laws of Concrete and Steel Bars in Membrane Elements under Cyclic Loading,' *Journal of Structural Engineering*, ASCE, **127** (12) 1402–1411.

Mansour, M. and Hsu, T.T.C. (2005a) 'Behavior of Reinforced Concrete Elements Under Cyclic Shear: Part I – Experiments,' *Journal of Structural Engineering*, ASCE, **131** (1) 44–53.

Mansour, M. and Hsu, T.T.C. (2005b) 'Behavior of Reinforced Concrete Elements Under Cyclic Shear: Part II – Theoretical Model,' *Journal of Structural Engineering*, ASCE, **131** (1) 54–65.

Mattock, A.H. and Kaar, P.H. (1961) 'Precast-Prestressed Concrete Bridges 4: Shear Tests of Continuous Girders,' *Journal of the PCA Research Development Laboratories*, pp. 19–47.

Mattock, A.H. (1995) Private communication from A.H. Mattock to T.T.C. Hsu concerning torsion in box girders, 19 March 1995.

Mo, Y.L., Zhong, J.X. and Hsu, T.T.C. (2008) 'Seismic Simulation of RC Wall-Type Structures,' *Engineering Structures*, **30** (11) 3167–3175.

Modjeski and Masters Inc. (2003) 'Comprehensive Design Example for Prestressed Concrete (PSC) Girder Super-structure Bridge with Commentary,' *Technical Report FHWA NHI-04-043*, submitted to the Federal Highway Administration, 381 pp.

Morsch, E. (1902) Der Eisenbetonbau, seine Anwendung und Theorie, 1st edn, Wayss and Freytag, A.G., Im Selbstverlag der Firma, Neustadt, a. d. Haardt, 118 pp.; 2nd edn, Verlag Von Konrad Wittmer, Stuttgart, 1906, 252 pp.; 3rd edn (English Translation by E.P. Goodrich), McGraw-Hill, New York, 1909, 368 pp.

Navier, C.L. (1826) Resume des Leçons Donnees a l'Ecole des Ponts et Chaussées sur l'Application de la Mécanique a l'Etablissement des Constructions et des Machines, Firmin Didet, Paris.

Newmark, N.M. (1959) 'A Method of Computation for Structural Dynamics,' *Journal of the Engineering Mechanics Division*, ASCE, pp. 676–94.

Nielsen, M.P. (1967) 'Om Forskydningsarmering i Jernbetonbjaelker,' (On Shear Reinforcement in Reinforced Concrete Beams), *Bygningsstatiske Meddelelser*, **38** (2) 33–58.

Nielsen, M.P. and Braestrup, M.W. (1975) 'Plastic Shear Strength of Reinforced Concrete Beams,' *Bygningsstatiske Meddelelser*, **46** (3) 61–99.

Oesterle, R.G., Aristizabal-Ochoa, J.D., Shiu, K.N. and Corley, W.G. (1984) 'Web Crushing of Reinforced Concrete Structural Walls,' *ACI Journal*, **81** (3) 231–241.

Oesterle, R.G. (1986) 'Inelastic Analysis for In-plane Strength of Reinforced Concrete Shearwalls,' PhD Dissertation, Department of Civil Engineering, Northwestern University, Evantson, IL.

Okamura, H. and Maekawa, K. (1991) *Nonlinear Analysis and Constitutive Models of Reinforced Concrete*, ISBN4-7655-1506-0 C 3051, published in Japan.

Ou, Y.C. (2007) 'Precast Segmental Post-Tensioned Concrete Bridge Columns for Seismic Regions,'' PhD Dissertation, Department of Civil, Structural, and Environmental Engineering, State University of New York, Buffalo, NY.

Palermo, D. and Vecchio, F.J. (2003) 'Compression Field Modeling of Reinforced Concrete Subjected to Reversed Loading: Formulation,' *ACI Structural Journal*, (100) (5) 616–625.

Pang, X.B. and Hsu, T.T.C. (1995) 'Behavior of Reinforced Concrete Membrane Elements in Shear,' *ACI Structural Journal*, **92** (6) 665–679.

Pang, X.B. and Hsu, T.T.C. (1996) 'Fixed-Angle Softened-Truss Model for Reinforced Concrete,' *Structural Journal of the American Concrete Institute*, **93** (2) 197–207.

Paz, M. (1997) Structural Dynamics Theory and Computation, Fourth Edition, Chapman and Hall, New York, 824 pp.

Rangan, B.V. (1991) 'Web Crushing Strength of Reinforced and Prestressed Concrete Beams,' *ACI Structual Journal*, **88** (1) 12–16.

Rausch, E. (1929) *Design of Reinforced Concrete in Torsion*, (Berachnung des Eisanbetons gegen verdrehung), Technische Hochschule, Berlin, Germany, 53 pp. (in German). A second edition was published in 1938. The third edition was titled *Drillung* (Torsion) *Schub und Scheren in Stahlbelonbau*, Deutcher Ingenieur-Verlag GmbH, Dusseldorf, Germany, 1953, 168 pp.

Riks, E. (1972) 'The Application of Newton's Method to the Problem of Elastic Stability,' *Journal of Applied Mechanics*, **39**, 1060–1066.

Ritter, W. (1899) 'Die Bauweise Hennebique,' *Schweizerische Bauzeitung*, **33** (7) 59–61.

Robertson, I.N. and Durrani, A.J. (1987) 'Shear Strength of Prestressed Concrete T Beams with Welded Wire Fabric as Shear Reinforcement,' *PCI Journal*, **32** (2) 46–61.

Robinson, J.R. and Demorieux, J.M. (1972) 'Essais de Traction-Compression sur Modèles d'Ame de Poutre en Beton Arme,' *IRABA Report*, Institut de Recherches Appliquees du Beton de L'Ame, Part 1, June 1968, 44 pp; 'Resistance Ultimate du Beton de L'ame de Poutres en Double Te en Beton Arme,' Part 2, May 1972, 53 pp.

Schlaich, J., Schafer, K. and Jennewein, M. (1987) 'Toward a Consistant Design of Structural Concrete,' *PCI Journal*, Prestressed Concrete Institute, **32** (3) 74–150.

Scott, B.D., Park, R. and Priestley, M.J.N. (1982) 'Stress–Strain Behavior of Concrete Confined by Overlapping Hoops at Low and High Strain Rates,' *ACI Journal*, 13–27.

Shahawy, M.A. and Batchelor, B. (1996) 'Shear Behavior of Full-Scale Prestressed Concrete Girders: Comparison Between AASTHO Specifications and LRFD Code,' *PCI Journal*, Precast/Prestressed Concrete Institute, **41** (3) 48–62.

Sittipunt, C. and Wood, S.L. (1995) 'Influence of Web Reinforcement on the Cyclic Response of Structural Walls,' *ACI Structural Journal*, **92** (6) 745–756.

Swann, R.A. (1969) 'Flexural Strength of Corners of Reinforced Concrete Portal Frames,' *Technical Report TRA 434*, Cement and Concrete Association, London, England, 14 pp.

Tamai, S., Shima, H., Izumo, J. and Okamura, H. (1987) 'Average Stress–Strain Relationship in Post Yield Range of Steel Bar in Concrete.' *Concrete Library of JSCE, No. 11*, June 1988, pp. 117–129. (Translation from *Proceedings of JSCE*, No. 378/Vol. 6, February 1987).

Taucer, F.T., Spacone, E. and Filippou, F.C. (1991) 'A Fiber Beam–Column Element for Seismic Response Analysis of Reinforced Concrete Structures,' *Report UCB/EERC-91/17*, University of California, Berkeley.

Taylor, R.L. (1999) 'FEAP User Manual v7.1.' Department of Civil and Environmental Engineering, University of California, Berkeley, 1999, http://www.ce.berkeley.edu/~rlt/feap/.

Thurlimann, B. (1979) 'Shear Strength of Reinforced and Prestressed Concrete – CEB Approach,' ACI-CEB-PCI-FIP Symposium, Concrete Design: U.S. and European Practices, ACI Publication SP-59, American Concrete Institute, Detroit, MI, pp. 93–115.

Vecchio, F.J. and Collins, M.P. (1981) 'Stress–Strain Characteristic of Reinforced Concrete in Pure Shear,' IABSE Colloquium, Advanced Mechanics of Reinforced Concrete, Delft, Final Report, International Association of Bridge and Structural Engineering, Zurich, Switzerland, pp. 221–225.

Vecchio, F.J. and Collins, M.P. (1982) 'The Response of Reinforced Concrete to In-Plane Shear and Normal Stresses.' *Publication 82-03* (ISBN 0-7727-7029-8), Department of Civil Engineering, University of Toronto, Toronto, Canada, 332 pp.

Vecchio, F.J. and Collins, M.P. (1986) 'The Modified Compression Field Theory for Reinforced Concrete Elements Subjected to Shear,' *Journal of the American Concrete Institute, Proceedings*, **83** (2) 219–231.

Vecchio, F.J. (2000) 'Disturbed Stress Field Model for Reinforced Concrete: Formulation,' *Journal of Structural Engineering*, ASCE, **126** (9) 1070–1077.

Vecchio, F.J. (2001) 'Disturbed Stress Field Model for Reinforced Concrete: Implementation,' *Journal of Structural Engineering*, ASCE, **127** (1) 12–20.

Wang, J. (2006) 'Constitutive Relationships of Prestressed Concrete Membrane Elements,' PhD Dissertation, Department of Civil and Environmental Engineering, University of Houston, Houston Texas, USA.

Wilson, E.L., Farhoomand, I. and Bathe, K.J. (1973) 'Nonlinear Dynamic Analysis of Complex Structures,' *International Journal of Earthquake Engineering and Structural Dynamics*, **1**, 241–252.

Yokoo, Y. and Nakamura, T. (1977) 'Nonstationary Hysteretic Uniaxial Stress–Strain Relations of a Steel Bar,' *Transactions of the Architectural Institute of Japan*, **260**, 71–80.

Zhang, L.X. and Hsu, T.T.C. (1998) 'Behavior and Analysis of 100 MPa Concrete Membrane Elements,' *Journal of Structural Engineering*, ASCE, **124** (1) 24–34.

Zhong, J.X. (2005) 'Model-Based Simulation of Reinforced Concrete Plane Stress Structures,' PhD Dissertation, Department of Civil and Environmental Engineering, University of Houston, Houston, TX.

Zhong, J., Mo, Y.L., Jacob, P. and Gur, T. (2006) 'Finite Element Modeling of a Seven-story Reinforced Concrete Wall Building,' NEES/UCSD Seminar on Analytical Modeling of Reinforced Concrete Walls for Earthquake Resistance, University of California at San Diego, San Diego, CA, 16 December 2006, https://www.nees.org/images/uploads/4am/Wednesday/BlindPredictionContestPlenarySession/J_Restrepo.pdf.

Zhu, R.H., Hsu, T.T.C. and Lee, J.Y. (2001) 'Rational Shear Modulus for Smeared Crack Analysis of Reinforced Concrete,' *Structural Journal of the American Concrete Institute*, **98** (4) 443–450.

Zhu, R.R.H. and Hsu, T.T.C. (2002) 'Poisson Effect of Reinforced Concrete Membrane Elements,' *Structural Journal of the American Concrete Institute*, **99** (5) 631–640.

Zienkiewicz, O.C. (1971) 'Incremental Displacement in Non-Linear Analysis,' *International Journal for Numerical Methods in Engineering*, **3**, 587–592.

Index

Note: Italicized f and t refer to figures and tables

cyclic stress-strain curves of, 267–71
envelope curves of, 269
in shear, 247, 258
in tension, 247, 258
nonlinear constitutive relationship of, 90–3
smeared stress-strain of, 232–6
smeared tensile strength of, 234–5
strength, 230–1, 362
bending of, 299–301
strain distribution in, 301–2
strains and stresses in, 301 f
concrete constitutive matrix, 212, 257
concrete contribution V_c, 55, 143–5, 361, 364,
366, 378
in fixed angle theory, 144
in rotating angle theory, 143
in UH shear strength equation, 366–7
concrete shear stress, 144–5, 211, 232, 259
concrete shear stress-strain curve, 253, 264
concrete shear strain, 212
concrete struts, 2, 9–11, 13, 16, 25, 34, 45, 51–2,
55, 68, 139, 150, 167–8, 175, 188, 202,
210, 213, 216, 223, 227, 232, 264, 280,
282, 284, 299–302, 305, 307, 314, 320,
324, 343, 346, 348–9, 353, 356, 362, 390
constitutive law of concrete in compression, 247,
258, 308
constitutive law for steel, 95t
of compression steel, 115t
of mild steel, 308
of prestressing steel, 308–9
of tension steel, 115t
constitutive laws of materials, 247–8, 258
constitutive matrix
of smeared bars, 218–19
of smeared concrete, 217–18
constitutive relationships, 12, 13, 142–3, 212–13,
224–5, 234, 257, 266–7, 272, 385.
See also constitutive law
continuous orthotropic materials, 212–13
contribution of concrete, *See* concrete
contribution
contribution of steel, *See* steel contribution
conversion matrices, 224–5
crack control, 53, 156, 161–5, 339
steel strains for, 161–3
strain ratio, 163, 165
yielding of longitudinal steel, 163, 164 f
yielding of transverse steel, 164–5
cracked moment of inertia, 78, 81
cracked sections, bending rigidities of, 78–81
curvature, 72–3, 76, 78, 83–4, 88, 90, 96, 101,
110–11, 299, 301
cyclic shear, 216, 267, 269–70, 274, 277, 279,
279 f, 281, 284–7

cyclic softened membrane model (CSMM),
12–13, 266–93, 383–5, 385 f, 388–94,
396, 407, 409
2-D model for walls, 393–6
analysis procedures, 396
coordinate systems for concrete structures,
393–4
implementation, 394–6
basic principles of, 266–7
coordinate systems transformations, 393–4
demonstration panels, 284–7
Series A0, 286, 288–9, 292
Series A45, 286, 288–9, 292
Series P, 286–9
shear properties of, 287 f
steel bar angle, 285
steel percentage, 285
failure mechanism, 284
flow chart for, 275t
Hsu/Zhu ratios, 12–13, 212, 216–17, 219–25,
248, 253, 255, 264, 266–7, 274, 276, 395
hysteretic loops, 267, 269, 274, 276–81, 289,
292, 417, 417 f
deformation characteristics, 279–81
panel CA3 vs. panel CE3, 276–8
pinching effect, 278–9
Mohr circles, 283–4
physical visualization, 281–4
pinching mechanism, 281
shear ductility, 288–9
shear energy dissipation, 289–93
shear stiffness, 288
solution procedure, 274–6
installation of strain history, 274–6
modification of coordinates, 274
modifications of material laws, 276–81
cyclic stress-strain curves of concrete, 267–71
compressive envelope curves, 269
damage coefficient for compression envelope,
269–70
tensile envelope curves, 269
unloading/reloading curves of concrete,
270–1
cyclic stress-strain curves of mild steel,
272–4
envelope curves of mild steel, 272
unloading/reloading curves of mild steel,
272–4

D region, 5, 9, 14. *See also* local regions
damage coefficient for compression envelope,
269–70, 392
damping, 399, 400–1, 428, 470, 479
damping ratio, 401, 428, 432
deformation, 279–81